Modern Physical Chemistry

Modern Chemistry Series

Under the supervisory editorship of D. J. Waddington, BSc, ARCS, DIC, PhD, Professor of Chemical Education, University of York, this series is specially designed to meet the demands of the new syllabuses for sixth form, introductory degree and technical college courses. It consists of self-contained major texts in the three principal divisions of the subject, supported by short readers and practical books.

Major Texts

Modern Inorganic Chemistry
G. F. Liptrot, MA, PhD

Modern Organic Chemistry
R. O. C. Norman, MA, DSc, C Chem, FRSC, FRS
D. J. Waddington, BSc, ARCS, DIC, PhD

Modern Physical Chemistry
G. F. Liptrot, MA, PhD
J. J. Thompson, MA, PhD
G. R. Walker, BA, BSc

Readers

Organic Chemistry: a problem-solving approach
M. J. Tomlinson, BSc, C Chem, MRSC
M. C. V. Cane, BSc, PhD, C Chem, MRSC

Investigation of Molecular Structure
B. C. Gilbert, MA, D Phil, C Chem, FRSC

Mechanisms in Organic Chemistry: Case Studies
R. O. C. Norman, MA, DSc, C Chem, FRSC, FRS
M. J. Tomlinson, BSc, C Chem, MRSC
D. J. Waddington, BSc, ARCS, DIC, PhD

Practical Books

Inorganic Chemistry Through Experiment
G. F. Liptrot, MA, PhD

Organic Chemistry Through Experiment
D. J. Waddington, BSc, ARCS, DIC, PhD
H. S. Finlay, BSc

G. F. Liptrot
MA, PhD

Department of Chemistry
Eton College

J. J. Thompson
MA, PhD

Professor of Education
University of Bath

G. R. Walker
BA, BSc

Headmaster
The Cavendish School
Hemel Hempstead
Herts.

Modern Physical Chemistry

Bell & Hyman Limited
London

First published in Great Britain 1982
by Bell & Hyman Limited
Denmark House, 37–39 Queen Elizabeth Street,
London SEI 2QB

© G. F. Liptrot, J. J. Thompson and G. R. Walker, 1982

Reprinted 1982

ISBN 0 7135 2231 3

Liptrot, G. F.
Modern physical chemistry
1. Chemistry Physical and theoretical
I. Title II. Thompson, J. J. III. Walker, G. R. 541.3 QD 453.2

Printed by Times Printers Sdn. Bhd.,
Singapore.

Contents

List of Plates

Preface

The book has been divided into what, for convenience, may be termed the four major divisions of physical chemistry, namely structure, energetics, equilibrium and kinetics; the relevance of these four broad aspects of the subject is summarised in an introductory chapter entitled 'The Nature of Physical Chemistry'.

The science of physical chemistry lies at the root of most modern developments throughout chemistry, whether they be inorganic, organic, biochemical, geochemical etc., and scientists specialising in all branches of chemistry are increasingly using the methods of physical chemistry in their studies, e.g. spectroscopy, thermodynamics and kinetics. It is hoped that this picture will emerge to the reader of this book.

In order to include such topics as the wave-nature of the electron, spectroscopic methods and chemical thermodynamics, which now feature in some A-level syllabuses, some of the more traditional topics such as classical atomic and molecular theory and the nature of the colloidal state have been omitted. This has been done solely to keep the size of the book within bounds and not because it was felt that such items were unimportant.

The Stock nomenclature has been used throughout the book and so too have SI units, although it has been felt more appropriate in places to use the atmosphere as the unit of pressure rather than the pascal (newton per square metre).

The book deliberately goes beyond the immediate needs of existing A-level syllabuses and because of this the sections are numbered so that more advanced ones can be omitted on first readings; in addition, some of the more advanced material is in smaller type. This book contains all the physical chemistry that should be required for University Scholarship Examinations and National Certificate work. It is also designed for initial courses in advanced chemistry at Colleges and Polytechnics and will provide a useful foundation for University courses.

G. F. Liptrot
J. J. Thompson
G. R. Walker
1981

Acknowledgements

We are indebted to a number of organizations and individuals for permission to reproduce photographs and these are acknowledged on the pages where they appear. For the Table of Relative Atomic Masses, thanks are due to the International Union of Pure and Applied Chemistry. We have also drawn on many of the excellent diagrams used in other books and they are acknowledged with thanks. In particular we thank the following authors and publishers for allowing us to reproduce, or base new drawings on, the following:

An Introduction to Modern Chemistry (M. J. S. Dewar: Athlone Press), Plate 1; *Atomic Spectra* (G. Hertzberg: Dover Publications), Figs. 4.2, 4.3, 4.4; *Chemistry in Context* (G. C. Hill and J. S. Holman: Nelson), Fig. 16.1; *Chemistry—Experimental Foundations* (R. W. Parry *et al.*: Prentice-Hall), Fig. 2.6; *Chemistry in Modern Perspective* (G. Gordon and W. Zoller: Addison-Wesley), Fig. 13.6; *Covalent Bond* (H. S. Pickering: Wykeham Publications), Figs. 8.8, 8.9; *Fundamentals of Physical Chemistry* (S. H. Maron and J. B. Lands: Macmillan Publishing Co. Inc.), Fig. 3.2; *Investigation of Molecular Structure* (Bruce Gilbert: Bell & Hyman), numerous Figs. in Chapter 13; *Man-made Transuranium Elements* (G. T. Seaborg: Prentice-Hall), Fig. 21.8; *Physical Chemistry* (P. W. Atkins: Oxford University Press), Figs. 4.1, 4.5, 4.10, 7.9, 9.18, 14.4; *Physical Chemistry* (G. M. Barrow: McGraw-Hill), Fig. 13.9; *Physical Chemistry* (G. W. Castellan: Addison-Wesley), Fig. 12.1; *Physical Chemistry* (W. J. Moore: Longman), Figs. 4.8, 4.9; *University Chemistry* (B. H. Mahan; Addison-Wesley), Figs. 11.2(a)(b), 11.3(a)(b), 11.11, 11.12.

We also thank *Scientific American* (B. J. Alden and T. E. Wainwright) and the Royal Society (E. G. Cox, D. W. J. Cruickshank and J. A. S. Smith) for permission to reproduce Figs. 12.2 and 13.7 respectively.

The selected questions from recent examinations papers are reproduced by courtesy of the following examination boards: The Associated Examining Board; University of Cambridge Local Examinations Syndicate; University of Cambridge Registrar for Entrance Examinations; Joint Matriculation Board; University of London School Examinations Council (also administering Nuffield questions); Oxford and Cambridge Schools Examination Board; Oxford Delegacy of Local Examinations; University of Oxford Registrar for Scholarship and Entrance Examinations; Southern Universities' Joint Board for School Examinations; Welsh Joint Education Committee. Questions are individually identified by the appropriate initials.

Questions 4, 6, 7, 8 and 9 at the end of chapter 13 are reproduced from *Organic Chemistry: A Problem-solving Approach* by M. C. V. Cane and M. J. Tomlinson, published by Bell & Hyman Ltd., with the authors' permission. Information in Appendix I and Appendix II is also taken from the same source.

Finally mention must be made of the valuable help given by Professor D. J. Waddington in reading the book in typescript and proof and by Dr R. J. Mawby (University of York) and Mr H. S. Pickering (Uppingham School) who read the typescript in part, all of whom made numerous suggestions for improving it, and by Michael Liptrot for checking the answers to numerical problems.

1981

G. F. L.
J. J. T.
G. R. W.

The nature of physical chemistry

1.1
The physical chemist

The chemist studies the nature of the material world around him and the possibility of changing it to his advantage. This, in turn, requires a study of change: the energetics, the rate and the direction of change and ultimately the underlying reason for the change taking place. The physical chemist seeks explanations of these phenomena in terms of things which cannot be observed directly. Thus by using instrumental methods such as microwave spectroscopy and X-ray diffraction he is able to explore areas which cannot be investigated by the conventional methods of descriptive analytical chemistry.

It has long been realised that outward appearance alone gives little information about the nature and behaviour of materials. For instance glass and perspex are outwardly similar but the former material shatters readily whereas the latter does not. The physical chemist peels away the layers and probes more deeply until he reaches inside the atoms themselves. His starting point is the structure of the material and from this point he can begin to study and to predict the way in which the material will behave under a wide range of conditions.

A knowledge of **structure** will lead to an understanding of the energy changes that accompany a chemical reaction (**energetics**), the direction in which the reaction occurs (**equilibria**), and the rate at which it takes place (**kinetics**). The main sections of this book are therefore concerned with these four areas of physical chemistry: **structure**, **energetics**, **equilibria** and **kinetics**, as well as the relationships between them.

1.2
Structure

Appearance alone provides little information about the structure of materials but a variety of methods, of increasing sophistication, may be used to extend our limited perception. By way of illustration we shall consider the kind of information which leads to a greater knowledge of the structure of common salt, sodium chloride.

The fact that sodium chloride is a white, crystalline solid suggests a regular arrangement of particles in the crystal, but leaves unanswered the questions: what are the particles and how are they arranged? The physical chemist can now use some of the more simple techniques at his disposal. For example:

(a) Molten sodium chloride conducts an electric current. This implies the presence of ions and this is supported by the decomposition of the salt to give chlorine at the anode and sodium at the cathode.

(b) Sodium chloride dissolves readily in water to produce a solution which again conducts an electric current. The implication is that the solution contains ions.

(c) When small amounts of sodium chloride are dissolved in water, the solution freezes at a lower temperature than the pure solvent. Moreover the lowering of the freezing point can only be explained in quantitative terms by postulating that each unit of NaCl splits up into two parts on dissolving.

All the evidence infers the existence of ions in the original substance. However, we are no nearer a solution as to the internal arrangement of the ions within the solid crystal. A more sophisticated technique, known as X-ray diffraction, provides this information.

A beam of X-rays is passed through a crystal of sodium chloride to produce a diffraction pattern of the kind shown in Plate 7 (page 177). This, in turn, can be used to determine the arrangement of the ions within the crystal. A range of physical techniques can be used to investigate a variety of different structures.

We can go further still and probe the arrangement of the fundamental particles within the ions themselves. A trace of sodium chloride imparts an intense yellow coloration to a flame and spectroscopic analysis of the light energy gives important information about the distribution of the electrons within the sodium ion.

1.3 Energy changes

All chemical reactions are associated with changes in energy which may be transferred into a variety of forms. For example, coal may be burnt in an open fire to keep us warm (chemical energy transformed into thermal energy); on the other hand, the same basic chemical reaction carried out in a coal-fired power station produces electricity. The stages here involve the production of thermal energy, some of which is converted into work of the moving turbine wheels; finally some of this work is transformed into electricity. Using molten salts as electrolytes, progress is slowly being made towards the ideal of combining coal and oxygen in a fuel cell to give electrical energy directly, instead of having to go through the wasteful extra stage of raising steam in boilers to drive electrical turbines.

Since a chemical reaction must involve a change in structure, the magnitude of the energy transferred can provide information about the size and nature of the forces binding the structure together. For example, aluminium and iron both react with oxygen to form oxides of the same basic formula but one mole of aluminium oxide is formed from the elements with the transfer of considerably more energy than is the case for the formation of one mole of iron(III) oxide, when compared under the same conditions.

$$2Al(s) + \tfrac{3}{2}O_2(g) \longrightarrow Al_2O_3(s) \qquad \Delta H^\ominus = -1676 \text{ kJ mol}^{-1}$$
$$2Fe(s) + \tfrac{3}{2}O_2(g) \longrightarrow Fe_2O_3(s) \qquad \Delta H^\ominus = -822 \text{ kJ mol}^{-1}$$

Aluminium oxide is therefore energetically more stable than iron(III) oxide with respect to the elements from which each is formed, and we could seek an explanation of this in terms of the structures of the two materials. To the industrial chemist, the stability of oxides is crucial since oxides represent the major source of many of our most important metals. It is not surprising that the production of aluminium involves a massive transfer of energy (in the form of electrical energy) in order to convert aluminium oxide (Bauxite) to the metal.

1.4 Equilibrium

A chemical substance never exists as the absolutely pure isolated substance, despite all the precautions that we may take. More often than not it is surrounded by air, a rather damp reactive mixture of gases. Even in a vacuum the substance is at once surrounded by its own vapour. In other words, while we can get a long way by treating substances as they appear to be in isolation, we must eventually recognise that they are always in a particular environment. This might be an aqueous environment, or the atmosphere of another gas, or simply the vapour of the substance in question. In every case, however, it will affect its behaviour, just as the environment affects the behaviour of human beings. In a similar fashion, it might even be the determining factor.

It is no coincidence that man used copper as a metal before he used iron even though iron is in many ways the more useful metal. The interaction of iron with its normal environment of oxygen and moisture is more marked than that of copper. Once we have made iron from its oxide, we must fight a constant battle to prevent it returning to the oxide from which it was formed. The reaction is reversible and the choice of conditions will favour one or other direction. Under normal atmospheric conditions iron is continually 'on the move' changing to its oxide. Only when it has reached equilibrium with its surroundings will the change *apparently* stop. Such a state of equilibrium represents a position of stability towards which all chemical systems will move.

For iron, then, in conditions of oxygen and moisture, equilibrium is reached in a pile of rust. In many other chemical systems, however, the position of equilibrium is less one-sided. Water in a closed vessel is in an environment of water vapour. At a particular temperature the position of equilibrium is marked by the saturated vapour pressure. Appearances are deceptive, however, because the equilibrium state is not one of static stand-still but, at a molecular level, of dynamic but opposed processes. Increasing the temperature changes the position of the equilibrium until the saturated vapour pressure becomes equal to the atmospheric pressure and the water boils.

Since all systems *apparently* come to rest at equilibrium, it is not surprising that we shall be devoting a lot of attention to the study of chemical equilibria and to the factors which affect them.

1.5 Kinetics

The remaining factor in the study of chemical change is concerned with the rate at which the change takes place; this is a study of chemical kinetics. There are many reactions (a large number of them of great commercial importance) which, despite favourable energetic factors, move only slowly towards a position of equilibrium. The control of a reaction requires a study of what it is that determines the rate of the reaction and how we can affect it.

A single reaction can take place at many different rates. For example, a jet of hydrogen can be safely burnt in an atmosphere of chlorine but a mixture of the gases explode violently when irradiated with ultra-violet light.

1.6 The applications of physical chemistry

The concepts and techniques of physical chemistry have been developed in the past but their importance lies in the future, in the predictions which they enable us to make and in the problems that they allow us to solve.

The modern chemist is confronted by millions of different compounds and he must try to formulate basic laws that will rationalise their behaviour. These laws, which must apply to all existing systems and hopefully to those still to be discovered, are the laws of physical chemistry.

Sheer curiosity has in the past promoted many of the questions that have stimulated the progress of physical chemistry. Graham, in 1846, never imagined that his Law of Diffusion would provide the basis for the purification of the fuel that was contained in the first atomic bomb that exploded almost 100 years later. Nowadays, the problems to be solved are often more pressing and sometimes a matter of life or death.

The following four examples illustrate the contribution of physical chemistry in some important fields:

(a) Few scientific discoveries have captured the imagination like that of the 'double helix' of D.N.A. The context is a biochemical one, yet so much has depended upon the elucidation of its **structure** by physical methods and, in particular, by the use of X-ray diffraction techniques.

(b) Heart surgery has been the source of intense interest and controversy in recent years. Major surgery of this kind is liable to upset the delicately balanced state of biochemical **equilibrium** that exists in our bodies. To prevent this happening it is necessary to understand the complex equilibria that are set up, and especially those which control the level of acidity.

(c) Concern over the conventional 'fossil fuels' centres not only around the amount of pollution that they cause to the environment, but also around the problems posed by their rapid consumption by modern society. What shall we do when our fossil fuel supplies have become exhausted? One answer lies in the use of nuclear power, but any answer depends on the understanding and control of **energy**. Once again this is the concern of the physical chemist.

(d) Efforts to solve the world shortage of food depend to a large extent on increasing the rate at which atmospheric nitrogen can be 'fixed' or converted into nitrate. This is a problem of **kinetics** and physical chemists have already made some progress in improving the rate of the Haber process by the use of new catalysts such as titanium isopropoxide.

Many problems of increasing complexity remain to be solved. They will yield not to a hit-or-miss approach, but rather to an understanding of the behaviour of materials at the most fundamental level—the level of physical chemistry.

Structure

The structure of the atom

2.1
Historical development

From the earliest classical writings there have been references to the atomic nature of matter. There was much speculation on what would be left finally if a solid substance was continually divided in two. The word 'atomic' derived from an adjective in Greek meaning 'not divisible' represented the views of one school of thought, the principal advocates being the Greek philosopher Leucippus and his pupil Democritus (c. 400 BC). This idea was soon forgotten and it was not revived again until very much later by Boyle and Newton.

It was left to Dalton in 1808 to establish the atomic nature of matter based on experimental evidence. He produced a first list of relative atomic masses on a scale in which hydrogen was taken as 1 and other elements related to it. Despite inaccuracies in his experimental results he was able to explain the formation of new compounds by the combination of atoms and was thus able to understand the Law of Constant Composition (Proust 1799) and the Law of Multiple Proportions (Dalton 1803).

During the nineteenth century more accurate values of relative atomic masses became available and it soon became apparent that many were not whole numbers, a dilemma which remained unresolved until the nature of the atom itself had been more closely studied (p. 14). Another problem was posed when it became evident that 'atoms' could apparently combine together and yet produce the same number of particles. Nowadays we find it difficult to imagine the confusion that the simple reaction

$$H_2(g) + Cl_2(g) \longrightarrow 2HCl(g)$$

caused to chemists who, at that time, had no conception of a molecule. The problem was solved in 1811 by Avogadro but his solution was not appreciated until the work of Cannizzaro about 50 years later.

With the invention of the mercury vacuum pump in 1855 a piece of apparatus was available which allowed experiments to be carried out on the conduction of electricity through gases at low pressure in discharge-tubes. The following years saw a period of unprecedented scientific progress leading, first, to J. J. Thomson's 'plum-pudding' model of the atom. The most important landmarks in this period are summarised below.

2.2
Steps to the Thomson atom

1876   Goldstein passed a high-voltage electrical discharge through various gases at reduced pressure and discovered negatively charged particles (cathode rays as he called them). These were in fact electrons, produced by the splitting up of the gaseous atoms by the discharge. At low pressures the fragments did not recombine and the electrons were attracted towards the positive electrode.

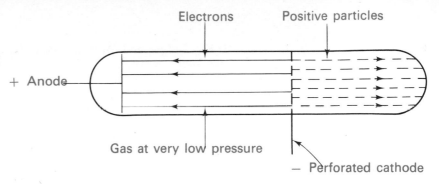

FIG. 2.1 .*Apparatus for producing negative and positive ions*

1897   Wien measured the charge/mass ratio of the positively charged fragments produced in discharge-tubes (positive rays as they were then called). He showed that the value depended upon the nature of the gas in the tube, and that the highest charge/mass ratio, i.e. the smallest positively charged particle, was obtained when the gas was hydrogen. This particle is now known to be the proton.

1897   Further work on gas discharge-tube reactions culminated in Thomson's measurement of the charge/mass ratio of the negatively charged particles (p. 8). He showed that the value was independent of the nature of the gas which suggested that the particles, called electrons, were constituents of all atoms. This was not finally proved until 12 years later when Millikan measured the charge on the electron (p. 9).

2.3
The Thomson atom

The results of the discharge-tube experiments described in the previous section may be summarised as follows:

Atoms may be split up into positively charged and negatively charged parts. The negatively charged part, the electron, is common to all atoms but the positively charged part differs from atom to atom. It is likely that the positively charged part has a much greater mass than that of the electron, but since only charge/mass ratios have so far been determined this cannot be said with certainty.

These findings were interpreted by Thomson in terms of a 'plum-pudding' model of the atom. This consisted of a positively charged 'pudding' of about 10^{-10} m in diameter and containing the main mass of the atom. Embedded in the pudding were just enough light, negatively charged electrons to make the whole atom electrically neutral.

2.4
Steps to the Rutherford atom

Thomson's model of the atom was based largely on the evidence from discharge-tube experiments. Rutherford's model, which was to replace Thomson's, depended on the newly-discovered phenomenon of radioactivity.

1896   Becquerel discovered a source of radiation present in all uranium salts which he called radioactivity. Two years later the Curies (husband and wife) followed up this work and succeeded in isolating two new elements, polonium and radium, from a uranium-containing ore called pitchblende.

1902   Rutherford showed that one type of radioactivity, α-radiation, was due to positively charged helium atoms. This suggested the existence of a fundamental, positively charged unit within the atom.

1910   Geiger and Marsden, two of Rutherford's pupils, bombarded a thin sheet of gold foil with a beam of α-particles. Some particles were deflected off-course and diverged, but the majority passed straight through the gold foil with little of no disturbance. The exciting fact that emerged was that an incredibly small number of the α-particles (about 1 in 20 000) were deflected backwards through angles greater than 90° (fig. 2.2). Rare though these deviations were, they occurred too frequently to be explicable on the Thomson model.

FIG. 2.2. *Representation of Geiger and Marsden's experiment. At A and B, α-particles collide with gold nuclei.*

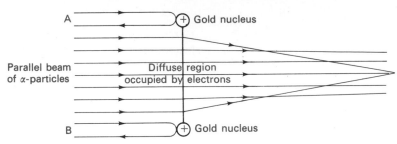

2.5
The Rutherford atom

Rutherford reasoned that the large angle scattering of the α-particles, observed by Geiger and Marsden, must be due to a collision or near collision with an incredibly small nucleus, which carried a positive charge approximately numerically equal to half the relative atomic mass of the atom. Furthermore virtually all the mass of the atom resided in this nucleus. The atom was visualised as containing a small positive nucleus where practically all the mass resided, and itself compounded from a number of positively charged particles (later called protons) together with a number of electrons. Around this nucleus there was a rather diffuse region containing sufficient electrons to maintain electrical neutrality. It has been estimated that nuclei of atoms have radii in the order of 10^{-14} to 10^{-15} m and that the outermost electron clouds are about 10^{-10} m from the centre of the nucleus, i.e. an atom magnified to about 60 cm in diameter would contain a nucleus no larger than 0·003 cm in diameter, which is about the same size as a very fine grain of sand. On this basis, it has been calculated that less than 10^{-12} of the volume of the atom is occupied by material particles.

2.6
The charge and mass of the electron

In 1897 Thomson measured the charge/mass ratio of the electron and also its velocity when produced in a discharge-tube. The apparatus is shown diagrammatically in fig. 2.3. Electrons from the cathode were passed through a slit in the anode and then through a second slit. They then passed between two aluminium plates spaced about 5 cm apart and eventually fell onto a fluorescent screen producing a well-defined spot. The position of the spot was noted and the magnetic field switched on, causing the electron beam to move in a circular arc while under the influence of this field.

An electric field was now applied in opposition to the magnetic field and gradually increased until the spot returned to its original position. If,

$$B = \text{magnetic flux density}$$
$$e = \text{charge on the electron}$$
$$v = \text{velocity of the electron}$$
$$m = \text{mass of the electron}$$
$$r = \text{radius of the arc in which the electron moves}$$

then the magnetic force, Bev, acting on each electron causes it to accelerate in the direction of the force, and thus to move along the arc of a circle.

FIG. 2.3. *Thomson's apparatus for producing negative and positive ions*

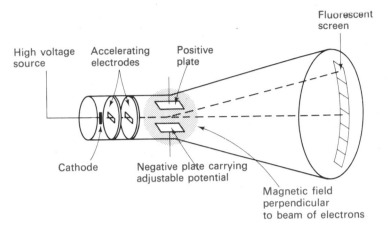

Thus:

$$Bev = \frac{mv^2}{r} \qquad \text{or} \qquad \frac{e}{m} = \frac{v}{rB} \tag{1}$$

When an electric field is applied in opposition (E) so that the electric force on each electron balances the magnetic force:

$$Ee = Bev \qquad \text{or} \qquad v = \frac{E}{B} \tag{2}$$

The velocity of the electrons can be calculated from equation (2). It is found that they travel at about 3×10^7 m s^{-1}, i.e. about $\frac{1}{10}$ the velocity of light. Substituting for v in equation (1) gives

$$\frac{e}{m} = \frac{E}{rB^2}$$

and since r can be determined from simple geometry knowing the dimensions of the apparatus, the value e/m can be evaluated. Thomson obtained a value in the order of 10^{11} C kg^{-1}, the accepted value for this ratio is now $1\cdot758\ 9 \times 10^{11}$ C kg^{-1}.

2.7
Conclusive proof that electrons are particles

Between 1910 and 1914 Millikan conducted a series of carefully controlled experiments to determine the value of the electronic charge; his apparatus is shown diagramatically in fig. 2.4. Using a telescope he was able to observe minute droplets of oil falling between parallel metal plates. Some of the droplets acquired a charge, either by friction on production in the spray, or from a beam of X-rays which could be specially used to charge the droplets. When a potential of several thousand volts was applied across the parallel metal plates, the charged droplets could be made to drift upwards towards the anode. Millikan was able to relate the rate of drift of a particular droplet to its charge, which could be varied by means of the X-ray beam.

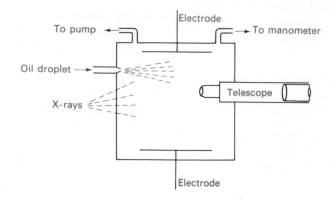

If an oil droplet of mass m and charge q_1 falls under gravity its terminal velocity, v_1, given by Stokes' Law, is proportional to the force acting on it:

$$mg = kv_1 \tag{3}$$

On applying an electric field, E, the droplet moves upwards towards the anode with a new constant velocity, v_2:

$$Eq_1 - mg = kv_2 \tag{4}$$

Equations (3) and (4) combine to give:

$$q_1 = \frac{mg(v_1 + v_2)}{Ev_1} \tag{5}$$

If the charge on the droplet is now changed by X-radiation to q_2 the new velocity becomes v_3 and

$$q_2 = \frac{mg(v_1 + v_3)}{Ev_1} \tag{6}$$

Equations (5) and (6) combine to give:

$$q_2 - q_1 = \frac{mg(v_3 - v_2)}{Ev_1}$$

Since mg/Ev_1 is constant, the change in charge is proportional to the change in velocity.

FIG. 2.5. *The forces acting on an oil droplet (a) uncharged, (b) charged q_1, (c) charged q_2. The upward and downward forces balance in each example, since a constant terminal velocity is achieved in each case*

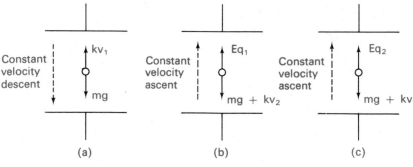

10

Millikan found a basic change in velocity fitted all other changes, i.e. all other changes were multiples of this basic value. He attributed this basic velocity change to a change in the charge carried by the oil droplet of just one unit, i.e. of one electron. From a knowledge of m (using Stokes' Law), E and the observed velocity v_1, Millikan was able to determine the value of this basic charge unit which he found to be independent of the size and material of the droplet and the source of the charge. The accepted value for the charge on the electron is now 1.602×10^{-19} C.

From Thomson's value of e/m and Millikan's value of e, the mass of the electron could be determined. The accepted value is now 9.11×10^{-31} kg.

2.8
The charge on the nucleus

Rutherford determined the charge on the nucleus of an atom from the α-particle scattering experiments (p. 8). Even his rather inaccurate values pointed to a relationship between nuclear charge and the 'serial number' of the element (the number given to the element when placed in ascending order of relative atomic mass). Moreover, experiments involving the scattering of β-particles (electrons emitted from the nucleus of certain radioactive elements) and of X-rays, carried out by Rutherford and Barkla respectively, suggested that the number of electrons in the atom, and thus the positive charge on the nucleus, was equal to about half the value of the relative atomic mass of the element concerned. In 1920, Chadwick carried out some β-particle scattering experiments of sufficient accuracy to show conclusively that the charge on the nucleus is equal to the serial number, or atomic number, of the element.

The importance of the atomic number of an element had already been made clear in 1913 by Moseley's experiments with X-ray spectra.

Atomic number

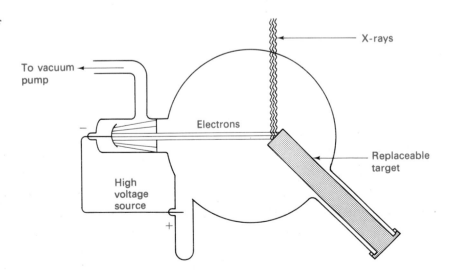

FIG. 2.6. *A schematic drawing of an X-ray tube*

To vacuum pump

Electrons

X-rays

Replaceable target

High voltage source

If a solid anode is bombarded by fast-moving electrons X-rays are produced with wavelengths characteristic of the particular element of which the anode is made (figs. 2.6 and 2.7). The spectral lines, which are now known to be caused by the displacement of electrons from the innermost orbits of the atoms appear in a series of groups called the *K*, *L*, *M*, *N* and *O* series, each of which resolves into a number of lines against a background of continuously varying wavelength.

FIG. 2.7. *Diagram of characteristic X-ray spectra*

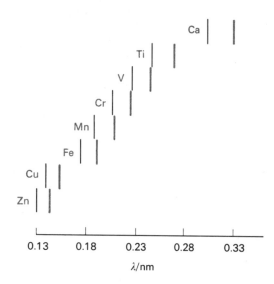

Moseley measured the wavelengths of corresponding lines within a particular series for different elements and showed that there was a linear relationship between the square root of the frequency of the line and a number which identified as the atomic number of the element. No such relationship exists with the relative atomic mass of the element (fig. 2.8) and thus it was realised that the atomic number, and not the relative atomic mass, was the fundamental property of the nucleus. As already mentioned above, subsequent scattering experiments identified the atomic number (until then just the serial number) with the positive charge carried by a particular nucleus.

FIG. 2.8. *Graph showing linear relationship between atomic numbers and the square root of the frequency for the K_α spectral lines. (a) Plot of relative atomic mass against $\sqrt{Frequency}$, (b) plot of atomic number against $\sqrt{Frequency}$*

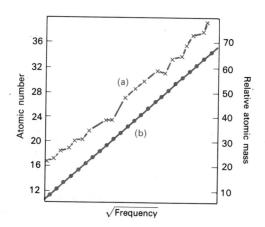

The Periodic Table

Since 1869, the elements in the Periodic Table had been arranged in order of increasing relative atomic mass. Moseley's measurement of atomic numbers from X-ray spectra enabled them to be put in a true order for the first time and thus many anomalies of the old Periodic Table were removed. For example, potassium with a relative atomic mass of 39·10 should immediately succeed argon, despite the fact that argon has a slightly larger relative atomic mass (39·95). Similarly iodine (relative atomic mass of 126·90) should immediately succeed tellurium (relative atomic mass of 127·60). Mendeléeff himself had been aware of examples of this kind and had carried out some corrections. Moseley's work resolved the difficulty; thus argon was found to have an atomic number of 18 and potassium one of 19. Similarly the atomic numbers of tellurium and iodine were found to be 52 and 53 respectively.

The elements lutetium and tantalum, hitherto shown as adjacent in the Periodic Table, were found to have atomic numbers of 71 and 73 respectively. Element 72, hafnium, was later identified in the ore of zirconium, an element which it resembles so closely that it is extremely difficult to separate the two. Elements 43 and 61 were also found to be missing and have since been identified in the products of nuclear piles (p. 60). They are technetium and promethium respectively.

The proton

Gas discharge-tube experiments (p. 7) had established that the positive fragment produced when hydrogen was used in the tube had the largest charge/mass ratio. From its position in the Periodic Table, the hydrogen atom was known to have a nuclear charge of +1 and it was logical to assume that the hydrogen nucleus, called the proton, was the fundamental unit of positive charge and the building block for all other elements.

The charge/mass ratio for the proton was found to be $9·57 \times 10^7$ C kg^{-1}. Electrolysis experiments confirm this value: thus 1 mole of protons (assuming each to have a single positive charge) carries a total of 1 Faraday, i.e. 96 500 C. The charge/mass ratio for the proton is therefore

$$96\,500/1·008 \times 10^{-3} = 9·57 \times 10^7 \text{ C kg}^{-1}$$

Assuming the charge on the proton to be the same size as that on the electron, i.e. $1·602 \times 10^{-19}$ C, the mass of the proton can be calculated to be $1·67 \times 10^{-27}$ kg. This means that the ratio of the mass of the proton to the mass of the electron is:

$$\frac{1·67 \times 10^{-27}}{9·11 \times 10^{-31}} = 1836$$

i.e. the mass of the proton is 1836 times the mass of the electron.

2.9
Isotopes

Even without Moseley's work on atomic numbers (p. 12), the chemist's faith in relative atomic mass as a fundamental quantity had been disturbed by the discovery that the same element could apparently exist in forms that had different atomic masses.

13

Since its discovery in 1896 the phenomenon of radioactivity had been the subject of increasing study (p. 49) and in 1910 scientists were puzzled by pairs of radioactive elements which seemed to be chemically identical, but which nonetheless had very different radioactive decay properties. Soddy invented the word **isotope** for these elements.

In 1913 Thomson had moved on from his study of electrons to a closer investigation of the positively charged ions that were produced in gas discharge-tubes. Using a technique similar to that which led to the determination of e/m for the electron (p. 8), he applied simultaneous electric and magnetic fields to a beam of positive ions produced from the gas neon. Particles of the same charge/mass ratio were focussed on the arc of a parabola and as well as detecting an arc corresponding to Ne^+ of mass 20 a faint trace at mass 22 was observed. However, the experiment was not conclusive since this trace could have been due to the presence of carbon dioxide impurity, i.e. doubly ionised carbon dioxide, CO_2^{2+}, rather than to an isotope of neon.

In 1919 Aston developed Thomson's technique into the mass spectrograph, the forerunner of the modern mass spectrometer (p. 198). This was a much more accurate instrument which focussed particles of the same charge/mass ratio on a line instead of on an arc. Using extremely pure neon, Aston still detected lines at 20 and 22 and he concluded that neon exists in two chemically identical forms having the same atomic number (10) but different atomic masses (20 and 22). The measured atomic mass of naturally-occurring neon was known to be 20·183 and this implied that the heavier atom is present with an abundance of about 9%. The two forms of neon are isotopes of the same element, being atoms of the same atomic number but different atomic mass. Aston actually separated neon-20 and neon-22 and showed that they had different atomic masses.

Isotopes of other elements were soon identified. Chlorine, for example, atomic mass 35·5, was shown to have a composition of 75% chlorine-35 and 25% chlorine-37. At last it had become evident why elements had non-integral atomic masses because they are, in many cases, mixtures of isotopes. Nevertheless, the origin of isotopes could not be understood until the composition of the nucleus had been further elucidated and, in particular, until the neutron had been discovered.

FIG. 2.9. *The mass spectrum of naturally occurring lead*

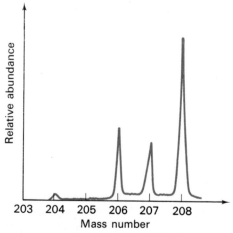

By 1920 the structure of the atom had been largely unravelled. The knowledge that had been gained in the previous 30 years may be summarised as follows.

The atom was known to consist of a tiny, positively charged nucleus, in which most of the atomic mass was concentrated, surrounded by negatively charged electrons. Most of the atom is 'empty space'.

The masses and charges of the electron and proton, the two fundamental particles in the atom, had been determined.

Moseley's work on X-ray spectra had led to the recognition of the atomic number as a fundamental atomic quantity, rather than atomic mass, Elements were now classified in the Periodic Table in increasing order of atomic number.

It was recognised that an element could exist in different isotopic forms each with the same atomic number but different atomic masses, but the reason for this was not then known.

2.10
The neutron

In the 1920s it was thought that the difference between the atomic mass and the atomic number of an isotope could be explained in terms of neutral (proton + electron) pairs (Table 2.1).

Table 2.1 The (proton + electron) pair explanation of atomic masses

	In the Nucleus	Around the Nucleus
Neon-20	10 protons + 10(proton + electron) pairs	10 electrons
Neon-22	10 protons + 12(proton + electron) pairs	10 electrons

The idea of neutral (proton + electron) pairs was appealing since it seemed to explain the emission of β-particles (high energy electrons) from the nuclei of certain radioactive elements and it suggested that the electrons might act as the 'nuclear glue' which held the protons together. However, as early as 1920, Rutherford suggested that a neutral particle might exist in the nucleus. He realised that its lack of charge would make its detection difficult, since it would be unaffected by electric and magnetic fields which had been the key to techniques hitherto used to study atomic particles.

In 1932 Chadwick discovered the neutron—the missing neutral particle. He knew that when beryllium is bombarded with α-particles it emits a particle of great penetrating power and which is unaffected by electric and magnetic fields. This new particle is the neutron and its formation can be represented thus,

$$\mathrm{^{9}_{4}Be} + \mathrm{^{4}_{2}He} \longrightarrow \mathrm{^{12}_{6}C} + \mathrm{^{1}_{0}n}$$

where the superscript refers to the atomic mass and the subscript to the atomic number.

It now became clear that two isotopes of an element have the same number of protons in their nuclei (and, therefore, electrons round them) but different numbers of neutrons. Table 2.2 lists the stable isotopes of sulphur.

Table 2.2 Stable isotopes of sulphur (atomic number 16)

| Isotope | Structure of the nucleus | | Abundance (%) |
	Number of protons	Number of neutrons	
^{32}S	16	16	95·0
^{33}S	16	17	0.76
^{34}S	16	18	4.2
^{36}S	16	20	0.021

The fact that neutrons have been observed outside the nucleus of an atom does not necessarily mean that they exist as fundamental particles inside the nucleus. Even a free neutron is unstable and decays into a proton, an electron and a neutrino. It is still possible that the neutron exists as a (proton + electron) pair within the nucleus.

It is now known that there are many more nuclear particles in addition to the proton and neutron. The most important of these are described in Chapter 5.

2.11
Important definitions

The **atomic number** of an element is the number of protons contained in the nucleus of the atom and is given the symbol Z. It is the same as the number of electrons surrounding the nucleus. The atomic number is a fundamental characteristic of an element, each element having a different atomic number.

The **mass number** is the sum of the protons and the neutrons in a particular nucleus. Unlike the atomic mass, it is a whole number and is given the symbol A.

A **nuclide** is any nuclear species of given mass number (A) and atomic number (Z). It is written as the element symbol with the mass number and atomic number as the left-hand superscript and subscript respectively, e.g.

$$^{16}_{8}O \quad \text{and} \quad ^{32}_{16}S$$

Although there are only 104 elements yet characterised, there are about 1400 different nuclides. Table 2.2 lists no less than 4 stable nuclides for the element sulphur.

The **relative atomic mass** is the ratio of the mass of a particular atom to $\frac{1}{12}$ of the mass of an atom of the nuclide $^{12}_{6}C$. This scale was adopted in 1961, previous scales being based on hydrogen and oxygen.

Table 2.3 illustrates the terms defined above for the case of the element chlorine.

Table 2.3 The atomic properties of chlorine

Nuclide	Atomic Number (Z)	Mass Number (A)	Relative Atomic Mass	Natural Abundance (%)
$^{35}_{17}Cl$	17	35	34·97	75·8
$^{37}_{17}Cl$	17	37	36·97	24·2

The atomic mass of an isotopically mixed element is the weighted mean of the isotopic masses; thus for the element chlorine the relative atomic mass of the naturally occurring element is

$$\frac{34\cdot97 \times 75\cdot8}{100} + \frac{36\cdot97 \times 24\cdot2}{100} = 35\cdot45$$

Table 2.4 lists some of the important properties of the fundamental atomic particles that have been discussed in this Chapter.

Table 2.4 Properties of the electron, proton and neutron

	Proton	Neutron	Electron
Symbol	p	n	e
Mass	$1\cdot672\ 52 \times 10^{-27}$ kg	$1\cdot674\ 82 \times 10^{-27}$ kg	$9\cdot109\ 1 \times 10^{-31}$ kg
Charge	$1\cdot602\ 10 \times 10^{-19}$ C	0	$1\cdot602\ 10 \times 10^{-19}$ C
Mass relative to the electron	1836	1839	1
Charge relative to the proton	+1	0	−1

2.12
Questions on chapter 2

1 (a) When the elements are arranged in order of increasing relative atomic mass, the 19th element is an inert (noble) gas. What are the characteristics of the 19th element when the elements are arranged in order of increasing atomic number? State your reasons and suggest an explanation for the difference.

(b) When listed in order of increasing atomic number, the 12th, 26th and 34th elements all exhibit some form of divalency. Give the electronic configurations of these elements and show how the different forms of divalency may be explained. (*N.B.* You are not required to identify these three elements.)
Give equations to represent the reactions that take place between
(i) the hydride of the 20th element and the hydride of the 8th element,
(ii) the hydride and oxide of the 16th   C

2 Outline the use of the mass spectrometer in the determination of relative atomic mass.
Explain briefly why H = 1 and O = 16 are no longer used as standards for the determination of relative atomic mass.
What do you understand by the term *isotope*? Illustrate your answer with examples drawn from **two** elements.
The mass spectrum of a certain element consists of a number of closely spaced lines. The intensity of one of these lines decreases over a period of time as two new lines, one of which corresponds to a relative mass of 4, simultaneously appear. Explain these observations as fully as you can.   C (Overseas)

3 (a) Outline the principles involved in determining relative atomic mass using a mass spectrometer.

(b) The isotope $^{211}_{83}$Bi undergoes radioactive decay becoming the isotope $^{207}_{83}$Bi. Trace a possible path for this conversion giving (i) the types of radiation emitted, and (ii) the relative atomic mass and atomic number of the intermediate isotopes.

(c) What is the exact nature of these radiations, and how does each of them behave when
  (i) directed towards a thin aluminium sheet
  (ii) influenced by a magnetic field
  (iii) passed through a gaseous element?   AEB

4 Fig. 1 represents one form of a mass spectrometer.

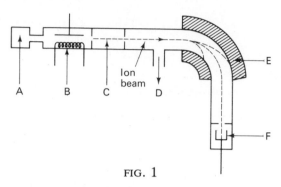

FIG. 1

(a) Identify the function of, or process occurring in, each of the parts of the instrument labelled A to F inclusive.

A mixture of $^2_1H_2$ and $^{81}_{35}Br_2$ was analysed in the mass spectrometer. The following pattern of lines due to singly charged ions was obtained:

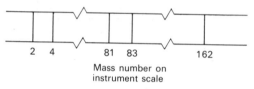

FIG. 2

(b) State which ions give rise to each of these lines.
(c) What apparent mass number would register on the instrument scale if the heaviest of these ions acquired a second charge?   C (Overseas)

5 Fig. 3 illustrates the ion paths in a mass-spectrometer, using bromine gas. (Bromine, atomic weight (relative atomic mass) 79·91, consists entirely of isotopes of mass numbers 79 and 81.)

Each of the groups of lines, A, B and C, is caused by one of the ions $Br^+(g)$, $Br_2^+(g)$ and $Br^{2+}(g)$. State which ion causes the lines in
(a) group A (b) group B (c) group C.

Identify each of the lines in group B and C, and state which line in each group is the most intense.   C

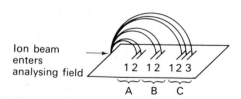

FIG. 3

The behaviour of electrons

3.1
The failures of the Rutherford atom

Rutherford's model of the atom (p. 8), proposed in 1911, consisted of a positively charged nucleus, in which was concentrated most of the mass of the atom, surrounded by negatively charged electrons whose orbits around the nucleus determined the size of the atom.

Rutherford's model agreed with the experimental evidence, but it contradicted some current theories, and in particular the theory of electromagnetic radiation. This predicted that charged particles, undergoing an acceleration, would continuously emit radiation and thus lose energy. According to this theory, the Rutherford model of the atom could not have a stable existence, since electrons are charged particles and, in circling the nucleus, are subjected to an acceleration. They should thus lose energy and spiral into the nucleus with the subsequent collapse of the atom.

A second contradiction was presented by the nature of atomic spectra, which consisted of a series of discrete lines in different parts of the electromagnetic spectrum. Plate 1 shows that part of the spectrum of atomic hydrogen which occurs in the visible region. In 1885 Balmer showed that there was a simple mathematical relationship between the frequencies of these lines, but at that time there was no reason to suppose that an atom represented by the Rutherford model should produce a line spectrum. A continuous spectrum would have been more likely.

In 1913 these contradictions were resolved by Bohr who commenced the study of the behaviour of the electrons within the atom.

Plate 1. The spectrum of atomic hydrogen—the Balmer series

3.2
The Bohr atom

The spectrum of atomic hydrogen (other atomic spectra are similar but more complex) consist of several series of discrete lines which converge in different parts of the electromagnetic spectrum. Three such series, named after their respective discoverers, are shown in fig. 3.1. In order to explain the origin of this spectrum Bohr adopted the principles of the quantum theory, then newly discovered, and applied them to the problem of the electron revolving around the nucleus of the hydrogen atom. He made the following assumptions:

(a) The Rutherford model of the atom is essentially correct.

(b) Each spectral line is produced by a single electron.

(c) Electrons can exist only in orbits of definite ('quantised') angular momentum and energy.

FIG. 3.1. *Series in spectra of atomic hydrogen*

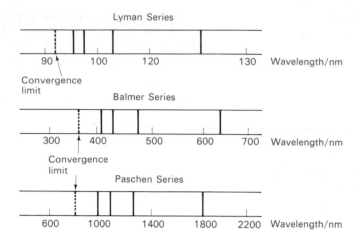

(d) The angular momentum J of these permitted orbits is given by

$$J = nh/2\pi \tag{1}$$

n = integer
h = Planck's constant
Planck's constant
$= 6.625\ 6 \times 10^{-34}$ J s

(e) An electron emits energy in the form of radiation when it moves from a higher to a lower permitted orbit; this produces a line in the atomic emission spectrum and this line will be of a particular energy, since the energies of the higher and lower orbits are fixed. Similarly, on moving from the lower to a higher orbit, the electron absorbs the same quantity of energy and the line is produced in the absorption spectrum.

(f) The difference in energy ΔE of the two orbits in question is related to the frequency of the radiation by Planck's equation:

v = frequency of radiation

$$\Delta E = hv \tag{2}$$

For example, the wavelength of the yellow D-line in the atomic spectrum of sodium is 589 nm; the difference in energy between the two orbits concerned in this change ΔE is given by

c = speed of light in vacuum
λ = wavelength of light

$$\Delta E = hv = hc/\lambda$$
$$= \frac{6.625\ 6 \times 10^{-34} \times 2.998 \times 10^{8}}{5.89 \times 10^{-7}} \text{ J}$$
$$= 3.37 \times 10^{-19} \text{ J}$$

The Principal Quantum Number

In its most stable state, known as the **ground state**, the single electron in the hydrogen atom is in an orbit of minimum energy, closest to the nucleus. Each orbit in the Bohr atom is denoted by the **principal quantum number**, n, (the integer in equation (1), above and in the ground state $n = 1$. Absorption of energy (for example, from an electrical discharge, a light source, or a flame) will promote the electron into an orbit of higher energy where $n = 2, 3$ or 4 etc. The atom is now said to be in an **excited state**. If the electron falls to the lowest orbit, the atom returns to its ground state and the energy emitted produces a line in the Lyman series of the atomic spectrum, in accordance with equation (2). The Lyman series occurs in the ultra-violet region of the electromagnetic spectrum.

20

In a similar manner, the Balmer, Paschen, Brackett and Pfund series in the atomic spectrum of hydrogen correspond to electrons in excited orbits dropping back, not this time to the lowest orbit, but to orbits where $n = 2, 3, 4$ and 5 respectively (fig. 3.2). In this manner Bohr was able to account for the general appearance of the spectrum of atomic hydrogen but, unlike Rutherford, he treated his model mathematically in order to derive results that could be much more rigorously tested.

FIG. 3.2. *Bohr orbits for the hydrogen atom (not to scale)*

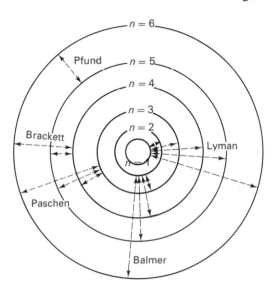

The mathematics of the Bohr atom

The most crucial of Bohr's assumptions (item (d) on p. 20) can be put in terms of the equation:

$$mvr = nh/2\pi \tag{3}$$

m = mass of electron
v = velocity
r = radius of orbit

The electrostatic force of attraction between the electron and the proton in the nucleus can be equated to the centripetal force necessary to hold the electron in a circular orbit:

$$ke^2/r^2 = mv^2/r \tag{4}$$

k = a constant $\left(\dfrac{1}{4\pi\varepsilon_0}\right.$ where

ε_0 is the permittivity of free space$\left.\right)$

Equations (3) and (4) can be combined to give:

$$r = n^2h^2/4\pi^2mke^2 \tag{5}$$

From equation (5) Bohr was able to calculate r, the radius of the hydrogen atom in its ground state, to be 0·052 9 nm, which agreed with the value obtained from the kinetic theory of gases.

He went on to calculate the total energy of the electron in a particular orbit:

$$\text{Total energy} = \text{Kinetic energy} + \text{Potential energy}$$
$$= \tfrac{1}{2}mv^2 - ke^2/r \tag{6}$$

(The negative sign appears in the above equation, since the zero of potential energy is reckoned as being when the electron is an infinitely

21

large distance from the nucleus). Combining equations (4) and (6) we have:

$$\text{Total energy} = ke^2/2r - ke^2/r$$
$$= -ke^2/2r \qquad (7)$$

Similarly combining equations (5) and (7) we have:

$$\text{Total energy} = -2\pi^2 mk^2 e^4/n^2 h^2 \qquad (8)$$

Using equation (8) Bohr calculated the energy change ΔE from an orbit of principal quantum number n_1 to another of principal quantum number n_2 where $n_2 > n_1$:

$$\Delta E = \frac{2\pi^2 mk^2 e^4}{h^2}\left[\frac{1}{n_1{}^2} - \frac{1}{n_2{}^2}\right] \qquad (9)$$

Thus, using equation (2), the frequency of the corresponding spectral line is given by:

$$v = \frac{2\pi^2 mk^2 e^4}{h^3}\left[\frac{1}{n_1{}^2} - \frac{1}{n_2{}^2}\right] \qquad (10)$$

In 1885 Balmer noticed that the frequencies of the lines in the visible region of the spectrum of atomic hydrogen fitted the relationship:

$$1/\lambda = R_H(1/2^2 - 1/n^2) \qquad \text{or} \qquad v = cR_H(1/2^2 - 1/n^2)$$

Bohr's work, culminating in equation (10), explained the basis of this relationship and he calculated R_H, the so-called **Rydberg Constant**, to be $1 \cdot 097\ 37 \times 10^7$ m^{-1}, the experimental value being $1 \cdot 096\ 78 \times 10^7$ m^{-1}. The excellent measure of agreement provided further support for Bohr's model of the atom.

Equation (10) can be used to calculate the wavelengths of lines in the different series of the spectrum of atomic hydrogen. For example, the first line in the Balmer series corresponds to an electron transition from $n = 3$ to $n = 2$. The frequency of this line is therefore given by the equation:

$$v = \frac{2\pi^2 mk^2 e^4}{h^3}\left[\frac{1}{2^2} - \frac{1}{3^2}\right]$$

The constants in this equation have the following values (correct to the second decimal place):

$$m = 9 \cdot 11 \times 10^{-31}\ \text{kg},$$
$$k = 8 \cdot 99 \times 10^9\ \text{N m}^2\ \text{C}^{-2},$$
$$e = 1 \cdot 60 \times 10^{-19}\ \text{C},$$
$$h = 6 \cdot 63 \times 10^{-34}\ \text{J s}.$$

Substitution of these values into the above equation gives a value for the frequency of the first line in the Balmer series. The frequency is related to the wavelength by the equation:

$$v = c/\lambda$$

The wavelength of the first line in the Balmer series is calculated to be 656 nm, in agreement with the experimental value and thus providing more support for the Bohr theory of the atom.

The Bohr theory of the atom is able to account for the observed spectral lines of atomic hydrogen with exceptional precision. However, it needs a good deal of modification when applied to other atoms, and the problem of how the electrons are arranged in other atoms is best approached by a consideration of ionisation energies.

3.3
Ionisation energies

Within a particular series, for example the Balmer series, the lines in the spectrum of atomic hydrogen converge as the frequency increases (wavelength decreases). Each successive line becomes closer to the previous one (fig. 3.1, p. 20). As equation (9) predicts, each series of lines converges towards a limit beyond which the spectrum is continuous (fig. 3.3). At this point the electron responsible for the spectral line has been excited into an orbit of such high energy ($n = \infty$) that it has effectively escaped from the influence of the nucleus. In other words, the atom has lost its electron and formed a positive ion:

$$H(g) \longrightarrow H^+(g) + e^-$$

The energy difference between the ground state of the atom and the excited state that corresponds to the limit of convergence of the spectral

FIG. 3.3. *Energy levels in the hydrogen atom*

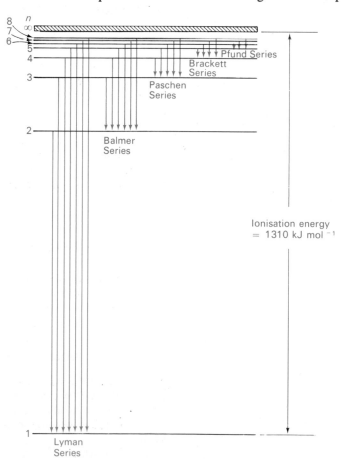

23

lines is called the **ionisation energy** of the atom. Note that ionisation energies always refer to the removal of an electron in the gas phase.

Equation (9) allows us to calculate the ionisation energy of the hydrogen atom, since $n_1 = 1$ (the ground state) and $n_2 = \infty$ (the limit of convergence of the Lyman series). The value so obtained is $2.178\ 5 \times 10^{-18}$ J for one electron; since ionisation energies are generally quoted in kJ mol^{-1}, multiplication of the above value by the Avogadro Constant gives a value of 1312 kJ mol^{-1} for the ionisation energy of the hydrogen atom.

Determination of ionisation energies

Ionisation energies can, therefore, be measured spectroscopically but they can also be determined by the method of electron bombardment, using a valve which is filled with the gas or vapour whose ionisation energy is to be determined. The simple device described below is suitable for measuring the ionisation energy of xenon (fig. 3.4).

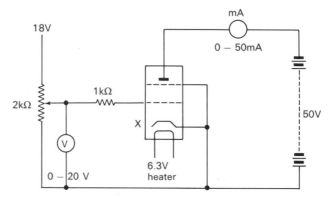

FIG. 3.4. *Determination of the first ionisation energy of xenon*
X—Mullard EN91 2D 21 Valve
mA—Milliameter reading 0–50 mA
V—Voltmeter reading 0–20 volts

A negative potential (50 volts) is applied to the anode of the valve, and the potential on the grid is gradually increased until a current is registered by the galvanometer. No current will flow until the potential on the grid is sufficient to produce singly ionised xenon Xe$^+$. Current should flow when the grid potential is 12·1 volts, from which the ionisation energy is calculated as 1170 kJ mol^{-1}.

$$Xe(g) + e^- \longrightarrow Xe^+(g) + 2e^-$$

A third method of determining ionisation energies is based on a similar principle. Radiation of varying frequency is passed through a sample of the gas or vapour until the energy of the radiation (calculated from the Planck equation (2), p. 20) is sufficient to cause ionisation. The ions are detected between two plates at a large potential difference containing the gas.

Successive ionisation energies

More than one electron may be removed from atoms other than hydrogen and we refer to first, second, third etc. ionisation energies. Successive electrons are removed with increasing difficulty, since the positive charge on the remaining ion will increase as each electron is removed.

It is important to remember that the second ionisation energy, for

24

example, is the energy required to remove the second electron from the singly charged positive ion and not the energy required to remove the first two electrons. For example:

$$Mg(g) \rightarrow Mg^+(g) + e^- \text{ 1st ionisation energy} = +738 \text{ kJ mol}^{-1}$$
$$Mg^+(g) \rightarrow Mg^{2+}(g) + e^- \text{ 2nd ionisation energy} = +1451 \text{ kJ mol}^{-1}$$

and so:

$$Mg(g) \rightarrow Mg^{2+}(g) + 2e^- \text{ Total energy required} = +2189 \text{ kJ mol}^{-1}$$

Successive ionisation energies may be determined spectroscopically, thus the second ionisation energy of magnesium can be measured from the line spectrum of the Mg^+ ion. They may also be measured using a mass spectrometer (p. 197); in the second technique the energy of the bombarding electrons can be measured at the exact moment when the species of the required mass/charge ratio appears in the mass spectrum.

Information from ionisation energies

Ionisation energies provide us with important evidence about the arrangement of the electrons in an atom. The magnitude of the ionisation energy measures the ease with which the particular electron may be removed from the gaseous atom or ion. This in turn is a measure of the stability of the electron in its orbit around the nucleus.

The stability of an electron in an atom depends upon:
(a) the charge carried by the nucleus of the atom
(b) the distance of the electron from the nucleus
(c) the extent to which the nucleus is 'screened' from the electron by other electrons in the atom.

Clearly, a large force of attraction will mean a large ionisation energy. Ionisation energies, therefore, tell us about the manner in which particular electrons are affected by the nucleus and by other electrons in the atom.

The electron volt

Ionisation energies are often quoted in energy units called the **electron volt**. The electron volt is not an S.I. unit, nevertheless it is still widely used and is especially useful in problems dealing with atoms and electrons.

An electron volt (eV) is the energy acquired by an electron when it moves through a potential difference of one volt. The charge carried by an electron is $1 \cdot 602 \times 10^{-19}$ C, so one electron volt (energy in joules is equal to charge in coulombs multiplied by the potential difference in volts) is $1 \cdot 602 \times 10^{-19}$ J. Multiplication by the Avogadro Constant gives a value of $96 \cdot 49$ kJ mol^{-1}.

3.4
Electronic energy levels

First ionisation energies

The atomic spectrum of hydrogen provides the evidence for the existence of different energy levels that are accessible to the electron in the hydrogen atom. Each level is characterised by the principal quantum number, given the symbol n.

In atoms of higher atomic number, atomic spectra confirm that similar energy levels still exist, but they are somewhat modified. The problem is to account for the way in which the many electrons of more complicated atoms occupy these energy levels. This pattern of occupation by the electrons of the energy levels available to them is known as the electronic configuration of the atom.

The first ionisation energies of elements 1 to 56 (hydrogen to barium) are plotted against atomic number in fig. 3.5. It is apparent that a definite pattern exists; thus the noble gases occupy peak positions and the Group 1A metals occur at minima. Further examination shows that there is a pronounced increase in ionisation energy from lithium to neon (subsidiary peaks occurring at positions occupied by beryllium and nitrogen which are discussed later). An exactly similar trend occurs in the portion of the graph from sodium to argon. The sharp decrease in ionisation energy from helium to lithium is understandable, if it is assumed that the two electrons in the helium atom occupy the same energy level (principal quantum number $n = 1$); in lithium, however, it appears as though two electrons occupy the same level, $n = 1$, with the third electron in the energy level of principal quantum number $n = 2$ and hence more readily removed. A logical extension of this reasoning would be to assume that from lithium to neon each additional electron

FIG. 3.5. *First ionisation energies of the first 56 elements*

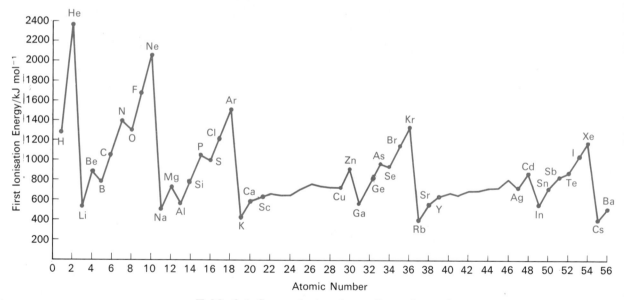

Table 3.1 Some electronic configurations of atoms

Element	Principal quantum number			
	$n = 1$	$n = 2$	$n = 3$	$n = 4$
Helium	2			
Lithium	2	1		
Fluorine	2	7		
Neon	2	8		
Sodium	2	8	1	
Chlorine	2	8	7	
Argon	2	8	8	
Potassium	2	8	8	1

goes into the energy level characterised by the principal quantum number $n = 2$, a new level, $n = 3$, being started with the element sodium.

Furthermore, for a particular periodic group, the first ionisation energies become progressively smaller, e.g. the ionisation energies of helium, neon and argon are respectively 2372, 2080 and 1520 kJ mol^{-1}; for lithium, sodium and potassium the respective values are 520, 496 and 419 kJ mol^{-1}. On the Bohr model, electrons in each successive element in a particular group occupy energy levels (orbits) that are increasingly distant from the nucleus; despite the increase in nuclear charge, the electrons become easier to remove as the group is descended.

Some electronic configurations of atoms are given in Table 3.1.

Successive ionisation energies

Very convincing evidence for the existence of energy levels is available from tabulated values of successive ionisation energies of atoms. Those for the potassium atom are shown in fig. 3.6; the ionisation energies cover a very wide range of values and for the convenience of graphical plotting their logarithmic values are used.

Clear-cut increases in ionisation energies are observed when the 2nd, 10th and 18th electrons are involved, which would seem to suggest that the nucleus of the potassium atom is surrounded by electrons grouped into a number of energy levels (sometimes called shells). Successive ionisation of the electrons becomes more difficult, since every time an

FIG. 3.6. *Successive ionisation energies (plotted logarithmically) for the potassium atom*

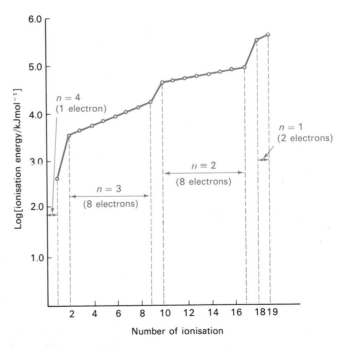

electron is removed the atom bears one more unit of positive charge. However, the large increase in ionisation energy for the removal of the 2nd, 10th and 18th electrons immediately suggests that the 2nd electron is closer to the nucleus than the 1st electron; similarly the 10th and 18th electrons are nearer to the nucleus than the 9th and 17th electrons respectively. The grouping of electrons into energy levels for the potassium atom is as shown.

27

Energy level	$n = 1$	$n = 2$	$n = 3$	$n = 4$
Number of electrons in each energy level	2	8	8	1

3.5
Subdivision of the main energy levels

More detailed study (mainly spectroscopic) shows that the main energy levels of an atom designated by the principal quantum number $n = 2, 3, 4, 5$ etc. and often referred to by their 'shell letters', L, M, N, O etc. are themselves capable of subdivision (note that the energy level $n = 1$, the K-level, is not subdivided). For instance, the 8 electrons in the L- and M-levels of potassium ($n = 2$ and 3 respectively) are distributed between two sub-levels containing 2 and 6 electrons respectively. It is possible to characterise the energy of an electron in an atom by four quantum numbers as follows:

(a) **The principal quantum number** n has integral values 1, 2, 3, 4 etc. As the electron with the largest value of n has the most energy, it is the one that requires the least input of energy to ionise it (i.e. it is the one most readily ionised).

(b) **The subsidiary or azimuthal quantum number** l has integral values ranging from 0, 1, 2, ... $(n - 1)$. For a given principal quantum number, an electron with the largest value of l is the one most readily ionised.

(c) **The third quantum number or magnetic quantum number** m has integral values ranging from $-l$, $-(l - 1)$, $-(l - 2)$ ... 0 ... $(l - 2)$, $(l - 1)$, l. This number arises because some levels which are normally of the same energy (degenerate), have slightly different energies when the atom is exposed to a strong inhomogeneous magnetic field.

(d) **The spin quantum number** has values of $-\frac{1}{2}$ and $+\frac{1}{2}$. The electron can be regarded as spinning on its axis, like a top, in a clockwise and anticlockwise direction.

It is convenient to refer to the electrons with different subsidiary quantum numbers by the letters s, p, d and f (the letters originally used to describe particular spectral series). Thus when the subsidiary quantum number $l = 0, 1, 2, 3$ the electrons are referred to as s, p, d and f electrons respectively.

Before it is possible to apply the quantum numbers to express the electronic configurations of atoms, it is necessary to state the **Pauli Exclusion Principle**:

No two electrons in the same atom can have the same values for all four quantum numbers.

This amounts to saying that no two electrons in any one atom behave in an identical manner. Thus consider the helium atom in its lowest energy state (ground state); the two electrons are assigned the quantum numbers $n = 1$, $l = 0$ and $m = 0$, and since their spins cannot be the same (Pauli Exclusion Principle), one electron has a spin quantum number of $-\frac{1}{2}$ and the other one a spin quantum number of $+\frac{1}{2}$. The electronic configuration of the helium atom in the ground state is written $1s^2$, the first numeral being the value of the principal quantum number n, the letter s denoting that the subsidiary quantum number l is 0, and the superscript indicating that there are two electrons in the same level with opposed spins. To indicate a pairing of electrons, as the above condition is called, the notation $\boxed{\uparrow\downarrow}$ is convenient.

3.6
The electronic configurations of the first ten elements

The electronic configurations of the elements are constructed by assuming that electrons occupy the lowest possible energy levels available, the number of electrons in any one level being determined by the four quantum numbers and the Pauli Exclusion Principle. The electronic configurations of the first ten elements are given in Table 3.2

Table 3.2 The electronic configurations of the atoms of the first ten elements

The Four Quantum Numbers				Maximum Number of Electrons in each sublevel	Maximum Number of Electrons in each principal level	Electronic Configuration of the atoms in their ground state
Principal	Subsidiary	Third	Spin			
$n = 1$	$l = 0$	$m = 0$	$\pm\frac{1}{2}$	2 s electrons	2	Hydrogen $1s^1$ Helium $1s^2$
$n = 2$	$l = 0$	$m = 0$	$\pm\frac{1}{2}$	2 s electrons	8	Lithium $1s^22s^1$ Beryllium $1s^22s^2$
	$l = 1$	$m = 1$	$\pm\frac{1}{2}$	6 p electrons		Boron $1s^22s^22p^1$ Carbon $1s^22s^22p^2$ Nitrogen $1s^22s^22p^3$ Oxygen $1s^22s^22p^4$ Fluorine $1s^22s^22p^5$ Neon $1s^22s^22p^6$
		$m = 0$	$\pm\frac{1}{2}$			
		$m = -1$	$\pm\frac{1}{2}$			

It can be seen that for $l = 1$ there are three levels corresponding to $m = 1, 0, -1$ which are degenerate (have the same energy) unless the atom is placed in a strong magnetic field (p. 28). These three levels can accommodate two electrons each with opposed spins (a maximum of 6 electrons). The question now arises: how will the electrons in these three levels of the same energy be arranged, i.e. if there are two, three or four electrons to be accommodated, will one level fill completely, holding two electrons with opposite spins, or will the electrons occupy each level singly before electron pairing takes place? The answer to this question is that electrons occupy each level singly before electron pairing takes place (because of their mutual repulsion), and only then does electron pairing occur. This principle is known as **Hund's rule**. Thus the electronic configuration of the nitrogen atom in the ground state can be represented as shown.

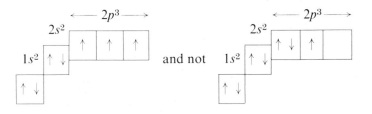

With this principle in mind, we can now display in a more detailed manner the electronic configurations of the atoms of the first ten elements in their ground states:

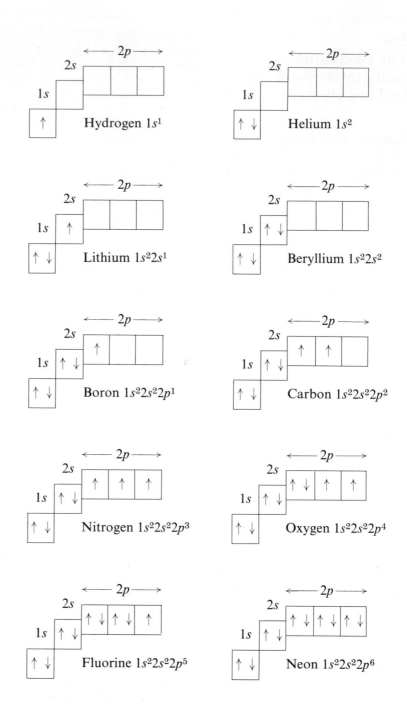

As we have seen above, the main energy levels of an atom are capable of subdivision (except for $n = 1$) and that, for example, the $2p$ level is of slightly higher energy than the $2s$ level. This is true for all atoms other than hydrogen; thus although sub-levels are available to the electron in the hydrogen atom, all sub-levels characterised by the same principal quantum number are degenerate. In this case the $2p$ and $2s$ levels are degenerate as are the $3d$, $3p$ and $3s$ levels.

3.7
The electronic configurations of some heavier atoms

The atom of sodium has one more electron than the atom of neon, and this electron is accommodated in the level characterised by the principal quantum number $n = 3$. It is found that the maximum number of electrons that can be accommodated in any given principal quantum level is $2n^2$ where n is the value of the principal quantum number, so that when $n = 3$, the maximum number of electrons that can enter this level is 18. In practice it is found that after 8 are present, the next 2 enter the principal quantum level $n = 4$ before the preceeding one expands from 8 to 18. This effect, which is due to the overlapping of energy levels, is discussed later (p. 43) and is responsible, for example, for the potassium atom having the electronic configuration $1s^2 2s^2 2p^6 3s^2 3p^6 4s^1$ and not $1s^2 2s^2 2p^6 3s^2 3p^6 3d^1$.

Table 3.3 The electronic configurations of atoms from sodium to zinc

The Four Quantum Numbers				Maximum number of electrons in each sublevel	Maximum number of electrons in each principal level	Atomic number	Electronic configurations of the atoms from sodium to zinc in their ground states	
Principal	Subsidiary	Third	Spin					
$n = 3$	$l = 0$	$m = 0$	$\pm\frac{1}{2}$	$2s$ electrons	18	11	Sodium	$1s^2 2s^2 2p^6 3s^1$
						12	Magnesium	$1s^2 2s^2 2p^6 3s^2$
	$l = 1$	$m = 1$	$\pm\frac{1}{2}$	$6p$ electrons		13	Aluminium	$1s^2 2s^2 2p^6 3s^2 3p^1$
		$m = 0$	$\pm\frac{1}{2}$			14	Silicon	$1s^2 2s^2 2p^6 3s^2 3p^2$
		$m = -1$	$\pm\frac{1}{2}$			15	Phosphorus	$1s^2 2s^2 2p^6 3s^2 3p^3$
						16	Sulphur	$1s^2 2s^2 2p^6 3s^2 3p^4$
						17	Chlorine	$1s^2 2s^2 2p^6 3s^2 3p^5$
						18	Argon	$1s^2 2s^2 2p^6 3s^2 3p^6$
	$l = 2$	$m = 2$	$\pm\frac{1}{2}$	$10d$ electrons		21	Scandium	$1s^2 2s^2 2p^6 3s^2 3p^6 3d^1 4s^2$
		$m = 1$	$\pm\frac{1}{2}$			22	Titanium	$1s^2 2s^2 2p^6 3s^2 3p^6 3d^2 4s^2$
		$m = 0$	$\pm\frac{1}{2}$			23	Vanadium	$1s^2 2s^2 2p^6 3s^2 3p^6 3d^3 4s^2$
		$m = -1$	$\pm\frac{1}{2}$			24	Chromium	$1s^2 2s^2 2p^6 3s^2 3p^6 3d^5 4s^1$
		$m = -2$	$\pm\frac{1}{2}$			25	Manganese	$1s^2 2s^2 2p^6 3s^2 3p^6 3d^5 4s^2$
						26	Iron	$1s^2 2s^2 2p^6 3s^2 3p^6 3d^6 4s^2$
						27	Cobalt	$1s^2 2s^2 2p^6 3s^2 3p^6 3d^7 4s^2$
						28	Nickel	$1s^2 2s^2 2p^6 3s^2 3p^6 3d^8 4s^2$
						29	Copper	$1s^2 2s^2 2p^6 3s^2 3p^6 3d^{10} 4s^1$
						30	Zinc	$1s^2 2s^2 2p^6 3s^2 3p^6 3d^{10} 4s^2$
$n = 4$ etc.	$l = 0$	$m = 0$	$\pm\frac{1}{2}$	$2s$ electrons	2	19	Potassium	$1s^2 2s^2 2p^6 3s^2 3p^6 4s^1$
						20	Calcium	$1s^2 2s^2 2p^6 3s^2 3p^6 4s^2$

FIG. 3.7. *Method by which the electronic structure of elements can be worked out*

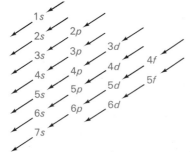

The electronic configurations of the elements from sodium to zinc are shown in Table 3.3. Reference to p. 28 will show how to determine the number of subsidiary levels etc., corresponding to a particular quantum number.

This table shows the subdivision of the principal quantum level $n = 3$ and part of the level $n = 4$. The $4s$ level is of lower energy than the $3d$ level for the potassium and calcium atoms. The chromium and copper atoms have the configurations . . . $3d^5 4s^1$ and . . . $3d^{10} 4s^1$ respectively, instead of the expected . . . $3d^4 4s^2$ and . . . $3d^9 4s^2$ configurations.

With few exceptions, which are not significant, the electronic structures of elements can be worked out from the sequence given in fig. 3.7. The electronic configurations of the elements are given in Table 3.4.

Table 3.4
The electronic configurations of the atoms of the elements

Element	Atomic Number	1s	2s	2p	3s	3p	3d	4s	4p	4d	4f	5s	5p	5d	5f	6s	6p	6d	6f	7s
H	1	1																		
He	2	2																		
Li	3	2	1																	
Be	4	2	2																	
B	5	2	2	1																
C	6	2	2	2																
N	7	2	2	3																
O	8	2	2	4																
F	9	2	2	5																
Ne	10	2	2	6																
Na	11	2	2	6	1															
Mg	12				2															
Al	13				2	1														
Si	14		10 electrons		2	2														
P	15				2	3														
S	16				2	4														
Cl	17				2	5														
Ar	18	2	2	6	2	6														
K	19	2	2	6	2	6		1												
Ca	20							2												
Sc	21						1	2												
Ti	22						2	2												
V	23						3	2												
Cr	24						5	1												
Mn	25						5	2												
Fe	26						6	2												
Co	27		18 electrons				7	2												
Ni	28						8	2												
Cu	29						10	1												
Zn	30						10	2												
Ga	31						10	2	1											
Ge	32						10	2	2											
As	33						10	2	3											
Se	34						10	2	4											
Br	35						10	2	5											
Kr	36	2	2	6	2	6	10	2	6											
Rb	37	2	2	6	2	6	10	2	6			1								
Sr	38											2								
Y	39									1		2								
Zr	40									2		2								
Nb	41									4		1								
Mo	42									5		1								
Tc	43									5		2								
Ru	44									7		1								
Rh	45									8		1								
Pd	46		36 electrons							10		0								
Ag	47									10		1								
Cd	48									10		2								
In	49									10		2	1							
Sn	50									10		2	2							
Sb	51									10		2	3							
Te	52									10		2	4							
I	53									10		2	5							
Xe	54	2	2	6	2	6	10	2	6	10		2	6							

Element	Atomic Number	1s	2s	2p	3s	3p	3d	4s	4p	4d	4f	5s	5p	5d	5f	6s	6p	6d	6f	7s
Cs	55	2	2	6	2	6	10	2	6	10		2	6			1				
Ba	56											2	6			2				
La	57											2	6	1		2				
Ce	58										2	2	6			2				
Pr	59										3	2	6			2				
Nd	60										4	2	6			2				
Pm	61										5	2	6			2				
Sm	62				| 46						6	2	6			2				
Eu	63				electrons						7	2	6			2				
Gd	64										7	2	6	1		2				
Tb	65										8	2	6	1		2				
Dy	66										10	2	6			2				
Ho	67										11	2	6			2				
Er	68										12	2	6			2				
Tm	69										13	2	6			2				
Yb	70										14	2	6			2				
Lu	71										14	2	6	1		2				
Hf	72										14	2	6	2		2				
Ta	73										14	2	6	3		2				
W	74										14	2	6	4		2				
Re	75										14	2	6	5		2				
Os	76										14	2	6	6		2				
Ir	77										14	2	6	7		2				
Pt	78										14	2	6	9		1				
Au	79										14	2	6	10		1				
Hg	80										14	2	6	10		2				
Tl	81										14	2	6	10		2	1			
Pb	82										14	2	6	10		2	2			
Bi	83										14	2	6	10		2	3			
Po	84										14	2	6	10		2	4			
At	85										14	2	6	10		2	5			
Rn	86	2	2	6	2	6	10	2	6	10	14	2	6	10		2	6			
Fr	87	2	2	6	2	6	10	2	6	10	14	2	6	10		2	6			1
Ra	88															2	6			2
Ac	89															2	6	1		2
Th	90															2	6	2		2
Pa	91														2	2	6	1		2
U	92														3	2	6	1		2
Np	93														5	2	6			2
Pu	94														6	2	6			2
Am	95				| 78										7	2	6			2
Cm	96				electrons										7	2	6	1		2
Bk	97														7	2	6	2		2
Cf	98														9	2	6	1		2
Es	99																			
Fm	100																			
Md	101																			
No	102																			
Lr	103																			

3.8
Summary

Atomic spectra provide evidence for electrons occupying discrete energy levels in atoms. The energy of each level is determined essentially by the principal quantum number. The first ionisation energies of the elements, plotted against atomic number, show a periodic variation which can be interpreted in terms of the filling of successive energy levels in the atoms. Successive ionisation energies of particular elements show a pattern which confirm this picture.

More detailed study (mainly spectroscopic) shows that the main energy levels of principal quantum number greater than 1 are divided into sub-levels, whose energy is distinguished by the subsidiary quantum number. In fact four quantum numbers are required to characterise the energy of an electron in an atom.

The electronic configurations of the elements are constructed by placing the electrons in the lowest possible energy levels available, as dictated by the four quantum numbers and the Pauli Exclusion Principle. Degenerate sub-levels are filled singly before pairing of electron spins occurs (Hund's rule).

Unfortunately, our model of the atom cannot explain why electrons start to occupy the main energy level of principal quantum number 4 before the level of principal quantum number 3 is completely filled. To seek an answer to this problem it is necessary to stop treating the electron as a particle and begin to treat it as though it behaved as a wave. This is done in the following chapter.

3.9
Questions on chapter 3

1   (a) Construct a schematic diagram representing the number and relative energies of the orbitals of the first three principal quantum numbers.
    (b) The electron configuration of the gaseous carbon atom might be represented in one of the following ways:

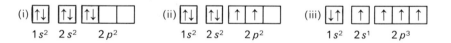

Which representation most nearly describes the configuration of the gaseous carbon atom?
    (c) Sketch the shapes of all the occupied orbitals in representation (b) (ii) above. C

2   (a) A metal ion M^{3+}, in the ground state, has the following electronic configuration.

$$1s^2 2s^2 2p^6 3s^2 3p^6 3d^3$$

Name the metal ion.
    (b) Give the equations which represent the changes taking place when the following ionisation energy measurements are made:
       (i) the first ionisation energy of sodium
       (ii) and the second ionisation energy of sodium.
    (c) Explain
       (i) why helium has the highest first ionisation energy of any element
       (ii) and why the first ionisation energy of potassium is less than the first ionisation energy of lithium. JMB (Syllabus B)

3   Account for the formation of the line spectrum of hydrogen and outline the relationship between this spectrum and the ionisation energy of hydrogen.
    The first six ionisation energies of an element A are as follows:
       762 (1st), 1540, 3300, 4390, 8950 and 11 900 kJ mol^{-1}.

State and explain what you would predict about (a) the structure, (b) the reaction (if any) with water, of the principal chloride of A. C (Overseas)

4 With the aid of a simple sketch, describe an experimental arrangement by means of which you could observe the emission spectrum of a gas such as hydrogen. What changes in the apparatus would be required to observe the absorption spectrum of the gas?
  (a) How does the appearance of an emission spectrum differ from that of the absorption spectrum of the same gas?
  (b) Discuss the relationship between the frequency of a line in the atomic (emission) spectrum and electron energy levels.
  (c) How are ionisation energies of elements calculated from spectroscopic measurements?
  (d) What information do ionisation energies yield about electronic energy levels? O and C

5 The atomic spectrum of hydrogen is given by the following relationship,

$$\frac{1}{\lambda} = R_H \left(\frac{1}{n_1{}^2} - \frac{1}{n_2{}^2} \right)$$

  (a) (i) What does λ represent?
      (ii) What do the terms n_1 and n_2 represent?
      (iii) What are the units of the constant R_H?
  (b) The spectrum comprises a number of lines which may be divided into a number of series.
      (i) Why does the spectrum consist of lines?
      (ii) Why is there a small number of series in the spectrum?
      (iii) Explain why each series converges and in what direction it converges.
  (c) What method is used to generate the light source for observing the atomic spectrum of hydrogen?
  (d) Name the instrument used to resolve the hydrogen spectrum.
  JMB (Syllabus B)

6 Discuss the atomic spectrum of hydrogen and its relation to our understanding of the electronic structure of atoms.
  Suggest explanations for the following observations:
  (a) The atomic spectrum of hydrogen contains lines in the radio-frequency region of the electromagnetic spectrum.
  (b) A line in the spectrum of atomic hydrogen on a distant object in the universe occurs at a wavelength of 300 nm though it is known to occur in the laboratory at 121·6 nm. O Schol. and Entrance

7 What experimental evidence suggests that electrons bound in atoms occupy well-defined energy levels?
  In atomic hydrogen these energy levels, referred to the situation where the electron and proton are at rest at infinite separation, are given by
  $$E_n = -R'/n^2$$
  where $n = 1, 2, \ldots$ and $R' = 2\cdot17 \times 10^{-18}$ J. Using this expression sketch the position of the first few energy levels of hydrogen and calculate:
  (a) the energy required to ionize a hydrogen atom in its ground state $n = 1$, in electron-volts;
  (b) the wavelength of the spectral line emitted in a transition from the lowest excited level, $n = 2$, to the ground state, $n = 1$;
  (c) the radius of the electron orbit in the level $n = 1$ given that the kinetic energy of the electron in this state is equal to $-\frac{1}{2}$ times its potential energy.
  $(e = 1\cdot60 \times 10^{-19}$ C; $h = 6\cdot63 \times 10^{-34}$ J s.)
  O Schol. and Entrance (Physical Science)

The wave properties of the electron

4.1
The electron as a wave

Although Bohr's model of the atom was remarkably successful in many ways, it was not long before its limitations became apparent. These were as follows:

(a) Attempts to predict the frequencies of the spectral lines for more complex atoms than hydrogen failed.

(b) There seemed little justification for the quantisation of the electron's angular momentum into units of $h/2\pi$; it just 'seemed to work'.

(c) Bohr's model did not really explain why Rutherford's atom did not collapse as the theory of electromagnetic radiation predicted.

In the 1920s new evidence came to light which cast doubt upon the validity of treating the electron as a revolving particle as Rutherford and Bohr had done. This evidence laid the foundations for what is now called wave-mechanics because it suggested that the electron should be treated as a wave rather than as a particle.

In 1924 de Broglie, making an analogy with light, predicted that electrons should show wave-like properties. Moreover, he recognised the link between Bohr's permitted orbits and the stationary-state solutions of wave equations. He related the wavelength λ of the predicted 'electron wave' to the momentum of the electron:

h = Planck's constant
m = mass of electron
v = velocity of electron

$$\lambda = h/mv \qquad (1)$$

Equation (1) allows us to calculate the wavelength of the wave associated with any moving object. Table 4.1 shows why wave properties of large bodies are not discernible—the wavelength is too small.

Table 4.1 The wavelength associated with different moving objects

Object	Mass/kg	Velocity/m s^{-1}	Wavelength/m
100 V electron	$9 \cdot 1 \times 10^{-31}$	$5 \cdot 9 \times 10^{6}$	$1 \cdot 2 \times 10^{-10}$
Golf ball	$4 \cdot 5 \times 10^{-2}$	30	$5 \cdot 0 \times 10^{-34}$
Bullet	$5 \cdot 0 \times 10^{-3}$	350	$4 \cdot 0 \times 10^{-34}$
Car	900	25	$2 \cdot 9 \times 10^{-38}$

Table 4.1 shows that an electron accelerated through a potential difference of 100 V should have a wavelength of about 10^{-10} m or 0.1 nm. This is similar to the wavelength of X-radiation (p. 174) and suggests that a beam of electrons might be capable of diffraction in a similar manner to a beam of X-rays.

Confirmation of de Broglie's predictions came in 1927 when Davisson and Germer, and G. P. Thomson (son of J. J. Thomson) demonstrated independently that an electron beam could be diffracted by thin metal foils (Plate 2). Accurate measurements of this diffraction pattern showed that equation (1) was correct.

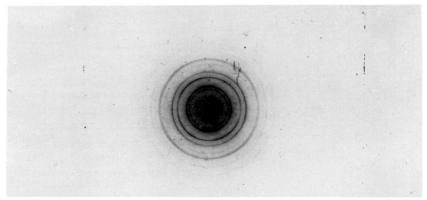

Two theories of electron behaviour

Two models are needed to account for the behaviour of electrons. The particulate model (which is easier to visualise, and for many purposes adequate) treats the electron as a tiny particle of definite mass, charge and momentum which obeys the conventional laws of moving objects.

The wave model associates the electron with a wave which has a certain wavelength and amplitude and which obeys the laws of wave motion. Little is to be gained by arguing for and against either of these two models and it must be accepted that electrons, like light, show a dual behaviour; the two theories stand side by side.

The wave properties of the electron make it a very elusive particle. If we try to isolate one by making it pass through a minute hole, this hole must be so small (to cut out all other electrons) that diffraction will take place. The subsequent direction, and hence the momentum of the electron then becomes uncertain.

The degree of uncertainty which is associated with a particular electron is summed up in the famous **Heisenberg Uncertainty Principle** (1927):

$$\Delta x \cdot \Delta p \approx h$$

Δx = uncertainty in position
Δp = uncertainty in momentum

This means that if we measure the momentum of an electron accurately then we cannot know its position with certainty. Similarly if we know its exact position, the momentum becomes uncertain.

The effects of the Uncertainty Principle are directly observable in the laboratory. For example, spectroscopists find that there is a definite limitation to the sharpness of spectral lines.

Fitting waves into atoms

If the electron is to be treated as a wave, then the first problem is that of fitting the wave into the confines of an atom. This can be done in the following approximate way for the hydrogen atom—approximate, since the problem is treated as a two-dimensional one only.

An integral number of wavelengths are fitted into the distance traced out by the orbiting electron. Two such situations are depicted in fig. 4.1.

FIG. 4.1. *Fitting electron waves into the hydrogen atom*

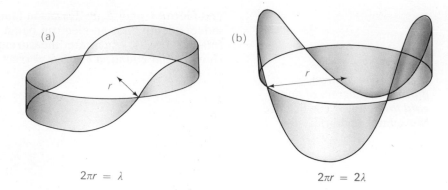

(a) (b)

$2\pi r = \lambda$ $2\pi r = 2\lambda$

In fig. 4.1(a) one complete wavelength is fitted into this distance, while in fig. 4.1(b) there are just two wavelengths in this distance. In the general situation the expression is:

r = radius of orbit
n = integer

$$2\pi r = n\lambda$$

Applying the de Broglie relationship (equation (1)) we have:

$$2\pi r = nh/mv \quad \text{or} \quad mvr = nh/2\pi$$

This latter expression, for the quantisation of angular momentum, was the most crucial assumption Bohr made in his theory of the hydrogen atom and for which there was originally no theoretical justification.

4.2
Atomic orbitals

The Schrödinger Equation

In 1927 Schrödinger expressed the wavelike characteristics of the electron in terms of a differential equation. This equation, given below, cannot be derived and Schrödinger arrived at it mainly by mathematical intuition. It can be solved exactly for one-electron systems, e.g. for the hydrogen atom and the helium ion He^+, and for such systems the predicted energy levels are in complete agreement with experimental results. The equation is:

m = mass of electron
h = Planck's constant
V = potential energy of electron
E = total energy of electron
ψ = wavefunction

$$\frac{\partial^2 \psi}{\partial x^2} + \frac{\partial^2 \psi}{\partial y^2} + \frac{\partial^2 \psi}{\partial z^2} + \frac{8\pi^2 m}{h^2}(E - V)\psi = 0$$

The details of solving the equation need not concern us here, the important point being that the equation will only give solutions for certain values of E, the total energy. These permitted values of the total energy are called **eigenvalues** and the corresponding values of ψ are called **wave functions**, or **atomic orbitals**.

ψ, the wave function or atomic orbital of the electron, has no physical reality, but ψ^2 is related to the probability of finding the electron within a specified volume of space. An important consequence of treating the electron as a wave is that the certainty of Bohr's orbits must be replaced by the probability description of atomic orbitals. No longer can we say that the electron in the hydrogen atom is at a distance of 0·052 9 nm from the nucleus (p.21); the best we can do is to state that the electron is most likely to be found at a distance of 0·052 9 nm from the nucleus, but there is a finite chance that it might be found elsewhere.

Three quantum numbers

Solution of the Schrödinger equation for the hydrogen atom requires that the eigenvalues are characterised by three quantum numbers n, l and m which, together with the spin quantum number, were used in Chapter 3 to construct the electronic configurations of the elements. The wave equation cannot be solved exactly for a system which contains more than one electron because, in addition to electron–nucleus attractive forces, we must consider electron–electron repulsion and the resulting mathematics is complex and not amenable to exact solution. Nevertheless, approximate solutions are possible which indicate that the electrons can be characterised by the three quantum numbers mentioned above. The significance of these numbers is described below.

Principal quantum number n takes integral values 1 to ∞. It is the main determinant of the energy of the electron and it determines how the probability of locating the electron varies with the distance from the nucleus.

Subsidiary quantum number l takes integral values from 0 to $(n - 1)$ but the numbers are more commonly known by their spectroscopic letters s, p, d, f etc. It determines the way in which the probability of locating the electron varies with direction from the nucleus.

Magnetic quantum number m takes integral values ranging from $-l$ through 0 to $+l$. Thus if $l = 1$, m can have values of -1, 0 and $+1$. The energy of the electron is unaffected by the value of m, except when it is placed in a magnetic field. The splitting of spectral lines in a magnetic field (Zeeman effect) provides the experimental justification for this quantum number.

It is quite easy to show that there is one wave function or atomic orbital with a principal quantum number of 1, four different atomic orbitals with a principal quantum number of 2, nine different atomic orbitals with a principal quantum number of 3, and no less than sixteen different atomic orbitals with a principal quantum number of 4. Study Tables 3.2 and 3.3 (p. 29 and 31) if you are in doubt.

4.3 Depicting atomic orbitals

It is quite easy to visualise and to depict the electron orbits in the Bohr model of the atom. The same is unfortunately not true for atomic orbitals. However, if instead of the wavefunction ψ, we consider the probability function ψ^2 then we can go some way towards a useful method of depicting the orbital.

ψ^2 is a measure of the probability of finding an electron in a particular volume of space. Two questions are of particular importance:

How does the probability, ψ^2, of finding an electron vary with the distance from the nucleus? What is the **radial distribution** of the probability?

How does the probability, ψ^2, of finding an electron vary with the direction from the nucleus? What is the **angular distribution** of the probability?

Radial distribution function

In its ground state the single electron in the hydrogen atom occupies an orbital with quantum numbers of $n = 1$ and $l = 0$, i.e. a $1s$ orbital. Fig.

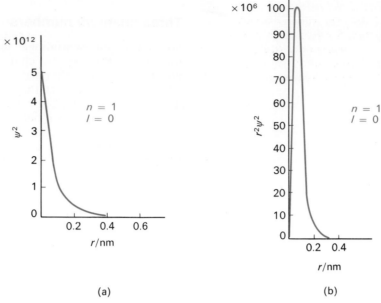

FIG. 4.2. *(a) Plot of* ψ^2 *against* r *for 1s orbital of hydrogen atom (b) Plot of* $r^2\Psi^2$ *(factor 4π omitted) against* r *for 1s orbital of hydrogen atom*

4.2(a) shows how the probability of finding the electron increases as the distance between the electron and the nucleus decreases, and is a maximum when the electron has reached the nucleus. However, when the probability of finding the electron in a given spherical shell around the nucleus, i.e. in the volume bounded by two spheres of radii r and $r + dr$, which is $4\pi r^2 dr$, is plotted against the distance of the electron from the nucleus, the maximum occurs at some distance (equal to the Bohr radius—0·052 9 nm in this case) from the centre of the atom (fig. 4.2b). This arises because, although the probability of finding the electron decreases with distance from the nucleus, the volume of the spherical shell containing the electron increases with the square of the distance. Although fig. 4.2b indicates that the maximum probability occurs at the Bohr radius it should be appreciated that there is a finite chance that the electron will be located at distances other than this. Figs. 4.3a and 4.3b show respectively the radial distribution functions for the 2s orbital ($n = 2$, $l = 0$) and the 3s orbital ($n = 3$, $l = 0$). The most probable region of locating the electron in these orbitals is further away from the nucleus. However, the smaller peaks in each of the graphs indicate a significant probability of locating the electron much nearer the nucleus. Electrons in s orbitals (loosely known as 's electrons') are said to **penetrate** towards the nucleus.

FIG. 4.3. *(a) Radial distribution function for the 2s orbital of the hydrogen atom (b) Radial distribution function for the 3s orbital of the hydrogen atom*

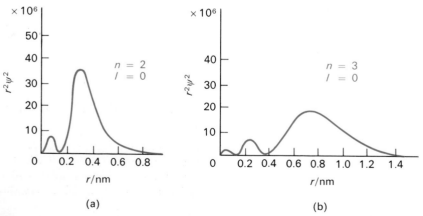

40

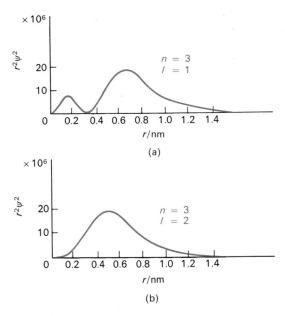

(a)

(b)

Penetration is less important in the case of electrons in *p* orbitals ($l = 1$) and in *d* orbitals ($l = 2$) as shown in fig. 4.4, and in general the extent of penetration follows the sequence:

$$s > p > d$$

Penetration is important because it allows an electron to experience more than its share of the positive nuclear charge. This will make the electron energetically more stable and thus more difficult to remove; hence we have a reason why, for example, the *s* level fills before the *p* level for a given principal quantum number; but note again, however, that in the hydrogen atom all sub-levels characterised by the same principal quantum number are degenerate.

FIG. 4.5. *The different penetrations of electrons in* 3s *and* 3p *orbitals through the inner shells*

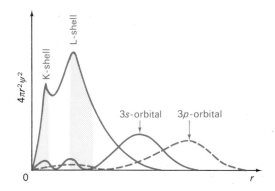

Angular distribution functions

The way in which the probability of locating an electron varies with the direction from the nucleus depends critically upon the subsidiary quantum number of the orbital which the electron occupies.

s orbitals ($l = 0$) are spherically symmetric (fig. 4.6(a)) which implies that the probability of locating the electron is independent of its direction from the nucleus. By contrast, p orbitals ($l = 1$) have a marked directional character and their angular distribution function is shown in fig. 4.6(b); the three orbitals correspond to the values of m, the magnetic quantum number, -1, 0 and $+1$. d orbitals ($l = 2$) and f orbitals ($l = 3$) have more complex angular distribution functions and their details need not concern us here.

FIG. 4.6. *(a) The 1s atomic orbital (angular distribution function) (b) The 2p atomic orbitals (angular distribution function)*

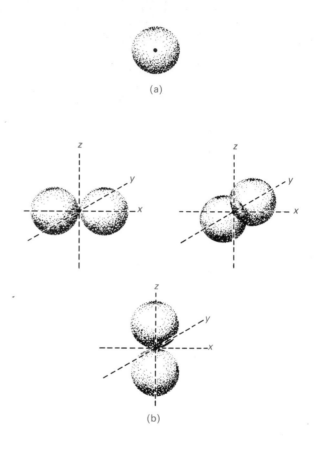

(a)

(b)

FIG. 4.7. *The 2p$_z$ atomic orbital showing electron probability dependence on the angle θ*

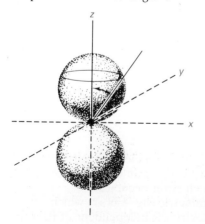

It is important to appreciate that the angular distribution function does not provide informaion about distance from the nucleus. If we consider the $2p_z$ orbital depicted in fig. 4.7 we can see that the electron in this orbital is most likely to be located along the z axis. However, this figure contains no information about how far along the axis the electron is most likely to be found. As the angle θ is increased, so the probability of locating the electron diminishes until at θ = 90° the probability is zero. Strictly speaking the diagram should be in three dimensions and two angles are needed to specify the position of the electron within the orbital.

It is possible to combine pictorially the radial and the angular distribution functions. The directional properties of orbitals are important when discussing the formation of bonds between atoms; it is, therefore, not generally necessary to consider the radial distribution function, although it should be appreciated that an approximation is being made. This simplification is of considerable help.

42

4.4
Atomic orbitals in multi-electron atoms

In the hydrogen atom the energy levels available to the electron are determined entirely by the principal quantum number. In the ground state, the electron occupies the $1s$ orbital. Excitation may promote the electron to the $n = 2$ orbitals but, as we have mentioned before, the $2s$ and $2p$ orbitals are of identical energy. Similarly the $3s$, $3p$ and $3d$ orbitals are degenerate.

In the case of heavier atoms the orbitals are not so clearly distinguished in their energies. The presence of other electrons in neighbouring orbitals affects relative orbital energies. Increasing nuclear charge also affects orbitals in different ways according to their powers of penetration (p. 41). The more penetrating the orbital, the more its energy is lowered with increasing nuclear charge relative to a less penetrating orbital.

Fig. 4.8 shows the calculated energies of the $3s$, $3p$ and $3d$ orbitals as a function of atomic number (note that the scale is a logarithmic one). At $Z = 1$ their energies are the same but as Z increases the orbitals split until at $Z \sim 20$ they are most widely spaced. At high values of Z the orbital energies converge since the charge on the nucleus is now so great that if far outweighs the difference in penetration between the three orbitals.

FIG. 4.8. *The relative energies of the* 3s, 3p *and* 3d *orbitals as a function of atomic number*

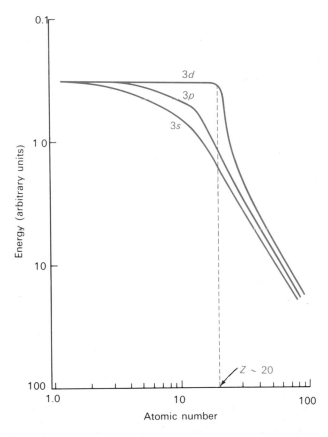

Fig. 4.9 shows the energy of orbitals of principal quantum number 1 to 4 plotted against atomic number. The most significant feature is the way in which the $4s$ orbital crosses the $3d$ orbital at an atomic number of about 7. This means that for elements of atomic number higher than 7 the $4s$ orbital is of lower energy than the $3d$ orbital, despite its higher

43

principal quantum number—such is the effect of the penetrating character of the *s* orbital.

FIG. 4.9. *The relative energies of orbitals of principal quantum numbers 1 to 4*

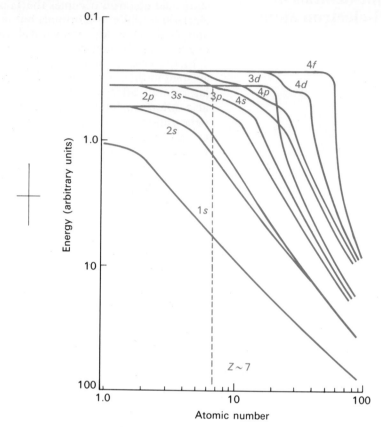

4.5
Some comments on first ionisation energies

FIG. 4.10 *The variation in electron density in the lithium atom. Note the formation of a shell structure*

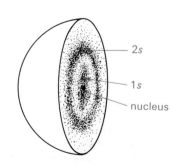

We are now in a position to offer some comments on the relative values of a few first ionisation energies plotted in fig. 3.5 (p. 26). The hydrogen and helium atoms have the electronic configurations $1s^1$ and $1s^2$ respectively. There is a doubling of the nuclear charge in proceeding from hydrogen to helium but, at the same time, the two electrons in the $1s$ shell of helium mutually repel each other. This repulsion, however, does not offset the large increase in nuclear attraction, and the first ionisation energy of helium is very much larger than that of the hydrogen atom.

The next element lithium has the electronic configuration $1s^2 2s^1$. In this case the two $1s$ electrons spend a considerably longer time closer to the nucleus than does the $2s$ electron. In fact, the $2s$ electron is effectively screened from the full effect of the nucleus by the $1s$ electrons and this explains the dramatic decrease in first ionisation energy in proceeding from helium to lithium.

The first ionisation energy rises from lithium to beryllium and this can be accounted for in terms of an increase in nuclear charge of $+1$. This more than offsets the repulsion that occurs owing to there being 2 electrons in the $2s$ shell of the beryllium atom. The next element boron has a lower first ionisation energy than beryllium. At this point the $2p$ level is occupied by one electron, thus the electronic configurations of the beryllium and boron atoms are respectively $1s^2 2s^2$ and $1s^2 2s^2 2p^1$. As we have seen previously a $2s$ electron has a greater probability of being very close to the nucleus than a $2p$ electron, thus a $2p$ electron does not experience as much attraction by the nucleus as does a 2s electron and is consequently more readily ionised.

There is a steady increase in first ionisation energy for the next two elements carbon and nitrogen. In each case the added electron(s) enter a 2p shell, but the added electrons remain unpaired hence reducing electron–electron repulsion (respective electronic configurations are $1s^2 2s^2 2p_x^1 2p_y^1$ and $1s^2 2s^2 2p_x^1 2p_y^1 2p_z^1$). The increasing nuclear charge ensures a steady increase in first ionisation energy. The electronic configuration of the next element oxygen is $1s^2 2s^2 2p_x^2 2p_y^1 2p_z^1$, i.e. the additional electron is forced to pair and, in this case, electron–electron repulsion must outweigh the increased attraction by the nucleus, since the first ionisation energy of oxygen is lower than that of nitrogen.

The first ionisation energies rise for the fluorine atom to a maximum for neon. In each case the added electron(s) pair in the 2p shell and, presumably here, increasing nuclear attraction is more important than electron–electron repulsion.

4.6
Questions on chapter 4

1 Why is it impossible to determine both the position and velocity of an electron at any instance with high precision? How has this led to a modification of the Bohr theory of the hydrogen atom?

2 Using de Broglie's equation determine the wavelength associated with an electron whose mass is 9×10^{-31} kg and whose velocity is 3×10^7 m s^{-1}.
(Planck's constant is $6 \cdot 625 \times 10^{-34}$ J s)

3 In the hydrogen atom the 2s and 2p orbitals are of the same energy, whereas in the lithium atom the 2p orbital is of higher energy than the 2s orbital. Explain why this should be so.

4 Explain the general features of the first ionisation energies of the first ten elements in terms of nuclear charge, screening, electron penetration and the Pauli principle.

The behaviour of the nucleus

5.1
Introduction

Through Rutherford's work on α-particle scattering (p. 8) and the study of radioactivity, the behaviour of the nucleus provided the key to the structure of the atom. When it became clear that it is the behaviour of the electrons which determine chemical properties, the nucleus was left for the physicists to study.

Work since the 1940s on nuclear fusion and fission and, in particular, the creation of a whole new family of man-made elements has reawakened the chemist's interest in the nucleus and in the way in which it behaves. This interest has been reinforced by the chemist's increasing use of radioisotopes in studying the pathways of chemical reactions.

5.2
The size and composition of the nucleus

Nuclear density

The radius of the nucleus ranges from about 1.5×10^{-15} m for a proton to about 6.5×10^{-15} m for one of the heaviest elements. This is very much smaller than atomic radii, e.g. the radius of the hydrogen atom is 0.53×10^{-10} m, which are about 10^4 times greater than nuclear radii. Most of the atom is empty space, whose boundaries are determined by the outermost electron clouds.

While the electrons determine the size of an atom, it is the nucleus which determines its mass. Each of the protons and neutrons which constitute the nucleus are about 1840 times heavier than an electron, so only a negligibly small portion of the mass of an atom resides outside the nucleus. The mass of a proton is 1.67×10^{-27} kg which means that its density is about 1.4×10^{14} greater than that of water. A drop of nuclear material large enough to see would weigh about 10^{10} kg!

Protons and neutrons

The most important nuclear particles are the proton and the neutron, known collectively as **nucleons**. Their masses are almost the same (the neutron is in fact about 0.13% heavier than the proton) but whereas the proton is positively charged the neutron is electrically neutral.

The atomic number of an element is the number of protons in the nucleus. However, nuclei of the same element may contain different numbers of neutrons. This gives rise to isotopes (p. 15) and accounts for the fact that most elements do not have integral relative atomic masses. We shall see later (p. 61) that the stability of a nucleus depends upon its proton/neutron ratio.

Any nuclear species characterised by its atomic number and mass number is called a nuclide and is conventionally written with the chemical symbol, together with an upper index denoting mass number and a lower index the atomic number. The atomic number is, of course, fixed by the chemical symbol itself, but it is nonetheless convenient to show it. The three nuclides of hydrogen are written as follows:

$$^1_1\text{H} \qquad ^2_1\text{H} \qquad ^3_1\text{H}$$

Hydrogen   Deuterium   Tritium

Other elementary particles

Experiments since the 1930s have revealed the existence of a host of 'elementary' particles in addition to the proton and the neutron. Most of them have only a transitory existence before they are transformed into something more stable. In some cases their existence was predicted theoretically long before they were detected experimentally. Table 5.1 shows the most important of these particles.

Table 5.1 Some subnuclear particles

Particle	Mass	Charge	Year of discovery
Antiproton	Same as that of a proton	Negative	1955
μ-meson (muon)	210 times that of an electron	Positive and negative	1937
π-meson (pion)	276 times that of an electron 265 times that of an electron	Positive and negative Zero	1947
Neutrino	Very much less than that of an electron	Zero	1956
Positron	Same as that of an electron	Positive	1932

There is no reason to suppose that these particles, or indeed the proton and neutron, will prove to be any more 'fundamental' than did the atom itself. On the contrary, there is evidence that protons and neutrons themselves have well-defined structures.

5.3 Nuclear stability and binding energy

The origin of the forces that accounts for the stability of the nucleus is still very much of a mystery, although it is clear that proton-neutron interaction is of great importance since no atom, other than that of hydrogen, contains only protons.

Despite the fact that so little is known about the origin of nuclear forces, it is possible to calculate the energy needed to separate atomic nuclei into their isolated protons and neutrons; this quantity is called the **binding energy** of the nucleus. Consider the helium nucleus which contains 2 protons and 2 neutrons; the mass of the helium nucleus (on the $^{12}\text{C} = 12$ scale) is 4·0017 amu (atomic mass units) whereas the individual masses of the isolated proton and neutron are 1·0073 and 1·0087 amu respectively. The total mass of 2 protons and 2 neutrons is $(2 \times 1\cdot0073) + (2 \times 1\cdot0087)$ or 4·0320 amu, i.e. there is a loss in mass of $4\cdot0320 - 4\cdot0017$ or 0·0303 amu when 2 protons and 2 neutrons form the helium nucleus, and this difference is called the **mass defect**. This annihilation of mass corresponds to the release of energy according to the Einstein equation:

$$E = mc^2$$

E = energy
m = mass
c = speed of light

The loss in mass for the formation of 1 mole of helium atoms from isolated protons and neutrons is 0.0303 g or 3.03×10^{-5} kg. The energy released is calculated from Einstein's equation:

$$E = mc^2 = 3.03 \times 10^{-5} \times (3 \times 10^8)^2$$
$$= 2.727 \times 10^{12} \text{ J mol}^{-1}$$
$$= 2.727 \times 10^9 \text{ kJ mol}^{-1}$$

This enormous value gives some idea of the scale of the energy changes that are associated with nuclear reactions and, in particular, with nuclear fusion reactions, i.e. the fusing together of components to form a new nucleus.

Binding energies are generally quoted as energy changes in terms of the mega electron volt, which is one million times the energy acquired by an electron when it passes through a potential difference of one volt. 1 mega electron volt (MeV) is equivalent to 9.6×10^7 kJ mol^{-1}. The formation of one mole of helium nuclei from its isolated protons and neutrons results in the release of

$$\frac{2.7 \times 10^9}{9.6 \times 10^7} = 28 \text{ MeV (approx.)}$$

But the helium nucleus contains four particles (2 protons and 2 neutrons), thus the binding energy per nucleon is

$$28/4 \text{ or } 7 \text{ MeV (approx.)}$$

Binding energies of the nuclei of other atoms can be calculated in a similar manner. Figure 5.1 shows the binding energies of the nuclei of atoms plotted against their respective atomic numbers.

FIG. 5.1. *Binding energy per nucleon plotted against mass number*

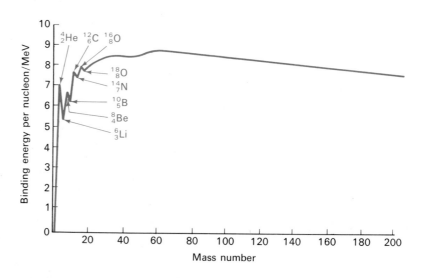

Inspection of the graph shows that elements whose mass numbers lie in the region of 60 have the most stable nuclei, e.g. iron, cobalt, nickel and copper. Elements with light nuclei, if they can be made to coalesce, should produce more stable and heavier nuclei with the release of energy, as in the hydrogen bomb explosion and in stellar reactions;

48

similarly, elements whose nuclei are very heavy should be capable of fission into lighter and more stable nuclei with the release of energy, as in the atomic bomb explosion and in the nuclear pile (p. 59).

5.4
Natural radioactivity

The spontaneous emission of particles and rays from the nucleus of the uranium atom was first noticed by Becquerel in 1896. This observation was followed up by the Curies who discovered that the ore pitchblende, from which uranium is extracted, was more radioactive than purified uranium oxide, and they eventually succeeded in isolating two new elements—polonium and radium—which were responsible for this increased activity.

α-particles, β-particles and γ-rays

Three types of radioactive emission have been characterised. α- and β-radiation is due to the emission of charged particles from the atomic nucleus but γ-radiation is a short wavelength electromagnetic radiation which accompanies α- and β-radiation, and represents the excess energy that a nucleus sheds on passing into a more stable arrangement.

α-radiation
α-particles are ejected at extremely high velocities from the nuclei of the heavier unstable elements. Rutherford measured the charge/mass ratio of these particles and suggested that they were helium nuclei. In a classic experiment, he sealed pure radon gas in a glass tube which was thin enough for the α-particles to pass through into a surrounding evacuated jacket. Several days later he examined the spectrum of the gas which collected in the jacket; he was able to confirm that it was helium and, furthermore, to establish that one atom of helium had originated from every α-particle that had penetrated into the evacuated jacket.

α-particles have a charge of +2 units and a mass of 4 units. The loss of an α-particle from the nucleus of an atom of an element results in the formation of a new element; for example, uranium (atomic number 92) changes into the element thorium (atomic number 90) and the atomic mass of the new element is 4 units smaller:

$$^{238}_{92}U \longrightarrow \ ^{234}_{90}Th \ + \ ^{4}_{2}He \text{ (an α-particle)}$$

Unstable nuclei emit α-particles of quite specific energies, though usually of more than one energy. For example, $^{238}_{92}U$ emits α-particles of 4·18 and 4·13 MeV. When an α-particle of energy 4·13 MeV is emitted the thorium nucleus is in an excited state and eventually sheds the excess energy, (4·18 − 4·13) or 0·05 MeV as γ-rays and drops down to its most stable state (see fig. 5.2).

FIG. 5.2. *The decay scheme of $^{238}_{92}U$ to $^{234}_{90}Th$*

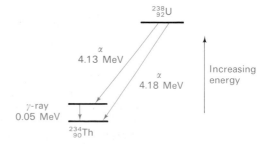

49

In view of their relatively large mass, α-particles will travel through gases in straight lines (Plate 3). They cause large amounts of ionisation, however, and this means that they do not travel far before they are stopped. In general, charged particles interact more strongly with other atoms and so they transfer their kinetic energy more readily and are quite easily stopped. This means that the health hazard from radiation of this kind, except through ingestion, is not great.

Plate 3. Cloud chamber photograph of α-particle tracks (photo: The Science Museum, London)

β-radiation

β-emission appears to be a more general kind of radioactivity than α-emission, which is confined to the heavier elements with their more highly charged nuclei. β-particles are electrons ejected from the unstable nuclei of atoms at velocities approaching that of light.

Since the nucleus does not contain electrons it must be supposed that they originate from the conversion of neutrons to protons:

$$\text{neutron} \longrightarrow \text{proton} + \text{electron}$$

50

The loss of a β-particle results in the formation of a new element whose atomic number is one unit greater than that of the decaying element. For example, thorium (atomic number 90) changes into protactinium (atomic number 91) by β-emission:

$$^{234}_{90}\text{Th} \longrightarrow {}^{234}_{91}\text{Pa} + {}^{0}_{-1}\text{e (a β-particle)}$$

β-emission provides the key to the production of new elements since the atomic number is one unit higher than that of the original nucleus (p. 56). Note that the mass number of the nucleus is unaffected by β-emission.

β-particles do not have great powers of penetration and they are stopped by a few mm of lead shielding; consequently they do not present a great hazard to health so far as external radiation is concerned. Because of their small mass their tracks in air, unlike those of α-particles, are generally meandering (Plate 4).

β-emission is quite distinct from α-emission in that the β-particles have a continuous spread of energies (fig. 5.3). At first this discovery puzzled nuclear physicists since it seemed to imply the breakdown of the principle of conservation of energy. In 1927 Pauli suggested that the emission of a β-particle was accompanied by the emission of a neutrino (see Table 5.1 p. 47). It was suggested that the total available energy was divided between the β-particle and the neutrino (negligible mass and no charge). Only in the case of E_{max} (fig. 5.3) would the total energy be carried away by the β-particle itself.

In view of its negligible mass and lack of charge the neutrino, discovered by scientists at Los Alamos in New Mexico in 1956, is able to penetrate matter to considerable distances (about 5×10^6 km in air before reacting with another nucleus). Neutrino detectors are therefore generally built deep in the earth, e.g. at the bottom of mine shafts, where interference from other particles is unlikely.

FIG. 5.3. *Distribution of energies among β-particles*

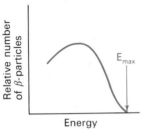

Plate 4. β-particle tracks with a fast β-track running from left to right across picture (photo: The Science Museum, London)

γ-*radiation*

The emission of an α-particle or a β-particle by an unstable nucleus is generally accompanied by the simultaneous emission of electromagnetic radiation of very short wavelength known as γ-radiation. This represents the excess energy that the nucleus sheds on passing into a more stable arrangement.

γ-rays are extremely penetrating and the more energetic ones will travel through several cm of lead shielding. They represent a severe health hazard to those who work with radioactive chemicals.

5.5
Detection of radiation

The detection of radiation associated with radioactivity depends on the detection of charged particles. γ-rays although uncharged themselves will, like α- and β-particles, ionise gas molecules through which they pass. Several types of detector are described below.

The cloud chamber

This device, invented by C. T. R. Wilson in 1911, consists of a chamber into which is introduced dust-free air saturated with water vapour (see fig. 5.4). Rapid expansion of the damp air, obtained by movement of the piston, causes a drop in temperature and the air becomes supersaturated with the water vapour. Any ions that happen to pass through the chamber act as nuclei onto which tiny droplets of water condense. The tracks of these ions thus become visible and may be photographed (see Plates 3 and 4 which show respectively the tracks of α- and β-particles).

FIG. 5.4. *Principle of the Wilson cloud chamber*

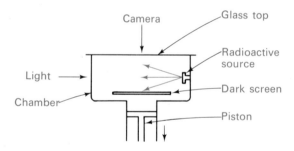

The bubble chamber

This piece of apparatus, invented by Glaser in 1951, involves the boiling of a liquid under pressure and then the sudden release of pressure in order to obtain a superheated liquid i.e. the absence of the usual gaseous phase. When ions pass through this superheated liquid they act as nuclei on which bubbles form and rapidly grow in size. Once again the tracks of any ions are made visible and are photographed.

This apparatus is superior to the cloud chamber since the tracks are much denser. Liquid hydrogen has been used in such chambers as the working fluid.

The Geiger-Müller tube

A Geiger-Müller (GM) tube consists of a metallized glass tube cathode sealed at one end by a thin mica window (fig. 5.5). Inside the tube, and insulated from it, is a tungsten wire anode. The tube may be filled with a variety of gases at low pressure, one such mixture being argon and bromine vapour. A potential difference of about 400 V is applied across the anode and cathode, and in the absence of ionising radiation the tube is a non-conductor of electricity. When an ionising particle enters the tube, the gas inside is ionised to give positive ions and electrons which travel to their respective electrodes, i.e. a current flows. This current produces a potential difference across the resistor, which is then amplified and recorded either as a click on a loud-speaker or as a current pulse. Modern GM tubes are capable of recording up to 10^5 counts per second.

FIG. 5.5. *The Geiger-Müller tube*

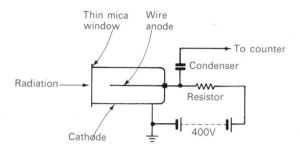

Photographic emulsion

Ionising radiation causes the blackening of photographic film. This observation, first noticed by Becquerel in 1896, has resulted in the development of specialised photographic emulsions which have proved of particular value in the study of cosmic rays, i.e. protons and nuclei of light elements which originate from outer space. In view of the rather high density of photographic emulsion the tracks produced by ionising radiation are very short range, so they have to be viewed under high magnification.

5.6
More about radioactive decay

The rate of decay

The decay of a radioactive element is a random process and is not influenced by external factors such as temperature changes. The rate of decay is directly proportional to the number of atoms present, following an exponential law, the rate of decay decreasing with time.

Kinetically the process is a first order reaction (p. 430) and can be expressed by the equation:

N = number of radioactive atoms
t = time
λ = constant

$$-\frac{dN}{dt} = \lambda N \qquad (1)$$

Rearrangement of equation (1) followed by integration gives:

$$-\int \frac{dN}{N} = \lambda \int dt$$

$$\text{or} \quad -\ln N = \lambda t + \text{const.}$$

If the number of radioactive atoms present at time $t = 0$ is N_0, then

$$-\ln N_0 = \text{const.}$$
$$\text{and} \quad -\ln N = \lambda t - \ln N_0 \quad \text{or} \quad \ln(N_0/N) = \lambda t$$

Conversion of Naperian logarithms to ordinary logarithms gives:

$$2 \cdot 303 \lg(N_0/N) = \lambda t \qquad (2)$$

Let $t_{1/2}$ represent the time needed for the original activity to decrease to a half, i.e. $N_0 = 2N$, then

$$2 \cdot 303 \lg(N_0/\tfrac{1}{2}N_0) = 2 \cdot 303 \lg 2 = \lambda t_{1/2}$$
$$\text{or} \quad 0 \cdot 6932 = \lambda t_{1/2}$$
$$\text{Thus,} \quad t_{1/2} = \frac{0 \cdot 6932}{\lambda} \qquad (3)$$

As equation (3) shows, the time taken for half the activity to disappear ($t_{1/2}$), or the **half-life** as it is called, is independent of the number of radioactive atoms. The half-life of a particular nuclide is a characteristic constant of that nuclide, and so too is the value λ which is termed the **decay constant**. In practice, half-lives of radioactive elements vary from thousands of years to milliseconds.

Equation (2) may be written in the form

$$N_0/N = e^{\lambda t}$$
$$\text{or} \quad N = N_0 e^{-\lambda t} \qquad (4)$$

A graph of N, the number of undecayed nuclei, against time produces the familiar exponential decay curve on which an equal increment of time accounts for the same fractional change in N (see fig. 5.6).

$-dN/dt$, measures the number of nuclear disintegrations per unit time. The unit of radioactivity is the becquerel (Bq), equal to one nuclear disintegration per second. $3 \cdot 7 \times 10^{10}$ becquerels equal 1 curie (Ci), which is the approximate activity of one gramme of radium.

FIG. 5.6. *Exponential decay of radioactive element*

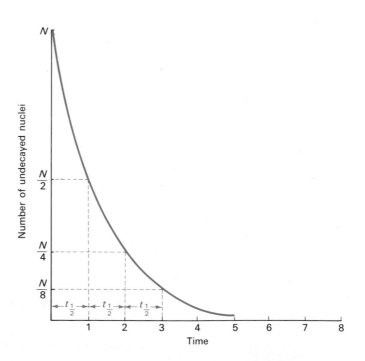

Decay products

If radioactive decay proceeds by a number of individual steps, i.e. if the first radioactive decay product is itself radioactive and thus decays further, then each stage in the process is represented by its own characteristic rate equation and thus by its own half-life and decay constant.

There are three naturally occurring radioactive decay series which start with uranium, thorium and actinium. In each case the final product of the series is a different stable isotope of lead. The uranium decay series is given in Table 5.2 below, together with the half-life of the intermediate elements and the type of emission which occurs at each stage in the series.

Table 5.2 The uranium decay series

Element	Symbol	Particle emitted	Half-life
Uranium	$^{238}_{92}$U	α	4.5×10^9 years
Thorium	$^{234}_{90}$Th	β	24·1 days
Protactinium	$^{234}_{91}$Pa	β	1·18 min
Uranium	$^{234}_{92}$U	α	2.48×10^5 years
Thorium	$^{230}_{90}$Th	α	8.0×10^4 years
Radium	$^{226}_{88}$Ra	α	1.62×10^3 years
Radon	$^{222}_{86}$Rn	α	3·82 days
Polonium	$^{218}_{84}$Po	α	3·05 min
Lead	$^{214}_{82}$Pb	β	26·8 min
Bismuth	$^{214}_{83}$Bi	β	19·7 min
Polonium	$^{214}_{84}$Po	α	1.6×10^{-4} s
Lead	$^{210}_{82}$Pb	β	19·4 years
Bismuth	$^{210}_{83}$Bi	β	5·0 days
Polonium	$^{210}_{84}$Po	α	138 days
Lead	$^{206}_{82}$Pb		Non-radioactive

Notice the wide range of half-lives from 1.6×10^{-4} s to 4.5×10^9 years.

5.7 Artificial breakdown of the atomic nucleus

So far we have been concerned with nuclei that are naturally unstable, for example, radium, uranium and thorium. Nuclei which are normally stable can be induced to undergo changes if they are stimulated in some way. The usual means of stimulation is to bombard the nucleus with a suitable 'projectile' particle.

The first artificial breakdown of the atomic nucleus was achieved by Rutherford in 1919. He bombarded nitrogen with α-particles and obtained oxygen and protons:

$$^{14}_{7}\text{N} + {}^{4}_{2}\text{He} \longrightarrow {}^{17}_{8}\text{O} + {}^{1}_{1}\text{H}$$

Other nuclear transformations rapidly followed, α-particles, protons and neutrons being used as the bombarding particles. Typical examples are:

$$^{9}_{4}\text{Be} + {}^{4}_{2}\text{He} \longrightarrow {}^{12}_{6}\text{C} + {}^{1}_{0}\text{n}$$
$$^{27}_{13}\text{Al} + {}^{1}_{1}\text{H} \longrightarrow {}^{24}_{12}\text{Mg} + {}^{4}_{2}\text{He}$$
$$^{16}_{8}\text{O} + {}^{1}_{0}\text{n} \longrightarrow {}^{13}_{6}\text{C} + {}^{4}_{2}\text{He}$$

Neutrons are particularly convenient particles to use since, being uncharged, they suffer no repulsion by atomic nuclei.

In all of the above examples the product nuclei are stable, but many nuclear reactions result in the formation of unstable, i.e. radioactive, isotopes. Thus in 1933 M. and Mme. Curie-Joliot succeeded in producing a radioactive isotope of silicon by bombarding magnesium with α-particles, the silicon decaying to aluminium by the process of positron emission, i.e. the loss of a positively charged electron:

$$^{24}_{12}Mg + {}^{4}_{2}He \longrightarrow \underset{\text{(radioactive)}}{^{27}_{14}Si} + {}^{1}_{0}n$$

$$^{27}_{14}Si \longrightarrow {}^{27}_{13}Al + {}^{0}_{+1}e$$

The above reactions are conveniently written in a shorthand fashion thus,

$$^{24}_{12}Mg(\alpha,n)^{27}_{14}Si \longrightarrow {}^{27}_{13}Al + {}^{0}_{+1}e$$

where α represents the projectile used in the transformation and n represents the particle emitted.

Other typical reactions which produce radioactive elements include the following:

α-particle-induced reaction $\quad {}^{14}_{7}N(\alpha,n)^{17}_{9}F \longrightarrow {}^{17}_{8}O \quad + {}^{0}_{+1}e$

Proton-induced reaction $\quad {}^{11}_{5}B(p,n)^{11}_{6}C \longrightarrow {}^{11}_{5}B \quad + {}^{0}_{+1}e$

Neutron-induced reaction $\quad {}^{23}_{11}Na(n,\gamma)^{24}_{11}Na \longrightarrow {}^{24}_{12}Mg + {}^{0}_{-1}e$

All the elements which have an atomic number greater than 92 (all those beyond uranium in the Periodic Table) are artificially produced by nuclear bombardment; they are called the transuranium elements. For example, uranium-238 can be converted into uranium-239 by neutron capture:

$$^{238}_{92}U + {}^{1}_{0}n \longrightarrow {}^{239}_{92}U$$

This uranium isotope is unstable, with a half-life of 2.4 days and it decays by β-emission to form a new artificial element called neptunium:

$$^{239}_{92}U \longrightarrow {}^{239}_{93}Np + {}^{0}_{-1}e$$

In several instances heavy bombarding particles such as the nuclei of nitrogen and oxygen atoms have been employed, particle accelerators, such as cyclotrons, synchrotrons and linear accelerators, being used to produce such particles with suffcient energy to induce reaction. Two typical reactions are given below, together with the half-life of the new element produced:

$$^{238}_{92}U + {}^{14}_{7}N \longrightarrow \underset{\text{(Einsteinium)}}{^{248}_{99}Es} + 4{}^{1}_{0}n \quad t_{1/2} = 25 \text{ min}$$

$$^{238}_{92}U + {}^{16}_{8}O \longrightarrow \underset{\text{(Fermium)}}{^{253}_{100}Fm} + {}^{1}_{0}n \quad t_{1/2} = 4 \cdot 5 \text{ days}$$

5.8
Uses of radioisotopes

The uses of radioisotopes are numerous and it is not possible to give more than a few examples of their application in widely divergent situations.

Medicinal uses

Ionising radiation, if of sufficient intensity, can destroy all living cells. However, developing cells are more susceptible to such radiation than those that are fully mature. In this context cobalt-60, which is a γ-ray emitter, has been used in carefully controlled situations to arrest the growth of cancerous cells.

Abnormal circulation of the blood can be investigated by injecting the patient with a solution of sodium chloride containing radioactive sodium-24. The radioactivity in the bloodstream can then be traced using a Geiger-Müller tube (p. 52).

Archaeological dating

The presence of carbon-14 in all organic matter provides a means of dating archaeological objects of a vegetable origin. The method depends upon the fact that the carbon dioxide in the atmosphere contains small amounts of the radioactive isotope $^{14}_{6}C$. This originates from the action of neutrons (produced by cosmic rays in the atmosphere) on the nitrogen in the air:

$$^{14}_{7}N + ^{1}_{0}n \longrightarrow ^{14}_{6}C + ^{1}_{1}H$$

Since plants and trees take up carbon dioxide and convert it into tissue material and since animals, including humans, eat plants, all living matter contains the same proportion of carbon-14 as occurs in the atmosphere. After death, the uptake of carbon dioxide ceases and the level of carbon-14 in the body and in plant tissues falls ($^{14}_{6}C$ has a half-life of 5600 years). For instance, a tree felled some 5600 years ago will only contain half the activity of a growing tree. Paper and papyrus have a vegetable origin and can thus be dated by the carbon-14 method, e.g. the Dead Sea scrolls.

Geological dating

Several methods employing radioisotopes have been used to date rocks. One of the most promising methods is based on the decay of potassium-40 to argon-40 by electron capture—a type of process not previously mentioned in this chapter. The basic process is:

$$^{40}_{19}K + ^{0}_{-1}e \longrightarrow ^{40}_{18}Ar$$

The abundance of potassium-40 in potassium-containing compounds is $0.0119 \pm 0.0001\%$ and any rocks containing potassium compounds would contain this percentage abundance of potassium-40 at the instant of formation. Over the course of millions of years such rocks will accumulate argon by the above process. Since the half-life of potassium-40 is 1.3×10^9 years this method can be used to cover a very wide geological time span. It is of course assumed that there is no serious escape of argon from rock samples over the geological time interval.

Industrial uses

The incorporation of a radioisotope of a metal into the piston rings of engines allows the accurate measurement of wear by friction. The engine is run under test for a specified period and the amount of radioactivity in the lubricating oil determined. Tests such as this have been useful in the development of more efficient engine oils.

β-particle emitters can be used to measure the thickness of metal plate, the amount of radioactivity penetrating the metal being measured with a suitable detector which is calibrated in units of length. This method allows the continuous monitoring of metal plate as it emerges from the rolling mill.

Some other uses

The uptake of phosphorus by plants has been investigated using phosphatic fertilizers labelled with radioactive phosphorus-32.

The mechanism of photosynthesis has been worked out using carbon-14.

A purely chemical application of radioisotopes has been the demonstration that the two sulphur atoms in the thiosulphate anion, $S_2O_3^{2-}$, are not equivalent. This has been achieved by first of all preparing the thiosulphate anion by boiling sulphur, containing the sulphur-35 isotope, with a solution of a sulphite which is left unlabelled:

$$^{35}_{16}S + SO_3^{2-} \longrightarrow [^{35}_{16}SSO_3]^{2-}$$

When the thiosulphate is treated with dilute acid it is found that the radioactivity appears in the precipitated sulphur and not in the sulphur dioxide:

$$[^{35}_{16}SSO_3]^{2-} + 2H^+(aq) \longrightarrow {}^{35}_{16}S + H_2O + SO_2$$

This experiment shows quite clearly that the two sulphur atoms in the thiosulphate anion do not occupy equivalent positions.

5.9 Nuclear energy

The graph of nuclear stability, measured in terms of the binding energy per nucleon, against mass number (fig. 5.1, p. 48) shows a maximum in the region of a mass number of 60. It is apparent, then, that energy should be released when very light nuclei coalesce to form heavier ones and when very heavy nuclei are broken up to form lighter ones. These processes, known respectively as nuclear fusion and nuclear fission, have both been used, with awesome effect, in the production of nuclear energy.

Nuclear fusion

We owe our existence to nuclear fusion, since the stars, including the sun, produce their energy by a nuclear fusion process. For example, in cooler stars like the sun, where internal temperatures are in the region of about 10^7 K, the following process is thought to occur:

Reaction	Energy released/MeV
$^1_1H + {}^1_1H \longrightarrow {}^2_1H +$ positron $+$ neutrino	0·4
$^1_1H + {}^2_1H \longrightarrow {}^3_2He +$ gamma radiation	5·5
$^3_2He + {}^3_2He \longrightarrow {}^4_2He + 2{}^1_1H$	12·8

The net reaction is therefore:

Energy released/MeV

$$4^1_1H \longrightarrow {}^4_2He + 2 \text{ positrons} + 2 \text{ neutrinos} \qquad 24.6$$

Neutrinos from the sun have been detected on the earth, but at a lower rate than predicted theoretically and this has cast some doubt upon the above scheme.

The barrier to nuclear fusion is the intense repulsive force which operates when nuclei approach each other to within collision distance. It is of no value to accelerate the nuclei in one of the machines mentioned on page 56 since the consumption of electrical energy by the machine exceeds the nuclear energy produced in the fusion process itself. Two methods have been used to produce a temperature high enough to overcome the repulsive force—a temperature of about 10^7 K is required.

The first method uses the process of nuclear fission as the triggering mechanism to provide a sufficiently high temperature to get nuclear fusion started. This is the principle behind the hydrogen bomb and, once fusion starts, enough energy is released for the process to become self-sustaining. This technique is inapplicable to the production of controlled nuclear fusion for peaceful purposes.

The second method involves heating the reaction mixture by rapid compression. The problem of containing the ionised gases (called plasma) in a vessel able to withstand the high temperatures involved has been solved by 'pinching' the plasma and containing it with a magnetic field—the so called 'magnetic bottle'. However, it has not yet proved possible to maintain a stable condition for sufficient time to reach 'break even' point, i.e. the point at which the nuclear energy released balances that put into the system. Once this problem is solved, an unlimited source of energy becomes available from the hydrogen atoms in water.

FIG. 5.7. *Uranium-235 fission-yield curve*

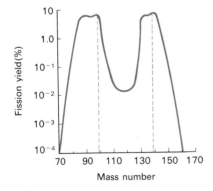

Nuclear fission

Nuclei of mass number greater than 200 are markedly less stable than those of mass number around 60. Nevertheless, fission does not occur spontaneously, since the activation energy (p. 443) of the process is too high.

Certain nuclei, notably uranium-235 and plutonium-239, are able to capture low-energy neutrons and become sufficiently excited for fission to occur. Uranium-235, the material that was used in the first atomic bomb, splits into two unequal fragments with mass numbers in the approximate range from 70 to 160; the most probable mass numbers are about 100 and 140 as shown in fig. 5.7. Approximately 200 MeV of energy is released in the process, together with 2 to 3 neutrons for every uranium atom split:

$$\text{Uranium-235} \longrightarrow \text{Fragment(1)-100} + \text{Fragment(2)-140} + 2$$
$$\text{or} \quad 3 \text{ neutrons} + \text{energy}$$

Most of the fragments are powerful β-emitters and many of them injurious to health through ingestion, e.g. strontium-90 is particularly potent since it can replace calcium in bone structures.

The importance of the fission of uranium-235 lies in the production of more neutrons, which may be absorbed by other uranium-235 nuclei and the fission process continued. Indeed, since one initial neutron will produce more than one from the fission process, a chain reaction can become established and this does occur in an atomic explosion.

In a nuclear pile, or nuclear reactor, the fission process is controlled so that the release of energy occurs gradually. The key to control lies in the absorption of neutrons in such a manner that just enough are available to maintain the fission process. This is achieved by using neutron-absorbing control rods made of boron- or cadmium-containing steel which can be lowered into the nuclear fuel should a dangerous situation arise.

A further problem in the production of nuclear energy for peaceful purposes is caused by the high energy of the neutrons which are emitted in the fission process. This makes them unsuitable for capture by another uranium-235 nucleus. The neutrons must be slowed down by a moderator, e.g. heavy water (deuterium oxide) or graphite, which surrounds the nuclear fuel and reduces the speed of the neutrons to about 10^3 m s^{-1}. These are called 'thermal neutrons' since they have energies of comparable magnitude to those of gas molecules.

One final problem in the control of nuclear fission lies in the nature of the fuel. Uranium-235 is present only to the extent of 0·7% in natural uranium. Over 500 tons of fuel were used in the first atomic pile built near Chicago in 1942. However, some of the neutrons released in the fission process can be used to convert uranium-238 (the predominant isotope) to plutonium-239 which has fissionable properties which are similar to those of the rare uranium-235. Thus, for every gramme of uranium-235 which is consumed, about one gramme of natural uranium-238 is converted into useful plutonium-239.

Reactor fuels may be enriched with plutonium-239 produced in **breeder reactors**. With enriched fuels, the control of the neutron energy becomes less important and the reactors using such fuels (fast reactors) are much reduced in size as a result. In 1971, a prototype reactor using only plutonium went into production at Dounreay in the United Kingdom. It makes use of high energy neutrons and no moderator is needed. The reactor is expected to be in full production in the 1980s.

5.10
Nuclear forces

There seems to be no apparent reason why positively charged protons and neutral neutrons should stick together and form a nucleus. Whatever the origin of this force, it is apparent that it acts over a short distance, i.e. nuclear radii are in the region of 10^{-15} m (p. 46), and over this sort of distance it is capable of acting on neutrons, in addition to overcoming the mutual electrostatic repulsion between protons. Over still shorter distances it must become repulsive, otherwise nuclei would collapse on themselves.

FIG. 5.8. *Binding energy curve for elements up to mass number 28*

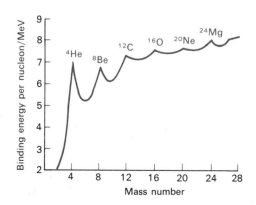

Important clues to the puzzle can be found in the binding energy curve (p. 48) which is shown on an expanded scale for elements up to a mass number of 28 in fig. 5.8. A definite series of maxima occurs at positions occupied by the following nuclei:

$$^4\text{He} \quad ^8\text{Be} \quad ^{12}\text{C} \quad ^{16}\text{O} \quad ^{20}\text{Ne} \quad ^{24}\text{Mg}$$

All these nuclei contain equal numbers of protons and neutrons, and they are all multiples of the helium nucleus. This nucleus is clearly especially stable and this is the reason why the α-particle features so prominently in nuclear chemistry.

If the stability of all the stable nuclei is considered the following facts emerge:
(a) 164 nuclei have even numbers of neutrons and protons
(b) 55 nuclei have an odd number of neutrons and an even number of protons
(c) 50 nuclei have an even number of neutrons and an odd number of protons
(d) 4 nuclei have odd numbers of neutrons and protons

It has been suggested that there is a series of energy levels for nucleons just as there is for the electrons in an atom. Each level can accommodate 2 neutrons and 2 protons, so the nuclei which are multiples of ^4He are the nuclear equivalents of the noble gases.

Nucleons, just like electrons, possess spin, and act as tiny magnets. This behaviour underlies the technique of nuclear magnetic resonance spectroscopy (chapter 13). However, the total spins of nuclei turn out to be very small and indeed in many cases zero (Table 5.3).

Table 5.3 Total spin of selected nuclei

Nucleus	Neutron number	Proton number	Spin
^1H	nil	odd	$\frac{1}{2}$
^4He	even	even	0
^{14}N	odd	odd	1
^{17}O	odd	even	$\frac{5}{2}$
^{41}Ca	odd	even	$\frac{7}{2}$

This suggests that nucleons 'pair up' so as to cancel out their spins, thus explaining the stability of the even/even nuclei where both neutrons and protons will be completely paired.

The apparent existence of energy levels and paired nuclear spins makes a satisfying analogy between the behaviour of the nucleus and the behaviour of electrons. However, it brings us no nearer to an explanation of why the nucleus sticks together at all. Yukawa has suggested that the nuclear 'glue' is provided by π-mesons (p. 47) which may be exchanged betweens neutrons and protons to produce strong 'exchange forces' which bind nucleons together

5.11
Questions on chapter 5

1 What is the nature of the three major radiations produced by radioactive elements?

How do these radiations differ in penetrating power?

Why are high energy particles usually used in bringing about nuclear reactions?

State and explain what changes you would anticipate in unstable nuclei which have (a) too high, (b) too low, a ratio of protons to neutrons.   C

2 (a) List the three main fundamental particles which are constituents of atoms, and give their relative masses and charges.
(b) Similarly, name and *differentiate* between the radiations emitted by naturally occurring radioactive elements.
(c) Complete the following equations using your Periodic Table to identify the elements X, Y, Z, Q, and R. Add atomic and mass numbers where these are missing:
    (i) $^{24}_{11}\text{Na} \rightarrow X + ^{0}_{-1}\text{e}$
    (ii) $^{14}_{7}\text{N} + ^{1}_{0}\text{n} \rightarrow ^{14}Y + _{1}Z$
    (iii) $\text{Si} \rightarrow ^{27}_{13}Q + ^{0}_{+1}\text{e}$
    (iv) $R + ^{4}_{2}\text{He} \rightarrow ^{13}_{7}\text{N} + ^{1}_{0}\text{n}$.

(d) Refer to c(i) and c(iii) above. For each of these two processes, briefly describe ONE chemical test which could be used to confirm that a change of chemical element has occurred.

(e) If lead(II) chloride is precipitated in the presence of thorium nitrate, using an aqueous solution of lead(II) nitrate and dilute hydrochloric acid, the lead(II) chloride contains radioactive ^{212}Pb atoms. (These radioactive lead atoms are a daughter product of the thorium.)

Show how you could experimentally use this information to establish that lead(II) chloride and its saturated solution are in dynamic equilibrium.　　　S

3　Name the three different types of radioactive emission. What is the nature of each of these three types of emission and how are they affected by a magnetic field?

Identify the particle A in the following nuclear equation:
$$^{32}S + A \longrightarrow {}^{32}P + {}^1_1H.$$

Define (a) *atomic number*, and (b) *mass number*. Explain what change you would expect to occur in an unstable nucleus with too high a mass number/atomic number ratio.　　　C (Overseas)

4　(a) What are α, β and γ emissions and how do they differ in their penetrating power and their behaviour in a magnetic field?

Explain the meaning of the two numbers before the symbol for uranium and identify P, Q, R and S in the following equations.

$$^{234}_{92}U \longrightarrow \alpha + P$$
$$^{239}_{92}U \longrightarrow \beta^- + Q$$
$$^{235}_{92}U \longrightarrow \gamma + R$$
$$^{238}_{92}U + {}^2_1H \longrightarrow {}^{239}_{92}U + S$$

(b) Deduce the nature of X in the following reaction.

$$^{235}_{92}U + {}^1_0n \longrightarrow {}^{95}_{42}Mo + {}^{139}_{57}La + 2{}^1_0n + 7X$$

There is approximately 0·1 per cent less mass on the right-hand side of the equation than on the left-hand side. What is the significance of this?

　　　JMB (Syllabus B)

5　(a) Give one example in each case of nuclear reactions brought about by bombardment with (i) protons, (ii) alpha-particles. Write equations for these reactions which show the identity, mass number and atomic number of each nucleus involved.

(b) Explain concisely the principles involved in the technique of radio-carbon dating.

(c) Indicate in outline how a mass spectrometer can be used
　　(i) to show that bromine has two (stable) isotopes,
　　(ii) to obtain an accurate value for the relative atomic mass of this element.
　　　O

6　(a) Describe in outline how a beam of neutrons can be produced.

(b) State briefly what experimental evidence led to the conclusion that a neutron has (i) no electrical charge, (ii) a mass approximately equal to that of the proton.

(c) Explain clearly the principles underlying the technique of radiocarbon dating.

(d) Name two isotopes (other than any isotope mentioned in your answer to (c)) which are used in industry, in medical research, or in some other aspect of everyday life, and briefly describe one use for each isotope.　　　O

7　Explain briefly the meaning of the terms *isotope* and *decay constant* (λ). Deduce the relationship between λ and $t_{1/2}$, the half-life of a radioactive nuclide.

Give the basic principles of any method that could be used to show whether a given naturally-occurring element consists of a single stable isotope (e.g. sodium) or several stable isotopes (e.g. magnesium).

The average human contains 140 g of potassium. This element consists of three isotopes, one of which is radioactive. Calculate the amount of natural radioactivity due to this isotope in the human body (in disintegrations second^{-1}). Data Book required.

Suggest how the total amount of radioactivity in a live human being could be determined.　　　O and C

8 (a) State the rate law governing radioactive decay and explain what is meant by the *half-life* of a radioactive isotope. Can a knowledge of the half-life be used to predict the lifetime of a single radioactive atom?

(b) The table below gives the number of undisintegrated atoms, N, present after different times measured in days, in a sample containing ^{32}P and ^{35}S but no other radioactive isotopes.

$N \times 10^{-12}$	Time/days
132	10
110	20
79·6	40
63·1	60
53·7	80
45·8	100
20·9	200
9·5	300

(i) The half-life of ^{35}S is about five times that of ^{32}P. Use the data in the table to obtain the half-life of ^{35}S.

(ii) Indicate how, in principle, the half-life of ^{32}P might also be obtained from these data.

(c) Hydrolysis of ethyl benzoate enriched with ^{18}O in the alkoxy group gives ethanol which contains all the ^{18}O. What method would be used to detect ^{18}O in the ethanol? What information does this particular experiment give about the mechanism of the hydrolysis? O (S)

9 Discuss the differences between radioactive decay and chemical reactions.

When meteorites travel through space they are bombarded by cosmic radiation which generates in them small amounts of radioactive nuclei. These generative processes cease when a meteorite strikes the earth. A recent stony meteorite was found to have a tritium, ^3H, to argon, ^{39}Ar, β-ray activity ratio of 40:1, compared with a ratio of 1:1 for an older meteorite of similar composition. Calculate the number of years that have elapsed since the older meteorite fell on the surface of the earth. State what assumptions you have made in working out your answer.

[Decay constant of ^3H = $5·8 \times 10^{-2}$ years^{-1};
Decay constant of ^{39}Ar = $2·1 \times 10^{-3}$ years^{-1}.] O (S)

10 You have discovered a new element. How would you determine its relative atomic mass and investigate its radiochemical properties? Could you distinguish different isotopes in it? Suggest experiments by which you might determine its atomic number. Oxford Schol. and Entrance

Ionic and metallic bonding

6.1
Introduction

Once the nuclear theory of the atom (Rutherford) and the structure of the hydrogen atom (Bohr) had been worked out, the time was ripe for an explanation of chemical bonding. Indeed, even before the general principles of atomic structure had been fully explained, Kossel and Lewis had independently put forward the view that elements tended to react together to attain the stable structure of the noble gases. While Kossel's work was mainly directed towards explaining the nature of the bonds in electrolytes (electrovalent or ionic bonding), Lewis was concerned with developing a theory to account for the bonding in non-electrolytes (covalent bonds). All modern refinements can be traced back to two important research papers published independently by them in 1916.

Chemists nowadays find it convenient to discuss bonding under five main headings. These are ionic bonding and metallic bonding, which form the subject matter of this chapter, and covalent bonding, hydrogen bonding and van der Waals' bonding which are discussed in chapter 7.

There is no such entity as a completely ionic compound or a completely covalent compound. Thus, whenever we say that a compound is ionic or that another one is covalent, we are implying that the bonding is predominantly of that particular type; we recognise that an ionic compound may have appreciable covalent character and that a covalent compound may possess appreciable ionic character.

6.2
The electrovalent or ionic bond

Although it has now been established that the noble gases krypton and xenon can be induced to form chemical compounds, it is still true to say that, as a group, the noble gases enter into chemical combination far less readily than other elements. It is therefore reasonable to assume that the unreactive nature of this group of elements is due to their particular electronic configurations (except for helium they all have an outer shell containing 8 electrons). Kossel pointed out that an alkali metal atom could attain this supposedly stable electronic configuration by the loss of one electron and an halogen atom by the gain of one electron. He visualised the hypothetical reaction between an atom of sodium and one of chlorine as a complete transfer of one electron from the sodium atom to the chlorine atom thus:

$$\text{Na (2.8.1)} + \text{Cl (2.8.7)} \longrightarrow \underset{\substack{\text{electronic} \\ \text{configuration} \\ \text{of neon}}}{\text{Na}^+(2.8)} + \underset{\substack{\text{electronic} \\ \text{configuration} \\ \text{of argon}}}{\text{Cl}^-(2.8.8)}$$

The resulting charged species are ions, which are held together by electrostatic attraction; the bonding is said to be electrovalent or ionic.

Electrovalent compounds are formed between the most reactive metals, e.g. Groups 1A and 2A of the Periodic Table, and the most reactive non-metals, e.g. Groups 6B and 7B. The following are a few more examples:

$$Ca\ (2.8.8.2) + O\ (2.6) \longrightarrow Ca^{2+}(2.8.8) + O^{2-}(2.8)$$

electronic configuration of argon   electronic configuration of neon

$$2K\ (2.8.8.1) + S\ (2.8.6) \longrightarrow 2K^{+}(2.8.8) + S^{2-}(2.8.8)$$

electronic configuration of argon   electronic configuration of argon

$$Li\ (2.1) \quad + F\ (2.7) \longrightarrow Li^{+}(2) \quad + \quad F^{-}(2.8)$$

electronic configuration of helium   electronic configuration of neon

In modern notation the latter example can be represented thus:

$$Li\ (1s^2 2s^1) + F\ (1s^2 2s^2 2p^5) \longrightarrow Li^{+}(1s^2) + F^{-}(1s^2 2s^2 2p^6)$$

There is a good deal of evidence which supports the idea of the existence of positive and negative ions in compounds such as those listed above, namely:

(a) Electron density maps of ionic compounds, determined by X-ray analysis, indicate isolated regions of charge associated with each ion (fig. 6.1). This contrasts markedly with the electron density map representation of a covalent bond (fig. 7.3, p. 81).

FIG. 6.1. *Electron density map for sodium chloride. Locations of equal electron density are linked with contour lines*

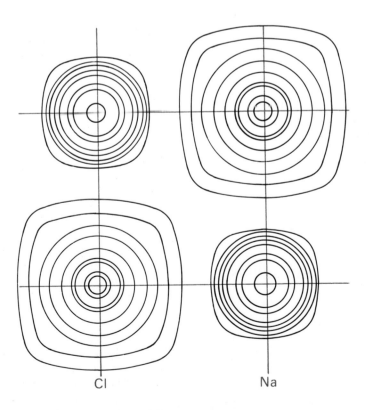

Cl   Na

(b) Molten ionic compounds conduct an electric current. This observation implies that the compounds are constructed from charged particles (called ions).

(c) X-ray analysis of potassium chloride, a typical ionic compound, shows that each species present has the same number of electrons associated with it. This suggests that the potassium atom (19 electrons) has donated an electron to a chlorine atom (17 electrons) to form the isoelectronic species K^+(18 electrons) and Cl^-(18 electrons).

(d) A particular ion shows characteristic chemical reactions which are independent of the other ion with which it is associated in an ionic compound. For example, all aqueous solutions of ionic chlorides react with aqueous silver nitrate to form a white precipitate of silver chloride:

$$Ag^+(aq) + Cl^-(aq) \longrightarrow AgCl(s)$$

The same is not true of covalently-bonded chlorine compounds.

As we have already hinted (see section 6.1, p. 64), it is difficult to draw a clear distinction between ionic and covalent compounds. Between the two extremes of the covalent hydrogen molecule and the predominantly ionic sodium chloride lies a spectrum of more or less polar bonds. The electron density map of lithium fluoride (fig. 6.2), for example, indicates a degree of covalent bonding. In our treatment of ionic bonding, therefore, we must be careful to see that any assumptions we make really are justified.

FIG. 6.2. *Electron density map of lithium fluoride*

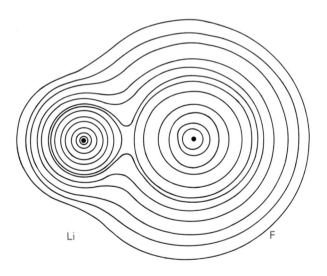

6.3
Cations and anions

In simple terms we have previously said that when a sodium atom is brought into the vicinity of a chlorine atom, the former loses an electron to the latter, since in this way both are able to achieve the electronic configurations of a noble gas as charged ions:

$$\begin{array}{ccc} Na & \longrightarrow & Na^+ & + & e^- \\ (2.8.1) & & (2.8) \end{array}$$

$$\begin{array}{ccc} Cl & + & e^- & \longrightarrow & Cl^- \\ (2.8.7) & & & & (2.8.8) \end{array}$$

However, this is not a very helpful approach since there are many examples of stable cations which do not possess the electronic configuration of a noble gas. A few examples are listed in Table 6.1 together with the electronic configurations of the parent atoms.

It is more useful to study the relative energetic stabilities of the atoms and the ions which are derived from them.

Table 6.1 Some examples of cations which do not possess the electronic configurations of a noble gas

Atoms	Cations
Mn(2.8.13.2)	Mn^{2+}(2.8.13)
Cu(2.8.18.1)	Cu^{2+}(2.8.17)
Zn(2.8.18.2)	Zn^{2+}(2.8.18)
Pb(2.8,18.32.18.4)	Pb^{2+}(2.8.18.32.18.2)

Cations

The ease with which an atom loses an electron to form a positive ion is measured by its ionisation energy (p. 23). Table 6.2 shows the decrease in the first ionisation energy of the alkali metals with increasing atomic number.

This decrease is due to a combination of size of the atom and shielding, both of which increase as the atomic number increases.

The removal of an electron from the atom of an element becomes increasingly difficult as a Period of elements is crossed from left to right as Table 6.3 indicates.

Table 6.2 First ionisation energies of the alkali metals

Element	First Ionisation energy/kJ mol^{-1}
Lithium	520
Sodium	496
Potassium	419
Rubidium	403
Caesium	376

Table 6.3 First ionisation energies of the elements of the second short period

Element	Na	Mg	Al	Si	P	S	Cl	Ar
Ionisation energy/kJ mol^{-1}	496	738	578	787	1012	1000	1250	1520

Except for some slight fluctuations, which can be explained in a similar manner to those for the first short period (lithium to neon) and which were discussed previously (p. 44), the general trend is for the first ionisation energy to rise. The extra nuclear charge experienced by each successive electron (in the same valence shell) binds the electron more tightly and thus makes it more difficult to remove than the previous one.

These vertical and horizontal trends in ionisation energies combine to make elements on the left and at the bottom of the Periodic Table the most electropositive, i.e. the most willing to form cations.

$$\uparrow \text{increasingly electropositive}$$

Li	Be				
Na	Mg	Al			
K	Ca	Ga			
Rb	Sr	In	Sn		
Cs	Ba	Tl	Pb	Bi	Po

←increasingly electropositive——

The first and second ionisation energies of the Group 2A elements are shown in Table 6.4. The removal of the second electron is more difficult than the removal of the first, since it is being pulled away from a positively charged ion. This considerable expenditure of energy will only be justified if the subsequent bonding in the ionic compound causes an even greater release of energy.

Table 6.4 First and second ionisation energies of the Group 2 elements

Element	First Ionisation energy/kJ mol^{-1}	Second Ionisation energy/kJ mol^{-1}
Beryllium	899	1757
Magnesium	738	1451
Calcium	590	1145
Strontium	549	1064
Barium	502	965

The removal of three electrons requires such a large input of energy that 3-valent cations are rare. The Al^{3+} ion is present in the fluoride, $Al^{3+}(F^-)_3$, and in the oxide, $(Al^{3+})_2(O^{2-})_3$, but not in the chloride, Al_2Cl_6, which is covalent. Metals may show high oxidation numbers (chapter 18) but this is not the same as the charge on an ion and, in any case, these states are generally shown in oxoanions, e.g. MnO_4^-, in which the oxidation number of manganese is +7.

The ionic radius of a cation is smaller than the atomic radius of the parent atom owing to the removal of the outermost electron(s), e.g.

Li	Na	K	Rb	Cs
0.133	0.157	0.203	0.216	0.235 nm

Li^+	Na^+	K^+	Rb^+	Cs^+
0.060	0.095	0.133	0.148	0.169 nm

Anions

The acceptance of an electron by an electronegative atom is generally an energy releasing process, in contrast to the ionisation of an atom. The energy change for the process:

$$X(g) + e^- \longrightarrow X^-(g)$$

is known as the **first electron affinity** for the element, and measures the ease with which it forms an anion in the gas phase. Electron affinities are not easy to measure thus, not surprisingly, there are some wide ranges in reported values.

Table 6.5 Some first electron affinities/kJ mol^{-1}

B	C	N	O	F
−15	−121	+31	−142	−333
Al	Si	P	S	Cl
−26	−135	−60	−200	−348
				Br
				−324
				I
				−295

There are a few anomalies in the above values of first electron affinities, thus the first row B to F values are less exoenergetic than corresponding values in the second row Al to Cl, and in fact the nitrogen atom has a positive value. An explanation may lie in the relative size of comparable atoms, for those in the first row are significantly smaller than those in the second. In general, however, the first electron affinity becomes more negative in moving from left to right in the Periodic Table.

Second electron affinities which correspond to the change

$$X^-(g) + e^- \longrightarrow X^{2-}(g)$$

are always positive, since the electron is now being added to a negative ion, e.g.

$$O^-(g) + e^- \longrightarrow O^{2-}(g) \qquad EA = +791 \text{ kJ mol}^{-1}$$
$$S^-(g) + e^- \longrightarrow S^{2-}(g) \qquad EA = +649 \text{ kJ mol}^{-1}$$

Unlike cations which are smaller than the parent atom from which they are derived, anions are always larger than the corresponding atom, since the acquisition of an extra electron results in electron–electron repulsion and thus an expansion of the electron clouds, e.g.

F	Cl	Br	I	
0.072	0.099	0.114	0.133	nm

F$^-$	Cl$^-$	Br−	I$^-$	
0.136	0.181	0.195	0.216	nm

6.4
Lattice energies

The energy required to ionise one mole of sodium atoms (496 kJ) is more than the energy released on the formation of one mole of chloride ions from chlorine atoms (348 kJ). It seems most unlikely then that the process:

$$Na(g) + Cl(g) \longrightarrow Na^+(g) + Cl^-(g)$$

will occur spontaneously, and yet we know that in a crystal of sodium chloride the species are present as sodium and chloride ions. The answer lies in the large amount of energy released when the ions arrange themselves into a regular ionic crystal structure. Some of the most common arrangements are discussed in chapter 11 and a model of the structure of sodium chloride is shown in Plate 5. The giant structure of ionic compounds accounts for their high melting-points, their crystalline appearance and their brittle nature (fig. 6.3). The bonding in an ionic compound is undirected in the sense that no single ion is associated with just one of the opposite charge; thus the formula for sodium chloride, Na^+Cl^-, for example, represents only the overall ratio of sodium and chloride ions within the structure and does not imply the existence of a discrete unit within the crystal.

Plate 5. A model showing the structure of sodium chloride. The small spheres represent the Na^+ ion and the larger spheres the Cl^- ion. Note that each ion has a co-ordination number of 6. (By courtesy of Catalin Limited, Waltham Abbey)

FIG. 6.3. *Two-dimensional representation of an ionic solid undergoing fracture*

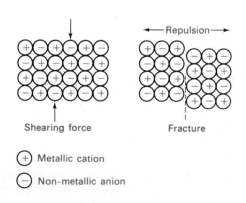

70

Calculation of lattice energies

The **lattice energy** for an ionic compound is defined to be the energy released when one mole of the ionic compound is formed from its constituent gaseous ions, e.g. for sodium chloride it refers to the process:

$$Na^+(g) + Cl^-(g) \longrightarrow Na^+Cl^-(s)$$

In the discussion that follows an expression is derived for the lattice energy of sodium chloride, but the general principles involved are applicable to any ionic solid. The following assumptions are made:

(a) that the geometrical arrangement of the ions within the crystal of sodium chloride is known (Plate 5) and so too is the interionic distance. This information is obtained from X-ray diffraction studies (p. 173).

(b) that the bonding within the crystal is completely ionic, the forces between the ions are electrostatic in nature and that the ions may be treated as 'point charges'.

The electrostatic force of attraction between a sodium ion and a chloride ion is given by the usual expression:

$$F = \frac{ke^2}{r^2} \tag{1}$$

k = **constant**

e = **charge on the electron since both ions are l-valent**

r = **distance apart of the ions**

Now imagine that the central ion is a sodium ion and that a chloride ion, from well outside the crystal (effectively at infinity) is brought up to an interionic distance, r, away from the sodium ion. The work done will be given by:

$$\text{Work done} = \int_\infty^r \frac{ke^2 dr}{r^2}$$
$$= \left[-\frac{ke^2}{r} \right]_\infty^r = -\frac{ke^2}{r} \tag{2}$$

FIG. 6.4. *Structure of sodium chloride*
Central sodium ion labelled 1 is surrounded by
(a) 6 chloride ions labelled 2 at distance r
(b) 12 sodium ions labelled 3 at distance $\sqrt{2}r$
(c) 8 chloride ions labelled 4 at distance $\sqrt{3}r$
(d) 6 sodium ions labelled 5 (only 4 shown) at distance 2r

The work done is equal to a decrease in the potential energy of the system. However, it is clear from fig. 6.4 that the central sodium ion is surrounded by six chloride ions at a distance r. The potential energy of this system (assuming a zero potential energy for the isolated ions) is given by:

$$\text{Potential energy} = -\frac{6ke^2}{r} \tag{3}$$

Further examination of fig. 6.4 shows twelve sodium ions at $\sqrt{2}r$ from the central ion; another eight chloride ions at $\sqrt{3}r$; six more sodium ions at $2r$ and so on. When all of these ions are brought together, the potential energy of the entire assembly of ions is given by:

$$\text{Potential energy} = -\frac{6ke^2}{r} + \frac{12ke^2}{\sqrt{2}r} - \frac{8ke^2}{\sqrt{3}r} + \frac{6ke^2}{2r} - \cdots \tag{4}$$

which may be rearranged thus:

$$\text{Potential energy} = -\frac{ke^2}{r}\left(6 - \frac{12}{\sqrt{2}} + \frac{8}{\sqrt{3}} - \frac{6}{2} + \cdots \right) \tag{5}$$

The series in the brackets may be summed into a constant, A, which is known as the **Madelung constant**:

$$\text{Potential energy} = -\frac{ke^2 A}{r} \tag{6}$$

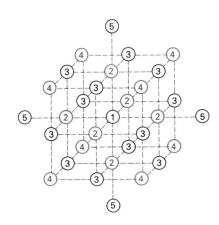

In one mole of sodium chloride there are $2L$ ions (where L is the Avogadro constant). The potential energy of one mole of sodium chloride is thus:

71

$$\frac{-\frac{1}{2}ke^2 A2L}{r} \qquad (7)$$

The factor of $\frac{1}{2}$ is introduced because the method of adding up the interactions between the ions which we have used has counted every pair of interactions twice. Thus, for one mole of sodium chloride

$$\text{Potential energy} = -\frac{kALe^2}{r}$$

$$= -\frac{ALe^2}{4\pi\varepsilon_o r} \qquad (8)$$

In fig. 6.5, the coulombic potential energy, calculated from equation (8) is plotted against interionic distance. It is apparent from the graph that, in the absence of some opposing force, the structure would collapse to a point.

$$k = \frac{1}{4\pi\varepsilon_0}$$
(see p. 21)

FIG. 6.5. *Coulombic potential energy plotted against interionic distance*

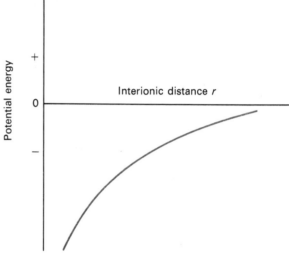

In the crystal of sodium chloride, the attractive coulombic force is balanced by a repulsive force which operates at short interionic distances owing to the overlap of the electron orbitals of the ions. This repulsive force contributes to the energy of the crystal according to the relationship:

$$\text{Repulsive energy contribution} = BL/r^n \qquad (9)$$

B = constant
n = integer

FIG. 6.6. *Potential energy curve showing attractive and repulsive contributions*

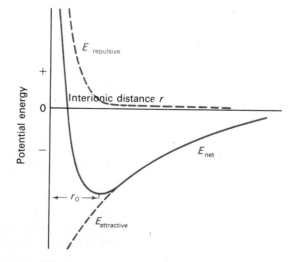

72

Fig. 6.6 shows the attractive and repulsive energy terms plotted against interionic distance, together with the net energy curve which shows a minimum at the equilibrium interionic distance in the crystal (0·281 nm in the case of sodium chloride). The net energy term is given by the expression:

$$\text{Net potential energy} = -\frac{ALe^2}{4\pi\varepsilon_\text{o}r} + \frac{BL}{r^n} \tag{10}$$

The Madelung constant, A, depends only upon the geometry of the crystal structure. Table 6.6 lists the Madelung constants for some of the most important crystal structures.

Table 6.6 Madelung constant values for some important crystal structures

Crystal structure	Coordination number	Madelung constant
Caesium chloride	8:8	1·76267
Sodium chloride	6:6	1·74756
Wurtzite (ZnS)	4:4	1·64132
Zinc blende (ZnS)	4:4	1·63806
Fluorite (CaF$_2$)	8:4	2·51939
Rutile (TiO$_2$)	6:3	2·408

The constant B in equation (10) can be determined since the net potential energy, U, is a minimum when $r = r_\text{o}$, the actual value of the Na^+ Cl^- equilibrium distance; thus dU/dr is zero at this point.

$$U = -\frac{ALe^2}{4\pi\varepsilon_\text{o}r} + \frac{BL}{r^n} \tag{11}$$

$$\frac{dU}{dr} = \frac{ALe^2}{4\pi\varepsilon_\text{o}r_\text{o}^2} - \frac{nBL}{r_\text{o}^{n+1}} = 0$$

thus,
$$\frac{ALe^2}{4\pi\varepsilon_\text{o}r_\text{o}^2} = \frac{nBL}{r_\text{o}^{n+1}} \quad \text{or} \quad B = \frac{Ae^2 r_\text{o}^{n-1}}{4\pi\varepsilon_\text{o}n}$$

Substituting this value of B and $r = r_\text{o}$ into equation (11) we have:

$$U = -\frac{ALe^2}{4\pi\varepsilon_\text{o}r_\text{o}} + \frac{ALe^2}{4\pi\varepsilon_\text{o}nr_\text{o}}$$

$$U = -\frac{ALe^2}{4\pi\varepsilon_\text{o}r_\text{o}}\left(1 - \frac{1}{n}\right) \tag{12}$$

The energy, U, is known as the lattice energy. The integer n in the equation can be determined from measurements of the compressibility of the crystal. In fact the repulsive energy term in equation (9) is not too sensitive to changes in n, so the value is not critical. A value of $n = 9$ is found for sodium chloride.

The general expression for the lattice energy of an ionic compound is:

$$U = -\frac{ALZ^+Z^-e^2}{4\pi\varepsilon_\text{o}r_\text{o}}\left(1 - \frac{1}{n}\right) \tag{13}$$

where Z^+ and Z^- are positive integers, showing the charges carried by the cation and anion respectively, e.g. for the fluorite structure, $Ca^{2+}(F^-)_2$, Z^+ is 2 and Z^- is 1.

Lattice energies may be determined experimentally and a comparison of the calculated and experimental values provides a check on the assumptions made in the calculations. Table 6.7 lists some of these values.

Table 6.7 Calculated and experimental values of some lattice energies/ kJ mol^{-1}

Compound	Expt. value	Calc. value*
LiF	−1033	−1021
NaCl	− 781	− 777
KBr	− 679	− 667
CaO	−3607	−3519
CaF$_2$	−2611	−2586
CdI$_2$	−2435	−1986
AgCl	− 890	− 769

* Calculated from a slightly modified version of equation (13).

The measure of agreement is good except for cadmium iodide and silver chloride. Here the discrepancy is due to the invalidity of one of our basic assumptions, namely that the bonding is completely ionic. In both these compounds there is an appreciable amount of covalent character; cadmium iodide, for example, has a layer structure intermediate between that of an ionic compound and a covalent compound.

In general, elements will only combine to form an ionic compound if there is a release of energy in the process of combination. This point will be developed more fully in a later chapter (p. 226) but, at this stage, it is worth noting that the energy released when sodium and chlorine atoms combine to give solid sodium chloride can be calculated from the energy processes mentioned in this chapter. For this reaction

$$Na(g) + Cl(g) \longrightarrow Na^+Cl^-(s)$$

the energy terms are as follows:

(a) first ionisation energy of sodium = +496 kJ mol^{-1}
(b) first electron affinity of chlorine = −348 kJ mol^{-1}
(c) lattice energy of sodium chloride = −781 kJ mol^{-1}

$$\text{Net energy change} = -633 \text{ kJ mol}^{-1}$$

Note that the value −633 kJ mol^{-1} (evolved) will not be the same as the energy change when solid sodium reacts with chlorine molecules, Cl$_2$, to give solid sodium chloride, since energy must be absorbed to give gaseous sodium and chlorine atoms. The reaction does result in the release of energy but it is, of necessity, less than the value given in the above calculation.

6.5
A simple model of metallic bonding

The most comprehensive definition of a metallic element is one which readily forms positive ions; metallic elements are electropositive and they have lower ionisation energies than non-metallic elements.

X-ray analysis shows that metals have a high co-ordination number; each atom is surrounded by a large number of neighbouring atoms (generally either eight or twelve). Metals tend to adopt close-packed structures (p. 154) which minimise the amount of empty space between the atoms.

These properties—the ease of positive ion formation and the close-packed nature of metallic structures—provide the clues to the nature of bonding in metals. The atoms contribute their valence electrons to a common 'sea' of electrons which cements together the resulting positive ions.

In magnesium, for example, each atom contributes its two $3s$ valence electrons to the common 'sea'. These electrons are free to move within the metal and help to provide the electronic 'glue' which binds each Mg^{2+} ion to its twelve close neighbours; these valence electrons are said to be **delocalised**. Metallic bonding thus represents the extreme case of undirected bonding, and we shall now examine if this simple model can account for the characteristic features of metals.

Thermal and electrical conductivity

Thermal and electrical conductivity are properties which distinguish metals from non-metals. In a metal, the loosely bound delocalised valence electrons can move freely within the framework of the positive ions and collide with them; their thermal motion is random just as the motion of molecules in a gas. On the application of an electric field a drift motion is superimposed on this random thermal motion of the electrons. The free movement of the valence electrons in a metal provides transport for thermal and electrical conductivity.

The electrical conductivity of metals decreases with increasing temperature. This is because the increased thermal motion of the metal ions restricts the drift motion of the delocalised electrons.

Malleability and ductility

The relative ease with which layers of close-packed atoms are able to slip over one another, in the absence of any rigid directed bonding, accounts for these two properties of metals.

Opaque, shiny and silvery appearance

Apart from the two metals copper and gold, all metals have the above three characteristics. There is a large number of very closely spaced energy levels into which the delocalised electrons can be excited, i.e. the energy levels are almost continuous. This means that light of a wide range of wavelengths is absorbed by metals. The excited electrons then re-emit all the light on returning to lower energy states, so effectively visible light is totally reflected, and hence metals are opaque, shiny and silvery in appearance.

Wide variation in melting-points

The melting-point of a metal is a measure of the strength of the bonding that holds the atoms in a rigid structure. This strength will depend largely on the number of valence electrons that each atom contributes to the delocalised 'sea'. Thus Group 1A metals, which contribute one valence electron per atom, have relatively low melting-points, while the melting-points of the Group 2A metals, which contribute two valence electrons per atom, are rather higher. The melting-points of the metals in the first transition series are higher still, since the $3d$ electrons make a contribution to the bonding. A definite correlation between melting-point and the maximum oxidation state of these transition metals can be observed. The unusually low melting-point of manganese is due to its distorted structure.

The melting-points of the Group 1A, the Group 2A and the first transition series metals are shown in fig. 6.7.

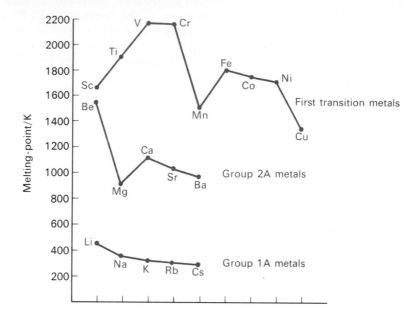

FIG. 6.7. *Melting-point of Group 1A, Group 2A and first transition series metals*

Large difference between enthalpies of fusion and vaporization

When a metal melts there is only a small increase in its volume and the delocalised bonding that existed in the solid is not totally destroyed. When the metal boils, however, there is a complete breakdown of the structure and the enthalpy of vaporization is much higher than the enthalpy of fusion as shown in Table 6.8 which includes, for comparison, two giant covalent structures of silicon and germanium.

Table 6.8 Some molar enthalpies of fusion and vaporization

Element	Molar enthalpy of fusion, $\Delta H^{\ominus}/kJ\ mol^{-1}$	Molar enthalpy of vaporization, $\Delta H^{\ominus}/kJ\ mol^{-1}$
Na	2·60	89·0
K	2·30	77·5
Mg	8·95	128·7
Ca	8·66	149·8
Fe	15·36	351·0
Cu	13·05	304·6
Si (non-metal)	46·44	376·8
Ge (non-metal)	31·80	334·3

The molar enthalpy of fusion is the energy absorbed when 1 mole of the solid is converted into liquid at its melting-point and at one atmosphere pressure (101 325 N m^{-2}). Similarly, the molar enthalpy of vaporization is the energy absorbed when 1 mole of liquid is converted into vapour at its boiling-point and at one atmosphere pressure. Enthalpy changes are dealt with in detail in chapter 14.

A more comprehensive theory of metallic bonding—the band model—is discussed in chapter 11.

Questions on this chapter are at the end of chapter 7.

Covalent, hydrogen and van der Waals' bonding

7.1
The covalent bond

An electrovalent or ionic bond is formed by the transfer of one or more electrons from a metal atom to a non-metal atom; in many instances the ions so formed have the electronic configurations of one of the noble gases. Lewis recognised that in simple molecules another form of bonding had to operate, and he suggested that electrons might be shared in pairs so that each atom could be said to have attained the stable electronic configuration of a noble gas. Thus consider the chlorine atom which has seven electrons in its outer shell; if one electron is provided by each atom and shared equally, then each chlorine atom can acquire a share in eight electrons—a completed octet as it is sometimes called.

$$: \overset{..}{\underset{..}{Cl}} \cdot \; + \; \cdot \overset{..}{\underset{..}{Cl}} : \; \longrightarrow \; : \overset{..}{\underset{..}{Cl}} : \overset{..}{\underset{..}{Cl}} :$$

The sharing of two electrons, one electron being provided by each atom constitutes a single covalent bond; it is usually represented by a single line joining the two atoms together. For example, the chlorine molecule can be represented thus:

$$Cl—Cl$$

The oxygen atom has six electrons in its outer shell, and in the oxygen molecule, O_2, a stable electronic configuration can be assumed to be attained by the sharing of four electrons, two being provided by each atom. The oxygen atoms are bound together by a double covalent bond and this is represented by a double line:

$$: \overset{..}{O} + \overset{..}{O} : \; \longrightarrow \; \overset{..}{O} \; \vdots \; \overset{..}{O} \qquad \text{or} \quad O{=\!\!=}O$$

Similarly, the nitrogen molecule contains a triple bond, which involves the sharing of six electrons, three being provided by each nitrogen atom:

$$\cdot \overset{.}{N} + \overset{..}{N} \cdot \; \longrightarrow \; : N \; \vdots \; N : \qquad \cdot \text{ or } \quad N{\equiv}N$$

When the two atoms that are bound together by covalent bonds are different the procedure is similar. Thus the combination of the hydrogen atom (one electron) with the chlorine atom (seven electrons in the outer shell) can be shown thus:

$$H \cdot + \overset{\times\times}{\underset{\times\times}{\times Cl \times}} \; \longrightarrow \; H \overset{\times\times}{\underset{\times\times}{\times Cl \times}} \qquad \text{or} \quad H—Cl$$

The covalent substances discussed above exist in the form of discrete molecules with little force of attraction between the individual molecules. This is true of many covalent substances and accounts for many of them being gases, liquids or easily fusible solids. There are some giant molecules or macromolecules, however, e.g. diamond and silicon dioxide in which directional covalent bonds extend throughout the whole structure. Compounds of this type are, of course, solids with high melting and boiling-points.

The structures of the covalent substances (that we have listed) have been drawn above with dots and crosses to represent the electrons in the valence shells of the atoms involved. This is a useful method of counting the electrons but it should not be supposed that electrons can be accurately located in a molecule any more than they can in an atom (p. 38). The best mental picture of an electron is that of a rather diffuse negative cloud. Sharing of electrons can then be thought of as a merging of electron clouds, resulting in a significant build-up of electron density between the bonded atoms.

7.2
The covalent bond in the hydrogen molecule

In order to gain some sort of appreciation as to why two atoms should covalently bond together, it is logical to examine the hydrogen molecule—the simplest possible system.

FIG. 7.1. *Potential energy as a function of internuclear separation for the hydrogen molecule*

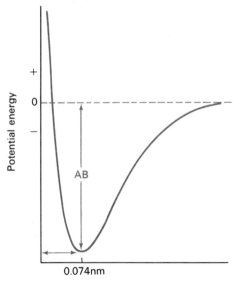

Hydrogen exists in the form of diatomic molecules commonly written as H_2, or H ⨯ H or H—H. The process

$$H_2(g) \longrightarrow 2H(g)$$

requires the input of 436 kJ mol^{-1} of energy, which indicates the marked stability of the molecule with respect to its constituent atoms.

The formation of the H—H bond can be summarised by a potential energy diagram, showing how the energy of the system changes as the two hydrogen atoms come closer together (fig. 7.1). Calculations show that at an interatomic distance of 0·074 nm, which is equal to the experimentally determined bond length, there exists a potential 'well' whose depth AB measures the bond energy of the molecule, i.e. the energy needed to pull the atoms completely apart from each other.

Why should a molecular unit of two electrons and two protons be more stable energetically than two separate atoms, each of one electron and one proton? The problem is really one of deciding on the most stable arrangement of two negative and two positive charges. Clearly we want to keep positive close to negative but not too close to positive; likewise we want to keep negative away from negative. The solution is a compromise which is achieved by the partial overlap of the two 1s orbitals (electron clouds) of the hydrogen atoms. The overlap is not sufficient to cause strong repulsion between like charges (fig. 7.2).

FIG. 7.2. *The overlap of hydrogen 1s orbitals to achieve a stable system*

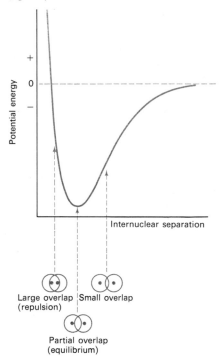

The overlapping of atomic orbitals forms **molecular orbitals** and the two electrons occupying it are no longer confined to their original nuclei. They are quite indistinguishable and, just like an atomic orbital, a molecular orbital can be described only in terms of the probability of finding an electron in a certain region of space. This may be represented by an electron density map on which locations of equal electron density are linked with contours. Fig. 7.3 shows such a map for the hydrogen molecule. This map, the result of detailed calculations using the technique of wave mechanics, shows a concentration of negative charge between the two nuclei and this acts as a 'glue' to keep the two atoms together in the molecule. The two electrons which occupy a molecular orbital have their spins opposed just as do any two electrons in the same atomic orbital (Pauli exclusion principle p. 28). This can be thought of as a means of achieving the most favourable spatial arrangement of the electrons.

Covalent bonds between atoms other than hydrogen are visualised in the same sort of way, but now the presence of more than one electron per atom make detailed calculations out of the question. We shall have more to say about molecular orbital theory in the next chapter.

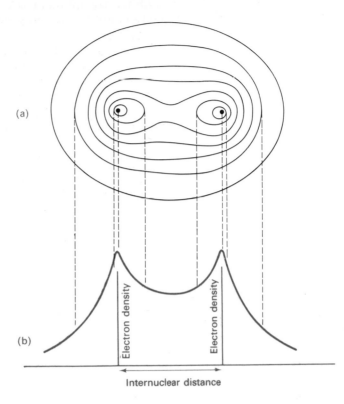

(a)

(b)

Electron density

Electron density

Internuclear distance

7.3
The co-ordinate or dative covalent bond

There is no real distinction between this and a single covalent bond, except that only one of the bonded atoms provides the two electrons which are shared. It is present in the stable complex $BCl_3.NH_3$ formed between boron trichloride and ammonia. The nitrogen atom in the ammonia molecule contains two electrons not involved in bonding (a lone pair of electrons) while the boron atom in the boron trichloride molecule is two electrons short of an octet (an example of a compound disobeying the octet rule, p. 82). The octet can be completed thus:

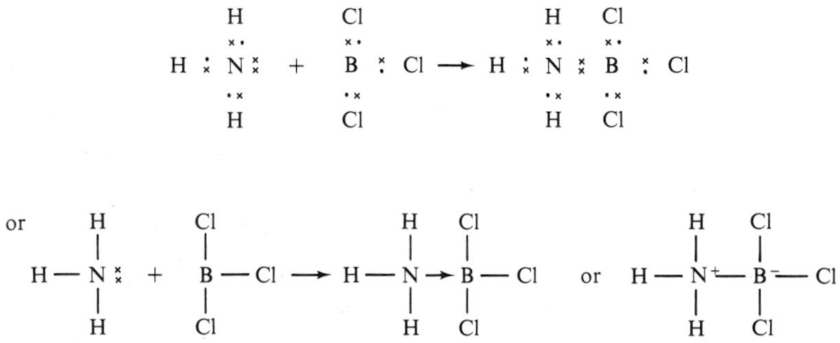

In molecular orbital language we should say that a filled atomic orbital in the nitrogen atom overlaps with an empty atomic orbital present in the boron atom.

A co-ordinate bond is represented by an arrow pointing from the donor to the acceptor atom, or by a single line with a negative charge on the acceptor atom and an equal but opposite charge on the donor atom.

81

Other examples of compounds containing co-ordinate bonds are carbon monoxide and nitric acid:

$$\times \overset{\cdot\cdot}{\underset{\cdot\cdot}{C}} \overset{\cdot\cdot}{\underset{\cdot\cdot}{\vdots}} \overset{\cdot\cdot}{O} \colon \quad \text{or} \quad C \equiv O \quad \text{or} \quad \bar{C} \equiv \overset{+}{O} \quad \text{carbon monoxide}$$

$$H \colon \overset{\times\times}{\underset{\times\times}{O}} \overset{\circ}{\underset{\circ}{\times}} \overset{\times\times}{\underset{\circ+\circ}{N}} \overset{\times\times}{\underset{\times\times}{O}} \overset{}{\underset{}{\quad}} \quad \text{or} \quad H-O-N \overset{\diagup O}{\underset{\diagdown\diagdown O}{}} \quad \text{or} \quad H-O-\overset{+}{N} \overset{\diagup \bar{O}}{\underset{\diagdown\diagdown O}{}} \quad \text{nitric acid}$$

Ammonium chloride is an ionic compound, $NH_4^+Cl^-$, whose formation from ammonia and hydrogen chloride can be considered to involve the formation of a co-ordinate bond thus:

| Initial state | Intermediate state | Final state |

7.4
Anions containing covalent bonds

Anions containing more than two combined atoms are common and include the carbonate, nitrite, nitrate, sulphite and sulphate anions. Consider the structure of the carbonate ion, CO_3^{2-}. Two electrons from a metal atom (or atoms) can be transferred to two oxygen atoms to give the configuration $O^-(2\cdot7)$; the carbon octet can now be completed by the formation of two $C-O^-$ covalent bonds and one $C=O$ double covalent bond thus:

e.g. as in $Ca^{2+}\ CO_3^{2-}$

The structures of other anions can be constructed in a similar manner and are represented as, for example,

| nitrite | nitrate | sulphite | sulphate |

Note that the sulphur atom in the sulphite ion is surrounded by 10 electrons, and that in the sulphate ion it is surrounded by 12 electrons. The octets could be preserved by writing the $S=O$ bonds as co-ordinate bonds, the sulphur atom being the donor atom. However, there is evidence to show that the former is the best representation of these structures, i.e. sulphur can expand its octet (see section 7.5).

7.5
Covalent compounds which violate the octet rule

Although the atoms of many elements which form covalent bonds can be considered to attain the electronic configuration of a noble gas, e.g. the structures of hydrogen, oxygen, hydrogen chloride, water etc. many exceptions exist and include $BeCl_2$, BF_3, PF_5 and SF_6, together with SO_3^{2-} and SO_4^{2-} (mentioned in the previous section). Modern valence theory focuses attention on electron pairing rather than on completion of the octet. Consider beryllium chloride: the beryllium atom has the configuration $1s^22s^2$ (no unpaired electrons) and the chlorine atom has the configuration $1s^22s^22p^63s^23p^5$ (one unpaired electron). When the beryllium atom enters into chemical combination an electron from the $2s$ level is considered to be promoted to the higher $2p$ level with the absorption of energy (promotion energy absorbed), and each single electron is paired off with the unpaired electrons of the two chlorine atoms.

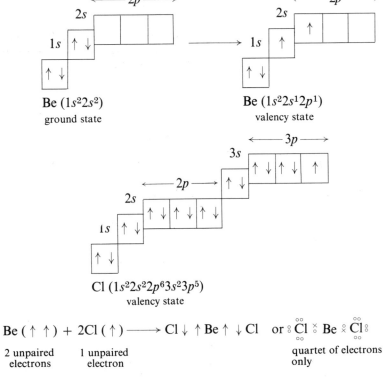

Be ($1s^22s^2$)
ground state

Be ($1s^22s^12p^1$)
valency state

Cl ($1s^22s^22p^63s^23p^5$)
valency state

Be ($\uparrow$ $\uparrow$) + 2Cl ($\uparrow$) $\longrightarrow$ Cl $\downarrow$ $\uparrow$ Be $\uparrow$ $\downarrow$ Cl   or $\overset{\circ\circ}{\underset{\circ\circ}{\text{Cl}}}\overset{\times}{\circ}\,\text{Be}\,\overset{\times}{\circ}\,\overset{\circ\circ}{\underset{\circ\circ}{\text{Cl}}}$

2 unpaired   1 unpaired   quartet of electrons
electrons   electron   only

The promotion energy is less than the energy released in forming bonds with the unpaired electrons of the chlorine atom; indeed if this were not so, then beryllium would be devoid of chemical reactivity like the noble gas helium.

The formation of boron trifluoride can be explained in a similar manner (the three equivalent p levels are designated p_x, p_y and p_z):

B ($1s^22s^22p^1$) $\longrightarrow$ B ($1s^22s^12p_x^12p_y^1$)
ground state   valency state
1 unpaired electron   3 unpaired electrons

F ($1s^22s^22p_x^22p_y^22p_z^1$)
valency state
1 unpaired electron

B ($\uparrow$ $\uparrow$ $\uparrow$) + 3F ($\uparrow$) $\longrightarrow$

3 unpaired   1 unpaired
electrons   electron

or

sextet of
electrons only

Phosphorus pentafluoride, PF_5, and sulphur hexafluoride, SF_6, are compounds in which the octet of electrons is exceeded around the central atom. The same reasoning as before applies; thus $3d$ levels are available to phosphorus and sulphur which can accommodate electrons promoted from the ground state:

$$P\,(1s^2 2s^2 2p^6 3s^2 3p^3) \longrightarrow P\,(1s^2 2s^2 2p^6 3s^1 3p^3 3d^1)$$

3 unpaired electrons   5 unpaired electrons
                                   highest valency state

$$S\,(1s^2 2s^2 2p^6 3s^2 3p^4) \longrightarrow S\,(1s^2 2s^2 2p^6 3s^2 3p^3 3d^1) \longrightarrow S\,(1s^2 2s^2 2p^6 3s^1 3p^3 3d^2)$$

2 unpaired electrons   4 unpaired electrons   6 unpaired electrons
                                                            highest valency state

Phosphorus pentafluoride and sulphur hexafluoride can be prepared because the energy released in bond formation with fluorine atoms is more than sufficient to produce a phosphorus atom with 5, and a sulphur atom with 6 unpaired electrons.

It is interesting to note that, whereas phosphorus has valencies of three and five, e.g. in PF_3 and PF_5, and sulphur has valencies of two, four and six, e.g. in H_2S, SF_4 and SF_6, nitrogen is limited to a valence of three and oxygen to one of two. This is because d levels are not easily accessible for the elements from hydrogen to neon in the Periodic Table, i.e. d levels first become available for atoms with a principal quantum number $n = 3$ (see Table 3.3 p. 31).

Generally only electron promotion is possible between sub-levels characterised by the same principal quantum number, e.g. $2s \to 2p$, $3s \to 3p \to 3d$ etc. are possible, since the energy can be provided by chemical reaction; but $2s \to 3s$ or $2s \to 3p$ etc. are impossible since the energy jumps are too high. One exception occurs in the chemistry of transition metals where $(n-1)d$ and ns levels are very nearly of the same energies.

7.6
The shape of covalent molecules—a simple theory

In contrast to the bonding in ionic solids, that operating between atoms in a covalent molecule is directional in character. A simple theory to account for the molecular shape of such molecules was first put forward by Sidgwick and Powell in 1940 and is based on electron-pair repulsion; it was refined and extended somewhat by Nyholm and Gillespie in 1957. The theory was developed as a result of studying the shapes of a wide variety of substances and can be summarised in the following rules.

(a) Electron-pairs, whether in bonding orbitals or lone-pair orbitals, arrange themselves in space so as to minimise their mutual repulsions. The resulting geometrical shapes for two to seven electron-pairs are shown in fig. 7.4.

(b) The repulsion between lone-pair/lone-pair is greater than that between lone-pair/bonding-pair which is, in turn, greater than that between bonding-pair/bonding-pair.

(c) Where there are more than four pairs of electrons to consider, the interactions where the electron-pairs make an angle greater than 90° at the central atom are ignored.

Table 7.1 illustrates the application of these rules to some simple molecules.

FIG. 7.4. *Arrangement of electron pairs*

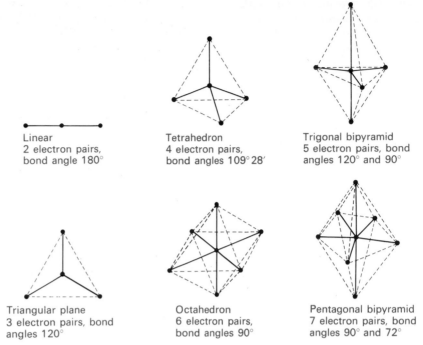

Linear
2 electron pairs,
bond angle 180°

Tetrahedron
4 electron pairs,
bond angles 109°28′

Trigonal bipyramid
5 electron pairs, bond
angles 120° and 90°

Triangular plane
3 electron pairs, bond
angles 120°

Octahedron
6 electron pairs,
bond angles 90°

Pentagonal bipyramid
7 electron pairs, bond
angles 90° and 72°

Table 7.1 Derivation of molecular shape from the electron-pair repulsion theory

Substance	Number of valence electrons round the central atom	Number of electron-pairs	Number of bonding-pairs	Number of lone-pairs	Basic shape
$BeCl_2$	4	2	2	0	Linear
BF_3	6	3	3	0	Triangular plane
CH_4	8	4	4	0	Tetrahedron
NH_3	8	4	3	1	
H_2O	8	4	2	2	
PF_5	10	5	5	0	Trigonal bipyramid
SF_4	10	5	4	1	
ClF_3	10	5	3	2	
SF_6	12	6	6	0	Octahedron
IF_5	12	6	5	1	
IF_7	14	7	7	0	Pentagonal bipyramid

The 'basic shape' listed in Table 7.1 will not be the actual shape of the molecule if lone-pairs are present; however, the lone-pairs are as important as the bonding-pairs in determining the shape. The way in which the actual shape of a molecule is related to the 'basic shape' is shown for four molecules, in fig. 7.5, which contain either one or two lone-pairs.

85

FIG. 7.5. *The actual shapes of some simple molecules. The shapes of ammonia, NH₃, and water, H₂O, are based on the tetrahedron. The shapes of sulphur tetrafluoride, SF₄, and chlorine trifluoride, ClF₃, are based on the trigonal bipyramid*

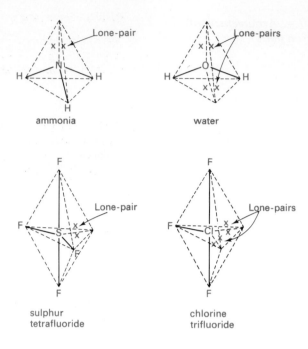

Rule (b) above accounts for the variation in the bond angles in the series of hydrides CH_4(approximately 109°), NH_3(approximately 107°) and H_2O(approximately 105°). The increasing repulsion from the lone-pairs (0, 1, and 2 respectively) forces the bonding-pairs together and decreases the bond angle.

A careful application of rules (b) and (c) explains why the lone-pair in SF_4 and the lone-pairs in ClF_3 adopt the equatorial rather than the axial positions. These facts have been confirmed experimentally.

7.7
Polar covalent bonds

So far our discussion of the covalent bond has contained the implicit assumption that the electron clouds 'cementing' the nuclei of the two atoms together are equally distributed between the two atoms. We have assumed that the probability of finding bonding electrons in the vicinity of one atom is the same as the probability of finding them in the same position with respect to the other atom. The electron density map for the hydrogen molecule (fig. 7.3) suggests that for this molecule the assumption is valid.

When two different atoms are covalently bonded together, the electrons forming the bond are not symmetrically distributed between them. The electron distribution becomes **polarised** and drawn preferentially towards one atom of the bond which acquires a partial negative charge, leaving the other atom with an equal partial positive charge. Such a distortion of charge produces a **polar covalent bond**, the extent of polarity depending upon the ease with which the two atoms accept, respectively, a partial negative and a partial positive charge.

Polarisation of a bond increases its bond strength owing to the increased electrostatic attraction between the two partial charges.

86

Dipole moments

If two charges of equal magnitude but opposite in sign, q^+ and q^-, are separated by a distance r, then the system is said to have a dipole moment of magnitude given by:

$$\text{charge} \times \text{distance} = q \times r$$

From what we have said above, any diatomic molecule which contains different atoms, e.g. HCl, must have a dipole moment.

Historically, dipole moments were measured in Debye units, D, chosen by the Dutch physical chemist Debye, such that

$$1\ D = 10^{-18}\ \text{esu cm}$$

This unit is still used today and conversion to S.I. can be made using the relationship:

$$1\ D = 3.335\ 6 \times 10^{-30}\ \text{C m}$$

Thus the dipole moment of a system comprising charges q^+ and q^- of magnitude $1.602\ 10 \times 10^{-19}$ C (the electronic charge) at a separation distance of 0.1 nm is given by:

$$
\begin{aligned}
\text{Dipole moment} &= 1.602\ 10 \times 10^{-19} \times 10^{-10}\ \text{C m} \\
&= \frac{1.602\ 10 \times 10^{-29}}{3.335\ 6 \times 10^{-30}}\ \text{D} \\
&= 4.80\ \text{D}
\end{aligned}
$$

Dipole moments are represented by the symbol μ and an arrow is drawn pointing in the direction of the negative end. Figure 7.6 shows the situation for the example discussed above.

FIG. 7.6. *Conventional representation of a dipole moment*

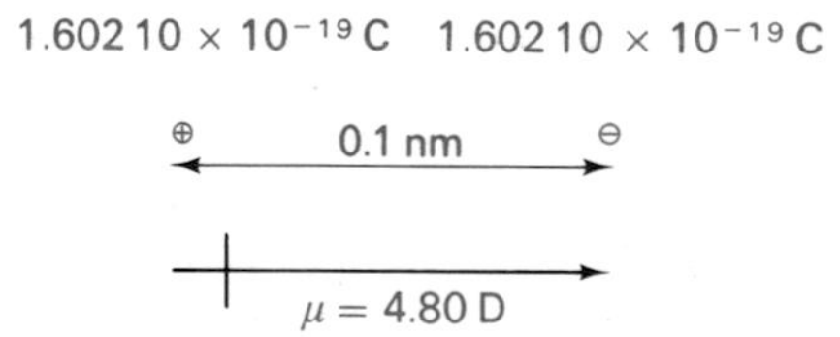

$$1.602\ 10 \times 10^{-19}\ \text{C} \qquad 1.602\ 10 \times 10^{-19}\ \text{C}$$

$$\oplus \xleftarrow{\hspace{1cm} 0.1\ \text{nm} \hspace{1cm}} \ominus$$

$$\mu = 4.80\ \text{D}$$

Table 7.2 lists the dipole moments of the hydrogen halide molecules.

Table 7.2 Dipole moments of the hydrogen halide molecules

Substance	Formula	Dipole moment/D
Hydrogen fluoride	HF	1·91
Hydrogen chloride	HCl	1·05
Hydrogen bromide	HBr	0·80
Hydrogen iodide	HI	0·42

Despite an increase in bond length from HF to HI the dipole moment, pointing in the direction of the halogen atom in each case, decreases in value. It would be convenient if the values could be related directly to the polarity of the hydrogen—halogen bond, but unfortunately this cannot be done for reasons given below.

FIG. 7.7. *The dipole moment of HCl*

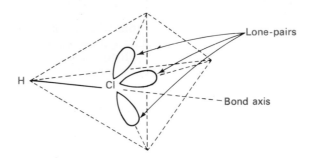

Suppose we consider the HCl molecule and assume that the bonding- and lone-pair electrons are pointing towards the apices of a regular tetrahedron (see fig. 7.7 and the theory of electron-pair repulsion), and there is some evidence that this is the approximate situation; then the lone-pair electrons would have a net dipole moment pointing along the bond axis and away from the molecule (dipole moments, like forces, are vectors and can be resolved). This dipole would reinforce the dipole due to the polarity of the bond, which may conveniently be called the 'ionic' dipole. There is yet another effect to consider; in establishing a covalent bond, atomic orbitals of hydrogen and chlorine overlap. The atomic orbital of hydrogen is smaller than that of chlorine, thus in the absence of any polar effects the bonding electrons would be closer to the hydrogen than to the chlorine nucleus. This so called overlap dipole would point in the direction of the hydrogen atom, in opposition to the previous two effects. The measured dipole moment can be expressed as:

$$\mu_{measured} = \mu_{overlap} + \mu_{ionic} + \mu_{lone\text{-}pair}$$

H—Cl $\xrightarrow{\ \ +\ \ }$ $\xleftarrow{\ +\ }$ $\xrightarrow{\ +\ }$ $\xrightarrow{\ +\ }$

It is possible to make allowance for the dipoles due to lone-pairs and overlap; when this is done, the ionic dipole moment which reflects the polarity of the bond can be used to calculate the ionic character of the bond. Typical values are 0·17 (or 17%) for hydrogen chloride and 0·05 (or 5%) for hydrogen iodide.

Table 7.3 Dipole moments of selected molecules in the vapour phase

Substance	Formula	Dipole moment/D
Carbon dioxide	CO_2	0
Boron trifluoride	BF_3	0
Benzene	C_6H_6	0
Ammonia	NH_3	1·48
Sulphur dioxide	SO_2	1·63
Chlorobenzene	C_6H_5Cl	1·67
Water	H_2O	1·84

Dipole moments can provide important structural information about a molecule. Consider, for example, the values listed in Table 7.3. The zero dipole moment for the carbon dioxide molecule confirms that the molecule is linear; each $C{=}O$ 'unit' has a dipole moment, but the two cancel when added vectorially. For similar reasons the zero moment of boron trifluoride confirms the planar triangular shape which was predicted by the theory of electron-pair repulsion (section 7.6). Clearly the dipole moments of the water and sulphur dioxide molecules rule out a linear structure. The symmetrical planar ring structure of benzene accounts for its zero dipole moment, but substitution of a chlorine atom into the ring destroys the symmetry and chlorobenzene does have a dipole moment.

Electronegativity

The ability of an atom to attract the electrons of a covalent bond is measured by the electronegativity of the atom. An atom of high electronegativity will attract electrons away from one of lower electronegativity (or higher electropositivity).

Two factors are clearly involved in determining the electronegativity of an atom: its ability to lose an electron (measured by its ionisation energy) and its ability to gain an electron (measured by its electron affinity). An atom with a high ionisation energy and a high electron affinity will be strongly electronegative. Mulliken proposed the relationship

$$\text{Electronegativity} = \frac{\text{First ionisation energy} + \text{electron affinity}}{2}$$

but a more widely used scale (due to Pauling) is based on bond energy values. In the absence of any polar character, Pauling argued, a covalent bond A—B would have a bond energy equal to the geometric mean of the bond energies of A—A and B—B. For example, if H—Cl were non-polar its bond energy, E, would be given by:

$$E = (E_{\text{H--H}} \times E_{\text{Cl--Cl}})^{1/2}$$
$$= (436 \times 242)^{1/2}$$
$$= 325 \text{ kJ mol}^{-1}$$

Table 7.4 The Pauling electronegativity values of some elements

			H 2·1			
Li 1·0	Be 1·5	B 2·0	C 2·5	N 3·0	O 3·5	F 4·0
Na 0·9	Mg 1·2	Al 1·5	Si 1·8	P 2·1	S 2·5	Cl 3·0
K 0·8	Ca 1·0	—	Ge 1·8	As 2·0	Se 2·4	Br 2·8
Rb 0·8	Sr 1·0	—	Sn 1·8	Sb 1·9	Te 2·1	I 2·5
Cs 0·7	Ba 0·9					

The measured bond energy of H—Cl is 431 kJ mol^{-1} and the difference in values is related, on the Pauling scale, to the **electronegativity difference** between the hydrogen atom and the chlorine atom. By allotting a value of 4 to the most electronegative atom, namely that of fluorine, values may be assigned to the atoms of other elements (Table 7.4). A definite increase in electronegativity can be observed on ascending a group of the Periodic Table and on crossing a period from left to right.

The concept of electronegativity is a somewhat artificial one, but it is nonetheless a valuable aid in predicting the polarity of a particular bond, which in turn may determine the way in which a bond will break during a chemical reaction. Various empirical relationships have been suggested in an attempt to relate electronegativity differences to percentage ionic character of a particular covalent bond. One set of suggested values is given in Table 7.5.

In view of the highly empirical approach to the concept of electronegativity, and the subsequent extension to percentage ionic character of covalent bonds, the above values should be viewed with caution.

Table 7.5 Percentage ionic character of some bonds

C—H	N—H	O—H	F—H
4%	19%	39%	59%

C—F	C—Cl	C—Br	C—I
43%	6%	2%	0%

7.8
The hydrogen bond

When a hydrogen atom is covalently bonded to a highly electronegative atom—commonly fluorine, oxygen or nitrogen—the bond is polarised to such a large extent that the hydrogen atom is able to form a weak, directional bond with a highly electronegative atom adjacent to it. It may be represented as follows:

$$A^{\delta-}\!\!=\!\!H^{\delta+}\text{-}B^{\delta-}\!\!=\!\!X^{\delta+}$$

The strength of a hydrogen-bond is much lower than that of a covalent bond (of the order 20–40 kJ mol^{-1} compared to a value of 150–900 kJ mol^{-1} for covalent bonds) but there is a good deal of evidence to support its existence.

FIG. 7.8. *The abnormal melting- and boiling-points of ammonia, water and hydrogen fluoride*

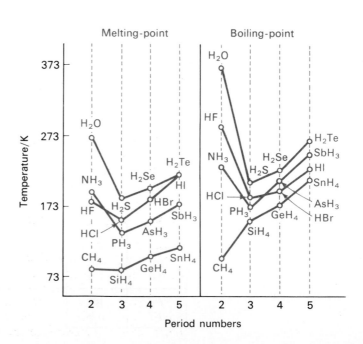

90

(a) The melting- and boiling-points of ammonia, water and hydrogen fluoride are anomalously high compared with those of the hydrides of other elements in Groups 5B, 6B and 7B respectively of the Periodic Table (fig. 7.8). This suggests the existence of bonding appreciably stronger than the normal type of weak bonding which exists between molecules (van der Waals' bonding which is discussed in section 7.9).

(b) Hydrogen fluoride forms acid salts, e.g. $K^+(HF_2)^-$, in which the anion has the hydrogen-bonded structure

$$F\text{—}H\text{---}F^-$$

Hydrogen-bonding in this ion is exceptional in that it is stronger than normal and the hydrogen atom is symmetrically situated between the two electronegative fluorines.

In the solid and liquid states hydrogen fluoride forms long hydrogen-bonded chains.

The zig-zag shape is a consequence of the orientation of the lone-pairs on the fluorine atoms.

(c) The infrared absorption spectrum (see chapter 13) of compounds containing an O—H group will show a reduction in the frequency of absorption for this group when it is hydrogen-bonded. For example, the infrared spectrum of ethanol vapour has a sharp band at 3700 cm^{-1} due to a 'free' O—H group (no hydrogen-bonding in the vapour state), but this absorption shifts to 3300 cm^{-1} for a solution of ethanol in an inert solvent (hydrogen-bonding between the ethanol molecules).

(d) Relative molecular mass determinations of carboxylic acids show that they exist as double molecules (dimers) which are held together by hydrogen-bonding, e.g. ethanoic acid, CH_3COOH, which is shown below:

The boiling-points of these compounds are unusually high, e.g. ethanoic acid boils at 391 K whereas propanone, CH_3COCH_3, which has approximately the same relative molecular mass as undimerised ethanoic acid, boils at 329 K.

Intermolecular and intramolecular hydrogen-bonding

Hydrogen-bonds may be formed between two molecules (**inter**molecular) or between atoms within the same molecule (**intra**molecular). For example, 2-nitrophenol possesses an intramolecular hydrogen-bond, whereas 4-nitrophenol forms intermolecular hydrogen-bonds. 2-Nitrophenol boils at 489 K whereas the 4-isomer, which is hydrogen-bonded into larger 'units', boils at 532 K. This large difference in boiling-points allows the two isomers to be separated by steam distillation.

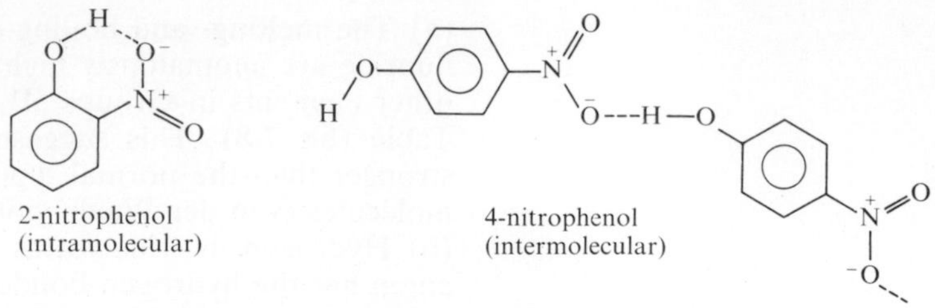

2-nitrophenol
(intramolecular)

4-nitrophenol
(intermolecular)

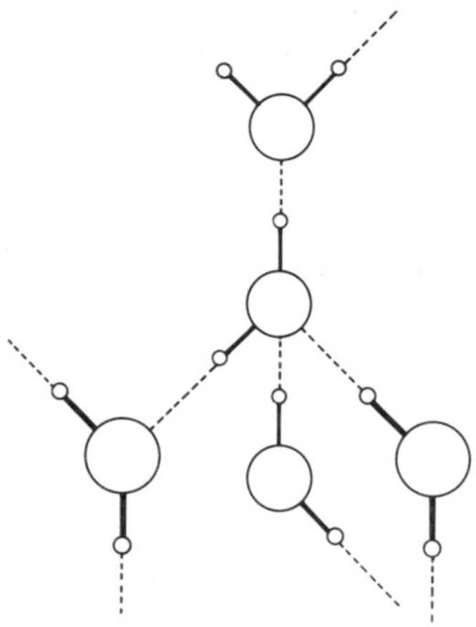

FIG. 7.9. *The partial structure of ice. The large circles represent oxygen atoms, and the smaller ones hydrogen atoms*

Hydrogen-bonding in ice and water

In ice, each oxygen atom is surrounded tetrahedrally through hydrogn bonds to four water molecules (fig. 7.9). The resulting structure is a very open one and this explains why the density of ice is lower than that of water. When ice melts, there is a contraction in volume of about 9% as the hydrogen-bonded structure breaks down. The closer packing of the molecules results in a maximum density at 277 K, after which there is an increase in intermolecular separation with increasing temperature. Nevertheless, the abnormally high boiling-point of water indicates the presence of appreciable hydrogen-bonding even at 373 K.

Hydrogen-bonding in DNA

Hydrogen-bonding plays a crucial role in the structure of deoxyribonucleic acid (DNA), since it holds together the two helical nucleic acid chains. Hydrogen-bonds are formed between specific pairs of bases, with an adenine unit in one chain bonding to a thymine unit in the other; similarly, a guanine unit in one chain bonds to a cytosine unit in another (fig. 7.10).

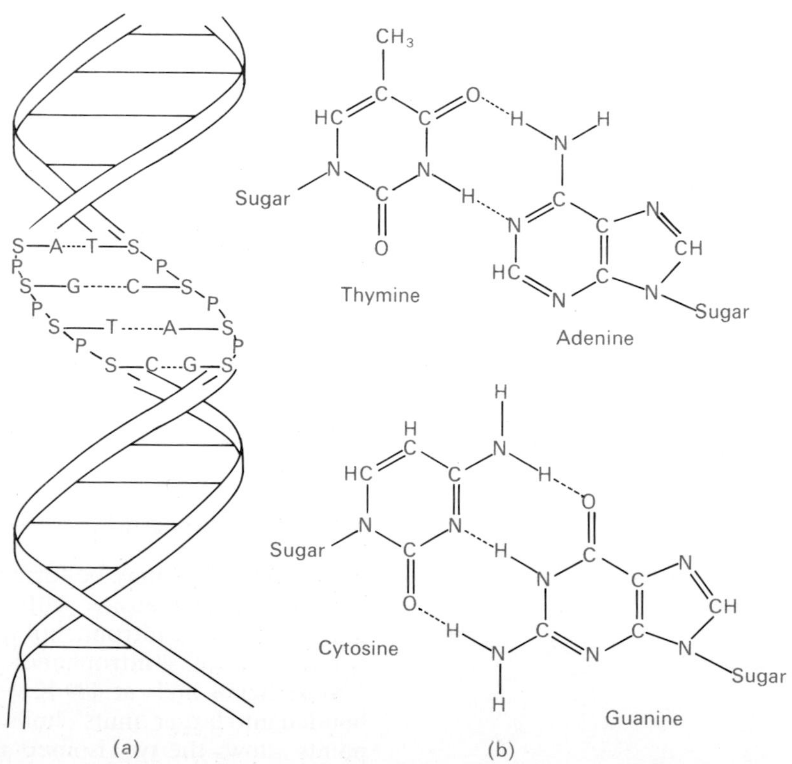

FIG. 7.10. *(a) The double helix structure of DNA. Each strand contains sugar (S) and phosphate (P) residues. The two helical chains are held together by the thymine/adenine (T/A) and cytosine/guanine (C/G) pairs (b) Hydrogen bonding between thymine/adenine and cytosine/guanine*

(a)

(b)

Some more comments on the hydrogen-bond

A hydrogen-bonded system, which may be represented as A—H---B, is generally linear with the hydrogen atom asymmetrically placed within the 'unit' even when the two outer atoms are the same, i.e. A = B; ice, for example, has an O—H bond length of 0·101 nm and an O---H bond length of 0·174 nm. However, there are some exceptions such as the symmetrical F—H---F⁻ ion and the 'bent' bond in the intramolecularly hydrogen-bonded 2-nitrophenol molecule (p. 92).

Although there is no universal agreement amongst chemists as to the best description of the hydrogen-bond, it is nevertheless agreed to be essentially electrostatic in origin. The fact that there are no bonds of a similar nature involving atoms other than hydrogen as the central atom of such a 'unit' is due to the rather special character of the hydrogen atom. In order to maximise the electrostatic energy of interaction, the atoms must be able to approach each other closely; this is possible with hydrogen, since its atomic radius is extremely small and, further, it does not have inner electron shells which would introduce repulsive terms. The small size of the other atoms in the bond, e.g. fluorine, oxygen or nitrogen, would allow of a close approach to the hydrogen atom, and this is clearly of importance as is their high electronegativities.

7.9
van der Waals' bonds

FIG. 7.11. *Plan of the crystal structure of iodine. The 'black molecules' lie in the plane of the paper; the 'blue molecules' lie above and below this plane*

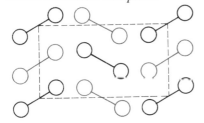

Substances which exist in the form of small, covalently-bonded molecules are generally gases or low-boiling liquids. The formation of covalent bonds within the molecule (intramolecular bonds) normally uses up all the valence electrons in the atoms concerned and prevents the formation of strong bonds between the molecules (intermolecular bonds).

It is clear, nevertheless, that weak forces of attraction do exist between discrete molecules. The presence of intermolecular attractions is one reason why real gases (even hydrogen) do not obey the ideal gas laws. When gases are cooled and the thermal motion of the molecules reduced, the intermolecular forces can impose some sort of ordered structure on the liquid which is formed. When the thermal motion of the molecules is low enough for the substance to solidify, the resulting molecular solid has a very ordered structure which is held together by intermolecular bonds. Iodine provides an example of such a structure at room temperature (fig. 7.11) and solid carbon dioxide provides another at a much lower temperature. The intermolecular bonds which exist between neutral molecules are known as van der Waals' bonds.

The nature of van der Waals' bonding is not obvious since it cannot be attributed to any of the bonding electrons in the molecules concerned. Any acceptable model of van der Waals' bonding must explain how the noble gases, which have completely filled valence shells, can be liquefied and eventually solidified, albeit at very low temperature, e.g. neon liquefies at 27 K and solidifies at 25 K.

van der Waals' bonding is evidently a very weak form of bonding. The molar enthalpy of vaporization of methane, which measures the strength of the intermolecular bonds in the liquid, is only 8.2 kJ mol^{-1} compared to the value of 435 kJ mol^{-1} for the strength of the C—H bond within the methane molecule.

Predictably, van der Waals' bonds are significantly longer than covalent bonds. In the molecular crystal of iodine, for example, (fig. 7.11) the molecules are 0·430 nm apart compared with the interatomic distance within the iodine molecule of 0·266 nm. Graphite provides a striking example of the contrast between van der Waals' and covalent bond lengths: the C—C bond length within the layers is 0·142 nm while the layers themselves, held together by van der Waals' bonds, are 0·335 nm apart (fig. 7.12). The lubricant properties of graphite are due to the weak inter-layer bonding which allows the layers to slip over each other.

FIG. 7.12. *The structure of graphite*

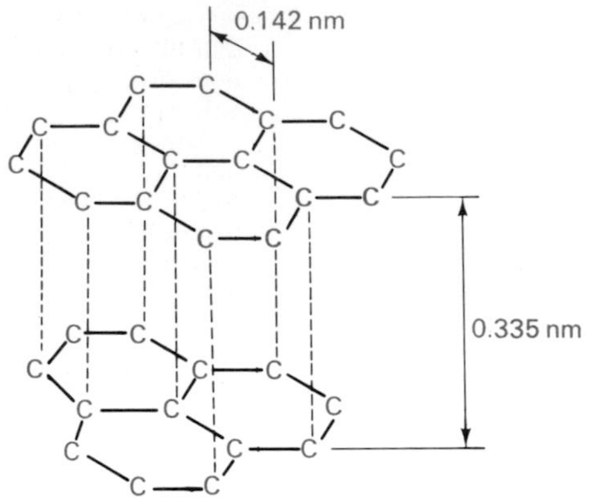

Three types of forces are responsible for van der Waals' bonds.

Dipole-dipole (orientation) forces (Keesom 1912)

If the covalent molecule has a dipole (p. 87), the positively charged end of the dipole of one molecule will attract the negatively-charged end of another molecule and the molecules will orientate themselves accordingly

The contribution made to the total bonding by this type of force is usually small.

Dipole-induced-dipole (induction) forces (Debye 1920)

A molecule with a permanent dipole will induce a dipole in another molecule and an attraction between the two will result.

The contribution to the total bonding from this effect is also very small.

Dispersion forces (London 1930)

Interatomic bonding in the noble gases cannot be explained in terms of the two forces described above. However, even in non-polar molecules momentary distortions of the electron orbitals can produce instantaneous fluctuating dipoles. These pulsations are synchronised to produce attractive forces between the molecules (or atoms in the case of the noble gases), known as **dispersion forces**.

Table 7.6 Boiling-points of the homologous series of alkanes

Compound	Formula	Boiling-point/K
Methane	CH_4	112
Ethane	C_2H_6	185
Propane	C_3H_8	231
Butane	C_4H_{10}	273
Pentane	C_5H_{12}	309
Hexane	C_6H_{14}	342

The larger the molecule, the greater the volume of space occupied by the electron orbitals and the larger the momentary distortion. Thus the dispersion forces between molecules (or atoms) increase with their relative molecular mass (or atomic mass). This effect is commonly seen in the increase, with increasing relative molecular mass, of the boiling-points of an homologous series (Table 7.6).

The van der Waals' bonding that is attributable to dispersion forces is undirected in the sense that, averaged over a period of time, it is impossible to distinguish attractions between particular pairs of atoms or molecules. Dispersion forces usually make by far the biggest contribution to van der Waals' bonding as Table 7.7 illustrates

Table 7.7 Approximate percentage contributions of the three main types of van der Waals' forces for a number of molecules

Molecule	Orientation effect	Induction effect	Dispersion effect
Ar	nil	nil	100
N_2	nil	nil	100
CH_4	nil	nil	100
CO	0·005	0·08	100
HCl	15	4	81

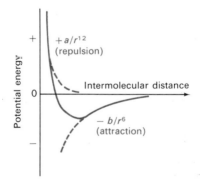

FIG. 7.13. *A total potential energy curve for van der Waals' attraction*

If there were not a balancing repulsive force operating between molecules, the attractive van der Waals' forces would cause the molecules to coalesce. In practice, a strong repulsive force comes into play at short range when the electron orbitals from the molecules (or atoms) begin to overlap.

Calculations show that the attractive potential energy term varies inversely as the sixth power of the distance between the molecules (all three contributions obey the same type of relationship) while the repulsive potential energy term varies inversely as the twelfth power of the intermolecular distance:

$$\text{Total potential energy} = a/r^{12} - b/r^6$$

This relationship, together with its component parts is shown in fig. 7.13.

7.10
Questions on
chapters 6 and 7

1 Using an appropriate example in each case, explain the essential features of the following types of bond:
(a) covalent bond
(b) ionic bond
(c) hydrogen bond
(d) van der Waals' bond

2 (a) Discuss the bonding in (i) calcium oxide, (ii) tetrachloromethane, (iii) ice, (iv) the molecule Al_2Cl_6.
(b) What are the spatial arrangements of the atoms in (i) boron trichloride, (ii) ammonia, (iii) the gaseous compound SF_6?　　　　O

3 (a) Describe and suggest reasons for the trends across and down the Periodic Table in (i) atomic radius, (ii) atomic volume and (iii) electronegativity.
   (b) Discuss the bonding in (i) sodium chloride and (ii) hydrogen chloride.
   (c) Describe and account for the bonding you would expect to be present in (i) rubidium chloride and (ii) iodine chloride (ICl).   JMB (Syllabus A)

4 (a) Interatomic bonds are of three *ideal* types—electrovalent, covalent and metallic.

     Describe these types and distinguish between them using the following information:
     (i) potassium conducts electricity in the solid state and in the molten state;
     (ii) potassium chloride is a crystalline solid of melting point 776°C; the solid does not conduct electricity, but the molten substance does;
     (iii) chlorine has a boiling point of −34°C; it does not conduct electricity.
   (b) Account for the following information in terms of the bonding present in the substances:
     (i) at room temperature and atmospheric pressure water is a liquid whilst hydrogen sulphide is a gas. At standard pressure the boiling points are, respectively, 100°C and −62°C;
     (ii) liquid hydrogen chloride does not conduct electricity, but an aqueous solution of hydrogen chloride does.   AEB

5 Give what explanation you can in terms of structure and bonding for the following data.
   (a) Ethanol boils at 78°C, dimethyl ether at −24°C (under 760 mm Hg).
   (b) 2,2-dimethylpropane, $(CH_3)_4C$, boils at 10°C, n-pentane at 38°C (under 760 mm Hg).
   (c) Yellow (white) phosphorus is soluble in carbon disulphide and ignites in air at 35°C, whereas red phosphorus is insoluble in carbon disulphide and ignites in air at 260°C.
   (d) Oxygen is a gas at room temperature whereas rhombic sulphur is a solid melting at about 113°C.   C (Overseas)

6 (a) The first seven ionisation energies of an element **B** are as follows:
     940 (1st); 2100; 3100; 4100; 7100; 7900; 15 000 kJ mol^{-1}.
    State, giving reasons, the group of the Periodic Table to which **B** is likely to belong and deduce (i) the empirical formula of its simplest chloride, and (ii) the effect of water on this chloride.
   (b) Explain the following observations in terms of structure and bonding.
     (i) The boiling points of the elements in the period sodium to argon are: sodium, 890°C; magnesium, 1120°C; aluminium, 2450°C; silicon, 2680°C; phosphorus, 281°C; sulphur, 445°C; chlorine, −34°C; argon, −186°C.
     (ii) Methane is almost insoluble in water but ammonia and hydrogen fluoride are readily soluble in water.   C (Overseas)

7 This question concerns the halogens bromine, chlorine and iodine.
   (a) Distinguishing between *s*, *p* and *d* electrons, give the electron configurations of (i) a chlorine atom, (ii) a bromide ion.
   (b) Explain why chlorine is more electronegative than bromine.
   (c) Why does iodine show some metallic properties? Give the name of one compound of iodine to illustrate this.
   (d) Place the following chlorides in order of increasing covalency, i.e. the least covalent first:
     aluminium (III) chloride, magnesium (II) chloride, phosphorus trichloride and sulphur dichloride.
     Briefly explain your reasoning and cite **one** piece of evidence for the order you give.
   (e) Give the structures of (i) ClO_3^-, (ii) PBr_5, (iii) CHI_3.   AEB

8 (a) For each of the following regular shapes, give the formula of one molecule which has that shape and say what bond angle(s) occur in each structure.
     (i) linear,
     (ii) triangular planar,
     (iii) tetrahedral,

96

(iv) octahedral.
  (b) Construct bond diagrams and so predict the bond angles in
      (i) the NO_2^- ion
      (ii) the CO_3^{2-} ion.
  (c) (i) Suggest, with some explanation, the geometrical shape of the PCl_5 molecule in the vapour phase.
      (ii) Give reasons why molecules often have bond angles different from those angles in the regular tetrahedral shape.   S

9 (a) For each of the following shapes, give the formula of one molecule or ion having the stated shape and give the bond angle(s) found in the regular structures listed.
      (i) linear (ii) trigonal planar (iii) tetrahedral (iv) trigonal bipyramid (v) octahedral.
  (b) The H-N-H bond angle in ammonia is 107·3°. Explain why this angle is not one of those found in the regular structures (i)–(v).   JMB (Syllabus A)

10 Describe the structure of any one ionic crystal, any one molecular crystal and any one giant molecular crystal. For each one of these solids, state three facts or properties which confirm or are derived from its structure, and explain briefly the connection between the property and the type of force in the crystal lattice.   O

11 (a) Describe the shape of each of the following molecules and account for each shape by considering the electronic repulsions within each molecule.
      BCl_3 PCl_3 XeF_4 H_2S
      For each compound, predict whether it would be likely to act as a Lewis acid, as a Lewis base, or as neither of these.
  (b) Arrange the following compounds in order of increasing boiling point. Justify and explain this order in terms of the intermolecular forces present.
      CH_3COONa $CH_3CH_2CH_3$ CH_3CH_2OH CH_3COCH_3
                                                                                              JMB (Syllabus B)

12 Explain the meaning of the statement 'that in its ground state the electron in a hydrogen atom is in a 1s orbital'. Further discuss the meaning of the term **covalent bond** illustrating your answer with reference to the hydrogen molecule and explaining why such a bond is stable.
   Explain the principles of electron repulsions in determining the shape of the ammonia (NH_3) molecule. Give the shape and electronic structures of the carbonate ion (CO_3^{2-}) and ethene (C_2H_4) and show how the principles can further be used to account for the shapes of the species present in phosphorus pentachloride.   W

13 Write an account of the influence of 'lone pairs' of electrons (non-bonded electrons) on the structure and properties of substances.   N (S)

14 Discuss the nature of the interaction between:
      (i) The hydrogen atoms in the hydrogen molecule.
      (ii) The argon atoms in liquid argon.
      (iii) The ions in crystalline LiI.
      (iv) Solute and solvent in aqueous $HClO_4$.   Camb. Entrance

15 Describe the experimental evidence which led to the suggestion of a helical structure for some protein molecules. What role is played in this helical structure by
      (a) the bonding in the peptide group, and
      (b) hydrogen bonding?   N

Chapter 8

More about the covalent bond

8.1 Introduction

We have seen in chapter 4 that the wave model of the electron leads to the inescapable conclusion that statements about the precise distance of a particular electron from the nucleus of an atom are invalid. Instead, we must use the language of probability and be content with saying that a particular electron is most likely to be found at a certain distance from a nucleus, at the same time recognising that there is a finite chance of finding it elsewhere. In order to get away from the rather abstract mathematical ideas, which a rigorous treatment of wave functions (atomic orbitals) demands, chemists have often found it sufficient and convenient to think in more concrete—but less precise— terms; thus an electron is visualised as a rather diffuse negative cloud.

A covalent bond has been visualised as the partial overlapping of atomic orbitals (electron clouds), the resulting molecular orbital, as its name implies, extending over both the bonded atoms. The overlap is sufficient to give a high electron density in the inter-bond region but not sufficient to cause strong internuclear repulsion (see section 7.2, p. 79).

It is the purpose of this chapter to develop these ideas more fully.

8.2 The covalent bond—molecular orbital theory

The molecular orbital theory of covalent bonding considers the ways in which electrons are influenced by the presence of two or more nuclei, the electrons in bonded atoms being said to occupy molecular orbitals. In the case of the hydrogen molecule there will be repulsive forces operating between the two nuclei and between the two electrons; in addition, there will be attractive forces operating between the nuclei and the two electrons. These complications mean that precise solutions of the energies and wave functions corresponding to molecular orbitals are out of the question, but approximate solutions are possible. We shall do no more than consider the molecular orbital theory in a purely qualitative sense.

Let us consider the hydrogen molecule. If the two atoms are widely separated, then each electron is only influenced by its own nucleus and occupies a 1s atomic orbital. As the hydrogen atoms approach one another both electrons come under the influence of the two hydrogen nuclei and eventually a chemical bond is established; under these conditions the separate atomic orbitals partially overlap and an electron cloud, consisting of two electrons, envelops both nuclei. A fundamental point in molecular orbital theory is that the number of possible molecular orbitals is equal to the number of atomic orbitals, thus for the hydrogen molecule there are two possible molecular orbitals. The reason for this is that electron waves, just like light waves, can combine constructively (in phase) or destructively (out of phase). The two molecular orbitals are thus obtained by adding and subtracting the two atomic orbitals (linear combination of atomic orbitals, abbreviated

LCAO, method). The two possible molecular orbitals are represented thus:

$$\psi_{bonding} = \psi_{A.1s} + \psi_{B.1s}$$
$$\psi_{antibonding} = \psi_{A.1s} - \psi_{B.1s}$$

where the letters A and B are used to distinguish the two hydrogen atoms. A bonding molecular orbital, obtained by addition of the separate atomic orbitals, results in an increase of electron density between the two nuclei. An antibonding molecular orbital, obtained by subtraction of the separate atomic orbitals, results in a decrease in electron density in the inter-bond region. A visual picture of what is involved can be gained by considering the overlap of the two atomic orbitals (fig. 8.1).

At this point it is necessary to explain the positive and negative signs attached to fig. 8.1. Solution of the Schrödinger equation (section 4.2., p. 38) gives an expression for the $1s$ wave function which is everywhere positive. Thus the bonding molecular orbital is also written with a positive sign attached to it. Since an antibonding molecular orbital is obtained by subtraction of the two atomic orbitals, this is conveniently represented by positive and negative signs as indicated. It must clearly be understood that these positive and negative signs are concerned only with the sign of the wavefunction and nothing else, i.e. they do not represent charge. In the case of hydrogen, the bonding molecular orbital is labelled $\sigma 1s$ and the antibonding molecular orbital $\sigma^* 1s$. This nomenclature will be explained later.

FIG. 8.1. *The combination of hydrogen* 1s *atomic orbitals to give a bonding and an antibonding molecular orbital*

Bonding molecular orbital ($\sigma 1s$)

Antibonding molecular orbital ($\sigma^* 1s$)

FIG. 8.2. *The energy difference between bonding and antibonding molecular orbitals*

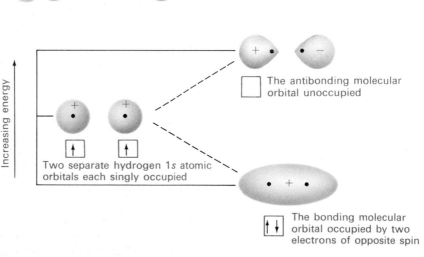

The antibonding molecular orbital unoccupied

Two separate hydrogen 1s atomic orbitals each singly occupied

The bonding molecular orbital occupied by two electrons of opposite spin

In general a bonding molecular orbital is of lower energy than the two separate atomic orbitals, while the reverse is true for an antibonding orbital. In fact, calculations show that an antibonding molecular orbital is slightly more antibonding than a bonding molecular orbital is bonding. Since the Pauli exclusion principle allows us to place two electrons with opposite spins in a molecular orbital, the hydrogen molecule contains both its electrons in a bonding molecular orbital. A chemical bond can be said to exist since the doubly occupied molecular orbital is of lower energy than the two singly occupied atomic orbitals of the hydrogen atoms (fig. 8.2).

Before proceeding to discuss the application of molecular orbital theory to more complex examples, it is necessary to list some conditions that must be met by the combining of atomic orbitals:

(a) The atomic orbitals must overlap to an appreciable extent. Otherwise there will be no build-up of charge between the nuclei and thus no chemical bond will be established.

(b) In the region of overlap the wave functions of the separate atomic orbitals must be of the same sign. Otherwise there will be no build-up of charge between the nuclei.

(c) The energies of the separate atomic orbitals must be of comparable magnitude.

8.3
The helium molecule and the helium molecule-ion

Helium atoms contain two electrons occupying the $1s$ atomic orbitals, giving an electronic configuration of $1s^2$. If these atomic orbitals are combined they will give one bonding and one antibonding molecular orbital and, since four electrons are involved, both of these orbitals will be doubly occupied. However, the lower energy of the bonding molecular orbital is effectively cancelled out by the higher energy antibonding molecular orbital and no chemical bond is established; helium molecules, He_2, cannot be formed (fig. 8.3).

FIG. 8.3. *The higher energy of the antibonding molecular orbital cancels out the lower energy of the bonding molecular orbital and no He_2 molecule is formed*

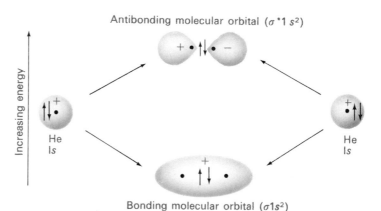

Antibonding molecular orbital ($\sigma^* 1s^2$)

Increasing energy

He
1s

He
1s

Bonding molecular orbital ($\sigma 1s^2$)

The entity He_2^+ molecule-ion can be obtained by subjecting helium gas, at low pressure, to a high energy discharge. It contains three electrons, two being in a bonding molecular orbital and one in an antibonding one; there is some overall bonding and the species exists. This is the first example we have met of a covalent species in which there is an unpaired electron. Its existence is explained quite naturally by molecular orbital theory, which demonstrates that electron pairing does not hold such a dominating role in bond formation as older theories have hitherto assumed. We shall discuss further examples of odd electron species later.

8.4
Molecular orbitals formed by the overlap of *p* atomic orbitals

It will be remembered from chapter 4 that the complete wave function for an electron consists of a radial part and an angular part. In the case of the $2p$ atomic orbital the radial wave function is everywhere positive but the angular wave function is both positive and negative; thus a positive sign would be attached to the upper lobe of fig. 4.7 (p. 42) (for values of θ from 0 to 90° and 270 to 360°) and a negative sign to the lower lobe (for values of θ from 90 to 270°). The implication of this is that p-type atomic orbitals can overlap in two distinct ways. Thus two $2p_x$ atomic orbitals can overlap along the x-axis and addition and subtraction of their wave functions give rise to a bonding and an antibonding molecular orbital. The two $2p_x$ atomic orbitals can also overlap laterally and again two molecular orbitals (one bonding and the other antibonding) can be obtained. These methods of overlap are shown in fig. 8.4 and fig. 8.5. In a σ bond the electron density is along the bond axis, while in a π bond the electron density is alongside the bond axis.

FIG. 8.4. *Molecular orbitals formed by the overlap of* $2p_x$ *atomic orbitals along the x-axis*

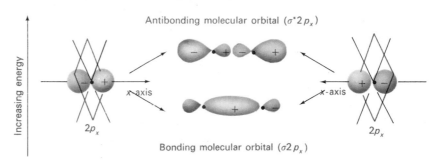

FIG. 8.5. *Molecular orbitals formed by the lateral overlap of* $2p_x$ *atomic orbitals*

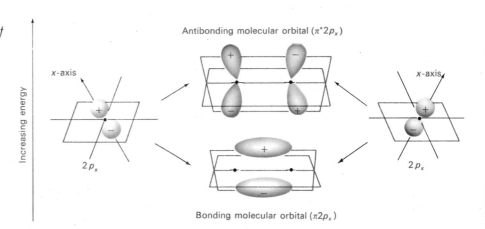

In general the energy of the $\sigma 2p_x$ molecular orbital is lower than that of a $\pi 2p_x$ molecular orbital (less effective overlap); similarly the energy of the $\pi^* 2p_x$ molecular orbital is lower than that of the $\sigma^* 2p_x$ molecular orbital. Thus the energy increases in the order:

$$\sigma 2p_x < \pi 2p_x < \pi^* 2p_x < \sigma^* 2p_x$$

101

FIG. 8.6. *A $2p_x$ and a $2p_z$ atomic orbital cannot overlap to form a molecular orbital*

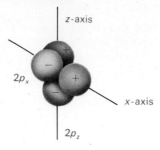

It is important to realise that some combinations of overlap of atomic orbitals are disallowed since the signs of the wavefunctions in the region of overlap mismatch. Thus a $2p_x$ atomic orbital cannot be combined with a $2p_z$ atomic orbital, for example, since the wavefunctions are not of the same sign in the region of overlap; the negative portion cancels the positive portion (fig. 8.6).

We have considered the formation of molecular orbitals by the overlapping of $2p$ atomic orbitals and it only remains to see how $2s$ atomic orbitals overlap. This type of overlapping is similar to that for the overlapping of two $1s$ atomic orbitals. Thus a bonding, $\sigma 2s$, and an antibonding, σ^*2s, molecular orbital are formed. The energy sequence of these molecular orbitals in a homonuclear diatomic molecule, i.e. a molecule containing two like atoms, is given in fig. 8.7.

FIG. 8.7. *The energy sequence for molecular orbitals (the bond direction is taken to be along the x-axis). In some instances the levels $\sigma 2p_x$ and $\pi 2p_y/\pi 2p_z$ are interchanged*

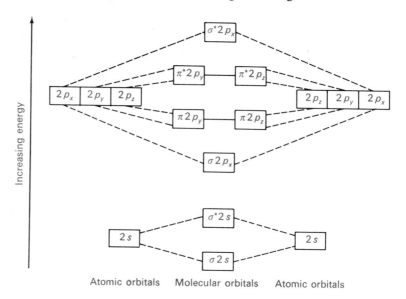

8.5
The nitrogen, oxygen and fluorine molecules

The electronic configurations of the atoms of nitrogen, oxygen and fluorine are respectively $1s^2 2s^2 2p^3$, $1s^2 2s^2 2p^4$ and $1s^2 2s^2 2p^5$. Consider two nitrogen atoms approaching each other; eventually a chemical bond is established and the fourteen electrons come under the influence of both nuclei. The four $1s$ electrons occupy a bonding and an antibonding molecular orbital, i.e. no net bond results, and this is the justification for neglecting inner shell electrons when discussing bond formation. There are now ten electrons to accommodate in the various molecular orbitals, and the four $2s$ electrons, like the four $1s$ electrons, occupy a bonding ($\sigma 2s$) and an antibonding (σ^*2s) molecular orbital; again no net bond results. Six electrons are now left, namely two $2p_x$, two $2p_y$ and two $2p_z$ atomic orbitals; since the bond is taken to be along the x-axis the two $2p_x$ atomic orbitals overlap along this axis and a bonding $\sigma 2p_x$ molecular orbital is formed and occupied by both electrons. The separate $2p_y$ and $2p_z$ atomic orbitals cannot overlap along their axes but they can overlap laterally to give respectively $\pi 2p_y$ and $\pi 2p_z$ molecular orbitals which are both filled. The nitrogen molecule thus contains one σ bond and two π bonds; the antibonding orbitals formed from the $2p$ atomic orbitals are left unoccupied and the net result is a triple bond, sometimes represented as N≡N.

The oxygen molecule contains two more electrons than the nitrogen molecule and these occupy the next higher energy molecular orbital. In fact there are two such orbitals having the same energy and they are antibonding (π^*2p_y and π^*2p_z, see fig. 8.7). Each of these two molecular orbitals is singly occupied (see Hund's rule p. 29). Since the higher energy of these antibonding orbitals effectively cancels out the lower energy of one of the $\pi2p$ bonding molecular orbitals (which is doubly occupied) the bond in the oxygen molecule can be considered to be a double bond. Molecular orbital theory is thus able to explain why the oxygen molecule is paramagnetic, i.e. attracted by a magnet, since this phenomenon is associated with the presence of unpaired electron spins. The formula $O{=}O$ is really not adequate for the oxygen molecule.

The fluorine molecule contains two more electrons than the oxygen molecule and thus the π^*2p_y and π^*2p_z antibonding molecular orbitals are now doubly occupied. There are, therefore, three bonding and two antibonding molecular orbitals occupied, and the fluorine molecule can be considered to contain a single covalent bond. In general the bond order is given by the formula:

$$\text{Bond order} = \tfrac{1}{2}(N_b - N_a)$$

N_b = number of bonding electrons
N_a = number of antibonding electrons

The molecular orbital diagrams for the three cases discussed above are given in fig. 8.8, together with some other relevant information.

8.6
Some other diatomic species

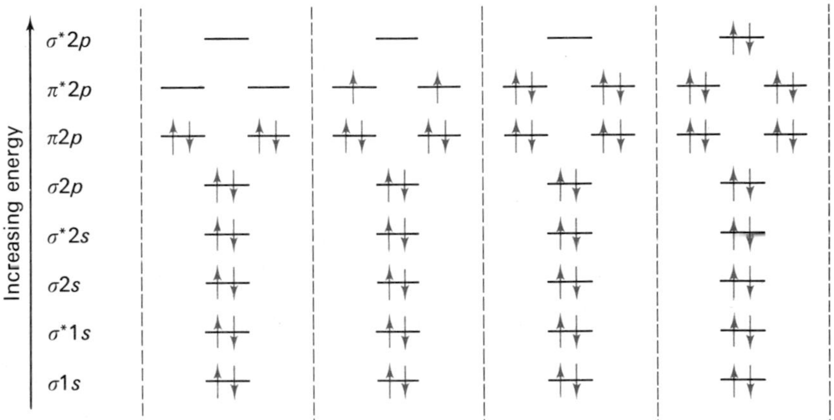

FIG. 8.8. *The occupation of the molecular orbitals for the nitrogen, oxygen and fluorine molecules. The molecule Ne_2 does not exist, because there is no net bonding*

Molecule	Nitrogen (N_2)	Oxygen (O_2)	Fluorine (F_2)	Neon (Ne_2) (Does not exist)
Bond order	3	2	1	0
Bond energy /kJ mol^{-1}	945	497	158	0
Unpaired electrons	0	2	0	0
Bond length /nm	0.110	0.121	0.142	—

Molecular orbital theory, even in its simplest form as we have presented it in this chapter, is able to account for the existence of unusual species, e.g. He_2^+, and the presence of unpaired electron spins in the oxygen molecule. Facts such as these were left unanswered by older theories of bonding, in which electron pairing was seen as the central idea; in molecular orbital theory there is nothing unusual in a molecule having an unpaired electron in a molecular orbital, and electron pairing is seen simply as a means of completely filling a particular molecular orbital.

The bonding in some more diatomic species is shown in fig. 8.9. It should be noted that we have included nitrogen oxide, NO, which contains two different atoms, and clearly the energies of the combining atomic orbitals of the nitrogen and oxygen atoms will be different. However, no great error is introduced here by assuming them to be the same.

It will be seen from figs. 8.8 and 8.9 that there is a decrease in bond length with increasing bond order, and that species which are isoelectronic have similar bond lengths, e.g. the nitrogen molecule and NO^+ ion have a bond order of 3 and bond lengths of 0·110 and 0·106 nm respectively, while the fluorine molecule and the superoxide ion, O_2^{2-}, have a bond order of 1 and bond lengths of 0·142 and 0·149 nm respectively. Bond strengths are seen to increase as the bond order increases as one would expect. Thus the sequence O_2^{2-}, O_2^-, O_2

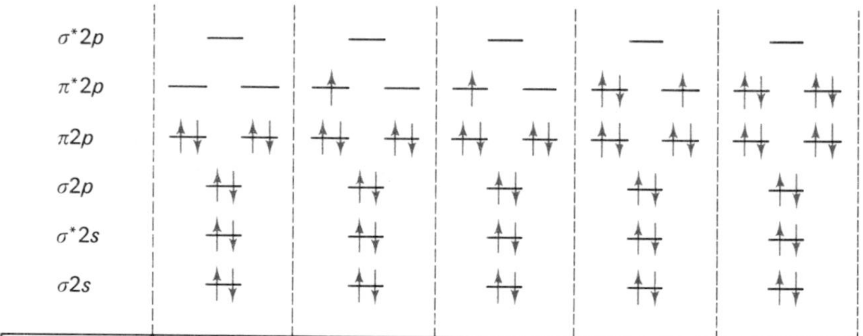

FIG. 8.9. *The occupation of the molecular orbitals for some other diatomic species. Note that the* $\sigma 1s$ *and* $\sigma^* 1s$ *orbitals are not included*

Species	NO$^+$	NO	O$_2^+$	O$_2^-$	O$_2^{2-}$
Bond order	3	$2\frac{1}{2}$	$2\frac{1}{2}$	$1\frac{1}{2}$	1
Bond energy /kJ mol^{-1}		670	~630	~350	~140
Unpaired electrons	0	1	1	1	0
Bond length /nm	0.106	0.115	0.112	0.126	0.149

and O_2^+ have respective bond orders of 1, $1\frac{1}{2}$, 2 and $2\frac{1}{2}$, and the bond strengths of these species increase in the order 140, 350, 497 and 630 kJ mol^{-1}.

The hydrogen fluoride molecule

So far we have discussed the bonding in diatomic molecules and ions which contain like atoms or, in the case of NO and NO$^+$, which contain atoms whose atomic orbitals are closely matched in energy. We must now go on to consider what happens when two different types of atom combine which have atomic orbitals that are dissimilar in energy; a case in point is the combination of the hydrogen and fluorine atoms to give hydrogen fluoride.

FIG. 8.10. *(a) Bonding between* H*(1s) and* F*(2p$_x$) orbitals (b) Zero overlap for* H*(1s) and either* F*(2p$_y$) or* F*(2p$_z$)*

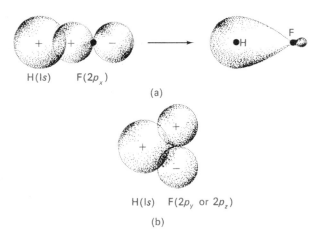

H(1s) F(2p$_x$)

(a)

H(1s) F(2p$_y$ or 2p$_z$)

(b)

The hydrogen atom can only contribute one electron which is in a 1s atomic orbital whereas, in theory, fluorine can provide several. However, it is only the 2p atomic orbitals which have an energy comparable to the 1s orbital of the hydrogen atom (see fig. 4.9, p. 44) which shows the relative energies of some atomic orbitals as a function of atomic number); in fact the energy of the 2p orbitals of the fluorine atom is lower than the energy of the 1s atomic orbital of the hydrogen atom. Since the x-axis is conventionally taken as the bond axis, combination takes place between H(1s) and F(2p_x) and this is represented in fig. 8.10(a). Note that any combination between H(1s) and either F(2p_y) or F(2p_z) is ruled out, since overlap from the positive lobe of these 2p orbitals is cancelled by that of the negative lobe; this is shown in fig. 8.10(b). In this example, the bonding molecular orbital is doubly occupied leaving the antibonding molecular orbital vacant. Since the bonding molecular orbital is closer in energy to that of the atomic orbital of the fluorine atom, it will resemble this orbital rather more than that of the hydrogen atom (see fig. 8.11).

FIG. 8.11. *Molecular orbital diagram for* HF

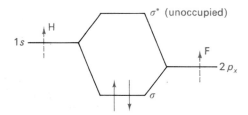

σ^* (unoccupied)

1s — H

F

2p_x

σ

The bonding in the other hydrogen halides is similar to that in hydrogen fluoride, thus in hydrogen chloride overlap occurs between H(1s) and Cl (3p_x) and the reason for this is again made clear in fig. 4.9, p. 44.

105

When we come to discuss polyatomic molecules we are immediately confronted with the problem of constructing molecular orbitals which extend over more than two nuclei. For example, methane contains ten electrons but, since two are in a closed shell (the $1s^2$ electrons of carbon), the problem is one of constructing eight molecular orbitals (four bonding and four antibonding) to accommodate these eight electrons and spread over five nuclei. Such **delocalised** molecular orbitals as they are called have been calculated, but it is clearly advantageous to look for possible ways of simplifying the problem. Fortunately this can be done in many cases, and the problem resolves into one of constructing **localised** molecular orbitals which embrace just two nuclei. In the case of methane we construct molecular orbitals for each C—H 'unit'. We discuss this sort of approach below.

The water molecule, H_2O

The possible atomic orbitals available for combination are the two hydrogen $1s$ orbitals and the $1s$, $2s$ and $2p$ orbitals of the oxygen atom (electronic configuration $1s^2 2s^2 2p_x^2 2p_y^1 2p_z^2$).* The $1s^2$ electrons of the oxygen atom are too tightly bound to be used for chemical combination and so too are the $2s^2$ electrons by themselves (but see later p. 108 in terms of hybridisation). Molecular orbitals are constructed from $H(1s)$ and $O(2p_x)$ and from $H(1s)$ and $O(2p_y)$ as depicted in fig. 8.12. In each case two localised bonding molecular orbitals are doubly occupied.

The bond angle in the water molecule would be expected to be 90°. In fact it is known to be 105° and we shall discuss this discrepancy later (p. 108).

FIG. 8.12. *Overlap of $2p_x$ and $2p_y$ atomic orbitals of the oxygen atoms with the $1s$ atomic orbital of hydrogen atoms*

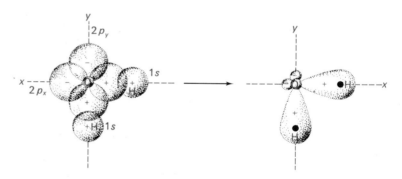

The ammonia molecule, NH_3

The bonding in the ammonia molecule can be discussed in a similar manner. Molecular orbitals are constructed from $H(1s)$ and $N(2p_x)$, from $H(1s)$ and $N(2p_y)$ and from $H(1s)$ and $N(2p_z)$. In this case the molecule would be expected to be pyramidal—which it is—with the N—H bonds at 90° to each other. Again there is some error in the predicted bond angle which is found experimentally to be 107°.

The methane molecule, CH_4

The electronic configuration of the outermost shell of the carbon atom is $2s^2 2p_x^1 2p_y^1$ and, by analogy with the water molecule, carbon would be expected to form two bonds with hydrogen giving a similar bond angle

---

* It is immaterial whether we write $2p_x^1 2p_y^1 2p_z^2$ or $2p_x^2 2p_y^1 2p_z^1$. We have chosen the former, simply because we have followed the convention of using the x- and y-axes as the bond directions.

FIG. 8.13. *An* sp³ *hybridised atomic orbital*

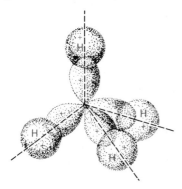

to that in the water molecule. However, one of the $2s$ electrons can be unpaired and promoted to a vacant $2p$ orbital, thus changing the electronic configuration to $2s^1 2p_x^1 2p_y^1 2p_z^1$ and producing the necessary four unpaired electrons which can now participate in covalent bond formation, by the overlap of these orbitals with the $1s$ orbital of each of the hydrogen atoms. The promotion energy of such a process is more than compensated for by the additional energy released in forming the two extra bonds.

We now have a problem: it is an experimental fact that all four bonds in the methane molecule are identical and that they are directed towards the corners of a regular tetrahedron, whereas the process of covalent bond formation, mentioned in the previous paragraph, would be expected to form two different types of bond: three equivalent (s—p bonds) and one other (s—s bond). To get over this difficulty we imagine that one s and three p atomic orbitals of carbon are mixed in such a manner that four equivalent, or hybridised, orbitals are obtained. When this process is followed through mathematically, four hybridised orbitals are constructed, which point towards the corners of a regular tetrahedron. Each has $\frac{1}{4}s$ character and $\frac{3}{4}p$ character. Like p orbitals, they are directional but unlike p orbitals one lobe is larger than the other (fig. 8.13). They are called sp^3 hybridised atomic orbitals.

The bonding in methane is visualised as the overlap of a hydrogen $1s$ atomic orbital with each of the four sp^3 hybridised atomic orbitals of the carbon atom (fig. 8.14).

8.8
Various types of hybridisation

Three types of hybridisation are important when discussing the chemistry of carbon compounds and these are given below. It is important to realise that the process of hybridisation is a purely mathematical technique, brought into operation when experimental results indicate that the bonding in compounds is not accurately reflected by the use of pure, or unmixed, atomic orbitals. Generally speaking, hybridisation is only a feasible technique when the participating atomic orbitals are energetically rather similar, i.e. the atomic orbitals have the same principal quantum number.

sp (linear) hybridisation

The mixing of an s and a p atomic orbital results in the formation of two hybridised atomic orbitals which are directed at an angle of 180° to each other. Each sp hybrid will have $\frac{1}{2}s$ and $\frac{1}{2}p$ character and unlike a p orbital has one lobe larger than the other (fig. 8.15). This type of hybridisation is important in accounting for the linear arrangement of the bonds in ethyne for example (p. 110).

FIG. 8.15. *Formation of two* sp *hybridised atomic orbitals*

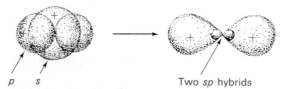

p   *s*   Two *sp* hybrids

sp² (trigonal) hybridisation

The combination of an s and two p atomic orbitals results in the construction of three coplanar hybridised atomic orbitals which enclose angles of 120°. Each sp^2 hybrid has $\frac{1}{3}s$ and $\frac{2}{3}p$ character and once again has one lobe larger than the other (fig. 8.16). This type of hybridisation is employed when discussing the shape of molecules such as ethene (p. 108), buta-1,3-diene (p. 111) and benzene (p. 112) for example,

FIG. 8.16. *Formation of three* sp^2 *hybridised atomic orbitals*

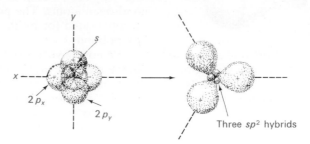

Three sp^2 hybrids

sp^3 (tetrahedral) hybridisation

We have already discussed this in connection with the bonding in methane. The angle between any two of the sp^3 hybridised atomic orbitals is 109°28′.

Some measure of sp^3 hybridisation in the water and ammonia molecules would account for the increase in bond angles from 90° to 105° and 107° respectively (p. 106).

The tetrahedral disposition of the valence bonds of carbon in the alkanes and in the giant molecular structure of diamond are explained in terms of sp^3 hybridisation.

It has been calculated that the relative overlapping abilities of atomic orbitals increase in the order:

$$s < p < sp < sp^2 < sp^3$$

By implication this would seem to suggest that bond strength should increase in this order. While this certainly appears to be true, it should be mentioned that in dealing with overlap we have restricted ourselves to angular functions only. We have neglected to consider radial functions, and hence the manner in which electron density varies with radial distance from the nucleus of the atom concerned (this simplification was noted previously, p. 42).

8.9
Some polyatomic molecules which contain multiple bonds

We have previously discussed the nature of multiple bonds in the context of homonuclear diatomic molecules, e.g. in the nitrogen and oxygen molecules (p. 102). We shall use those ideas, and further points developed in this chapter, in order to describe the bonding principally in multiple bonded carbon compounds.

The ethene molecule, C_2H_4

The ethene molecule is found experimentally to be flat with inter-bond angles very close to 120° (the actual values are given in fig. 8.17, which shows the basic skeleton structure of the molecule). We can account for these facts by imagining that electron promotion and sp^2 hybridisation occurs in the carbon atom prior to bond formation:

108

FIG. 8.17. *The skeleton structure of the ethene molecule showing the inter-bond angles*

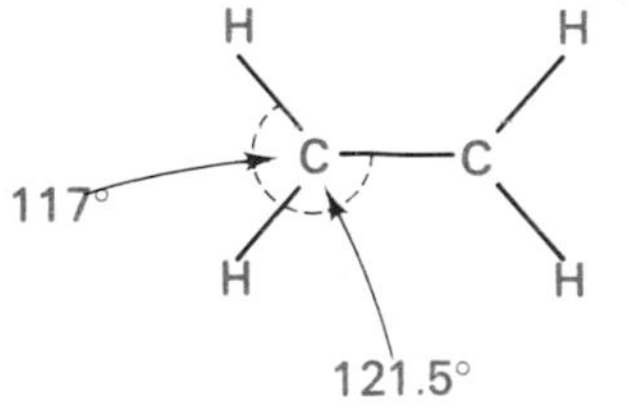

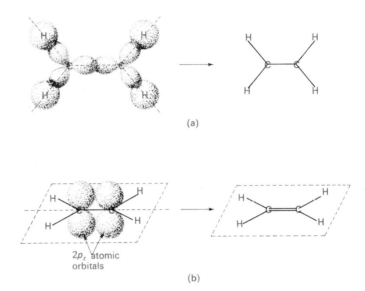

$$C(2s^2 2p_x{}^1 2p_y{}^1) \longrightarrow C(2s^1 2p_x{}^1 2p_y{}^1 2p_z{}^1) \longrightarrow C(2sp^2 \text{ hybrid} + 2p_z{}^1)$$

ground state   promoted state   hybridised state

Two of the sp^2 hybridised atomic orbitals from each of the carbon atoms form σ-bonds with each of the $1s$ atomic orbitals of the hydrogen atoms, the carbon atoms being bonded together by the remaining sp^2 hybridised atomic orbital on each atom; thus a total of five σ-bonds are constructed. This leaves one $2p_z$ atomic orbital on each carbon atom, directed at right angles to the plane of the ethene molecule; these overlap laterally to form a π-bond. The formation of five σ-bonds and one π-bond is depicted in fig. 8.18.

FIG. 8.18. *(a) The formation of five σ-bonds in the ethene molecule (b) The formation of one π-bond in the ethene molecule (the signs of the wave functions have been omitted for the sake of clarity)*

(a)

$2p_z$ atomic orbitals

(b)

The fact that the carbon atoms are bonded by a σ-bond and a π-bond means that these two carbon atoms are bound more tightly than in say ethane, C_2H_6, where a single σ-bond is present. This is reflected in a shortening of the carbon to carbon bond distance (C=C bond length 0·134 nm, C—C bond length 0·154 nm). There is also an increase in bond energy which would be expected (C=C bond energy 598 kJ mol^{-1}, C—C bond energy 346 kJ mol^{-1}). The presence of the π-bond in the ethene molecule prevents rotation about the C—C axis, since any tendency for rotation to occur would result in less effective overlap of the carbon $2p_z$ atomic orbitals, i.e. weakening of the π-bond.

Geometrical isomerism

The formation of a π-bond in addition to a σ-bond between two atoms has the effect of 'locking' the bond and preventing the free rotation of groups attached to it. This results in the existence of **geometrical isomers**, i.e. compounds that have the same structural formula but a different spatial arrangement of groups within the molecules. For example, the two structures

109

$$\underset{H}{\overset{H_3C}{>}}C=C\underset{H}{\overset{CH_3}{<}} \quad \text{and} \quad \underset{H}{\overset{H_3C}{>}}C=C\underset{CH_3}{\overset{H}{<}}$$

are geometrical isomers.

Geometrical isomerism is not confined to carbon compounds. For example, the structure N_2F_2 contains a double bond and two isomers exist with different physical properties.

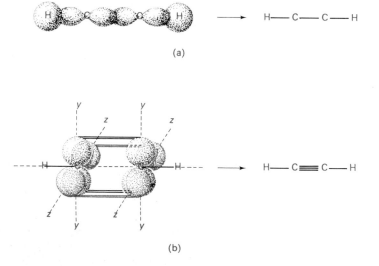

lone-pairs · lone-pairs

The ethyne molecule, C_2H_2

This molecule contains all four atoms lying along the same line. The bonding can be discussed in the following way: Both carbon atoms use *sp* hybridised atomic orbitals in bonding to each other and to the two hydrogen atoms. This leaves a $2p_y$ and a $2p_z$ atomic orbital on each of the carbon atoms and these overlap in pairs to form two π-bonds. The molecule thus consists of a total of three σ-bonds and two π-bonds and is depicted in fig. 8.19.

FIG. 8.19. *(a) The formation of three σ-bonds in the ethyne molecule (b) The formation of two π-bonds in the ethyne molecule*

H——C——C——H

(a)

H——C≡C——H

(b)

There is a further shortening of the carbon to carbon bond length (C≡C bond length 0.121 nm) and an increase in bond energy (C≡C bond energy 837 kJ mol^{-1}).

8.10
Systems containing delocalised molecular orbitals

In our discussion of the bonding in polyatomic molecules, we explained the construction of localised molecular orbitals, i.e. molecular orbitals confined to just two of the bonded atoms. There are situations, however, when this approach has to be modified in the light of experimental facts. In such instances we are forced to accept that some molecular orbitals are characteristic of the molecule as a whole, and, as such, will contain delocalised electrons. Examples of molecules which contain delocalised electrons are discussed below.

The molecule of buta-1,3-diene, C_4H_6

This molecule has interbond angles of 120°, so we begin by writing down the skeleton structure, which is based on the utilisation of sp^2 hybridised atomic orbitals of the four carbon atoms. This is shown in fig. 8.20(a) where the lines represent σ-bonds. Each carbon atom has a $2p_z$ atomic orbital at right angles to the plane of the molecule and these are shown in fig. 8.20(b). If these $2p_z$ orbitals now combine in pairs, C_1 with C_2 and C_3 with C_4, then two π-type molecular orbitals, similar to the one in ethene, would result (the π^* antibonding molecular orbitals would clearly be unoccupied). The molecule of buta-1,3-diene could thus be written as $CH_2{=}CH{-}CH{=}CH_2$ and it would be expected to have a C=C bond length of 0·134 nm as in ethene, and a C—C bond length of 0·154 nm as in ethane. The 'actual' molecule has indeed two different bond lengths but one is significantly shorter than a C—C bond (0·147 nm instead of 0·154 nm) and the other two are significantly longer than a C=C bond (0·137 nm instead of 0·134 nm).

FIG. 8.20. *(a) The skeleton structure of buta-1,3-diene indicating the σ-bonds (b) The four $2p_z$ orbitals on the carbon atoms of buta-1,3-diene*

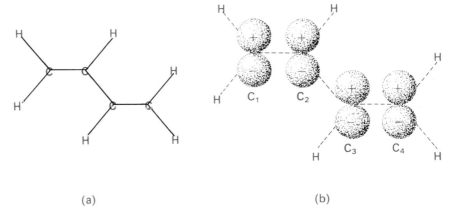

(a)   (b)

Another glance at fig. 8.20(b) indicates that the $2p_z$ atomic orbital on C_2 will clearly supply electron density not only to the bond between C_1 and C_2 but also to the bond between C_2 and C_3; a similar situation holds for the $2p_z$ atomic orbital on C_3. This means that we should expect the bond between C_2 and C_3 to be somewhat shorter than a normal C—C bond; similarly, the bonds between C_1 and C_2 and between C_3 and C_4 would be expected to be slightly longer than a normal C=C bond. We are thus able to account for the bond lengths in a qualitative manner.

The four $2p_z$ atomic orbitals can be combined to give four molecular orbitals (two bonding and two antibonding), and since we have four electrons they will enter the two bonding molecular orbitals with opposed spins. The energy levels of the four molecular orbitals are shown in fig. 8.21(a) but the details of their construction need not

111

concern us. The lowest energy molecular orbital extends over all four carbon atoms as a diffuse negative cloud as shown in fig. 8.21(b).

FIG. 8.21. *(a) The bonding (π-type) and antibonding (π*-type) molecular orbitals of buta-1,3-diene (b) The negative electron cloud of the lowest energy molecular orbital extending over the whole molecule of buta-1,3-diene*

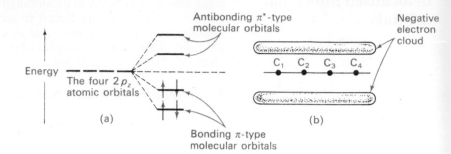

The molecule of benzene, C_6H_6

Benzene has a planar structure in which the six carbon atoms are bonded into a regular hexagonal ring. Each hydrogen atom is attached to carbon and, as in the case of buta-1,3-diene, all the interbond angles are 120°. The skeleton structure of benzene is shown in fig. 8.22(a), and again sp^2 hybridised atomic orbitals from carbon are employed in forming the twelve σ-bonds. The six $2p_z$ atomic orbitals, projecting at right angles to the plane of the molecule are combined to give six molecular orbitals (three bonding and three antibonding) and these six electrons doubly occupy the three bonding molecular orbitals (see figs. 8.22(b) and 8.23). The fact that all of the $2p_z$ atomic orbitals are

FIG. 8.22. *(a) The skeleton structure of benzene indicating the σ-bonds (b) The six $2p_z$ atomic orbitals on the carbon atoms of benzene*

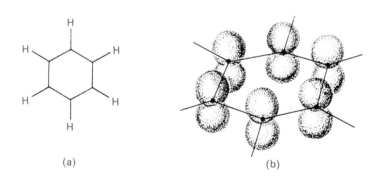

(a)   (b)

FIG. 8.23. *The molecular orbitals in the benzene molecule*

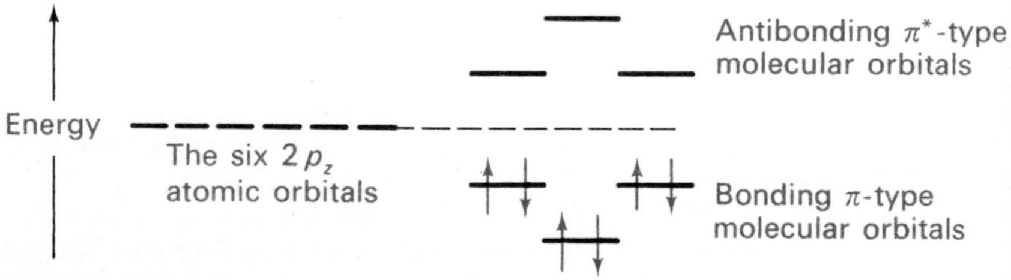

112

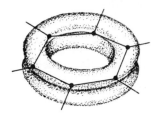

symmetrically disposed around the benzene ring means that all the carbon to carbon bond lengths are the same; the value is 0·139 nm, c.f. C—C (0·154 nm) and C=C (0·134 nm). Once again, the lowest energy molecular orbital extends over all the carbon atoms, rather like a double halo of negative charge (fig. 8.24).

The molecule of chloroethene, C_2H_3Cl

This molecule resembles that of ethene in basic structure and for most purposes is adequately represented by the formula $CH_2{=}CH{-}Cl$. However, the chlorine atom in the compound resists replacement by nucleophilic groups such as OH^-, unlike most organic chlorides which contain a single C—Cl bond. We can explain this experimental observation in terms of electron delocalisation.

The basic skeleton of the molecule is similar to that of ethene and is constructed from σ-bonds. Since there are singly occupied $2p_z$ atomic orbitals on each of the carbon atoms and a doubly occupied $2p_z$ atomic orbital on the chlorine, there is clearly the possibility of combining these three atomic orbitals into three molecular orbitals. The LCAO method of combining atomic orbitals allows for the formation of equal numbers of bonding and antibonding molecular orbitals, thus there is one bonding and one antibonding molecular orbital; the third is called a non-bonding molecular orbital. Since we have four electrons to accommodate in the molecular orbitals, two enter the bonding and two enter the non-bonding molecular orbital, leaving the antibonding orbital empty. Fig. 8.25(a) and (b) show respectively the $2p_z$ atomic orbitals, and the negative electron cloud of the bonding molecular orbital extending over the carbon and chlorine atoms. The relative energies of the three molecular orbitals are shown in fig. 8.26.

Since the formation of a bonding molecular orbital extending over the C—C—Cl system will amount to a strengthening of the C—Cl bond, we have an explanation of the enhanced stability of the chlorine atom in this molecule. In fact, the shortening of the carbon to chlorine bond is quite considerable; the experimental value of this bond length is 0·169 nm compared with a value of 0·177 nm for a 'normal' carbon to chlorine single bond.

FIG. 8.25. *(a) The three* $2p_z$ *atomic orbitals in the chloroethene molecule (b) The negative electron cloud of the molecular orbital extending over the chloroethene molecule*

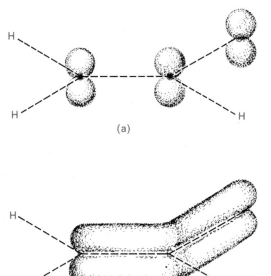

(a)

(b)

FIG. 8.26. *The molecular orbitals in the chloroethene molecule*

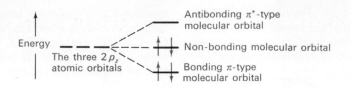

8.11
Questions on chapter 8

1 Draw diagrams to illustrate the shapes of s and p orbitals. In terms of orbitals, describe the bonding in methane and account for the shape of the molecule.

What would you predict about the shapes of the boron trifluoride molecule (the atomic number of boron is 5) and of the molecule that boron trifluoride forms with dimethyl ether, i.e. $(CH_3)_2O \rightarrow BF_3$?   C

2 (a) With the aid of simple sketches describe the shape and symmetry of (i) s and p orbitals, (ii) sp^3 and sp hybrid orbitals of carbon.

  (b) Using appropriate diagrams give the spatial arrangements of the atoms in the following species: (i) monomeric beryllium chloride; (ii) boron trifluoride; (iii) silicon tetrachloride (iv) phosphorus (v) chloride. For each species discuss briefly the spatial arrangement in terms of the Sidgwick–Powell theory of electron pair repulsion.   O

3 (a) Describe the **shape** and **symmetry** of (i) an s orbital, (ii) a p orbital, (iii) one sp^3 hybrid orbital.

  (b) What are the spatial arrangements of the atoms in the following molecular species and how may these be accounted for using the Sidgwick–Powell theory of electron pair repulsion: (i) $BeCl_2$; (ii) H_2O; (iii) BCl_3; (iv) NH_3?

For each of these molecular species state whether or not it has a dipole moment. [Your answers to parts (a) and (b) should include simple diagrams wherever appropriate.]   O

4 How are the two separate atomic orbitals of hydrogen atoms combined to give (a) a bonding molecular orbital and (b) an antibonding molecular orbital? How are the energies of the bonding and antibonding molecular orbitals related to the energies of the separate atomic orbitals?

5 Helium atoms do not combine to give the molecule He_2 but a helium atom can bond to a helium ion to give the entity He_2^+. How can these facts be explained using molecular orbital theory? Would you expect the H_2^+ species to exist under certain conditions? Give your reasons.

6 What justification is there for the statement: 'inner shell electrons do not contribute to chemical bonding'. Using molecular orbital theory, show that the fluorine molecule can be considered to contain a single covalent bond.

7 What is meant by a σ bond and a π bond? Show how molecular orbital theory is successful in explaining the presence of two unpaired electrons in the oxygen molecule, i.e. it is paramagnetic. Sulphur vapour contains some S_2 molecules; would you expect these molecules to be paramagnetic?

Chapter 9

Towards a theory of matter

9.1
The three states of matter

Under any given set of physical conditions most substances can be conveniently categorised into one of three distinct states of matter known as solid, liquid and gaseous states respectively. Some substances exist in all three states over a fairly short range of temperature (for example water, which exists in solid, liquid and gaseous states over a range of 100 K), whereas other substances exist in a single state within a fairly large range of temperature (for example diamond, which only melts when the temperature reaches 3823 K). Since a gas is formed from a liquid by increase in temperature, there is theoretically no upper limit to the range of temperature over which a substance can exist in the gas phase (although decomposition of the substance into elementary particles will occur at a high enough temperature). There is a lower limit to the range of temperature over which a solid can exist, since no substance can be at a temperature lower than 0 K.

Each of the states is characterised by particular properties which make it possible to classify any given material into one of these states. Thus the gaseous state is one in which the volume of the substance is affected to a large degree by changes in pressure and temperature, although it is also true to say that because a gas will always fill the complete volume of the container in which it is placed then a gas does not possess a surface of its own. A liquid, however, does have a finite volume although it too will tend to take up the shape of the vessel in which it is placed. That part of the liquid which is not in contact with the walls of the vessel constitutes the surface of the liquid and so sets a limit to the volume it can occupy. The existence of a fixed volume for a liquid makes it much less compressible than a gas. Solids may readily be distinguished from liquids and gases since they have a fixed shape, largely independent of pressure and only slightly affected by changes in temperature.

There are, however, some substances which tend to resist classification into one of the three states in spite of the specific characteristics referred to above. Amongst these are glasses, plasmas and liquid-crystals.

9.2
Early ideas on the nature of matter

It is natural to question the reason for the existence of three well-defined states of matter in terms of the theory of the atom as we know it at present. So far we have developed the atomic theory to the point where it is possible to account for the existence of discrete atoms and molecules, but we have not yet reached the point at which bulk properties of materials can be rationalised in terms of a basic hypothesis. To do so we can briefly retrace the development of the theory of matter from the earliest times. We have already referred to the contribution made by Democritus in establishing the existence of discrete entities known as atoms (p. 6), and there were others (Aristotle and Lucretius, and much later Gassendi) who gave valuable support to such an idea; but these early theoreticians did more than establish a particulate view of matter—they also attempted to account for the different states of matter. Thus Democritus taught that the particles of matter, being smooth and round, were unable to stick to one another so that in rolling over like small globules they produced a 'liquid' state; iron, on the other hand, being composed of jagged and uneven particles which were able to cling together existed in the 'solid' state. Lucretius explained how the atoms of all bodies were in continuous motion, colliding and rebounding from each other, and that when the distance between collisions was small the solid state resulted, whereas, when large distances were involved between collisions the gaseous state arose. This view was held for almost seventeen hundred years until Gassendi (1649) investigated some of the physical consequences of the atomic theory as it had stood for that period of time. He assumed that the atoms of different substances were similar in constitution although different in size and form, and further that they were able to move in all directions through empty space, and were completely rigid. On this basis he was able to account for a number of physical phenomena including the three states of matter, and the transitions from one state to another, in a manner which was not too different from the present day theory. He is therefore regarded by many as the father of the **kinetic theory** of matter. In the later part of the seventeenth century Hooke developed the ideas of Gassendi, as did Newton and Bernoulli in turn, but it was left to Clausius (1857) and Maxwell (1859) to propose what has become known as the modern kinetic theory of matter. It was Clausius, for example, who derived quantitative relationships between the pressure, volume and temperature of a gas in terms of particles of infinitesimal size, and it was Maxwell who presented his famous distribution law, as applied to the speeds of particles, at a meeting of the British Association for the Advancement of Science in Aberdeen.

9.3 Modern ideas

Although the **kinetic theory** was developed during the second half of the nineteenth century, it remains the basis of the present-day approach to the theory of matter and, in the case of the gaseous state, is the foundation for the quantitative interpretation of the properties of matter. The kinetic theory was based on certain assumptions concerning the nature and behaviour of the particles of matter, which are an essential prerequisite to the development of the theory as outlined in later sections of this chapter and also in chapter 10. They are as follows:

(a) All matter is made up of spherical particles (atoms and molecules) which are in constant motion.

(b) The particles are constantly colliding with each other, but when they do so there is no loss of energy on impact. Thus the collisions are said to be **perfectly elastic**.

(c) The particles exert no force upon each other (except during collision).

(d) Particles in a gas are separated from each other by distances which are large compared with the diameter of the particles themselves. The average distance travelled by a particle between collisions is known as the **mean free path**.

(e) The time spent in collision is extremely small compared with the time spent between collisions.

As we shall see, the assumptions above are justified to the extent that development based upon them lead to a reasonable interpretation of the properties of gases at an elementary level. However, the model is seen to be particularly inadequate in accounting for the existence of the so-called **condensed** states of matter (the liquid and solid states), for if assumption (c) had validity there would be **no** condensed states at all, since the particles would not be able to cohere! A satisfactory model of the three states of matter, based upon a particulate view of matter, can be derived, and this will be discussed now, before the ideas of the kinetic theory itself are further developed.

The three states of matter

Assuming the existence of individual particles of matter, the formation of three well-defined states may be conceived in terms of a balance between the attractive forces between the particles tending to bring them together in clusters, and the kinetic energy of the particles which keeps them in constant motion.

FIG. 9.1. *Potential energy curve for the interaction of two particles*

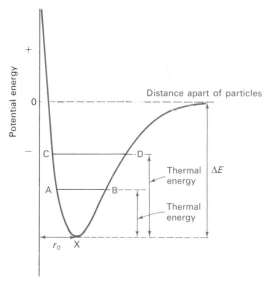

We have already seen (p. 80) how the potential energy of two atoms (or two molecules) changes as the particles approach each other from a large distance apart. This is illustrated in fig. 9.1. As the distance between the particles decreases the potential energy for the pair decreases until a minimum value is reached at point X, the equilibrium distance apart of the particles. Closer approach between the two would result in a rapid increase in potential energy due to the repulsive forces dominating the interaction between the particles. In the absence of any kinetic energy the particle-pair would remain a fixed distance, r_0, apart, a situation theoretically corresponding to the interaction of the particles at the absolute zero of temperature. At any temperature greater then absolute zero the particles will possess kinetic energy which will result in motion of the particles. In fig. 9.1 this is shown as a vibration of the pair of particles, such that their maximum distance apart is given by the horizontal line AB drawn across the potential energy 'well'; at higher temperatures, and therefore at greater thermal energies, the maximum distance apart during vibration will be greater, for example the distance

117

CD. It is obvious that if enough thermal energy is supplied (given by ΔE in fig. 9.1) the two particles will eventually separate from each other and interaction will have been effectively reduced to zero.

The thermal energy of the individual particles at a temperature T is of the order of kT, where k is the Boltzmann constant, (value $1.380\ 54 \times 10^{-23}$ J K^{-1}) and thus an approximate value for the average kinetic energy of the particles may be computed. It is clear that at high temperatures the value of kT will be large relative to ΔE and the particles will be moving about freely with very little interaction between them; in such conditions the substance will be a gas. At very low temperatures when kT may be very much less than ΔE, the particles will tend to cluster together, packing as closely as they can and motion will be confined to vibration about fixed positions. This situation describes the solid state. At some intermediate temperature the particles will have sufficient kinetic energy to break free momentarily from each other but there still remains a considerable degree of clustering. It is as if the individual particles exchange places in the separate clusters. The substance will then be in the liquid state, and there are good reasons for considering this state both as a very imperfect gas under some circumstances and as a modified solid under other circumstances. Table 9.1 summarises the balance between kinetic and potential energies leading to the three states of matter.

Table 9.1

Relative kinetic and potential energies	State of substance	Description
$kT \gg \Delta E$	gas	Little or no interaction between the particles, which are moving about continuously and freely
$kT \sim \Delta E$	liquid	Particles in small clusters, but exchange of particles between clusters takes place.
$kT \ll \Delta E$	solid	Particles vibrating about fixed positions in a three-dimensional cluster

Each of these principal energy terms, kinetic and potential will be examined in more detail.

9.4
The motion of particles

In describing the motion of the particles in the three states of matter there are various parameters which are of fundamental importance and which will be used in accounting for the bulk properties of substances. Amongst these quantities are the speed at which particles move, the distance particles move between collisions, the collision frequency and the energies of the particles. Most of these will be applied to the gaseous state, since the movement of the particles is very much restricted by intermolecular attractions in the liquid and solid states.

The speeds of particles

Since the particles in a gas are in constant motion they must be colliding with each other at a high frequency. In addition, any collision will be likely to change the velocities of the colliding particles and, since there are an extremely large number of particles in any sample of gas, the result will be a statistical distribution of the speeds of the particles. It is important that a distinction between the terms velocity and speed is made in connection with the distribution law as applied to a large number of particles. Since speed is a **scalar** quantity (it has magnitude but not direction) and velocity is a **vector** quantity (it has magnitude **and** direction), then for a large number of particles the average velocity must be zero (unless the whole sample of gas is moved bodily in a specific direction), whereas the average speed will always be finite.

FIG. 9.2. *Distribution of speeds of nitrogen molecules at a temperature of 273 K*

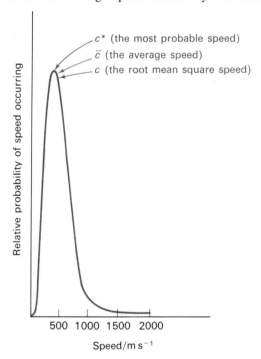

A typical distribution of molecular speeds in nitrogen gas at 273 K is shown in fig. 9.2, where the relative probability of a particular speed occurring is plotted against speed. It is immediately apparent that the curve is not symmetrical, spreading out more at high speeds than at low ones. The curve is a relatively narrow one, in that the majority of molecules lie within a fairly narrow range of speeds, although some molecules have very low and others very high speeds. Three different ways of quoting the 'mean' speed of molecules are important in the context of the kinetic theory.

The most probable speed, c^*, is that which the greatest fraction of molecules possess, and is consequently the speed corresponding to the maximum in fig. 9.2.

The average speed, $\bar{c}$, (sometimes called the mean speed), which is defined by the equation:

$$\bar{c} = \frac{1}{N}(c_1 + c_2 + c_3 + c_4 + \cdots + c_N)$$

119

i.e. it is the simple arithmetic mean of the individual speeds.

The root mean square speed, c, which is defined by the equation:

$$c = \left[\frac{1}{N}(c_1^2 + c_2^2 + c_3^2 + c_4^2 + \cdots + c_N^2)\right]^{1/2}$$

If the distribution curve was symmetrical then the most probable speed, c^*, would be equal to the average speed, $\bar{c}$; but because the curve spreads out more at high speeds the average speed is greater than the most probable speed. It will always be true, whatever the shape of the curve, that the root mean square speed, c, will be greater than the average speed, $\bar{c}$. It can be shown mathematically that for a large number of molecules the ratios of these speeds are:

$$c^* : \bar{c} : c = 1 : 1 \cdot 128 : 1 \cdot 225$$

thus

$$c^* \sim 0 \cdot 8c \qquad \text{and} \qquad \bar{c} \sim 0 \cdot 9c$$

A change in temperature would be expected to affect the distribution curve for molecular speeds. The distribution curves for nitrogen gas at temperatures of 273, 1273 and 2273 K are shown in fig. 9.3. It is important to note that it is not a proportionate increase in the speeds of all molecules which takes place when the temperature is raised; if this were so, then the curves for successive temperatures would simply be displaced to the right along the x-axis. It is the actual distribution, or spread, of the molecular speeds which changes with temperature, such that at a higher temperature the spread is greater. This is an important feature of the behaviour of molecules in gases, especially when the theory of the rates of chemical reactions is being studied, and we shall meet it again (chapter 25).

The values of c^*, $\bar{c}$, and c for nitrogen gas, at the temperatures corresponding to the three curves in fig. 9.3 are given in Table 9.2.

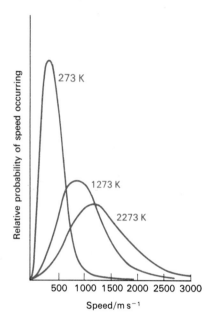

FIG. 9.3. *Distribution of speeds of nitrogen molecules at three different temperatures*

Table 9.2 Molecular speeds of nitrogen at various temperatures

Temperature/K	Most probable speed (c^*)/m s^{-1}	Average speed $(\bar{c})$/m s^{-1}	Root mean square speed (c)/m s^{-1}
273	403	454	493
1273	870	981	1066
2273	1162	1311	1424

Calculation of molecular speeds

It is possible to calculate the root mean square speeds, c, of molecules of gases at any given temperature if the relative molecular mass of the gas is known. Hence, it will be shown in chapter 10 that for an ideas gas

$$pV = \tfrac{1}{3}Nmc^2$$

120

p = pressure
V = volume
N = number of molecules
m = mass of molecule

which, on rearranging gives

$$c^2 = \frac{3pV}{Nm}$$

thus:

$$c = \left(\frac{3pV}{Nm}\right)^{1/2}$$

Now for an ideal gas, applying the equation $pV = nRT$ we·have

n = number of moles
R = gas constant with a value
   of 8·314 J K^{-1} mol^{-1}
T = Temperature (K)

$$c = \left(\frac{3nRT}{Nm}\right)^{1/2}$$

But $Nm/n = M_r$ (relative molecular mass in kg), thus

$$c = \left(\frac{3RT}{M_r}\right)^{1/2} \qquad (1)$$

This expression relates the root mean square speed with the temperature of the gas, and will be of importance in the study of chemical kinetics (chapter 25). Using the equation, a number of root mean square speeds of various gases at 273 K have been calculated and are shown in Table 9.3, together with the average speeds, $\bar{c}$.

Table 9.3 Molecular speeds (root mean square and average speeds) of selected gases at 273 K

Gas	Formula	Root mean square speed (c)/m s^{-1}	Average speed $(\bar{c})$/m s^{-1}
Hydrogen	H_2	1845	1698
Helium	He	1305	1202
Methane	CH_4	652	600
Nitrogen	N_2	493	454
Oxygen	O_2	461	424
Carbon dioxide	CO_2	393	362

Note that if the root mean square speed of the molecules of a gas is in S.I. units (m s^{-1}) then the value of M_r must be in terms of the kilogramme; thus for methane, CH_4, we have a relative molecular mass of 16 which, if expressed in terms of the kilogramme, becomes 16×10^{-3} kg, giving a root mean square speed of c where

$$c = \left(\frac{3 \times 8·314 \times 273}{16 \times 10^{-3}}\right)^{1/2} = 652 \text{ m s}^{-1}$$

Measurement of the distribution of molecular speeds in a gas

The distribution of molecular speeds in a gas, first deduced theoretically by Maxwell, has been verified by a series of experiments successively refined in accuracy over the past fifty years. Many such experiments have been based on that originally conceived by Zartmann (1931), and which is illustrated in fig. 9.4. The source of particles was a platinum filament coated with the metal whose atoms were under investigation (silver, bismuth and caesium were used). When the filament was heated electrically the metal 'boiled off' and streams of atoms were ejected in all directions. One such stream passed through a series of collimating slits to produce a narrow beam which then passed through a fine slit in a rotating drum. Once during every revolution of the drum the beam entered through the fine slit, but owing to the atoms in the beam having different speeds they were deposited on the opposite side of the drum over a range of distance, rather than in a single position. By measuring the intensity of the atoms at various places on the inside surface of the drum, the distribution of atomic speeds in the beam was evaluated. Close agreement with the distribution predicted by the Maxwell curve was found. Experiments conducted more recently have confirmed the shape of the distribution curve.

FIG. 9.4. *Determination of the distribution of molecular speeds*

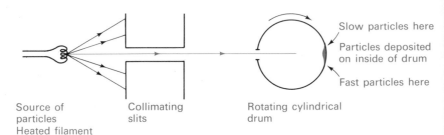

Slow particles here

Particles deposited on inside of drum

Fast particles here

Source of particles
Heated filament

Collimating slits

Rotating cylindrical drum

Collisions between particles in a gas

From the simple fact that one cubic metre of air at s.t.p. contains approximately 3×10^{25} particles with an average speed of about 450 m s^{-1} it must be obvious that in a gas the particles are not only in constant motion, but are constantly colliding with one another. An estimate of the number of collisions occurring each second may be made as follows.

Suppose we consider a gas whose molecules are assumed to be spherical and have a diameter d, then no molecule will be able to approach closer to another molecule than a distance d between their centres without a collision taking place. The molecule is said to have a **collision diameter** of d. (It should be appreciated that even if the molecules were infinitesimally small they would still possess a collision diameter, since at a finite distance they would mutually repel each other and therefore effectively collide.)

Now let us consider the purely hypothetical situation of one molecule moving with an average speed $\bar{c}$ with all the other molecules at rest. The 'moving' molecule will sweep out an 'effective' volume of $\pi d^2 \bar{c}$ in unit time (see fig. 9.5), colliding with other molecules within this volume. Now if there are n molecules per unit volume of the gas (m^3) the number of collisions made by this 'moving' molecule will be given by the expression:

$$\frac{\text{collision frequency}}{\text{for a single molecule}} = \frac{\text{volume swept out by 'moving' molecule}}{\times \text{number of molecules in unit volume}}$$
$$= \pi d^2 n \bar{c}$$

FIG. 9.5. *Effective volume swept out by a molecule in unit time*

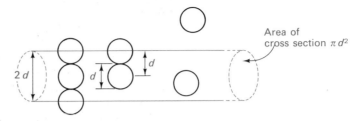

Area of cross section πd^2

A more exact formula, which takes into account the fact that all the molecules are moving and that it is the relative average speeds of the colliding molecules which matter, leads to the expression:

$$\text{Collision frequency for a single molecule} = \sqrt{2}\pi d^2 n \bar{c} \qquad (2)$$

Since there are n such molecules per unit volume making collisions, the total collision frequency Z_{AA} is given by:

$$Z_{AA} = \sqrt{2}\pi d^2 n^2 \bar{c} \times \tfrac{1}{2}$$
$$= \frac{\sqrt{2}\pi d^2 n^2 \bar{c}}{2} \qquad (3)$$

The necessity to divide by 2 in the above equation stems from the fact that it takes two molecules to make a collision, and therefore it is necessary to avoid counting any given molecule twice. The final expression, equation (3), is of particular value in the study of the theory of rates of reaction (chapter 25). A typical value for the collision frequency is that for oxygen at s.t.p. which is approximately 10^{35} collisions $\mathrm{m^{-3}s^{-1}}$

The average distance travelled between collisions by any given molecule is known as the mean free path, λ. It may be obtained by dividing the average speed, $\bar{c}$, by the number of collisions made in unit time by a single molecule (equation 2). Thus

$$\lambda = \frac{\bar{c}}{\sqrt{2}\pi d^2 n \bar{c}}$$
$$\lambda = \frac{1}{\sqrt{2}\pi d^2 n} \qquad (4)$$

Table 9.4 Mean free paths for various gases at 273 K and 1 atmosphere pressure (101 325 N m^{-2})

Gas	Formula	Mean free path λ/nm
Hydrogen	H_2	112·3
Helium	He	179·8
Nitrogen	N_2	60·0
Oxygen	O_2	64·7
Carbon dioxide	CO_2	39·7

123

It will be seen from equation (4) that the mean free path varies inversely with the number of molecules per unit volume of gas. This means that when values for the mean free paths of various gases are quoted, the pressure of the gas should also be given. Mean free path values may be obtained from measurements of the viscosities of gases, some typical values being given in Table 9.4.

It is clear from the values given in Table 9.4 that the molecules do not travel far before collision takes place. From equation (4) it is clear that as the pressure decreases (n decreases) the mean free path will increase and the collision frequency will correspondingly decrease. This reduction in collision frequency can account, in part, for the lower rate of gaseous chemical reactions as the pressure is reduced.

Experimental evidence for collisions between particles

Since the particles of matter are too small to be observed directly, evidence for the motion which they display in any given state must necessarily be indirect. The earliest evidence was provided by Brown in 1828, when he observed that pollen grains suspended in water were in a constant state of agitation when viewed through a microscope. Although he immediately jumped to the conclusion that he had discovered a peculiar property of pollen grains, he subsequently found that many small particles, when similarly suspended, described the same sort of motion. Later workers concluded that the pollen grains were set in motion by collision with the particles of water moving at very high speeds, and the phenomenon became known as **Brownian motion**.

Particles suspended in gases will also exhibit Brownian motion, and this is usually demonstrated by observing the light scattered from smoke contained in a glass cell. The arrangement is shown in fig. 9.6. Here the smoke particles experience constant bombardment from the fast-moving air molecules, the speed of the smoke particles being much less than that of the air molecules since they have a much larger mass.

FIG. 9.6. *Brownian movement in smoke*

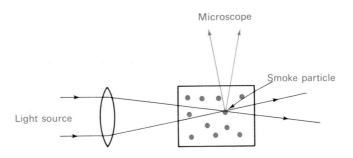

If it were possible to observe the actual motion of the molecules in a liquid or gas, it would be seen to be completely random (fig. 9.7). The displacement, d, of a particle from any given 'starting' position was originally shown by Einstein to be related to the value of the Avogadro constant, L, (p. 126) and careful observation of the displacements led to some of the earliest evaluations of this constant. The values obtained are in close agreement with subsequent more accurate determinations of a widely differing nature, and this lends further support to the validity of the basic kinetic theory of matter.

A further consequence of Brownian motion concerns the rate at

FIG. 9.7. *Random motion of particles in a gas*

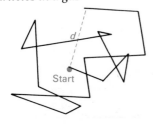

which particles, suspended in a liquid, fall under the influence of gravity. It was Perrin in 1909 who first noticed that when gamboge particles (yellow in colour, made from the gum resin of Siamese trees) were suspended in a liquid of slightly lower density all the particles did not sink to the bottom of the vessel. A distribution of particles occurred throughout the liquid, in such a manner, that the Maxwell distribution law (sometimes called the Boltzmann distribution law or the Maxwell-Boltzmann distribution law since both men were involved in its derivation) was obeyed when equilibrium had been established. Perrin chose gamboge particles which were uniformly spherical and of equal mass and, by estimating the number of particles at two different levels using a microscope (fig. 9.8), was able to confirm the general nature of the distribution law and make an estimate of the value of the Avogadro constant. The distribution law as applied to the solid suspended particles may be expressed as follows:

$$n_2 = n_1 e^{-mgh/kT} \tag{5}$$

from which it follows that

$$\ln \frac{n_1}{n_2} = \frac{mgh}{kT} \tag{6}$$

FIG. 9.8. *Distribution of small particles of solid suspended in a liquid*

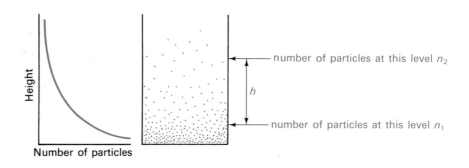

Perrin actually measured the radius of the gamboge particles and from a knowledge of the density of gamboge was able to determine their mass. Substitution of the measured quantities into equation (6) allowed a value of k, the Boltzmann constant, to be deduced.

Since $k = R/L$, an estimate was also made of the Avogadro constant L.

Consequences of Brownian motion occur in other fields of science.

FIG. 9.9. *Broadening of spectral lines with increased temperature*

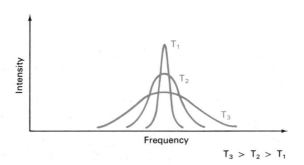

Thus the so-called line spectrum of gaseous elements (which is a very narrow curve in reality) gets broader as the temperature increases, owing to the increased motion of the gaseous atoms (fig. 9.9). There exists a limit to the sensitivity of a mirror galvanometer because of the quivering of the suspension due to molecular bombardment (fig. 9.10). Random electron movement in electrical circuits results in the so-called electrical noise, detected in amplification, which sets a limit to the sensitivity of electrical equipment. The effects diminish with a reduction of temperature.

FIG. 9.10. *Movement of the suspension in a mirror galvanometer*

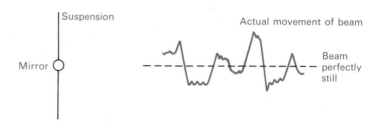

9.5
The mole and the Avogadro constant

We have referred to the Avogadro constant in several places in this chapter and since its definition is linked to that of the mole we shall define both terms now.

The mole

The mole, as agreed by the International Union of Pure and Applied Chemistry, is defined as **the amount of substance of a system containing as many elementary units as there are carbon atoms in 12 g (0·012 kg) of carbon-12**.

At first sight this seems a most complicated definition but we shall explain it with an example or two. The relative atomic mass of carbon-12 is 12, so there will be a certain number of atoms of carbon-12 in 12 g of carbon-12. The relative atomic mass of sodium is 23, so there will be this same number of **atoms** of sodium in 23 g of sodium. Similarly the relative molecular mass of carbon dioxide is 44 so again there will be this same number of **molecules** of carbon dioxide in 44 g of carbon dioxide. In each case, 12 g of carbon-12, 23 g of sodium and 44 g of carbon dioxide is one mole of that particular substance.

The Avogadro constant, *L*

The Avogadro constant, L, is the number of atoms of carbon-12 in 12 g of carbon-12. It follows that this same number is also the number of atoms of sodium in 23 g of sodium and the number of molecules of carbon dioxide in 44 g of carbon dioxide. The most accurate value of the Avogadro constant is

$$L = 6\cdot022\ 52 \times 10^{23}\ \text{mol}^{-1}$$

although for most purposes it is accurate enough to use $L \sim 6 \times 10^{23}\ \text{mol}^{-1}$

9.6
The kinetic energy of particles

In chapter 10 it will be shown that for an ideal gas

$$pV = \tfrac{1}{3}Nmc^2$$

This equation was previously used in the chapter (p. 120). On the assumption that the absolute temperature of an ideal gas is directly proportional to the average translational kinetic energy of the gas particles, $\tfrac{1}{2}Nmc^2$, it is possible to derive another equation relating to an ideal gas (p. 137), namely

$$pV = nRT$$

which again we have met previously in this chapter. Equating these two expressions we have:

$$\tfrac{1}{3}Nmc^2 = nRT$$

Since the average translational kinetic energy of the gas particles, E_k, is given by $\tfrac{1}{2}Nmc^2$, if follows that

$$\tfrac{2}{3}(\tfrac{1}{2}Nmc^2) = nRT$$
$$\text{or} \qquad E_k = \tfrac{3}{2}nRT$$

If we are dealing with one mole of gas instead of n moles the average translational kinetic energy, E_k, becomes

$$E_k = \tfrac{3}{2}RT \qquad\qquad (7)$$

Alternatively, the average translational kinetic energy per particle of gas is given by

$$E_k = \tfrac{3}{2}kT \qquad\qquad (8)$$

FIG. 9.11. *The rotation of a single atom about an axis through its centre*

It is important to realise that both equations (7) and (8) have been derived on the assumption that the kinetic energy is entirely due to the translation of the particles. For a gas composed of single atoms, i.e. a noble gas, this will be approximately true, since the only other kinetic energy possible in a single atom is the rotation of the atom about an axis passing through its centre (fig. 9.11). It turns out that the energy associated with this rotation is very small compared with that of the translation of the particle and it therefore makes no significant contribution to the total kinetic energy of the particle. In the case of gases composed of several atoms the situation may be very different due to the contributions from other rotational and vibrational modes, and such cases will be considered later.

Degrees of freedom

When energy is supplied to a substance in the form of heat, elementary kinetic theory supposes that this energy is absorbed by the substance in increasing its total kinetic energy. We have just noted that the total kinetic energy may be made up of several parts, due to translation and rotation of the particles; in addition, vibration of the atoms constituting the molecules of the substance will occur. Since a substance may therefore absorb the heat energy supplied to it in several different ways, it is said to possess a number of degrees of freedom. Examples of these degrees of freedom are as follows:

(a) Translational degrees of freedom

Because a particle travelling in three dimensions may be described in terms of its speed in three mutually perpendicular directions, then the total translational kinetic energy of the particle may be expressed in terms of three kinetic energy components, i.e.

$$\tfrac{1}{2}mc^2 = \tfrac{1}{2}mc_x^2 + \tfrac{1}{2}mc_y^2 + \tfrac{1}{2}mc_z^2$$

Thus any particle in the gas phase will possess three degrees of translational freedom.

Since the particles in the solid state do not undergo translation then a solid substance does not possess any translational degrees of freedom.

(b) Rotational degrees of freedom

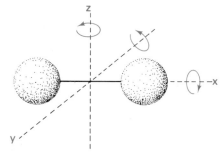

FIG. 9.12. *The rotation of a diatomic molecule about three orthogonal axes*

It has already been noted that the rotational energy of a single atom makes no significant contribution to the total kinetic energy of the particle. The three principal ways in which the rotation of a diatomic molecule may be described are shown in fig. 9.12. The rotations about the axes designated y and z will each be such that the molecule will possess a moment of inertia about these axes. The kinetic energy of rotation, in each case, will be given by

$$E_k = \tfrac{1}{2}I\omega^2$$

The rotation about the x-axis, however, is one for which the molecule has a negligible moment of inertia since both atoms lie on the axis of rotation (for point masses, this moment of inertia would be zero). Thus this rotation will make little contribution to the total kinetic energy of rotation of the molecule. Since the molecule possesses two modes of rotation which do contribute to the total kinetic energy, it is said to possess two rotational degrees of freedom.

In the case of a polyatomic molecule, which is non-linear, the rotation about the x-axis will involve a moment of inertia and there will therefore be three rotational degrees of freedom (fig. 9.13).

Rotational degrees of freedom exist for substances in both the gaseous and liquid states, but not for solids in which rotation is severely restricted.

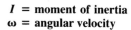

I = moment of inertia
ω = angular velocity

FIG. 9.13. *The rotation of a non-linear triatomic molecule about three orthogonal axes*

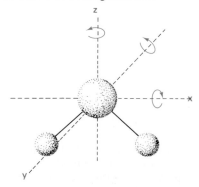

(c) Vibrational degrees of freedom

It is obvious that a gas composed only of single atoms can possess no

FIG. 9.14. *The vibration of a diatomic molecule about the centre of mass*

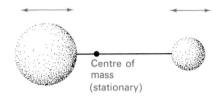

Centre of mass (stationary)

vibrational degrees of freedom, but a gas containing diatomic molecules will possess vibrational energy. Each of the atoms in a diatomic molecule can be considered to vibrate about the stationary centre of mass (fig. 9.14). During the vibration, energy absorbed as kinetic energy of motion is converted into potential energy of position. For example, at the two extreme positions of vibration, which amount to compression and extension of the chemical bond between the two atoms, the potential energy will be a maximum; while the kinetic energy will be a maximum when the two atoms are at their equilibrium distance apart. There is thus a constant interchange of kinetic and potential energy but the total vibrational energy is constant:

$$\text{Total energy of vibration} = \text{kinetic energy of vibration} + \text{potential energy of vibration}$$

There is clearly only one way in which a diatomic molecule can vibrate, so it possesses one vibrational degree of freedom.

For a polyatomic molecule, such as carbon dioxide or water, there will be several such vibrational modes. Some will involve stretching of bonds, others bending and yet others both stretching and bending. Some of these modes for a linear molecule, e.g. carbon dioxide, and for a non-linear molecule, e.g. water, are shown in fig. 9.15.

FIG. 9.15. *Some vibrational modes of (a) carbon dioxide and (b) water*

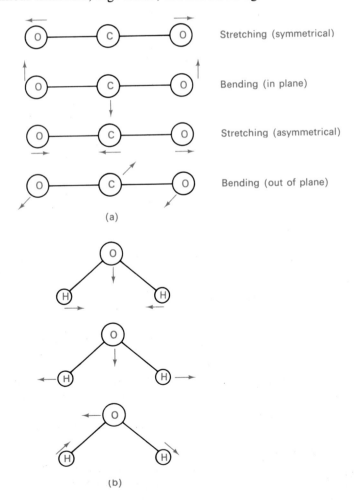

Stretching (symmetrical)

Bending (in plane)

Stretching (asymmetrical)

Bending (out of plane)

(a)

(b)

Total degrees of freedom

In general, a molecule consisting of N atoms will have $3N$ degrees of freedom associated with it. Thus for a diatomic molecule there will be six degrees of freedom; of these three will be translational, two will be rotational and one will be vibrational. For a monatomic gas there will be three degrees of freedom, all of which are translational. For a triatomic gas, there will be nine degrees of freedom altogether, of which three will be translational. This means that in the case of a triatomic gas there will be six degrees of freedom for the rotational and vibrational modes (these are known as the **internal** degrees of freedom). In the case of a linear triatomic gas, two of the internal degrees of freedom will be rotational and four vibrational (see fig. 9.15(a) for carbon dioxide). For a non-linear triatomic gas there will be three rotational and three vibrational degrees of freedom (see fig. 9.15(b) for water). It is possible to summarise the number of vibrational degrees of freedom of a molecule containing N atoms as follows:

linear molecule has $3N - 5$ vibrational degrees of freedom
non-linear molecule has $3N - 6$ vibrational degrees of freedom

Each degree of freedom is associated with an energy expression containing at least one 'squared-term'; thus the translational energy is given by the expression $\frac{1}{2}mc^2$ and the rotational energy by the expression $\frac{1}{2}I\omega^2$. The vibrational energy can be expressed in terms of two 'squared-terms', one for the kinetic energy contribution and the other for the potential energy. For each energy expression that can be written in such a form one degree of freedom may be counted for the molecule as a whole.

Equipartition of energy

It is possible using classical mechanical ideas to determine the manner in which the energy supplied to any molecule is distributed between the various possibilities of translation, rotation and vibration. The result is the Maxwell-Boltzmann principle, which may be summarised as follows:

'For a large number of particles in thermal equilibrium, the total energy is divided equally between all the squared energy terms; each squared term thus contributes the same average value which only depends upon the absolute temperature'.

We already know from equation (8) (p. 127) that for a monatomic gas each molecule has an average energy of $\frac{3}{2}kT$. Since there are only three 'energy terms' for such a gas each 'energy term' (or 'squared term') therefore corresponds to an average energy of $\frac{1}{2}kT$. In general we may therefore suppose that each 'energy term' will contribute $\frac{1}{2}kT$ to the total average energy of the molecule.

For a diatomic molecule having three translational, two rotational and one vibrational degrees of freedom, the average energy of the molecule will be given by:

$$3(\tfrac{1}{2}kT) + 2(\tfrac{1}{2}kT) + (\tfrac{1}{2}kT + \tfrac{1}{2}kT) = \tfrac{7}{2}kT$$

In general, therefore, the kinetic theory predicts that at a temperature T (kelvin) the average energy of a molecule may be given by the expression:

130

$$\text{Average energy of molecule} = (\text{number of 'squared-terms'}) \times \tfrac{1}{2}kT$$

Molar heat capacities

Since it is possible to calculate theoretically the energy absorbed by a molecule when its temperature is raised through a given interval, then by measurement of the heat capacities of different substances it ought to be possible to test the validity of the assumptions which have been made in the treatment given above.

The **molar heat capacity** of a substance is defined to be **the quantity of heat that must be supplied to one mole of the substance in order to raise its temperature by one unit amount (one kelvin)**.

For a gaseous substance, energy may be supplied so that the volume of the gas remains constant or it may be allowed to expand at constant pressure. A gas, therefore, has two different molar heat capacities, one at constant volume, C_v, and the other at constant pressure, C_p. In the discussion which follows we shall assume that the gas is not allowed to expand when energy is supplied to it, i.e. C_v will be used.

According to the kinetic theory so far developed, a monatomic gas has an average energy of $\tfrac{3}{2}RT$ per mole of gas (see equation (7), p. 127). Thus

$$U = E_k = \tfrac{3}{2}RT \text{ and, therefore, } \frac{dU}{dT} = \tfrac{3}{2}R = C_v$$

Taking the value of R as $8\cdot314$ J K^{-1} mol^{-1}, this gives a value for the molar heat capacity (at constant volume) for an ideal gas of $12\cdot47$ J K^{-1} mol^{-1}. Table 9.5 shows the calculated molar heat capacities of a number of gases, together with their measured values at constant volume.

Table 9.5 Some calculated and experimental molar heat capacities, C_v, of a selection of gases

Gas	Molar heat capacity, C_v/J K^{-1} mol^{-1}				
	Calculated values	Experimental values			
		298 K	600 K	1000 K	2000 K
He	12·47	12·47	12·47	12·47	12·47
Ne	12·47	12·47	12·47	12·47	12·47
Ar	12·47	12·47	12·47	12·47	12·47
H$_2$	29·10	20·52	21·01	21·89	25·89
N$_2$	29·10	20·81	21·80	24·39	27·68
H$_2$O	54·04	25·25	27·98	32·89	42·77
CO$_2$	54·04	28·81	39·00	45·98	52·02

There is excellent agreement between the calculated and experimental values of C_v for monatomic gases, although for diatomic and polyatomic gases in general the agreement is far from close. Another point concerning the values of molar heat capacities of gaseous substances is that whereas the values for monatomic gases are independent of temperature, all other gases have values which increase with increase in temperature. For example, nitrogen and carbon dioxide have values of C_v at 2000 K which are approaching the calculated values of 29·10 and 54·04 J K^{-1} mol^{-1} respectively.

A rationalisation of these observations was made by the proponents of classical kinetic theory by supposing that only the translational motions of the molecules absorb the applied heat energy at low temperatures, and that as the temperature increases the rotational motions are activated, and subsequently at still higher temperatures the vibrational modes begin to absorb the energy. This idea is supported by the observation that the molar heat capacities of gases (other than those which are monatomic) decrease with a fall in temperature to a limiting value of approximately 12·50 J K^{-1} mol^{-1}, a value corresponding to the absorption of energy by the three translational modes (fig. 9.16).

FIG. 9.16. *Variation of the molar heat capacity, C$_v$, of a selection of gases with temperature*

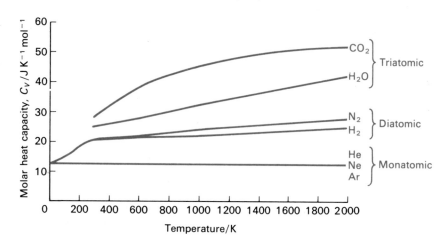

Further support came from the molar heat capacities of solid elements for which translation and rotation do not occur and which would be expected to have a value of $3R$ (since each particle in the solid will have one vibrational degree of freedom in three perpendicular lattice directions, the total vibrational energy will be $6 \times \frac{1}{2}RT$, i.e. molar heat capacity of $3R$). It was as long ago as 1819 that Dulong and Petit suggested that most solid elements at room temperature had a molar

FIG. 9.17. *Variation of molar heat capacity, C$_v$, for solid elements with temperature*

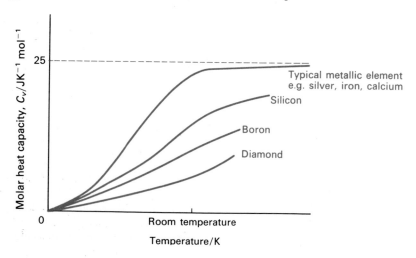

heat capacity of approximately 25 J K^{-1} mol^{-1} which is $3R$. The variation of molar heat capacity with temperature for solid elements is illustrated in fig. 9.17. It can be seen that there are important exceptions to the Dulong and Petit law, e.g. silicon, boron and diamond, and the decrease in the value of the molar heat capacity of metallic elements at low temperature cannot be explained in terms of the classical kinetic theory.

The idea that the two atoms in a diatomic molecule, such as hydrogen, do not vibrate at room temperature, in spite of the fact that the atoms have a finite mass and are held together by an elastic interaction, is clearly unacceptable. Thus the classical theory is unable to account satisfactorily for the variation of the molar heat capacities of both gases and solids with temperature.

9.7
Quantum theory and molar heat capacities

The resolution of the problem of the failure of classical theory to account for the variation of molar heat capacity with temperature came through an application of the quantum theory. It was Planck, in 1900, who proposed that the radiation from an emitting body is not propagated continuously but in discrete packets, or **quanta**. Bohr (1914) applied this idea to the energy associated with the position of an electron in the atom, and an essential feature of his theory of the hydrogen atom is that an electron may have only certain 'allowed' energy values, depending upon its distance from the nucleus (p. 21). The general idea of the quantum theory is that the energy of a body can only change from a particular value by an integral number of units (quanta). This applies to energy of all types, whether it is the result of a translation through space, of rotation about an axis, or of vibration about a fixed point; but what distinguishes the quanta of energy of different types are their **magnitude**. Thus the quanta associated with a moving car are so small that the energy changes during its passage along a road appear to be continuous. This 'apparent' continuous nature of energy arises when the energy supplied to a body is very large compared with the size of the quanta concerned. Of course, at the other extreme, if the energy supplied is extremely small compared with the size of the quanta then no change at all can take place.

So far as the molecule of a substance is concerned, we have already seen that the total energy is made up of contributions from several sources. There is the energy of the electrons arising from their positions relative to the nuclei in the molecule (electronic energy), the energy of the molecules as they move through space (translational energy) and as they rotate and vibrate (rotational and vibrational energies respectively). Hence we may say, to a good approximation, that the energy of a molecule can be expressed in terms of several components, as shown below.

$$\text{Energy of a molecule} = E_{\text{elec}} + E_{\text{vib}} + E_{\text{rot}} + E_{\text{trans}}$$

In general, the order of magnitude of the quanta of electronic energy (E_{elec}) is ~400 kJ mol^{-1}; that for quanta of vibration (E_{vib}) is ~20 kJ mol^{-1} and for rotational energy (E_{rot}) ~0·05 kJ mol^{-1}. So far as the translational energy (E_{trans}) is concerned, the size of the quantum is so small as to be undetectable in normal laboratory experiments. Thus

$$E_{\text{elec}} \gg E_{\text{vib}} > E_{\text{rot}} \gg E_{\text{trans}}$$

As a rough and ready rule we have previously used the value of kT to represent the average thermal energy of a molecule at room temperature (p. 118) or RT if we are considering one mole. Hence thermal energies are in the region of about 2·5 kJ mol^{-1} at room temperature. Since electronic quanta far exceed this value, it may be concluded that the majority of electrons in a molecule are in the lowest energy state and would not be promoted to a higher level until a very high temperature is reached. At the other extreme, translational quanta are so small that for all practical purposes the supply of energy for translation may be considered to be continuous. At room temperature energy may be absorbed by the molecule in rotation, since the rotational quanta are quite small. However, only those molecules with very low vibrational energy levels are likely to absorb energy of vibration, since vibrational quanta are generally greater than RT at room temperature. Thus the molar heat capacity of a molecule would be expected to vary in a discontinuous manner with temperature, as the increasing value of RT enables energy to be absorbed by the molecule in various ways. This is illustrated in fig. 9.18 for a diatomic molecule. As the graph shows, if the temperature is high enough, then the value of RT will be such that the energy will be absorbed in all four ways, and it would appear that the molecule can take in energy continuously. Thus at high enough temperatures the classical theory and quantum theory find their equivalence.

FIG. 9.18. *Variation of molar heat capacity, C_v, of a diatomic gas showing discontinuities, when different modes become active*

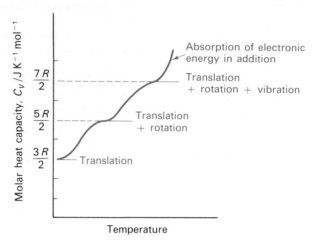

Generally speaking, the electronic contribution to the molar heat capacity of a gas (shown in fig. 9.18) is negligible, and often the molecule dissociates into atoms before any significant contribution from this source is made to the molar heat capacity.

It is possible to estimate the proportion of energy absorbed by the various modes from the value of the molar heat capacity at a given temperature. For a diatomic gas the translational contribution will always be 12.5 J K^{-1} mol^{-1} and a full rotational contribution will add a further 8.3 J K^{-1} mol^{-1}. The remainder will be the contribution to the total energy from the vibrational quanta. Hence the value of C_v for chlorine at 273 K is 25.79 J K^{-1} mol^{-1}, and we may therefore estimate the proportion of translational, rotational and vibrational energy as follows:

$$\text{Translational energy} = \frac{12.5 \times 100\%}{25.79} = 48.5\%$$

$$\text{Rotational energy} = \frac{8.3 \times 100\%}{25.79} = 32.2\%$$

$$\text{Vibrational energy} = \frac{(25.79 - 20.8) \times 100\%}{25.79} = 19.3\%$$

Hence almost one fifth of the absorption of energy by chlorine at 273 K is through the vibration of the atoms in the molecules.

Questions on this chapter are at the end of chapter 10.

Properties of gases

10.1
Pressure of a gas

This chapter is concerned with the way in which the theoretical ideas expressed in the kinetic theory of matter (as outlined in the previous chapter) can be used to account for the bulk properties of gaseous substances. Thus the microscopic properties associated with substances at the molecular level are related to the macroscopic properties of these substances.

One of the most obvious properties of a gas is its pressure which is exerted on the walls of the vessel which confines it. Since this property is also the most accessible for measurement, it can be used to test the validity of the kinetic theory as applied to gases.

FIG. 10.1. *The impact by a gas molecule on one side of a box is calculated by first resolving the speed, c', along three mutually perpendicular axes*

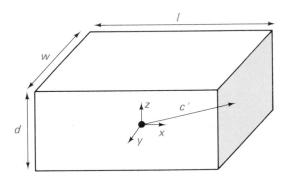

Suppose a gas is contained in a rectangular box of dimensions length l, width w, and depth d (fig. 10.1). Let us assume that the single particle of mass m has a speed c' with components c'_x, c'_y and c'_z in the three perpendicular directions. On impact with one end of the box (see shaded area in fig. 10.1) the particle will experience a change of momentum given by

$$mc'_x - (-mc'_x) = 2mc'_x \tag{1}$$

assuming an elastic collision.

The length of time which will pass before the same particle makes a second collision with the shaded face is given by:

$$\text{time taken} = \frac{\text{total distance travelled}}{\text{velocity of particle}}$$

$$= \frac{2l}{c'_x} \tag{2}$$

The number of collisions between the particle and the shaded face, taking place in unit time, will be the above expression inverted, i.e.

$$\frac{c'_x}{2l} \tag{3}$$

Now the force exerted by the single particle on the shaded face is given by the rate of change of momentum experienced by the particle (Newton's second law) and from equations (1) and (3) we may therefore write

$$\text{force} = \text{rate of change of momentum}$$

$$= 2mc'_x \frac{c'_x}{2l} = \frac{mc'^2_x}{l} \tag{4}$$

For the whole gas sample in which there are a total of N particles, the force exerted over the shaded face will be:

$$\frac{m}{l}[(c'_x)^2_1 + (c'_x)^2_2 + (c'_x)^2_3 + \cdots + (c'_x)^2_N]$$

The corresponding pressure (force per unit area) will be:

$$\frac{m}{lwd}[(c'_x)^2_1 + (c'_x)^2_2 + (c'_x)^2_3 + \cdots + (c'_x)^2_N]$$

$$= \frac{mNc^2_x}{lwd} \text{ where } c^2_x = \left\{\frac{(c'_x)^2_1 + (c'_x)^2_2 + \cdots}{N}\right\} \tag{5}$$

If the number N of particles is large (as in practice it will be), and if the particles are moving with random motion, we may write

$$c^2_x = c^2_y = c^2_z = \tfrac{1}{3}c^2$$

where c^2 is the mean square speed (or c is the root mean square speed). Therefore,

$$\text{pressure} = \frac{mNc^2}{3lwd}$$

$$p = \frac{1}{3}\frac{Nmc^2}{V} \tag{6}$$

The final expression for the pressure of a gas may be written in an alternative form; thus mN is the total mass of the gas and V is its total volume, hence the density, ρ, is mN/V giving:

$$\text{pressure} = \tfrac{1}{3}\rho c^2 \tag{7}$$

It is important to note that these final expressions are valid for all shapes of container; they are employed to derive some of the fundamental gas laws in terms of pressure, volume and temperature.

136

The Ideal Gas

In the previous chapter one of the basic assumptions made was that the temperature of a gas is proportional to the average translational kinetic energy of its particles. This can be expressed as follows:

$$T(\text{kelvin}) \propto \tfrac{1}{2}mc^2$$

Now from equation (6) it follows that

$$p = \frac{1}{3}\frac{Nmc^2}{V}$$

$$pV = \frac{1}{3}Nmc^2$$

$$= \frac{2N}{3}\left(\frac{1}{2}mc^2\right)$$

But since $T \propto \tfrac{1}{2}mc^2$ (assumed) we have:

$$pV \propto NT$$

$$\text{or} \quad pV = \text{const.}NT \tag{8}$$

This last relationship indicates that the product pV is proportional to both the absolute temperature and the amount of gas present (represented by the number of molecules N). Expressed in its more usual form, it becomes the equation of state for an ideal gas (or, more simply, **the ideal gas equation**):

n = number of moles of gas

$$pV = nRT \tag{9}$$

The constant R is known as the universal molar gas constant and has a value of $8\cdot314$ J K^{-1} mol^{-1}.

It is from the ideal gas equation that three frequently encountered gas laws may be deduced. Thus at constant values of n and T (that is, for a fixed number of moles of gas at constant temperature) the product pV will be constant. Expressed in the form

$$pV = \text{constant (for fixed } n \text{ and } T) \tag{10}$$

this relationship is known as **Boyle's Law**, after the scientist who discovered it in 1662 as a result of experiments on the compressibility of air. In terms of the kinetic theory, if the volume of the vessel containing a fixed amount of gas at constant temperature is halved there will be the same number of molecules of gas in half the volume. The number of collisions made by the gas on the walls of the vessel will increase by a factor of two, with a consequent increase in the rate of change of momentum by a factor of two. The pressure is, therefore, doubled.

The second law, that formulated by Charles in 1787, concerns the behaviour of a fixed amount of ideal gas under conditions of constant pressure. Under such conditions

$$\frac{V}{T} = \text{constant (for fixed pressure)} \tag{11}$$

that is, **the volume of a fixed amount of gas at constant pressure is directly proportional to its absolute temperature (Charles' law)**. On the kinetic theory, the effect of increasing the temperature of a fixed amount of gas would be to increase the average kinetic energy of the molecules, resulting in a greater rate of change of momentum. This would normally result in an increase of pressure, but in a container with flexible walls the gas will expand to occupy a larger volume and the original pressure will be maintained.

The third law, simply known as the **pressure law**, describes the manner in which the pressure of a fixed amount of gas at constant volume varies with the absolute temperature.

$$\frac{p}{T} = \text{constant (for fixed volume)} \qquad (12)$$

Thus the **pressure law** states that **for a fixed amount of gas at constant volume the pressure varies directly as its absolute temperature**. Again, increase of temperatrue increases the average kinetic energy of the molecules, leading to an increase in the rate of change of momentum at the walls of the container of fixed volume. This results in an increase in pressure. This effect is commonly noticed in the tyre of a car in which the air is contained in an almost fixed volume. Through the flexing of the rubber the air inside the tyre becomes warmer and the pressure builds up.

Real gases

Since the formulation of the ideal gas laws in the seventeenth and eighteenth centuries many scientists have devoted a life's work to an investigation of the behaviour of real gases. Without exception, it has been found that all gases deviate in some way or other from the relationships given in equations (9), (10), (11) and (12); indeed, an ideal gas is defined in terms of these relationships but, since no gas has been found

FIG. 10.2. *The compression factor* $Z = pV/nRT$ *plotted against pressure for a variety of gases*

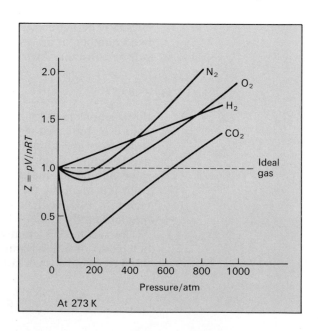

which obeys these laws under all conditions of concentration, pressure and temperature, the concept of an ideal gas is an abstract one. However, some gases under certain conditions of temperature and pressure approach 'ideal behaviour' fairly closely and, for practical purposes, such gases are taken to be ideal. The conditions under which real gases approach ideal behaviour can be deduced by inspection of figs. 10.2 and 10.3, in which the results of experiments, designed to test the validity of the ideal gas equation for different gases at varying temperatures and pressures, are illustrated. In each case the **compression factor** $Z = pV/nRT$ has been plotted as a function of pressure, p, for different gases (fig. 10.2) and as a function of pressure, p, at different temperatures, T, fig. 10.3. The most obvious conclusion which can be drawn from even a cursory inspection of both graphs is that real gases do not obey the ideal gas laws over a range of pressure and temperature. Further examination of both graphs leads to the conclusion that the behaviour of a gas deviates widely from ideality as the pressure increases. In the case of nitrogen at a temperature of 273 K and a pressure of about 800 atmospheres the product pV has a value at least twice that expected for ideal behaviour, i.e. the compression factor $Z \sim 2$ (fig. 10.2). At lower pressures the deviations are less and in the region of atmospheric pressure are negligible for nitrogen. In practice, if a gas obeys the ideal gas equation within an error of $\pm 1\%$ it is usually considered ideal. Most gases commonly encountered in the laboratory behave 'ideally' in terms of such a criterion, and it is a fairly safe assumption that at pressures less than atmospheric and temperatures greater than 273 K any gas will obey the ideal gas laws closely.

FIG. 10.3. *The compression factor* $Z = pV/nRT$ *plotted against pressure, at a series of temperatures, for nitrogen*

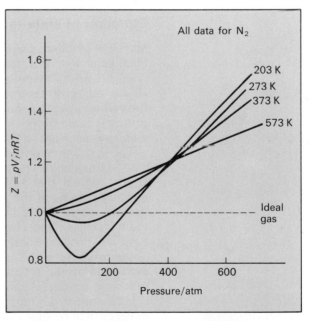

It is interesting to note that the value of the compression factor for hydrogen is always greater than unity at 273 K and increases as the pressure is increased, whereas for the other gases listed in fig. 10.2 there is an initial decrease in compression factor followed by a subsequent increase, i.e. there is a minimum in the relevant curve. The behaviour of hydrogen, however, is not exceptional, since it too shows a minimum in its compression/pressure curve at a low enough temperature.

From an examination of fig. 10.3 it is clear that at some temperature between 273 K and 373 K the minimum in the curve for nitrogen will lie on the 'ideal gas' line (fig. 10.4). At such a temperature the gas will obey the gas laws over quite a wide range of pressure; this temperature, which is a characteristic constant for each gas, is known as the **Boyle temperature**, e.g. oxygen has a Boyle temperature of 406 K whereas that for helium is 23 K. In general, the more easily liquefiable gases have a high Boyle temperature, and those which are difficult to liquefy have a low Boyle temperature.

FIG. 10.4. *Isothermals for a real gas, illustrating the Boyle temperature*

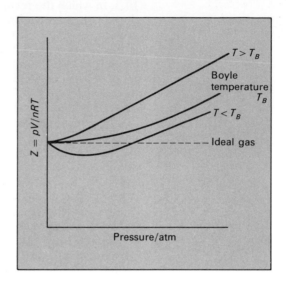

Equation of state for a real gas

Sufficient evidence has been given to show that real gases deviate from ideal behaviour and, from the earliest of experiments with real gases, scientists have attempted to discover an equation of state which more closely relates to the behaviour of the gaseous state of matter. One of the earliest attempts was made by **van der Waals** who suggested that the principal cause of the deviations from ideality arose through two of the assumptions made in the kinetic theory of gases (p. 116).

The first concerns the assumption that the volume of the molecules of a gas are negligible compared with the total volume occupied by the gas. It was argued that since the molecules have a finite size then the effective volume of the gas will be less than the measured volume, and that the magnitude of this discrepancy will be related to the volume of the molecules themselves. van der Waals simply replaced the volume term V in the gas equation by the expression $(V - b)$, where b is known as the **covolume** of the gas. It might be thought that the covolume of the gas would be the actual volume of the molecules themselves, but the following arguments shows that this is not so. Suppose that each molecule (assumed to be spherical) has a radius a; the actual volume of one molecule will be $\frac{4}{3}\pi a^3$. However, because it is impossible for two molecules to approach each other closer than the distance between their centres, the effective radius of a molecule becomes $2a$ (fig. 10.5). The volume involved in the collision between two molecules is therefore $\frac{4}{3}\pi(2a)^3$, i.e. $\frac{32}{3}\pi a^3$. Since two molecules are involved in the collision the effective volume of a molecule is $\frac{16}{3}\pi a^3$, which is four times the actual volume. After taking into consideration the covolume of the molecules of a gas the ideal gas equation is modified to

FIG. 10.5. *The effective volume of a gas molecule*

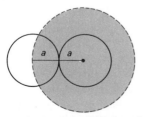

Volume of sphere of radius $2a = \frac{4}{3}\pi(2a)^3$

$\qquad\qquad\qquad\qquad\quad = \frac{32}{3}\pi a^3$

140

n = number of moles of gas
b = covolume of gas

$$p(V - nb) = nRT$$

The second assumption made in the kinetic theory of gases is that the attractive forces between the molecules are negligible, except at the closest approach during collision. In fact the attractive forces will be finite over appreciable distances, since the potential energy term for the attraction between two molecules varies as the inverse of the sixth power of the distance apart of the molecules (p. 95). Consider a molecule of gas travelling towards the wall of a container and about to collide with it (fig. 10.6). There will be two principal attractive forces exerted on such a molecule—one from the other molecules in the bulk of the gas and the other from the particles which constitute the walls of the container. Whereas a molecule in the interior of the container experiences no net attractive force since it is uniformly surrounded by similar molecules, that near the wall of the container is not uniformly surrounded and will therefore experience an attractive force towards the centre of the vessel. Thus at the instant the molecule is about to collide with the wall, its velocity will be decreased and the consequent change of momentum will be less than that derived in the simple kinetic theory treatment (p. 135). The measured pressure will be less than that expected for an ideal gas under the same conditions. The attraction of the wall of the vessel for the gas molecules has no net effect on the change in momentum, since it speeds up the molecules before impact and slows them down after collision.

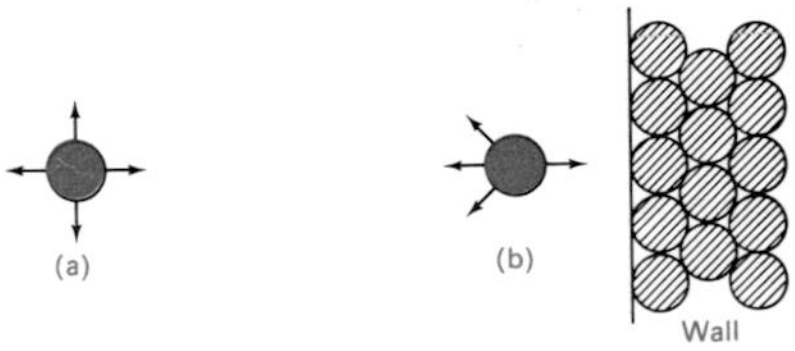

The inward attractive force on the molecules about to strike the wall of the container will be proportional to the number of particles in the bulk of the gas, and therefore to the density of the gas (n/V), and the number of molecules striking the wall at any given instance will also be proportional to the density of the gas. Thus we may write:

FIG. 10.6. *Attractive forces on a molecule of gas (a) in the bulk of the gas (b) near the wall of a vessel*

(a)

(b)

Wall

Force on gas molecules near the wall $\propto \dfrac{n}{V}\dfrac{n}{V}$

Hence the pressure defect $\propto \dfrac{n^2}{V^2}$

$$= \frac{an^2}{V^2} \qquad \text{(where } a \text{ is a constant)}$$

The equation of state, modified to take into account the so-called pressure defect, now becomes

$$\left(p + \frac{an^2}{V^2}\right)(V - nb) = nRT \qquad (13)$$

and this is known as the van der Waals' equation. For one mole of gas it becomes:

$$\left(p + \frac{a}{V^2}\right)(V - b) = RT \qquad (14)$$

The values of the constants a and b for a particular gas may be obtained by substituting known values of p, V and T into equation (14); two values for each of p, V and T will give a pair of simultaneous equations which can then be solved for a and b.

141

Table 10.1 Values of van der Waals' constants for some gases

Gas	Formula	a/atm dm^6 mol^{-2}	b/dm^3 mol^{-1}
Helium	He	0·0341	0·0237
Hydrogen	H$_2$	0·244	0·0266
Nitrogen	N$_2$	1·39	0·0391
Oxygen	O$_2$	1·36	0·0318
Chlorine	Cl$_2$	6·49	0·0562
Carbon dioxide	CO$_2$	3·59	0·0427

As can be seen from Table 10.1 the constant b is of the same order of magnitude for each gas. However, the value of the constant a varies widely from a numerical value of 0·0341 atm dm^6 mol^{-2} for helium (the most difficult gas to liquefy) to 6·49 atm dm^6 mol^{-2} for chlorine (an easily liquefied gas). This trend is expected since the constant a is a measure of the force of attraction between gas molecules.

Although the van der Waals' equation represents the behaviour of real gases in a fairly satisfactory qualitative manner it is, at best, only an approximation. For this reason many other workers have attempted either to modify van der Waals' equation or to propose different solutions.

The most general equation of state is that due to Kammerlingh-Onnes and is written:

$$pV = RT\left[1 + \frac{B}{V} + \frac{C}{V^2} + \frac{D}{V^3} + \cdots\right]$$

B, C, D etc. vary with temperature and are known as the second, third, fourth etc. virial coefficients respectively. The equation is known as the **virial equation of state**, and is generally used to account for the behaviour of real gases especially at high pressures. The equation is tedious to use but one of its merits is that once virial coefficients are determined the behaviour of any real gas can be predicted.

10.3 Pressure in a mixture of gases

So far we have been concerned only with the pressure exerted by a single gas. We now turn our attention to the pressure resulting from a mixture of two or more gases. Consider the case of two ideal gases which completely mix in a container of volume V, without chemical combination of any kind. We may therefore write

$$p_1V = \tfrac{1}{3}N_1m_1c_1^2 = \tfrac{2}{3}(\tfrac{1}{2}N_1m_1c_1^2) = \tfrac{2}{3}E_1$$

where the subscript 1 refers to one of the gases and E_1 is the average kinetic energy of this gas. For the other gas it likewise follows that

$$p_2V = \tfrac{2}{3}E_2$$

Since chemical combination does not occur there will be no change in temperature on mixing the two gases. Hence

$p_fV = \tfrac{2}{3}E_f = \tfrac{2}{3}(E_1 + E_2)$ $(E_f = E_1 + E_2)$
or $p_fV = p_1V + p_2V$
or $p_f = p_1 + p_2$ (15)

p_f = **final pressure after mixing**

Thus the final pressure is simply the sum of the two original pressures of the gases. The derivation can easily be extended to cover the mixing of any number of ideal gases which do not interact with each other. The result is summed up in Dalton's law of partial pressures, which states that **when two or more gases which do not react together chemically are present in the same container, the total pressure exerted by the mixture is the sum of the partial pressures of each constituent gas** (the partial pressure of each gas being that which each gas would exert if it alone occupied the whole volume).

Real gases do not obey Dalton's law exactly and, as would be expected, the deviations become greater as the total pressure increases. These deviations are due not only to attractions between molecules of the same kind, for clearly there will also be attractions between different kinds of molecule. In addition, the space in which the molecules are contained in the mixture is partly occupied by molecules of a different kind thus reducing the effective volume of the gases present. In practice, however, Dalton's law may be used without serious error at pressures in the region of one atmosphere where deviations are very small.

10.4
Molar volume of a gas

For some practical purposes it is more convenient to measure volume changes than pressure changes during the course of gaseous reactions, and it is therefore of importance to establish some standard volume which may be taken as a reference point for such measurements. For an ideal gas this standard is chosen, quite arbitrarily, as the volume occupied by one mole of gas under standard conditions of a temperature and a pressure (s.t.p.) which are 273 K and 101 325 N m^{-2} (one atmosphere). This volume is known as the **standard molar volume** of the gas V_m.

From the ideal gas equation we have

$$pV = nRT$$

thus

$$V = \frac{nRT}{p}$$

Under standard conditions:

$$V_m = \frac{1 \times 8 \cdot 314 \times 273}{101\ 325} \text{ m}^3 \text{ mol}^{-1}$$

$$= \mathbf{2 \cdot 24 \times 10^{-2}\ m^3\ mol^{-1}} \text{ (approx.)}$$

One mole of any ideal gas under standard conditions of temperature and pressure has a volume of $2 \cdot 24 \times 10^{-2}$ m^3 (or 22·4 dm^3, a unit more frequently used in chemical calculations). The extent to which real gases behave in an ideal way may therefore be determined by measuring the standard molar volume of the gas in question, and comparing it with the accepted value of **22·414 dm^3 at s.t.p.** Table 10.2 lists the values of the standard molar volumes for a selection of gases.

It can be seen that, although the standard molar volume is not independent of the nature of the gas, there is very good agreement between the values given in Table 10.2. The largest deviations occur for gases which are the most easily liquefied, e.g. chlorine and carbon dioxide. The values may be regarded as constant enough to enable relative molecular masses of gaseous substances to be evaluated from experimental measurements based on standard molar volumes (p. 149).

Table 10.2 Some standard molar volumes, $V_m/dm^3 \, mol^{-1}$, of gases

Gas	Formula	Standard molar volume $V_m/dm^3 \, mol^{-1}$
Helium	He	22·396
Hydrogen	H_2	22·432
Nitrogen	N_2	22·403
Oxygen	O_2	22·392
Chlorine	Cl_2	22·063
Carbon dioxide	CO_2	22·263

This is possible because there is a simple relationship between the volumes of two gases (at the same temperature and pressure) and the relative numbers of molecules present.

Consider two ideal gases, then

$$p_1 V_1 = \tfrac{1}{3} N_1 m_1 c_1^2$$
$$p_2 V_2 = \tfrac{1}{3} N_2 m_2 c_2^2$$

If the two gases have the same pressure, temperature and volume, then $p_1 = p_2$ and $V_1 = V_2$ and since the average kinetic energy of a molecule is proportional to the absolute temperature, then

$$\tfrac{1}{2} m_1 c_1^2 = \tfrac{1}{2} m_2 c_2^2 \quad \text{or} \quad m_1 c_1^2 = m_2 c_2^2$$

It therefore follows that

$$N_1 = N_2$$

This result is usually expressed in words such as:

Equal volumes of all ideal gases measured under the same conditions of temperature and pressure contain equal numbers of molecules.

This statement is known as **Avogadro's law**, after the scientist who first formulated it in 1811. The significance of the law is that it enables changes in the number of molecules taking place during a chemical reaction in the gas phase to be followed by measurement of changes in the total volume (or pressure).

10.5
Diffusion and effusion of gases

The phenomenon of self-diffusion of gases was used as direct evidence for a particulate theory of matter. The interpretation of the mixing of two gases, when placed in contact with each other, can be made quite simply in terms of the random motions of the molecules of the gases concerned; the molecules of one gas move freely into the spaces between the molecules of the other gas. We shall see later that this simple picture of the spontaneous mixing of gases can be used to interpret the factors responsible for the movement of all chemical systems towards a state of equilibrium (chapter 15).

It is a fundamental property of all gases that they will spontaneously diffuse into one another when they are brought into contact, thus any gas is completely soluble in any other.

We have already seen that the molecules of a gas move very rapidly, but it is worth recalling that the average distance travelled by any molecule between collisions is very small. Thus diffusion into a vacuum will take place much more rapidly than diffusion into another gas. The relative rates at which two gases diffuse into a third gas (or into a vacuum) depends upon the nature of the two gases concerned as the following argument shows:

Since the molecules of two gases will have the same average translational kinetic energy at the same temperature, then

$$\tfrac{1}{2}m_1c_1^{2\cdot} = \tfrac{1}{2}m_2c_2^{2}$$

where the subscripts 1 and 2 refer to the two gases. If the molecules of gas (1) have a larger mass than the molecules of gas (2), then it follows that the root mean square speed of gas (1) will be less than that of gas (2). Hence we might expect an inverse relationship between relative masses of gas molecules and the relative rates at which they diffuse under the same conditions. This relationship was first established experimentally by **Graham** (1829) and is summarised in the statement which has become known as **Graham's law: the rate of diffusion of a gas is inversely proportional to the square root of its relative molecular mass**.

The kinetic theory confirms this statement; hence for any ideal gas we know that

$$pV = \tfrac{1}{3}Nmc^2$$

$$\text{thus} \quad c^2 = \frac{3pV}{Nm}$$

$$\text{and} \quad c = \sqrt{\frac{3pV}{Nm}}$$

But from the ideal gas equation $pV = nRT$ we have:

$$c = \sqrt{\frac{3nRT}{Nm}}$$

This last equation can be simplified by recognising that Nm/n is the total mass of the gas divided by the number of moles of gas present, i.e. it is the relative molecular mass of the gas, M_r, thus:

$$c = \sqrt{\frac{3RT}{M_r}} \tag{16}$$

Equation (16) not only demonstrates the statement of Graham's law, namely that $c \propto 1/\sqrt{M_r}$, but it enables an estimate to be made of the root mean square speed of the molecules of a gas in terms of its temperature and relative molecular mass.

It is experimentally difficult to compare the relative rates of diffusion of two gases under the same conditions of temperature and pressure, and in practice it is easier to compare the relative rates of effusion of the two gases. Effusion is the process by which a gas is allowed to escape through a fine hole in the containing vessel into a surrounding gas (fig. 10.7). If the hole in the side of the container is small enough (ideally the diameter of the hole should be small compared with the mean free path of the gas) a molecule which would otherwise strike the wall of the container will pass through the hole into the surrounding gas.

FIG. 10.7. *Effusion of a gas*

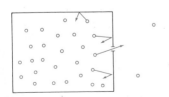

It is to be expected that the rate of effusion would be related to the number of molecules striking unit area of wall in unit time, and calculation shows that this is given by the expression $\frac{1}{4}n\bar{c}$, where n is the number of molecules per unit volume and $\bar{c}$ is the average speed of the molecules (**not** the root mean square speed, c). Comparison of experimental results with those calculated from the kinetic theory show close agreement when the pressure of the gas is low and the size of the hole, through which effusion takes place, is small. At low pressures the mean free path of the gas will be quite large and therefore the possibility of molecular collisions in the region of the hole is less than at higher pressures. Similarly, if the diameter of the hole is relatively large, then the molecules will have a greater chance of colliding with each other in passing through the hole and the basic assumptions are no longer valid. Since the rate of effusion is proportional to the mean or average speed, $\bar{c}$, which in turn is proportional to the root mean square speed, c, then Graham's law of diffusion applies equally to the phenomenon of effusion.

There are two main applications of gaseous effusion. The first is for the determination of the relative molecular mass of a gas. If two gases under the same conditions of temperature and pressure are allowed to effuse through the same hole, in separate experiments, then:

$$\frac{\text{Rate of effusion of gas A}}{\text{Rate of effusion of gas B}} = \sqrt{\frac{M_B}{M_A}} \qquad (17)$$

The experiment is generally carried out by allowing the same volume of each gas to effuse, under the same conditions of temperature and pressure, and noting the time taken in each case. Since the times taken are inversely proportional to the rates of effusion we may write:

$$\frac{\text{Time taken for certain volume of gas A to effuse}}{\text{Time taken for the same volume of gas B to effuse}} = \sqrt{\frac{M_A}{M_B}} \qquad (18)$$

If the relative molecular mass of one gas is known, then the value for the other gas may be determined.

The second application of gaseous effusion is in the physical separation of gases which are otherwise difficult to separate. The most familiar example is in the enrichment of uranium to give a higher proportion of ^{235}U to ^{238}U than is found in naturally occurring uranium. Uranium hexafluoride, UF_6, which is quite volatile is prepared and will contain $^{235}UF_6$ and $^{238}UF_6$ with relative molecular masses of 349 and 352 respectively. Although the difference in relative molecular mass is small, repeated effusion through a permeable barrier composed of very small pores leads to the eventual enrichment of the volatile fluoride $^{235}UF_6$.

An application of gaseous diffusion is in the field of medicine, where mixtures of oxygen and helium have been substituted for normal air to help relieve respiratory difficulties. Patients breathe more easily owing to the greater rate of diffusion of helium than of nitrogen in ordinary air. Another useful application of gaseous diffusion is in high vacuum techniques, through the use of so-called diffusion-pumps.

10.6
Liquefaction of gases

The ideal gas equation $pV = nRT$ predicts pressure/volume relationships at constant temperature of the type shown in fig. 10.8, a series of rectangular hyperbolae, known as the isothermals of an ideal gas. The shape of the isothermals for real gases was thoroughly investigated by Andrews (1869) using carbon dioxide, and he confirmed the hyperbolic nature of the curves at high temperatures. However, at relatively low temperatures deviations from the predicted shapes were noticed, and these are illustrated in fig. 10.9 for 2-methylbutane, $CH_3CH(CH_3)CH_2CH_3$.

FIG. 10.8. *Isothermals for an ideal gas*

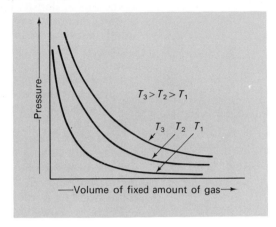

At a temperature of 553 K the isothermal corresponds approximately to that of an ideal gas. As the temperature of the gas falls to 473 K the curve becomes distorted until at the temperature 461 K the pressure/volume curve is the lowest continuous isothermal. At still lower temperatures the curves become discontinuous (so far as experimental investigation is concerned, at least) and the extent of the discontinuity increases with a decrease in temperature, as indicated by the area within the dotted-line. The isothermal for a temperature of 433 K represents such a situation.

At the point *A* the substance is in the gas phase and as the pressure is gradually increased the volume decreases approximately according to

FIG. 10.9. *Isothermals for 2-methylbutane,* $CH_3CH(CH_3)CH_2CH_3$

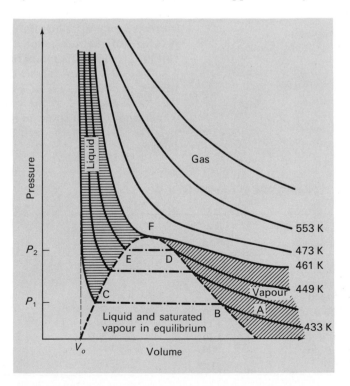

Boyle's law. At the point B the gas begins to liquefy, and there is a consequent rapid decrease in volume with very little increase in pressure as the higher density liquid phase is formed. At the point C liquefaction is complete and any further application of pressure produces only a small decrease in volume of the liquid, since liquids are much less compressible than gases. Application of even the highest pressures result in only a very small decrease in volume until the volume V_o is reached—this being the smallest volume which can be occupied by the molecules of the substance when they are theoretically in contact. All isothermals when extrapolated to high pressure tend to this point on the volume axis.

At a higher temperature, 449 K, the behaviour of the gas follows a similar path to that described for a temperature of 433 K, except that the gas begins to liquefy at the higher pressure (point D) and is complete at a point E, the volume of the liquid being larger that for the corresponding point C at a lower temperature.

There is, however, an isothermal at a temperature of 461 K which defines the highest temperature at which liquefaction can occur for the gas. This temperature is known as the **critical temperature**, T_c, for the gas and above this temperature the substance cannot exist as a liquid no matter how large a pressure that is applied to it. At the point F the corresponding pressure and volume are known as the **critical pressure**, p_c, and the **critical volume**, V_c, respectively. Strictly speaking, the critical volume is the **molar** volume at the critical point for the substance. Above the critical temperature the substance behaves more nearly like an ideal gas, the closeness to ideality increasing as the temperature increases.

We have seen that for the isothermal at 433 K, the substance is in the gaseous phase up to the point B and beyond point C is entirely in the liquid state. Therefore between the points B and C the substance consists of an equilibrium mixture of the liquid in contact with its saturated vapour. Thus the pressure p_1 represents the saturated vapour pressure of the substance at 433 K and similarly p_2 is the saturated vapour pressure at a temperature of 449 K. It is clear from fig. 10.9 that a saturated vapour cannot obey the gas laws, but that the saturated vapour pressure increases with an increase in temperature. The area of the graph enclosed by the dotted line $CEFDB$ represents the region in which liquid and vapour are in equilibrium.

Table 10.3 lists the values of the critical constants for some substances.

Table 10.3 Some critical constants

Substance	Formula	Critical press. p_c/atm	Critical vol. V_c/dm³ mol⁻¹	Critical temp. T_c/K
Helium	He	2·26	0·0578	5·21
Hydrogen	H_2	12·8	0·0650	33·2
Nitrogen	N_2	33·5	0·0901	126·3
Oxygen	O_2	50·1	0·0780	154·8
Chlorine	Cl_2	76·1	0·124	417·2
Carbon dioxide	CO_2	72·7	0·0940	304·2
Ammonia	NH_3	111	0·0725	405·5
Water	H_2O	218	0·0553	647·4

As Table 10.3 shows, the range of critical temperatures for the substances listed is quite large. It can be seen that some of the substances have critical temperatures above room temperature, i.e. above 290 K. These substances, which include chlorine, carbon dioxide and ammonia can be liquefied at room temperature simply by applying sufficient pressure. Others, like helium and hydrogen, have to be cooled below their critical temperatures (by using a liquid boiling below room temperature to cool them or by rapid expansion of the gas or by a combination of both methods).

The fact that all real gases have a critical temperature is a direct consequence of the attractions between the molecules of the substance; thus an ideal gas could not have a critical temperature. In terms of the balance between the kinetic energy of the molecules of the gas and the potential energy tending to bring the molecules together, it is clear that if the thermal energy is large enough the molecules will always escape from their neighbours no matter how great a pressure is applied to force them into closer proximity. At all temperatures above the critical temperature the thermal energy is sufficient to keep the molecules apart.

10.7 Relative molecular masses of gases and volatile liquids

The ideal gas equation $pV = nRT$ provides the basis for the determination of the relative molecular mass of a gas. Thus measurements of p, V and T for any gas will enable the number of moles of gas to be determined:

$$n = \frac{pV}{RT}$$

If the mass of the gas is m and its relative molecular mass (to be determined) is M_r, then $n = m/M_r$, hence:

$$M_r = \frac{mRT}{pV} \qquad (19)$$

Determination of the relative molecular mass of a gas

The principle of the method is simply to determine the mass of a known volume of gas at a given temperature and pressure and apply equation (19). In order to gain as much accuracy as possible, however, it is best to work at fairly low pressures. A suitable piece of apparatus is shown in fig. 10.10, and the stages involved in the experiment are given below.

(a) With the liquid air container removed, tap B closed and taps A, C, D, E, F and G open, the whole system is evacuated to give a high vacuum. Taps F and G are now closed.

(b) The gas-bulb (of known volume) is now detached and weighed on a high precision balance.

(c) The gas-bulb is now re-attached to the apparatus and taps F and G opened again to remove the small amount of air between F and G.

(d) With taps A, D, E and F closed and tap C open, some of the gas is condensed into the cold-trap, through tap B, by placing the liquid air container round it. With the liquid air container still in position, tap B is closed and tap A opened to remove any air that may have entered the apparatus.

(e) Tap A is now closed, taps C, D and F opened and some of the gas is introduced into the gas-bulb by lowering the liquid air container. Once a suitable pressure of gas is reached (see diagram) tap *C* is closed and the liquid air replaced round the cold-trap. The gas is now allowed to attain room temperature, which is recorded, and the pressure reading taken using a travelling telescope which is focussed onto the mercury menisci (as indicated in the diagram).

(f) Taps *F* and *G* are now closed, the gas-bulb detached and reweighed.

(g) The gas may be recovered after the experiment but the details need not concern us.

FIG. 10.10. *Apparatus used to determine the relative molecular mass of a gas*

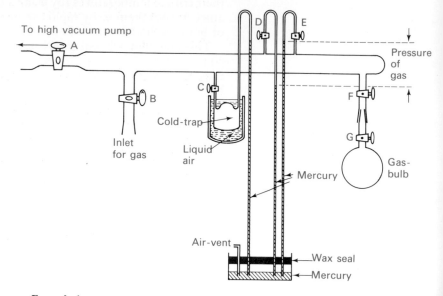

Example 1

0·0725 g of a gas has a volume of 204·6 cm³ and a pressure of 0·0726 atm at a temperature of 290 K. What is the relative molecular mass of the gas?

If we are going to use equation (19) and the gas constant $R = 8·314$ J K^{-1} mol^{-1} we must convert the pressure into N m^{-2} and the volume into m³. 1 atmosphere is 101 325 N m^{-2}, and 1 cm³ is 10^{-6} m³, thus

$$M_r = \frac{0·0725 \times 8·314 \times 290 \times 10^6}{0·0726 \times 101\ 325 \times 204·6} = 116·1$$

Relative molecular mass, M_r, = 116

Determination of the relative molecular mass of a volatile liquid

This method determines the relative molecular mass of the substance above its boiling point, i.e. as a vapour, thus there is no guarantee that this is the same as that of the liquid itself. The principle of the method is exactly similar to that discussed for a gas, and a simple apparatus which enables an approximate value to be obtained (and this is generally sufficient) is shown in fig. 10.11.

FIG. 10.11. *Apparatus to determine the relative molecular mass of a volatile liquid (as a vapour)*

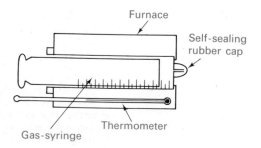

The apparatus consists of a glass syringe with the nozzle fitted with a self-sealing rubber cap. The syringe is placed in a furnace, whose temperature is above the boiling point of the volatile liquid in question. After allowing sufficient time for the syringe to reach thermal equilibrium, the volume of any air remaining in it is read. A much smaller syringe, fitted with a hypodermic needle, is weighed and reweighed containing a small amount of the volatile liquid. Some of this liquid is now injected into the larger gas syringe where it vaporises and pushes back the piston. The volume of vapour produced and its temperature are recorded and the small glass syringe is reweighed so that the mass of injected liquid may be determined. The pressure of the vapour is simply the atmospheric pressure which can be read off a barometer.

The calculation of the relative molecular mass is exactly similar to that given in the example above.

Relative molecular masses by the method of limiting densities

Relative molecular masses obtained by applying equation (19) (p. 149) are only approximate, since the quantity pV varies with pressure (at a fixed temperature) for all gases. However, a gas approaches ideal behaviour as the pressure is reduced, and the dilemma is resolved by determining a series of pV values for the gas at 273 K as the pressure is reduced; the pV values are plotted graphically against p and extrapolated to zero pressure. Suppose the extrapolated value of pV is denoted by p_0V_0 and its value at s.t.p. is p_1V_1, then

Relative molecular mass of gas at 1 atmosphere, $M'_r = \dfrac{mRT}{p_1V_1}$

Accurate relative molecular mass of gas at zero pressure, $M_r = \dfrac{mRT}{p_0V_0}$

$$M_r = M'_r \times \frac{p_1V_1}{p_0V_0}$$

or since $M_r/22{\cdot}414$ and $M'_r/22{\cdot}414$ are respectively the limiting density and the normal density of the gas (g dm^{-3} at 273 K):

$$\text{Limiting density} = \text{Normal density} \times \frac{p_1V_1}{p_0V_0}$$

Example 2
The normal density of hydrogen chloride is 1·639 15 g dm^{-3}. The values p_1V_1 and p_0V_0 (found by extrapolation of a $pV - p$ graph) are respectively 548 03 and 552 13. Calculate the relative molecular mass of hydrogen chloride.

$$\text{Limiting density} = \frac{1{\cdot}639\ 15 \times 548\ 03}{552\ 13}\ \text{g dm}^{-3}$$

Relative molecular mass of hydrogen chloride
$$= \text{limiting density} \times 22{\cdot}414$$
$$= \frac{1{\cdot}639\ 15 \times 548\ 03 \times 22{\cdot}414}{552\ 13} = 36{\cdot}467$$

Variable relative molecular masses

In some instances the variation in the values of relative molecular masses for a given substance under different experimental conditions is too great to be accounted for in terms of deviations from an ideal gas. For example, relative molecular mass determinations on ethanoic acid vapour (carried out at reasonably high temperatures since its b.p. is 391 K) give values ranging from about 60 to 120. The value tends to about 60 as the pressure is reduced and the temperature increased, while a value approaching 120 is found for higher pressures and lower temperatures. This behaviour is taken as evidence that ethanoic acid

151

vapour can exist as the monomer, CH_3COOH, $(M_r = 60)$ and as the dimer, $(CH_3COOH)_2$, $(M_r = 120)$. At intermediate experimental conditions an equilibrium mixture of monomer and dimer exists:

$$2CH_3COOH \rightleftharpoons (CH_3COOH)_2$$

Another example of such behaviour is afforded by aluminium chloride, which can exist in the vapour phase as the dimer, Al_2Cl_6, but which breaks down into the monomer, $AlCl_3$, at high temperatures.

10.8
Questions on chapters 9 and 10

1 What is meant by an ideal gas? Discuss the ways in which the behaviour of real gases deviates from ideality, giving reasons for such deviations.
   How does the kinetic theory of gases lead to an explanation of (a) the relative rates of diffusion of gases, (b) the increased rate of reaction at higher temperatures?
   Calculate the root-mean-square velocity of oxygen molecules at 298 K.   W

2 In the kinetic theory of gases, what basic assumptions are made in the derivation of the expression $pV = \frac{1}{3}Nmc^2$ for an ideal gas?
   Use this expression (a) to derive Graham's law of diffusion, (b) to calculate the root-mean-square speed $(\sqrt{c^2})$ for argon at s.t.p. and (c) to calculate the kinetic energy for one mole of an ideal gas in terms of the gas constant, R, and the temperature, T.
   Sketch the curve showing the distribution of molecular speeds in a gas and mark on the curve the position of the *average speed*, the *most probable speed* and the *root-mean-square speed*.
   Describe how the distribution of molecular speeds changes when the temperature is increased.
   Explain how the gas equation $pV = nRT$ has been modified to take account of the behaviour of *real* gases.   O and C

3 (a) How does the pressure of a given mass of gas in a fixed volume change as the temperature is raised? How is this change interpreted in terms of the behaviour of the molecules of the gas?
   (b) At 0°C and 1 atm pressure (1 atm $= 1·013 \times 10^5$ Pa), a certain mass of gas has a volume of $100·0$ cm^3. The volume of the same mass of the same gas at the same temperature is $2·412$ cm^3 at 40 atm, and $1·170$ cm^3 at 80 atm pressure. Show that the behaviour of the gas deviates from that of the ideal gas, and explain the reasons for this.
   (c) Why does the rate of a reaction such as the decomposition of gaseous hydrogen iodide into hydrogen and iodine vapour increase rapidly with rising temperature?   O

4 The times taken for equal volumes of three gaseous elements to diffuse under identical conditions were determined. The results are shown in the following table.

Gas	oxygen	A	E
Time/s	28·3	10·0	10·0

   (a) Calculate the relative molecular mass of the gases A and E.
      A was unreactive towards common laboratory reagents, whereas E exploded when ignited with air.
   (b) Identify the gases A and E.   C (Overseas)

5 Gases exert a pressure on a container, they may be compressed, they interdiffuse and they diffuse at different rates through a membrane. How may the kinetic theory of gases be used to explain these properties?
   Explain, **concisely**, how real gases deviate in their behaviour from that expected on simple kinetic theory.
   Under comparable conditions, 200 cm^3 of oxygen diffused through a membrane in 600 seconds and 60 cm^3 of an unknown gas diffused through the same membrane in 300 seconds. Find (a) the mean molecular mass of the gas, (b) the temperature at which the gas has the same root-mean-square velocity as oxygen at 273 K, both gases being at the same pressure.   W

6 (a) (i) Explain the significance of the symbols in the equation $PV = nRT$ and state any **two** of the gas laws it summarizes.
   (ii) Calculate the value of R in appropriate units.
   (iii) Why, and how, did van der Waals modify this equation?
(b) (i) State Graham's Law of Diffusion.
   (ii) Write the molecular formulae of three hydrocarbons which diffuse at the same rate as nitrogen, nitrogen oxide (nitric oxide), dinitrogen oxide (nitrous oxide) respectively.
   (iii) Calculate the ratio of the rates of diffusion of carbon monoxide and carbon dioxide under identical conditions.
      [1 mole of a gas occupies $22{\cdot}4 \times 10^{-3}$ m^3 ($22{\cdot}4$ dm^3) at 101 kPa and 273 K (s.t.p.).]   AEB

7 Discuss briefly, but critically, the basic postulates underlying the kinetic theory of gases and explain
   (a) gas pressure,
   (b) thermal expansion at constant pressure,
   (c) the use of $\overline{c^2}$, the mean square velocity, in the equation
$$pV = \tfrac{1}{3}mn\overline{c^2}$$
   where n molecules of ideal gas, each of mass m, occupy a volume V at a pressure p.
   Calculate
   (i) the kinetic energy (in joules) of the molecules in one mole of ideal gas at 47°C;
   (ii) the *root* mean square velocity of hydrogen iodide molecules, HI, at 47°C in the gaseous phase; (in this calculation the molar mass of hydrogen iodide must be expressed in the appropriate SI unit, i.e. the *kilogram*).
   (iii) the *ratio* of the *root* mean square velocities of oxygen, O$_2$, and hydrogen iodide at 47°C;
   (iv) the time expected to be taken for a given volume of hydrogen iodide at 47°C to effuse (or diffuse) through a pin-hole if the same volume of oxygen under the same conditions takes 60 seconds.
   (v) Compare this with that predicted by Graham's Law of Diffusion, which should be stated.
$$\text{H} = 1, \text{O} = 16, \text{I} = 127; R = 8{\cdot}314 \text{ J K}^{-1} \text{ mol}^{-1};$$
$$0°\text{C} = 273 \text{ K}; \text{J} = \text{kg m}^2 \text{ s}^{-2}.   \text{S}$$

8 Give a brief account of the basic assumptions of the kinetic theory of gases. Show that the pressure exerted by a gas containing n molecules per unit volume, each of mass m, is given by
$$P = \tfrac{1}{3}mn\overline{c^2}$$
where $\overline{c^2} = \overline{u^2} + \overline{v^2} + \overline{w^2}$ and u, v, w are the components of the molecular velocity along the x, y, z directions respectively.
   In molecular oxygen the root mean square velocity, $\sqrt{c^2}$, is $6{\cdot}82 \times 10^2$ m s^{-1} at 600 K. Calculate the root mean square velocity in hydrogen gas at 300 K.
                                                        Oxford Schol and Entrance

9 Discuss the assumptions on which the kinetic theory of ideal gases is based and state the equation which governs the behaviour of such gases.
   The van der Waals' equation for one mole of a gas.
$$(P + a/V^2)(V - b) = RT   (a \text{ and } b \text{ are constants})$$
is often used to describe the behaviour of real gases. Suggest reasons for the terms a/V^2 and b in this equation.
   Explain why, under the same conditions, different gases diffuse at different rates.
   A wad saturated with aqueous ammonia is placed at one end of a narrow tube 1 m long and one saturated with concentrated hydrochloric acid at the other. Describe and explain what happens.   JMB (Syllabus B)

The structures of some solids

11.1
Introduction

It is convenient to discuss the structures of solids under several major headings, as follows:

(a) **Metallic solids** in which the bonding is undirected and in which, by and large, the atoms are packed together as closely as possible.

(b) **Ionic solids** in which the bonding is again undirected and whose crystal structures are dictated largely by the relative sizes of the anions and cations concerned.

(c) **Atomic and covalent solids**, frequently called giant or macromolecular structures, in which the bonding is essentially covalent and thus directed.

(d) **Molecular solids** in which relatively small covalent molecules are bonded together by weak van der Waals' forces.

(e) **Layer structured solids**, which may be essentially ionic or covalent or, indeed, may have bonding of an intermediate character. The individual layers are generally held together by weak van der Waals' forces.

11.2
The structures of metals

FIG. 11.1. *The close packing of metal atoms*

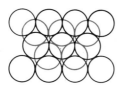

The manner in which the atoms of a metal are packed together in the crystal can be determined by the technique of X-ray diffraction (chapter 13). For most metals it is found that the atoms pack together as closely as possible in much the same kind of way that ping-pong balls would, if they were allowed to pack together as efficiently as possible. Thus spheres of equal radius can be packed so that one sphere is in contact with six others. A second layer can be built up so that each sphere in this layer is in contact with three spheres in the first layer. This second layer is shown in a different colour in fig. 11.1. A third layer can be constructed in one of two ways. If the spheres in this layer are arranged to be directly over those in the first layer, the resulting structure is said to

FIG. 11.2. *Hexagonal close packing; (a) space filling model (b) exploded view (c) unit cell*

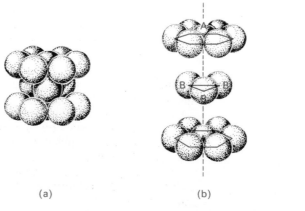

(a)　　　　　　(b)　　　　　　(c)

be hexagonal close-packed (hcp) and the basic structure is repeated every two layers (ABABABAB . . .). If, however, the spheres in the third layer are arranged in an alternative manner, the arrangement is said to have a cubic close-packed (ccp) structure (or face-centred cubic) if the basic unit is repeated every three layers (ABCABCABC . . .). These two basic close-packed structures are illustrated in figs. 11.2 and 11.3; both are common structures adopted by metals.

FIG. 11.3. *Face-centred cubic packing (cubic close packing); (a) space filling model (b) exploded view (c) unit cell*

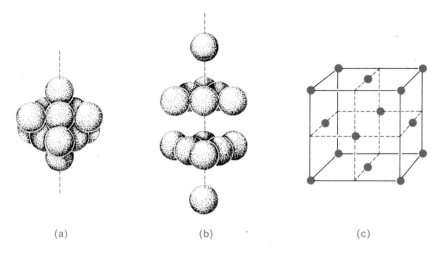

(a)   (b)   (c)

FIG. 11.4. *Body-centred cubic packing (a) space filling model (b) unit cell*

(a)

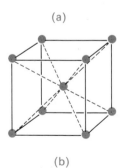

(b)

A third type of structure is adopted by some metals but it is not as close-packed as the two described above. It is called body-centred cubic packing (bcc) but instead of each atom having twelve nearest neighbours, i.e. a co-ordination number of 12, there are only eight with another six only slightly further distant; it is illustrated in fig. 11.4.

The unit cell

It is clearly advantageous to have a way of representing the internal structure of crystalline material in as simple a manner as possible. This is done in terms of the **unit cell** which is the **smallest portion of the crystal which contains all the fundamentals without repetition**. There are three kinds of unit cell based on the cube and these are:

(a) A simple cube in which atoms are placed one at each corner.

(b) A face-centred cube in which there is an atom at each corner and one in the centre of each face of the cube. This is shown in fig. 11.3(c).

(c) A body-centred cube in which there is an atom at each corner and another at the centre of the cube. This is shown in fig. 11.4(b).

The unit cell in hexagonal close-packing is not as simple as the three above and is shown in fig. 11.2(c).

If a particular unit cell is repeated in three dimensions then the pattern produced reproduces the arrangement of the atoms in the structure. This means that each atom at the corners of a face-centred cube is shared by eight such unit cells, while the atoms at the centre of each face are shared by two. Thus the number of atoms 'belonging' to a particular unit cell is:

155

$$8 \times \tfrac{1}{8} + 6 \times \tfrac{1}{2} = 4$$

In the case of a body-centred cube the relevant number of atoms in the unit cell is:

$$8 \times \tfrac{1}{8} + 1 = 2$$

The volume occupied by atoms in some unit cells

FIG. 11.5. *One face of a face-centred cube*

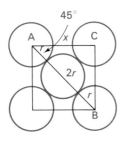

We have already said that in terms of packing as many atoms together in as small a volume as possible hexagonal close-packing and cubic close-packing are more efficient than any other sort of arrangement. We can easily determine the percentage of space occupied by atoms in a cubic close-packed structure (face-centred cubic) and a body-centred cubic structure as follows.

Fig. 11.5 represents one face of a face-centred cube. Along the diagonal AB the atoms, of radius r, are in contact thus this diagonal distance is $4r$. Since we have seen that 4 atoms 'belong' to this type of unit cell, the volume of the unit cell occupied by atoms will be:

$$4 \times \text{volume of one atom} = 4 \times \frac{4}{3}\pi r^3 = \frac{16\pi r^3}{3}$$

If we let x represent the length of the cube edge, then the volume of the cube is x^3, and

$$\frac{AC}{AB} = \frac{x}{4r} = \cos 45° = \frac{1}{\sqrt{2}}$$

$$\text{or} \quad x = \frac{4r}{\sqrt{2}} = 2\sqrt{2}r$$

$$\text{volume of cube} = x^3 = (2\sqrt{2}r)^3 = 16\sqrt{2}r^3$$

Hence

$$\frac{\text{volume occupied by atoms}}{\text{volume of the cube}} = \frac{16\pi r^3}{3} \Big/ 16\sqrt{2}r^3$$

$$= \frac{\pi}{3\sqrt{2}} = 0{\cdot}74 \text{ or } 74\%$$

In this close-packed structure 74% of the space is occupied by atoms and the result is exactly the same for hexagonal close-packing.

Fig. 11.6 represents part of the unit cell for a body-centred cubic structure. Along the diagonal AB the atoms are in contact and this diagonal distance is $4r$. Since we have already seen that 2 atoms 'belong' to this unit cell, the volume of the unit cell occupied by atoms will be:

$$2 \times \text{volume of one atom} = 2 \times \frac{4}{3}\pi r^3 = \frac{8\pi r^3}{3}$$

FIG. 11.6. *Part of a body-centred cube*

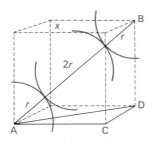

If we let x represent the length of the cube edge, then the volume of the cube is again x^3. From fig. 11.6 it will be seen that

$$AD^2 = AC^2 + CD^2 = x^2 + x^2 = 2x^2$$
$$AB^2 = AD^2 + BD^2 = 2x^2 + x^2 = 3x^2$$

hence

$$AB = \sqrt{3}x$$

but $AB = 4r$, thus $\sqrt{3}x = 4r$ and $x = \dfrac{4r}{\sqrt{3}} = \dfrac{4\sqrt{3}r}{3}$

$$\text{volume of cube} = x^3 = \left(\frac{4\sqrt{3}r}{3}\right)^3 = \frac{64\sqrt{3}r^3}{9}$$

Hence

$$\frac{\text{volume occupied by atoms}}{\text{volume of the cube}} = \frac{8\pi r^3}{3} \bigg/ \frac{64\sqrt{3}r^3}{9}$$

$$= \frac{\sqrt{3}\pi}{8} = 0\cdot68 \text{ or } 68\%$$

Thus we see that this type of structure is about 6% less efficient, in terms of utilisation of space, than the two closest forms of packing.

Metallic radii

Strictly speaking, an atom does not have a definite radius since this would imply a sharp cut-off point to the electron density of the valence electrons and we know, from the probability nature of the electron, that this is not so. In addition, when two atoms 'come together' there is an overlapping of their electron clouds. However, there is a definite value that can be attached to the internuclear distance between two bonded atoms and, for a metal, half this distance is taken as the atomic radius of the atom.

If we know the density of a metal and the type of structure which it adopts, we can readily calculate the atomic radius of the metal atom. The following example illustrates the method.

Example

Copper has a relative atomic mass of 63·5 and a density of 8·94 g cm^{-3}. It is found by X-ray analysis to adopt the face-centred cubic structure. Calculate the atomic radius of the copper atom (the Avogadro constant, L, is 6·02 $\times$ 10^{23}).

For a face-centred cubic structure the unit cell contains 4 atoms, hence

$$\text{Mass of copper in one unit cell} = \frac{63\cdot5 \times 4}{6\cdot02 \times 10^{23}} \text{ g}$$

If x is the length of the cube edge, the volume of the cube is x^3 and since volume = mass/density we have

$$x^3 = \frac{63\cdot5 \times 4}{6\cdot02 \times 10^{23} \times 8\cdot94}$$

thus
$$x = \sqrt[3]{\frac{63\cdot5 \times 4}{6\cdot02 \times 10^{23} \times 8\cdot94}} = 3\cdot613 \times 10^{-8} \text{ cm}$$

We have previously shown (p. 156) that $x = 2\sqrt{2}r$, where r is the atomic radius, thus

$$r = \frac{3 \cdot 613 \times 10^{-8}}{2\sqrt{2}} = 1 \cdot 28 \times 10^{-8} \text{ cm}$$
$$= 1 \cdot 28 \times 10^{-10} \text{ m}$$
$$= 0 \cdot 128 \text{ nm}$$

The structures of some metallic elements are shown in Table 11.1.

Table 11.1 The structures of some metallic elements

Li	Be										
c	b										
Na	Mg										
c	b										
K	Ca	Sc	Ti	V	Cr	Mn	Fe	Co	Ni	Cu	Zn
c	a,b	a,b	b,c	c	c	—	a,c	a,b	a,b	a	b
Rb	Sr	Y	Zr	Nb	Mo	Tc	Ru	Rh	Pd	Ag	Cd
c	a	b	b,c	c	b,c	b	a,b	a	a	a	b
Cs	Ba	La	Hf	Ta	W	Re	Os	Ir	Pt	Au	Hg
c	c	a,b	b,c	c	c	b	a,b	a	a	a	—

a = cubic close-packed (face-centred cubic)
b = hexagonal close-packed
c = body-centred cubic

Manganese has a distorted structure and is not included. It will be noticed that a number of metals can adopt two different structures; for example, iron has a body-centred cubic structure below a temperature of about 910°C, but above this temperature it becomes more compact in changing to a cubic close-packed structure (face-centred cubic).

The band theory of metallic bonding

In chapter 6 the bonding in metals was described in terms of positive ions held together by mobile valence electrons. This very simple picture is able to explain some of the characteristic properties of metals (p. 75). Having developed the simple qualitative ideas involved in molecular orbital theory in chapter 8 it is relevant, at this stage, to apply these ideas to the problem of bonding in metals. Consider the bringing together of two sodium atoms; at a sufficiently close distance of separation their 3s atomic orbitals begin to overlap and, in molecular orbital language, a bonding and an anti-bonding molecular orbital are constructed. With three sodium atoms being brought up together there will be three molecular orbitals, one bonding, one non-bonding and one antibonding. In general for N atoms there will be N molecular orbitals, the energy separation between adjacent molecular orbitals decreasing as the value of N increases.

FIG. 11.7. *The formation of a band by the combination of N atomic orbitals*

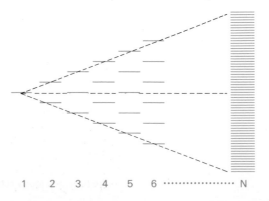

In a piece of metal, N will be exceedingly large and the consequence of this is that a series of closely spaced energy levels is available, which can be described in terms of an almost continuous band (fig. 11.7). A band constructed from s-orbitals is called simply an s-band.

By the Pauli exclusion principle we are allowed to place two electrons, but no more than two, in a molecular orbital; hence in the case of sodium these N electrons will occupy the lower $N/2$ molecular orbitals leaving the rest vacant.

Since metals have p-orbitals they may also be combined in a similar fashion to give a similar system of closely spaced energy levels called the p-band, which may or may not overlap partially with an s-band. The two possible situations are shown in fig. 11.8.

From the point of view of this general discussion it matters little whether the alkali metals have overlapping s- and p-bands since the s-band is only half filled. On the application of a potential difference, electrons in the occupied levels are easily moved into some of the unoccupied levels and the electrons are able to move in one direction. Thus we can account for electrical conductivity. Overlapping of bands is, however, important for the Group 2A metals since their atoms have two 's' electrons and hence the s-band is completely filled. However, the presence of an overlapping p-band makes energy levels easily accessible and these elements are electrical conductors.

FIG. 11.8. *The energy levels in s- and p-bands. In (a) the two bands are quite distinct but in (b) partial overlap occurs*

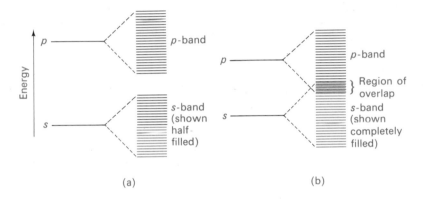

In chapter 6 we explained the opaque, shiny and silvery appearance of metals in terms of the absorption of a wide range of wavelengths of visible light, and the re-emission of this light as the excited electrons dropped back to their original energy levels. The net effect is total reflection of light and, for this to be possible, it is necessary to postulate the existence of an almost continuous band of energy levels in a metal. We now have an explanation of these extremely closely spaced levels.

11.3
The structures of ionic solids

A crystal of sodium chloride consists of a symmetrical arrangement of sodium and chloride ions, each sodium ion being surrounded by six chloride ions and vice-versa. The interionic distance (the distance from the centre of one ion to the centre of its nearest neighbour of opposite charge) was found to be 0·281 nm for sodium chloride. Other alkali metal halides adopt the sodium chloride structure (p. 164) but with differing interionic distances, e.g.

K^+Cl^- 0·314 nm	Na^+Cl^- 0·281 nm	Difference = 0·033 nm
K^+F^- 0·266 nm	Na^+F^- 0·231 nm	Difference = 0·035 nm
Difference 0·048 nm	0·050 nm	

The almost constant differences are explained if we assume that each ion acts as a sphere of constant radius, the measured interionic distance being represented by two spheres that are just in contact. Before proceeding further, we ought to make it clear that the actual structure and the interionic distance can be established quite conclusively for an ionic compound by X-ray diffraction (p. 173). However, the fact that no ionic compound is completely 'ionic', i.e. they all possess some covalent character in small but varying degrees, means that there will be some merging of the electron clouds of the ions. Furthermore, the sudden cut-off point to the electron density of an ion, which a definite ionic radius implies, is at variance with the probability nature of the electron. Despite these drawbacks, it is still useful to think in terms of a particular ion behaving like a hard sphere, and thus being possessed of a definite radius.

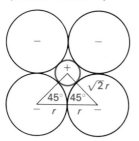

FIG. 11.9. *Calculation of ionic radii (Landé's method)*

Determination of ionic radii

The determination of the radius of an ion involves deciding how much of the interionic distance between two ions of opposite charge should be allocated to each of the two ions concerned. Several methods of approach have been used, one of the earliest and most readily understood being that introduced by Landé (1920).

Landé's method may be illustrated by the following example. Lithium iodide clearly contains a small cation and a very much larger anion, and Landé reasoned that in the crystal of lithium iodide the small lithium ion would not prevent the very much larger iodide ions from coming into contact with each other (see fig. 11.9). Since the interionic distance in lithium iodide is 0·302 nm, an ionic radius of 0·213 nm for the I^- ion ($\sqrt{2}r = 0\cdot302$ nm or $r = 0\cdot213$ nm) is now easily deduced by simple geometry (fig. 11.9).

Once the ionic radius of one ion has been fixed, it becomes possible to determine the values for other ions. For example, the interionic distance in sodium iodide is 0·323 nm so the ionic radius of the sodium ion is $(0\cdot323 - 0\cdot213) = 0\cdot110$ nm.

Pauling used more refined but more complex methods to calculate ionic radii from interionic distances. Thus for alkali halide compounds containing isoelectronic ions (ions with the same number of electrons), e.g. NaF, KCl and RbBr, he computed values for the effective nuclear charges of the isoelectronic ions, i.e. he corrected for the screening of the nucleus by the electrons. He now assumed that the interionic distance could be divided in the inverse ratio of these computed nuclear charges, and so arrived at values for the ionic radii of some alkali metal cations and some halide ions. Other ionic radii were now derived from these, and it is the Pauling crystal radii which are used in this book.

The arrangement of ions in crystals

Ionic compounds do not adopt the hexagonal close-packed and the cubic close-packed structures which are a feature of many metals, and hence do not have co-ordination numbers as high as twelve. However, the crystal structures adopted by many simple ionic compounds are related to the close-packed structures of metals and it is worth pursuing this point in more detail.

Consider a close-packed system of equal spheres with another close-packed arrangement stacked on top (as shown in two dimensions in fig. 11.10). These two layers of spheres enclose holes of two distinct types. Three spheres in the lower layer in contact with each other will have their centres at the corners of an equilateral triangle. A sphere in the second layer placed on top, and in the depression, will form a tetrahedral arrangement with the three lower spheres, i.e. their centres will be at the corners of a regular tetrahedron (see figs. 11.10 and 11.11).

160

FIG. 11.10. *The tetrahedral and octahedral holes in a close-packed structure*

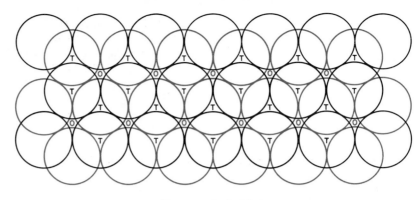

T = tetrahedral hole
O = octahedral hole

FIG. 11.11. *Generation of a tetrahedral hole*

FIG. 11.12. *Generation of an octahedral hole*

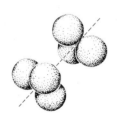

Each sphere in a close-packed system will give rise to two tetrahedral holes, and not one as shown in fig. 11.10, since each layer of close-packed spheres is sandwiched between two other close-packed layers and not one as drawn in the fig. 11.11. The second type of hole is generated by six enclosing spheres, three from each of the two layers and are marked in fig. 11.10, where it can be seen that they are half as numerous as the tetrahedral holes, i.e. one sphere gives rise to one octahedral hole. They are not so easy to visualise as the tetrahedral holes but reference to fig. 11.12 should help. Octahedral holes are larger than tetrahedral holes as would be expected, since the former are enclosed by six spheres whereas the latter are only enclosed by four.

Since an ionic bond is non-directional, an ionic solid will consist of a symmetrical array of positive and negative ions. As many positive ions surround the negative ion and vice-versa, as is consistent with
(a) maintaining electrical neutrality,
(b) ensuring a maximum lattice energy (p. 69).
In the case of sodium chloride, which has a co-ordination number of 6, the structure can be visualised in terms of a cubic close-packed array of the larger chloride ions with the smaller sodium ions occupying the octahedral holes. In practice the basic close-packed structure of the chloride ions is expanded to accommodate the sodium ions, so that the chloride ions are not in contact with each other and the lattice energy is maximised. Plate 5 (p. 70) shows the structure which can be described as two expanded interpenetrating cubic close-packed (or face-centred) lattices of sodium ions and chloride ions.

There are two basic structures for zinc sulphide, ZnS. It does not adopt the sodium chloride structure since the much smaller zinc ion cannot surround itself with six larger sulphide ions without causing a large measure of repulsion between these ions. One structure (**zinc blende**) consists of an expanded cubic-close packed arrangement of sulphide ions with zinc ions occupying half the tetrahedral holes. The other form of the compound is known as **wurtzite**, which consists of an expanded hexagonal close-packed arrangement of sulphide ions with the zinc ions once again occupying half the tetrahedral holes. Only half the tetrahedral holes are occupied in each case, since this is consistent with the empirical formula ZnS. The two structures are shown in fig. 11.13(a) and (b), where it is clear that the co-ordination number of each ion is four.

161

FIG. 11.13. *(a) Structure of zinc blende (b) Structure of wurtzite*

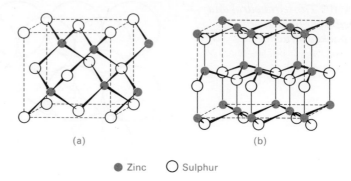

(a)   (b)

● Zinc   ○ Sulphur

Caesium chloride, CsCl, has a co-ordination number of eight and is thus not based on either the expanded cubic-close packed or the hexagonal close-packed arrangements. It is a body-centred cubic structure which is the arrangement adopted by some metals (p. 158). Its higher co-ordination number than sodium chloride is consistent with the principle of maximum co-ordination number where this works in favour of a maximum lattice energy. This arrangement is possible since the caesium ion is appreciably larger than the sodium ion. The structure of caesium chloride is shown in fig. 11.14, where it is compared with that of sodium chloride.

FIG. 11.14. *The structure of (a) sodium chloride (b) caesium chloride*

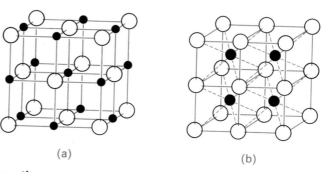

(a)   (b)

Radius ratio

Let us consider the sodium chloride structure where each ion has a co-ordination number of six; by considering the situation depicted in fig. 11.15, where the size of the anion is gradually increased, it is possible to determine the critical ratio (radius of cation/radius of anion), or the **radius ratio** as it is called, when a lower co-ordination number must be adopted. The critical situation is shown in fig. 11.15(b) where an anion is just in contact with other anions and with a cation. By simple geometry it can be seen that:

$$ab = cb \cos 45°$$

or

$$\text{radius of anion} = \frac{\text{sum of radii of cation and anion}}{\sqrt{2}}$$

or

$$\frac{\text{radius of cation}}{\text{radius of anion}} = \frac{r_+}{r_-} = \sqrt{2} - 1 = 0.414$$

If the radius ratio is less than 0.414 a binary ionic compound must adopt a structure with a co-ordination number less than six.

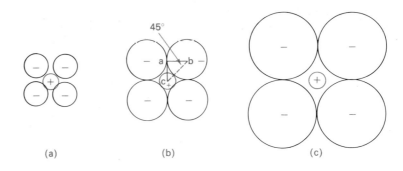

(a) (b) (c)

Let us now calculate the critical radius ratio for a compound of the type A^+B^- which assumes a co-ordination number of four, i.e. either the zinc blende or the wurtzite structures (see fig. 11.13). In fig. 11.16 the cation is situated at the centre, D, of the cube with four anions arranged tetrahedrally around it at points A, B, E and F. The critical situation occurs when all four anions are in contact with the cation and are in contact with each other.

FIG. 11.16. *The limiting condition for 4:4 co-ordination. The cation is at point D and is surrounded by four anions at A, B, E and F*

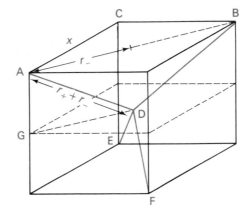

Two anions are in contact along AB and half this distance is equal to the radius of the anion, r_-. An anion and a cation are in contact along AD and this distance is equal to the sum of the radii of a cation and an anion, $r_+ + r_-$. If the length of a cube edge is x, we have

$$AB^2 = AC^2 + BC^2$$
$$= x^2 + x^2 = 2x^2$$

thus $AB = \sqrt{2}x = 2r_-$

$$r_- = \frac{AB}{2} = \frac{\sqrt{2}x}{2}$$

Now $AD = r_+ + r_-$

From fig. 11.16 it can be seen that:

$$AG = \frac{x}{2} \text{ and } GD = \frac{\sqrt{2}x}{2} \text{ (radius } r_-)$$

but $AD^2 = AG^2 + GD^2 = \frac{x^2}{4} + \frac{2x^2}{4} = \frac{3x^2}{4}$

$$AD = r_+ + r_- = \frac{\sqrt{3}x}{2}$$

hence $\dfrac{r_-}{r_+ + r_-} = \dfrac{\sqrt{2}x}{2} \Big/ \dfrac{\sqrt{3}x}{2} = \dfrac{\sqrt{2}x}{2} \cdot \dfrac{2}{\sqrt{3}x} = \dfrac{\sqrt{2}}{\sqrt{3}}$

163

thus
$$\sqrt{3}r_- = \sqrt{2}r_+ + \sqrt{2}r_-$$
$$r_-(\sqrt{3} - \sqrt{2}) = \sqrt{2}r_+$$
$$\frac{r_+}{r_-} = \frac{\sqrt{3} - \sqrt{2}}{\sqrt{2}} = \frac{1 \cdot 732 - 1 \cdot 414}{1 \cdot 414} = 0 \cdot 225$$

A compound of the type A^+B^- will have a co-ordination number of four if the radius ratio is between 0·225 and 0·414. If the radius ratio is less than 0·225 it must assume a lower co-ordination number.

We now complete this section by finding the critical radius ratio for a compound of the type A^+B^- to have a co-ordination number of eight.

FIG. 11.17. *The limiting condition for 8:8 co-ordination. The cation is at the centre of the cube and is surrounded by eight anions whose centres are at the apices of the cube*

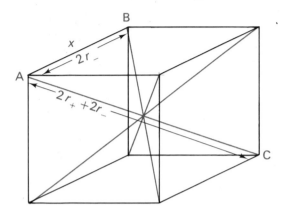

In fig. 11.17 a cation is at the centre of the cube and is surrounded by eight anions situated at the apices of the cube. The critical situation arises when all eight anions are in contact with the central cation, and two anions are in contact with each other along each cube edge. If x is the length of the cube edge, we have

and
$$AB = x = 2r_-$$
$$AC = \sqrt{3}x = 2r_+ + 2r_-$$

(the condition $AC = \sqrt{3}x$ has been derived on p. 157)

hence
$$\frac{r_-}{r_+ + r_-} = \frac{x}{\sqrt{3}x} = \frac{1}{\sqrt{3}}$$
$$\sqrt{3}r_- = r_+ + r_-$$
$$r_-(\sqrt{3} - 1) = r_+$$
$$\frac{r_+}{r_-} = \sqrt{3} - 1 = 1 \cdot 732 - 1 = 0 \cdot 732$$

We can now say that a compound A^+B^- will have a co-ordination number of eight if the radius ratio exceeds 0·732 (upper limit not specified) but this reduces to a co-ordination number of six if the radius ratio is in the range 0·414 to 0·732. Table 11.2 lists some simple compounds having co-ordination numbers of six and eight. The radius ratio criterion is a good approximation to the truth in many cases but it has some

Table 11.2 Some simple ionic structures

Type of structure	Examples	Radius ratio	Co-ordination number
Caesium chloride	Cs^+Cl^-, Cs^+Br^- Cs^+I^-	>0·732	8
Sodium chloride	Na^+Cl^-, Na^+Br^- K^+Cl^-, K^+Br^-	<0·732 >0·414	6

failures; this is not surprising, since the simple geometrical arguments we have used have been based on the assumption that ions can be treated like hard spheres with definite ionic radii. Exceptions to the simple rules are also expected when a compound has appreciable covalent character.

For a compound such as calcium fluoride, $Ca^{2+}(F^-)_2$, it is clearly impossible for both ions to have the same co-ordination number; the co-ordination number of the cation must be twice that of the anion in order to maintain electrical neutrality, in this case eight and four respectively. Thus in calcium fluoride each Ca^{2+} ion is surrounded octahedrally by eight F^- ions, and each F^- ion is surrounded tetrahedrally by four Ca^{2+} ions.

11.4
The structures of some giant covalent solids

FIG. 11.18. *Part of the structure of diamond, showing the tetrahedral arrangement of bonds*

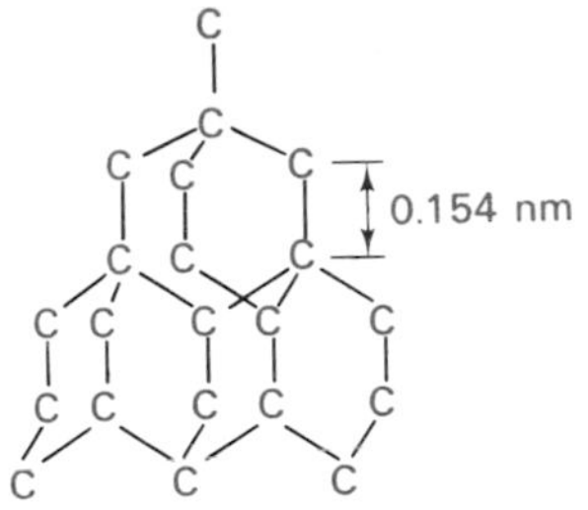

FIG. 11.19. *The structure of silicon dioxide represented in two dimensions*

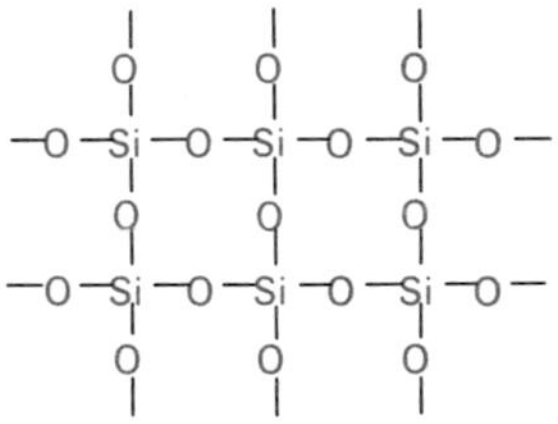

Typical three-dimensional covalent structures include those of diamond, silicon carbide, boron nitride and silicon dioxide.

The diamond allotrope of carbon contains a three-dimensional array of carbon atoms with the bonds directed towards the apices of a regular tetrahedron (electron-pair repulsion (p. 84) or alternatively sp^3 hybridisation (p. 108). The whole structure is one giant molecule and, because the bonding is strong and extended in three dimensions, diamond is exceptionally hard (see fig. 11.18 for the structure of diamond).

Silicon carbide, $(SiC)_n$, is nearly as hard as diamond itself and has the same type of structure with silicon and carbon atoms occupying alternate positions in the crystal lattice. One form of boron nitride, $(BN)_n$, adopts the diamond structure and this substance is also extremely hard. Since one atom of boron and one atom of nitrogen are isoelectronic with two atoms of carbon, the number of electrons on average per atom is the same as that in diamond.

Silicon dioxide is also a three-dimensional giant molecule and one form of the compound (cristobalite) has a structure in some ways similar to that of diamond. Each silicon atom is surrounded tetrahedrally by four oxygen atoms, each of these oxygen atoms being linked to other silicon atoms, i.e. each oxygen atom is shared equally between two silicon atoms giving the empirical formula SiO_2. A two dimensional representation of this structure is shown in fig. 11.19.

11.5
The structures of some molecular solids

FIG. 11.20. *The molecular unit, P_4, in white phosphorus*

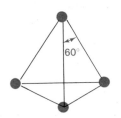

In these structures the basic units are small covalent molecules, except in the case of the solidified noble gases where they are single atoms, with weak van der Waals' forces (p. 93) providing the bonding between atoms in different molecules. The structures of molecular crystals are determined by the shape of the basic molecular units and they tend to pack together as efficiently as possible, since van der Waals' forces are undirected. Thus the solidified noble gases have a cubic-packed structure and therefore a co-ordination number of twelve like many metals, and the structure of iodine (p. 93) is very similar.

Black phosphorus is macromolecular but the reactive allotrope, white phosphorus, contains discrete P_4 units (fig. 11.20) held together by weak van der Waals' forces. The molecular unit in sulphur is a puckered ring containing eight sulphur atoms linked together by single covalent bonds (fig. 11.21). These basic units are preserved intact in both the rhombic and monoclinic allotropes but the actual crystal structures, which are clearly different in the two allotropes, are quite complex.

165

FIG. 11.21. *The molecular unit, S_8, present in both rhombic and monoclinic sulphur*

11.6
Layer structured solids

Molecular solids have low melting points, since only weak forces have to be overcome, unlike the very much higher melting-points of structures like diamond. The liquid state of these compounds contains the simple molecules, thus liquid sulphur just above its melting-point contains S_8 molecules, although the breaking of covalent bonds may occur as the temperature is increased.

A typical example of a layer structure is provided by graphite. It contains layers of carbon atoms, each carbon atom in a particular layer being covalently bonded to three others giving C—C—C bond angles of 120°, i.e. the carbon atom is sp^2 hybridised. Each carbon atom has a singly occupied p-orbital directed at right angles to the plane of the layer and these combine to form molecular orbitals, each layer being sandwiched between a double cloud of negative charge, i.e. the structure of one layer is rather like an extended version of that found in the benzene molecule (p. 113). The individual layers are held together by van der Waals' forces, hence the slippery nature of the substance is adequately explained. Graphite has appreciable electrical conductivity within each layer (but not at right angles to them) and this finds a ready explanation in terms of an extensive delocalisation of electrons within a layer.

It is interesting to note that boron nitride also exists in a form resembling the graphite structure; however, it differs from graphite in having no 'metallic lustre' and in being a semiconductor of electricity only. Its structure involves co-ordinate bonds from nitrogen to boron. Parts of the structures of graphite and boron nitride are shown in fig. 11.22.

FIG. 11.22 *Parts of the structures of graphite and boron nitride*

0.142 nm

0.335 nm

(a)

(b)

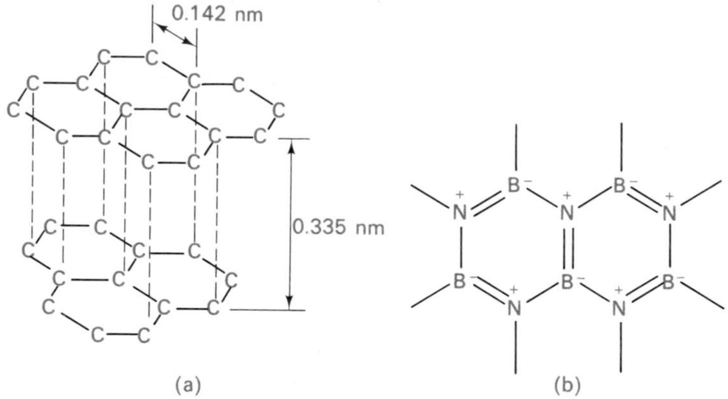

A number of chlorides, bromides, iodides and sulphides of the less electropositive metals have layer structures, a typical example being cadmium iodide, CdI_2. In this compound each cadmium atom, in a particular layer, is surrounded octahedrally by six iodine atoms and each iodine atom has three cadmium atoms as nearest neighbours. The bonding within a particular layer is intermediate in character being neither predominantly ionic nor covalent; weak van der Waals' forces hold the layers loosely in position and once again this substance has a greasy texture. Part of the layer structure of cadmium iodide is shown in fig. 11.23.

FIG. 11.23. *Part of the layer structure of cadmium iodide*

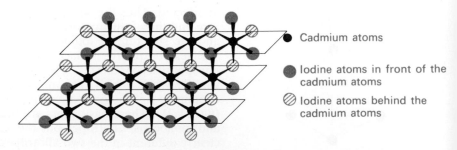

● Cadmium atoms

● Iodine atoms in front of the cadmium atoms

⊘ Iodine atoms behind the cadmium atoms

Some one-dimensional structures

Related to the layer structures are some one-dimensional structures in which the atoms are strongly bonded together to give extensive chains, which are again loosely held together by van der Waals' forces. Typical examples are silicon disulphide, $(SiS_2)_n$ and one form of sulphur trioxide. The chain-like structures of these compounds (fig. 11.24) explain why they crystallise in the form of long silky needles.

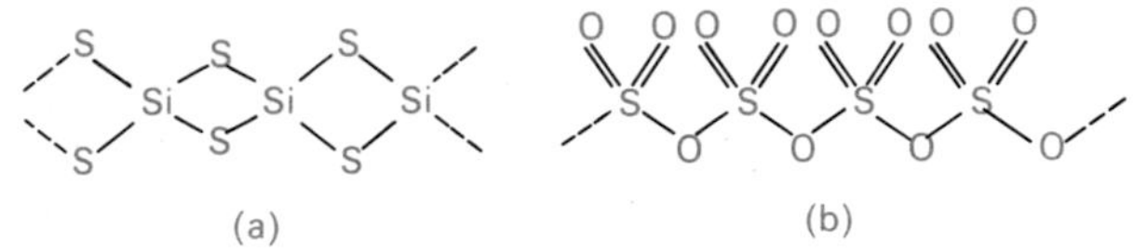

FIG. 11.24. *The structure of (a) silicon disulphide and (b) sulphur trioxide*

(a)

(b)

11.7
Questions on chapter 11

1 Metals crystallize into three main structures: hexagonal close packed, face-centred cubic, and body-centred cubic. Describe these three structures in detail, pointing out the differences between them.

Use a Book of Data to investigate the relationship between the crystal structure of a metal and its position in the Periodic Table. Comment on your findings. **N**

2 Metal lattices are *close-packed*. Explain what is meant by this term, giving examples of **two** *close-packed* structures. Give **one** way in which a metal may not have a *perfect close-packed* structure.

Describe the way in which the atoms in a metal are held together, and show how its characteristic thermal, mechanical and electrical properties can be accounted for, using this description.

Tin and lead melt at 505 K and 600 K respectively. They form a single eutectic containing 63% tin, the eutectic temperature being 455 K. Use these data to sketch, (roughly to scale), the tin-lead phase diagram, and use it to explain what happens when a molten mixture of 80% lead and 20% tin, which has a melting point of 550 K, is slowly cooled to room temperature. **W**

3 Write a brief survey of the types of bonds which exist in solids giving a specific example of each type of bond you mention.

The crystal of potassium chloride has a face-centred cubic structure. Calculate the theoretical density (in g cm^{-3}) of potassium chloride crystals. You will need a Data Book. **O and C**

4 Give a molecular description of (a) gases, (b) liquids, (c) solids.

In the case of solids, discuss, giving specific examples, the different types of bonding by which the constituent species may be held together and the influence this has upon the physical properties of the solids. **W (S)**

5 Determine the limiting value of the radius ratio when both anion and cation have a co-ordination number of 3.

6 Using the radius ratio criterion, determine the co-ordination numbers of both anions and cations in the following compounds: Cs^+Br^-, $Sr^{2+}(F^-)_2$ and $Ca^{2+}O^{2-}$. The radii of the ions (nm) are as follows:
$Cs^+ = 0.169$, $Br^- = 0.195$, $Sr^{2+} = 0.113$, $F^- = 0.136$, $Ca^{2+} = 0.097$, $O^{2-} = 0.140$.

7 (a) Explain the conditions under which an expression for the potential energy $U(r)$ of two particles (or planes of particles), separated from each other by a distance r, of the form:

$$U(r) = -\frac{A}{r^n} + \frac{B}{r^m}$$

may lead to a stable solid structure. (Assume A, B, m and n to be positive.) Sketch a typical curve of $U(r)$ for such a situation.

Indicate how the parameters A, B, m and n are related to the properties of the particles concerned in forming solids of (i) an inert gas such as neon and (ii) an ionic salt such as potassium chloride.

(b) Strontium oxide crystallises in a similar structure to that of sodium chloride.

Calculate a value for the Avogadro constant, given: Density of strontium oxide = 4·70 g cm^{-3}; Atomic masses Sr = 87·6, O = 16·0; dimension of a side of the unit cell = 0·52708 nm.

Oxford Schol. and Entrance (Physical Science)

8 A metal, M, crystallizes in a face centred cubic lattice with a unit cell length of 0·352 nm (3·52 Å). The metal which has a density of 8·94 g cm^{-3} is resistant to attack by caustic alkalis and liquid HF but is attacked by dilute acids to give a green solution, X. A green gelatinous precipitate, N, forms when an excess of KOH solution is added to X. N does not redissolve in excess alkali but does in excess ammonia solution to form a blue solution, Y. Addition of dimethylglyoxime (butanedione dioxime) to a solution of X which has been neutralized gives a bright red precipitate, O. Calculate the relative atomic mass of M. Identify the metal M, the solids N and O, and the solutions X and Y. Write down equations for the reactions that have been described above. Calculate the radius of the metal atom.

Oxford Schol. and Entrance (Physical Science)

The liquid state

12.1 Introduction

In discussing the kinetic theory of gases it was assumed that the interactions between the individual molecules were negligible by comparison with their translational kinetic energies. This led to some useful results which explained many of the bulk properties of gases in a satisfactory manner. Only when the discussion turned to consider the effect of increasing pressure and lower temperatures and, in particular, the liquefaction of gases was the treatment modified to take into account the finite size of the molecules and their interactions with one another. At the other extreme, the solid state was seen as an orderly arrangement of particles packed together into a crystal structure. The only motion available to the particles in a solid is a vibratory one about fixed lattice positions. Molecules in a liquid, on the other hand, are free to rotate as well as vibrate and, in addition, are able to move from point to point, albeit in a much more restricted manner than they are in a gas. Molecules in a liquid then are neither completely pinned down as they are in a solid nor do they enjoy anything like the complete freedom of action as in a gas. It is thus not very surprising that a comprehensive theory of the liquid state has yet to emerge, and we shall do no more than offer a few comments.

12.2 Some comments on the liquid state

It is possible to form a simple, if incomplete, mental picture of the liquid state by considering three pieces of evidence.

Volume changes associated with phase changes

Pure solids melt at a sharp temperature and at this point there is generally a sudden increase in volume in the order of about 10% (water is exceptional in this respect, having a smaller volume than the same mass of ice). This implies that a solid retains its ordered structure intact up to the instant of melting, but the rather small increase in volume suggests that the liquid also retains a high degree of order. This idea seems to be confirmed by the fact that there is a sudden and dramatic increase in volume when a liquid boils (roughly a 100–1000-fold increase) and the highly disorganised gaseous state is formed. Once again the change of phase takes place at a sharp temperature.

Molar latent enthalpies of fusion and vaporization

The amount of heat needed to convert one mole of solid into liquid at its melting-point and at one atmosphere pressure is called the molar latent heat of fusion (or more generally the enthalpy of fusion). Similarly the amount of heat needed to convert one mole of liquid into vapour at its boiling-point and one atmosphere pressure is called the molar latent heat of vaporization (enthalpy of vaporization). Some values of these two quantities for a selection of substances are given in Table 12.1.

Table 12.1 Enthalpies of fusion and vaporization for some selected substances

Substance	Formula	Enthalpy of fusion/kJ mol^{-1}	Enthalpy of vaporization/kJ mol^{-1}
Aluminium	Al	10.7	284
Mercury	Hg	2.30	58.2
Water	H_2O	6.02	41.1
Ethanol	C_2H_5OH	4.60	43.5
Benzene	C_6H_6	9.83	30.8

It is clear from the above values that it requires much less energy to convert a solid to a liquid than it does to convert a liquid into a gas. Intuitively it seems reasonable to argue that absorption of a large amount of energy is associated with the creation of a large amount of disorder. If this is a correct interpretation of these values, then once again we have some evidence that liquids possess a considerable measure of order in the arrangement of their constituent particles (but see chapter 15 for a discussion of order/disorder in terms of entropy).

X-ray diffraction by liquids

One characteristic property of a solid is its ability to produce sharp diffraction effects with X-rays (p. 174). The sharpness of the diffraction lines produced on photographic film is indicative of an orderly arrangement of the atoms or ions in the crystal. Gases, on the other hand, do not exhibit diffraction effects with X-rays, since the arrangement of the atoms or molecules in a gas is a completely random one. Liquids have been observed to give diffraction patterns with X-rays, very like those obtained with powdered solids; but there is one important difference in that the diffraction lines are rather diffuse. Since sharpness of lines is associated with a constant spacing between the particles, then clearly there is no such thing as a constant spacing between the particles in a liquid. At some instance two particles may be as close together as they would be in the solid state and at another they may have moved appreciably apart.

A simple model of a liquid

FIG. 12.1. *Two-dimensional models of (a) a solid (b) a liquid (c) a gas*

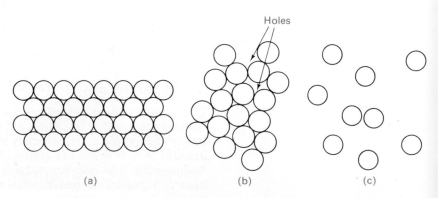

170

A two dimensional model of a liquid is compared with those of a solid and a gas in fig. 12.1. Whereas each particle in a solid is shown surrounded symmetrically by six more (in two dimensions) there are, on average, fewer nearest neighbours for any one particle in the liquid. The variable number of nearest neighbours means that in some regions of the liquid there are 'holes' which can move rapidly through the structure and thus appear in random positions in the liquid. The presence of holes precludes the possibility of a liquid having long range order like a solid; melting-points of solids are sharp because once a hole has been created it inevitably means an abrupt end to the long range order of the solid.

FIG. 12.2. *Computer simulation of the transition from solid to liquid, based on calculation of trajectories of molecules. Upper half shows solid behaviour; lower half liquid behaviour*

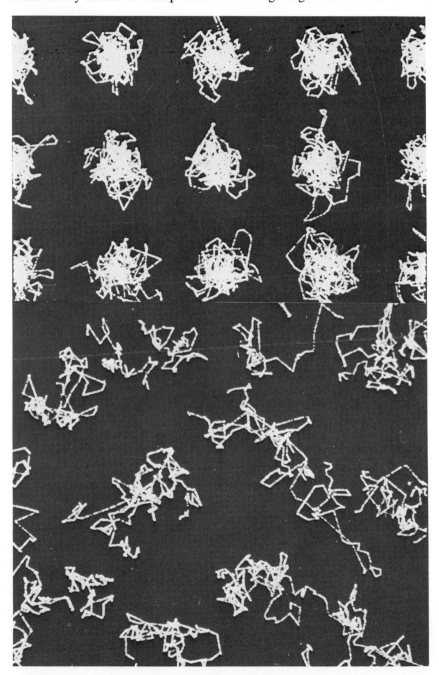

Chapter 13

Determination of structure

13.1
Introduction

It is the intention in this chapter to describe some of the methods which are in use by the chemist to determine the structures of a wide variety of substances, and thereby to present some degree of validity to the models of bonding which we have employed in previous chapters. The methods chosen are those which give some direct indication of the chemical structure of the substance under investigation, and as such represent only some of the diverse range of methods available for the analysis of materials. In each case a brief theoretical background is included, together with the essential experimental features and applications of the technique.

13.2
From chemical to physical methods

Until relatively recently the general method employed for the determination of chemical structure consisted of the preparation of the substance in as pure a condition as possible, followed by the systematic analysis of the substance according to well-defined and well-tried chemical reactions. This process would often be accompanied by the synthesis of the substance from starting materials of known structure, and a comparison would be made of the physical and chemical properties of the unknown and synthesised substances. Even after such a careful process a high degree of uncertainty could still exist about the internal structure of the substance concerned. This method is both time consuming and extravagant on materials, since large quantities are often involved. The application of physical methods to the study of chemical structure arrived with the developments in electronic and computer engineering over the past thirty years or so. As will be seen from the description of the methods which follow, the physical techniques are economic both in the amount of material which is required for analysis and also in the relatively short time necessary for a complete analysis. The material is rarely destroyed in the course of the investigation and may be recovered at the end. An additional advantage has been gained in the concurrent development of purification techniques such as zone-refining and gas-liquid chromatography (p. 365).

13.3
A summary of physical methods

Table 13.1 lists those methods which are described in this chapter, together with brief notes on the principal items of chemical information which can be obtained from each one of them. For convenience, the methods have been grouped into three main classes: diffraction methods, spectroscopic methods and others. Among the physical techniques listed that of X-ray diffraction is one of the oldest, and the method was developed as long ago as 1912. As this technique was one of the first and one which has had an enormous impact on the advance in structural chemistry it will be the first to be considered.

Table 13.1 Summary of some physical methods for investigation of chemical structure

Class	Technique	Information obtained
Diffraction methods	X-ray diffraction	Arrangement of atoms and ions in crystals, distances between atoms and ions in crystals
	Neutron diffraction	Location of hydrogen atoms in crystals
	Electron diffraction	Bond angles and bond distances in gaseous molecules
Spectroscopic methods	Visible and Ultra-violet Spectroscopy	Structure of unsaturated organic molecules, and of transition metal complexes
	Infra-red Spectroscopy	Characteristic groups in molecules and ions; bond properties
	Nuclear Magnetic Resonance Spectroscopy	Arrangement of hydrogen atoms, in particular, in molecules
Others	Mass Spectrometry	Relative atomic and molecular masses; molecular structure of organic compounds
	Magnetic properties	Electronic structures of transition metal ions

13.4
X-ray diffraction

The phenomena of interference and diffraction of light have been known to physicists for a very long time, and an explanation can readily be given in terms of the wave theory of light. There are many familiar examples of the interference and diffraction of light, amongst them

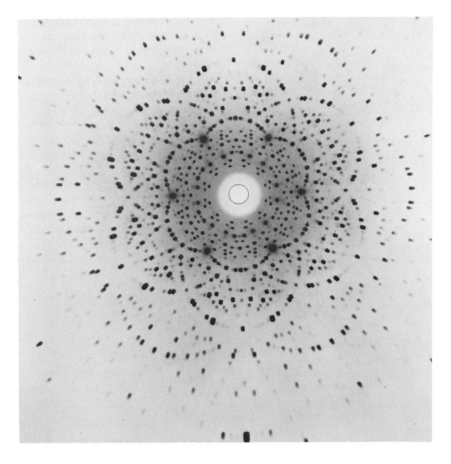

Plate 6. X-ray diffraction of beryl by the von Laue method (by courtesy of Kodak Research Laboratories)

being the variety of colours seen in thin films of oil, and the effects on the wall of a room when strong sunlight enters through a narrow slit in a door or window. Diffraction gratings depend for their effect on having transparent spaces between the ruled lines a distance apart comparable with the wavelength of light to be diffracted.

Since a crystal is composed of particles (atoms, ions or molecules) which are generally arranged in a regular pattern, the possibility arises that it may behave as a three-dimensional diffraction grating for radiation of suitable wavelength. Since the distance apart of the particles in a crystal is in the order of about 10^{-10} m, this is in the range of X-radiation.

It was von Laue in 1912 who first suggested the possibility of using a crystal to diffract a beam of X-rays, and his suggestion was taken up in the same year by two of his research students Friedrich and Knipping. They passed a beam of X-rays through a crystal of hydrated copper (II) sulphate onto a photographic plate. Development of the film resulted in a photograph composed of a dark central area corresponding to the undeviated beam, together with a pattern of spots arranged around the central area in a symmetrical manner. The surrounding spots corresponded to the diffraction of the X-ray beam in various directions. A diffraction photograph taken by a similar method (known as the von Laue method) is shown in Plate 6).

The analysis of von Laue type photographs obtained from sodium chloride, potassium chloride and zinc sulphide was soon undertaken by W. H. and W. L. Bragg, father and son, at the Royal Institution in England. In addition, they refined the experimental technique by using monochromatic X-radiation in place of the 'white radiation' (consisting of a range of wavelengths) used in the original work. At the same time they developed a spectrometer which measured the intensity of the diffracted X-ray beam by monitoring the amount of ionisation which it produced, thus isolating a characteristic X-ray spectrum and employing it for the determination of crystal structure.

The Braggs pointed out that the interaction of X-rays with atoms in a crystal could be treated in much the same manner as for the analogous reflection of light. Thus when a beam of X-rays is incident upon a crystal structure some of the radiation will pass straight through the substance, but a fraction will be reflected from the ions or atoms. The frequency and wavelength of the reflected beam are identical with those of the incident beam, and the angles of incidence and reflection are equal, by analogy with the optical situation. However, the beam of X-rays penetrates the crystal in such a manner that atoms or ions in many successive rows also become sources for the reflection of the X-rays. Now whereas all the waves reflected by a single row of atoms or ions will necessarily

FIG. 13.1. *Reflection of X-rays from two consecutive parallel planes in a crystal*

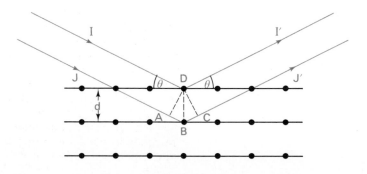

174

be in phase, only those waves reflected from successive rows will be in phase provided a single condition is fulfilled. The reflection of a beam of X-rays from consecutive planes in a crystal structure is shown in fig. 13.1.

The rays I and J are parts of the same X-ray wavefront and are incident upon consecutive planes distance d apart in the crystal, the angle of incidence being θ. The difference in path length between the two reflected rays I' and J' is given by the distance $AB + BC$, and in order that I' and J' are in phase this path difference must be an integral number of wavelengths. Thus, if reinforcement is to occur for radiation of wavelength λ, the condition

$$AB + BC = n\lambda$$

must be obeyed, where n is a whole number. From the geometry of fig. 13.1 it is clear that angles ADB and BDC are both equal to θ, and we may therefore write

$$AB = DB\sin\theta \text{ and } BC = DB\sin\theta$$
and
$$AB + BC = 2DB\sin\theta = 2d\sin\theta$$
hence
$$2d\sin\theta = n\lambda \tag{1}$$

This final relationship, known as the Bragg equation, is the condition for reinforcement to occur for X-rays reflected from parallel rows of atoms or ions spaced a distance d apart in the crystal. In principle it is possible to determine the spacing d if the angle of reflection is determined experimentally for X-rays of known wavelength as is shown in the example below.

Example
First-order reinforcement, i.e. $n = 1$, was observed for X-rays of wavelength 0·0576 nm at an angle $\theta = 6°54'$ for a single crystal of a certain solid. Calculate the distance apart of consecutive planes.
From the Bragg equation we have:

$$d = \frac{n\lambda}{2 \sin \theta} = \frac{1 \times 0·0576}{2 \sin 6°54'} = 0·236 \text{ nm}$$

There are several angles for which reinforcement occurs and these correspond to values of $n = 1, 2, 3$ etc.

In practice the task of determining the spacing between planes in a crystal is far from easy and it is a most painstaking and time-consuming one. The use of computers for the calculations involved has now been incorporated with great success. The relatively simple structure shown in two dimensions in fig. 13.2 illustrates some of the problems which are encountered. Thus there may be several sets of planes from which reflections can take place, and it is the assignment of particular planes to a given set of reflected beams that is a difficult operation for all but the most simple structures. As the intensity of a reflected beam of X-rays depends upon the distance apart of the planes of particles (and hence on the density of the population of particles in the planes), it is possible to calculate the position of the particles in the structure by a measurement of the position and intensity of the reflected beam.

FIG. 13.2. *Two-dimensional representation of some parallel planes within a simple crystal*

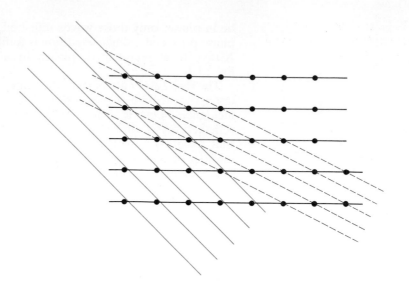

Experimental technique

FIG. 13.3. *Equipment used for X-ray diffraction*

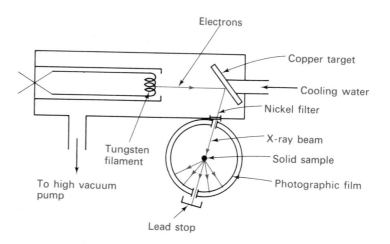

The general arrangement for the production of an X-ray diffraction photograph of a solid is illustrated in fig. 13.3. The beam of X-rays is produced from an X-ray tube in which electrons are accelerated from an electrically-heated tungsten filament through a potential difference in the order of 10 000 V in vacuum. When the electrons impinge on a copper target X-rays are produced with several different wavelengths, and these are then allowed to strike a sheet of nickel which filters out all the X-rays except those within a very narrow range of wavelength. This monochromatic beam of X-rays is now allowed to strike the solid sample which can take the form of either a single crystal or a compact powder. The X-ray beam is reflected at definite angles, as illustrated, and detected either on a photographic film wrapped around the inside of the equipment, or by an ionisation chamber attached to a Geiger counter which can rotate around the solid sample. Using the photographic method, an exposure time up to several hours might be employed before the film is removed and developed. In the case of a single crystal, which is mounted at the centre of the film and rotated about a major axis, the resultant photograph appears as a series of spots of different intensities, distributed along layer lines, as illustrated in Plate 7 which is an X-ray diffraction photograph from a single crystal of sodium chloride. For a sample of powdered crystals packed into a cylindrical form using a suitable adhesive, the photographs produced are of a different type. An X-ray diffraction photograph from powdered sodium chloride is shown in Plate 8.

Plate 7. X-ray diffraction photograph from a single crystal of sodium chloride

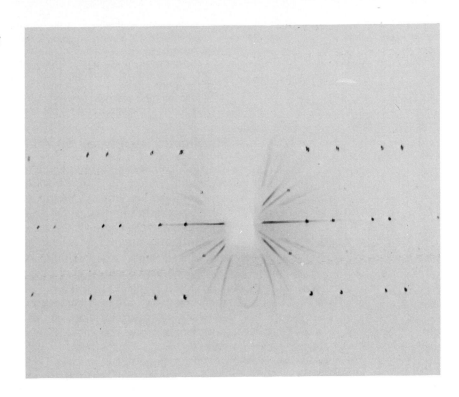

Plate 8. Powdered sample of NaCl, X-ray diffraction photo

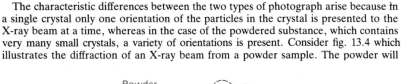

The characteristic differences between the two types of photograph arise because in a single crystal only one orientation of the particles in the crystal is presented to the X-ray beam at a time, whereas in the case of the powdered substance, which contains very many small crystals, a variety of orientations is present. Consider fig. 13.4 which illustrates the diffraction of an X-ray beam from a powder sample. The powder will

FIG. 13.4. *The origin of the curved lines of an X-ray diffraction photograph from a powder sample*

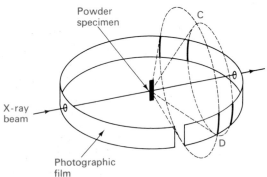

contain many small crystals at the correct angle θ to satisfy the Bragg equation and these crystals will be distributed uniformly about the axis of the X-ray beam. Thus for a particular spacing d the diffracted X-ray beam will trace out a cone with diameter say CD, which will cut the photographic film symmetrically in two positions. Since there are various parallel planes in crystals with different spacings (different values of d) there will be other values of θ which satisfy the Bragg equation. A whole family of diffraction cones will therefore be produced from a powder sample, and each cone will give rise to two symmetrical curved lines on the developed film.

Applications of X-ray diffraction

In practice it is much more convenient to use powdered samples for X-ray diffraction since the substance does not need extensive preparation, but the amount of information which can be obtained from such samples is rather limited. A large library of recorded X-ray photographs is available to the chemist, thus impurities in samples may be detected in a quantitative manner, provided a suitable calibration has been carried out with the camera beforehand. An early example of this analytical use concerned the constitution of bleaching powder, which was shown to be a mixture of $Ca(OCl)_2$ and the basic chloride $CaCl_2 . Ca(OH)_2 . H_2O$. In a similar way, the powder method may also be used to distinguish between polymorphic forms of the same compound. The applicability to both organic and inorganic compounds in the powdered state is very wide.

If the symmetry of the crystal is low, or the sample contains very small crystals, the powder method does not easily lend itself to determinations of dimensions of the constituents of the unit cell. For such purposes the single crystal specimens are much more suitable, although it is much more difficult to obtain a perfectly-formed crystal of sufficiently large size to be mounted in the camera (in practice they need to be at least 0·5 mm in length along each of the principal axes). Once a crystal specimen has been obtained then a great deal of information can be gathered from it. Analysis can lead to the measurement of the unit cell dimensions and thus to the establishment of the atomic and ionic radii of different atoms and ions. The technique has also been applied to the investigation of molecular crystals. In such structures, which may contain many different atoms in the unit cell, the molecules are symmetrically spaced with respect to each other but the X-ray diffraction pattern will arise from the individual atoms within the molecules; these atoms may therefore not occupy positions at the centre or corners of the unit cell. In such cases the analysis is even more complex, and the relating of the multiple reflections and different intensities to the arrangement of the atoms in the molecules is an extremely difficult one. The basic structure for the penicillin group of compounds was found by X-ray diffraction to be that shown in fig. 13.5, when exhaustive chemical analysis had failed to distinguish between the various possibilities in 1943. The extremely complicated structure of vitamin B_{12} (fig. 13.6) was obtained in a similar way.

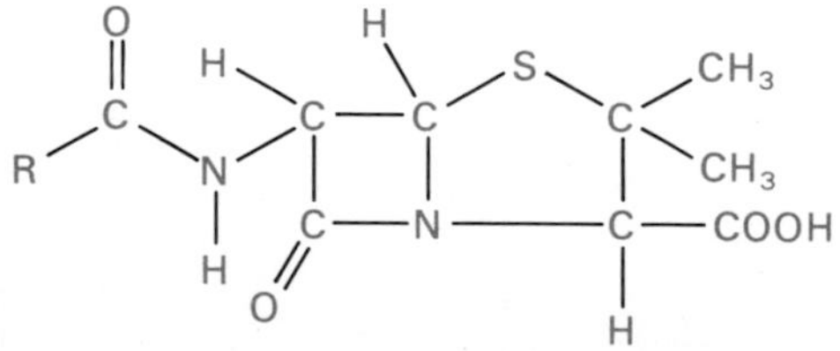

FIG. 13.5. *The basic structure for the penicillin group of compounds*

178

FIG. 13.6. *The structure of vitamin B$_{12}$*

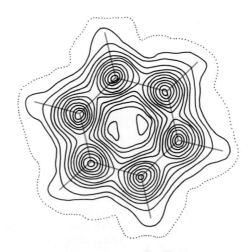

Since the degree of scattering of X-rays is dependent upon the number of electrons around a given atom, then it is possible to depict molecular structures in **electron density contour maps**, such as is shown for benzene in fig. 13.7. The positions of the hydrogen atoms do not show up on such maps because of the weak scattering from the lone electron of the hydrogen atom. Other methods must be employed to locate this atom in structures and one method is by the technique of neutron diffraction which is described in the next section.

FIG. 13.7. *Electron density map for benzene, derived from a single crystal X-ray diffraction analysis*

Recent X-ray work has concentrated on the investigation of giant molecules which are of importance in the science of molecular biology and also on the investigation of polymeric materials such as man-made fibres.

13.5 Neutron diffraction

In the previous section we have seen how the diffraction of X-rays is caused by the electrons surrounding the atoms in a crystal. It is also possible for other types of beam to be diffracted, and a neutron beam from an atomic pile is capable of exhibiting similar effects to those of an X-ray beam. Again, the fundamental requirement for neutron diffraction to occur is that the wavelength of the neutron beam should be comparable with the interatomic or interionic dimensions in a crystal. A beam of neutrons travelling with a velocity of 4×10^3 m s^{-1} has a wavelength of about 0·1 nm ($\lambda = h/mv$) and is therefore suitable for undergoing diffraction in crystalline solids. The major difference between diffraction of X-rays and neutrons arises from the source of diffraction; thus for X-rays it is located in the electrons which surround the atoms while for neutrons it is found in the interaction between the neutrons and the atomic nucleus (known as **nuclear scattering**).

Since neutrons readily pass through the electron clouds of a molecule there is much less variation in the amount of nuclear scatter from different atoms than there is for X-rays which are scattered more by 'heavy' atoms which contain many electrons. The technique may therefore be used for the location of hydrogen atoms in solids and is complementary to X-ray diffraction methods. It has been used to investigate hydrogen bonding (p. 90) and the accurate determination of the C—H bond lengths in organic compounds.

Because the neutrons are generated in an atomic pile the technique is both inconvenient and expensive. It is generally used only when the locations of the 'heavier' atoms in a structure have been determined by X-ray analysis.

13.6 Electron diffraction

The wave properties of an electron were first predicted by de Broglie in 1924 and confirmed independently by Davisson and Germer, and G. P. Thomson, when they succeeded in showing that an electron beam could undergo diffraction by thin metal foils (p. 37 and see also Plate 2).

At an accelerating potential difference of about 40 000 V the wavelength of an electron beam is approximately 0·006 nm and this is the sort of wavelength which is employed in the diffraction studies of molecules. Electrons are readily absorbed by liquids and solids and cannot be used to investigate the internal structures of substances in either of these two forms. However, an electron beam produces clear diffraction effects with gases at low pressures and it is possible to determine bond lengths and interbond angles of simple gaseous molecules.

Although a gas consisting of single atoms would scatter an electron beam in all directions, a gas composed of diatomic molecules has the component atoms in a molecule in fixed positions relative to each other. There will be many such molecules in a gas sample which have their bond axis at the correct angle to an approaching electron beam, at any given moment, for constructive interference to occur (the condition here is that $r\cos\theta = n\lambda$, where r is the bond distance, θ is the inclination of the axis of the molecule to the incoming electron beam and n is a whole number, see fig. 13.8). Since these molecules will be symmetrically disposed about the axis of the electron beam, a cone of diffracted electrons will be produced. The effect will show up on photographic film as a ring. Since the condition for constructive interference is satisfied for more than one angle, i.e. for $n = 2,3$ etc. a series of cones, and thus of concentric rings on photographic film, will result (fig. 13.9).

FIG. 13.8. *Diffraction of an electron beam by a diatomic molecule*

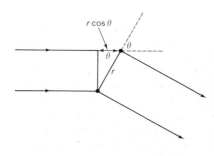

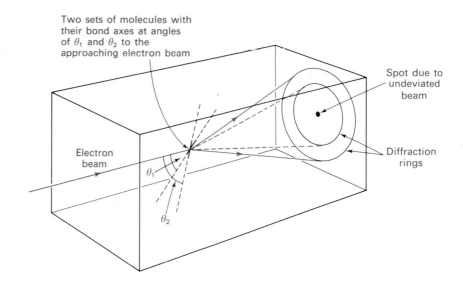

For a molecule such as tetrachloromethane, CCl_4, there will be two different types of source from which scattering of an electron beam can occur: one from the unit C—Cl and one from the non-bonded unit Cl--Cl, so the ring system will be more complex than is the case for a diatomic molecule. Analysis of the concentric rings allows bond lengths to be determined with considerable accuracy and, for a molecule such as tetrachloromethane, also bond angles. For a basically simple molecule such as $CHFCl_2$ the number of parameters is increased and the analysis becomes more complex.

An electron beam can be employed to investigate the surface of a solid and for such purposes electrons of lower energy and consequently higher wavelength are employed (typically 0·1 nm). The electron beam is focussed by a suitable arrangement of electric and magnetic fields and is then directed onto the surface of the solid under investigation. Reflection occurs from the top layer of atoms and the reflected beam is detected photographically. In this way, surface features of solids can be examined under very high resolution. The method is particularly useful for investigating the surface of solid catalysts.

13.7 Absorption spectroscopy

We have seen previously (chapter 3) that hydrogen atoms exhibit an atomic spectrum when electromagnetic radiation of the correct range of frequency is employed. This observation was explained in terms of discrete energy levels in the atom. An electron can pass from one energy level to another by emission or absorption of energy from the incident electromagnetic radiation; thus a transition of an electron from an energy level E_2 to another of energy E_1 will occur with a change of energy, ΔE, given by

$$\Delta E = h\nu$$

where ν is the frequency of the emitted or absorbed radiation. In a similar way molecules exhibit spectra, and several main types of molecular spectra can be identified, each associated with a characteristic type of energy change within the molecule. Each is also observed in a characteristic region of the electromagnetic spectrum; they are generally observed as **absorption** spectra, so that as a result of the absorption of incident radiation the total energy of the molecule is increased. **Absorption spectroscopy**, as the technique is known, is a most useful one for the determination of molecular structure. The different types of absorption

181

spectra which may be observed in the different regions of the electromagnetic spectrum are shown in fig. 13.10, together with an indication of the nature of the transitions involved. The boundaries between the various regions are not rigid but the types of molecular process associated with each region are quite different.

FIG. 13.10. *Spectroscopic processes associated with the various regions of the electromagnetic spectrum*

Wavelength/m	10^{-8}			10^{-6}		10^{-4}	10^{-2}
X-rays and γ-rays		Ultra-violet	Visible	Infra-red		Microwave	Radio-waves
Atomic electronic changes		Atomic and molecular electronic changes		Molecular vibrations		Molecular rotations	Nuclear magnetic energy levels

When a molecule has absorbed radiation it might appear that it should re-emit all of it again in dropping back to its original energy level. For a collection of such molecules the net result would therefore be zero, since on average as many molecules would be absorbing energy, at a given instant, as were emitting radiation of the same wavelength. Fortunately many so-called **excited** molecules (those in an upper energy level state) return to the **ground** state (the lowest energy level) not by a single drop in energy but by several, involving the transfer of the excess energy in absorption by collision with neighbouring molecules in the ground state. Eventually the sample under investigation becomes warm.

Table 13.2 General features of absorption spectrometers

	Ultra-violet	Visible	Infra-red	Microwave	Nuclear magnetic resonance
Source	Hydrogen discharge lamp	Tungsten filament lamp	Electrically heated rod of rare earth oxides	Klystron valve	Radiofrequency source and powerful magnet of stable field strength
Sample	Generally dilute solution but gases can be used. Quartz cells used in u.v. region		Gases, liquids, dilute solution Solids ground with KBr and pressed into discs	Gases or vapours	Solutions
System of Analysis	System of mirrors and rotating grating to select appropriate wavelength		System of mirrors and rotating grating	Frequency varied electronically	Magnetic field strength varied. Fixed radiofrequency used
Detector	Photographic device or more generally a photomultiplier tube		Heat sensor or photo-conductivity device	Crystal detector of radiofrequencies	Radiofrequency detector

13.8
General features of absorption spectroscopy

In essence the essential features of any absorption **spectrometer** are fairly straightforward; they consist of a suitable **source** of electromagnetic radiation, the presentation of the substance in a suitable form, a system of **analysis** of the radiation employed, and an appropriate **detector** to discover the wavelength of the radiation absorbed. In practice there may be difficulties arising in all of these features but, as it is not the present purpose of the book to give a complete account of the methods, these features are summarised in Table 13.2 for the methods discussed in this chapter.

The final absorption spectrum is displayed either on an oscilloscope screen or, more generally, on a chart recorder, and shows the manner in which absorption of the particular radiation varies with the wavelength.

13.9
Visible and ultra-violet spectroscopy

Radiation in the visible and ultra-violet regions of the electromagnetic spectrum is of sufficient energy to cause transitions of the outermost electrons in molecules from a lower to an upper level. Very often the outermost electrons are involved in bonding within a molecule, and the characteristic absorption wavelengths provide information about the types of bond present in the molecule. For example, the electronic absorption spectrum of propanone, CH_3COCH_3 (see fig. 13.11) shows that there is a maximum in the absorption at about 280 nm. One can also see in the diagram that there will be another maximum at a lower wavelength. It occurs at 188 nm. These absorptions are typical of the carbonyl group, $>C{=}O$. An absorbing centre such as this is called a **chromaphore**.

FIG. 13.11. *Electronic absorption spectrum of propanone, CH_3COCH_3*

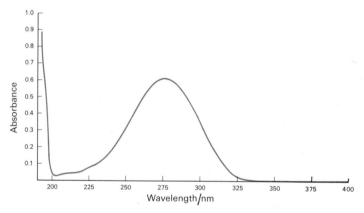

In organic compounds the transitions responsible for absorption of radiation are generally due to the promotion of σ and π-electrons of double bonds, or lone-pair electrons, to unoccupied antibonding molecular orbitals (p. 99). The wavelength corresponding to maximum absorption, λ_{max}, by a particular chromophore is to some extent dependent upon the solvent employed and the particular compound containing the chromophore, but the variation is generally quite small. For example, all simple aldehydes and ketones have a λ_{max} value in the range 275–295 nm. The values of λ_{max} shift significantly towards longer wavelength, however, when two chromophores are conjugated with each other, i.e. on adjacent atoms. Thus the values of λ_{max} shift from 188 and 279 nm for propanone, $CH_3{-}\underset{\underset{O}{\|}}{C}{-}CH_3$, to 219 and 324 nm respectively for butenone, $CH_2{=}CH{-}\underset{\underset{O}{\|}}{C}{-}CH_3$, and this affords a useful means of detecting such conjugated systems.

The extent to which radiation is absorbed by a substance depends upon the concentration of the substance, the distance the radiation travels in passing through the sample (the path length) and on the particular wavelength. For a fixed wavelength absorbed, it is found that dilute solutions obey the Beer-Lambert law fairly closely:

I_o = intensity of incident radiation
I = intensity of transmitted radiation
l = path length (m)
c = concentration (mol m^{-3})
ε = molar absorption coefficient (m^2 mol^{-1})

$$\lg \frac{I_o}{I} = \varepsilon l c \qquad (1)$$

Lg (I_0/I) is known as the **optical density** or **absorbance** and it is this quantity which appears along the y-axis of recorded visible and ultra-violet spectra. Molar absorption coefficients ε, measure the intensity of the absorption, and some values of ε_{max} (molar absorption coefficient for maximum absorption) together with λ_{max} values are given for some chromophores in Table 13.3.

Table 13.3 λ_{max} and ε_{max} for some chromophores

Substance	Formula showing chromophore	Solvent	λ_{max}/nm	ε_{max}/m^2 mol^{-1}
(a) Unconjugated system				
Propanone	CH$_3$—C—CH$_3$ ‖ O	Hexane	188 279	90 1·5
Pent-1-ene	CH$_3$—CH$_2$—CH$_2$—CH=CH$_2$	Hexane	190	1000
Ethyl ethanoate	CH$_3$—C—O—CH$_2$—CH$_3$ ‖ O	Water	204	6·0
(b) Conjugated system				
Buta-1,3-diene	CH$_2$=CH—CH=CH$_2$	Hexane	217	2100
Butenone	CH$_2$=CH—C—CH$_3$ ‖ O	Ethanol	219 324	360 2·4

Visible and ultra-violet absorption spectroscopy is also useful for investigating inorganic compounds which show absorption in this region of the electromagnetic spectrum. The technique is particularly applicable to the study of transition metal complexes which are often highly coloured and hence absorb in the visible region. For example, an

FIG. 13.12. *Electronic absorption spectrum of [Ni(H$_2$O)$_6$]$^{2+}$*

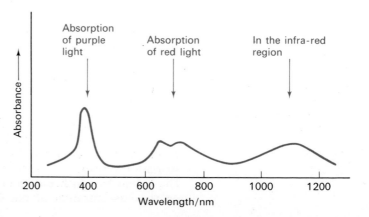

184

aqueous solution of a nickel salt is green and shows an absorption in the purple region of the visible spectrum, together with two absorption maxima in the infra-red region (fig. 13.12).

Visible and ultra-violet spectra of compounds are useful in the identification of compounds. Thus the spectra of an unknown compound can be compared directly with known spectra or measured values of λ_{max} and ε_{max} can be compared with tabulated values for known chromophores. The Beer-Lambert law (equation (1), p. 184) relates absorbance with concentration and this enables the concentration of a known species in solution to be determined accurately; the method is particularly useful for measuring very low concentrations which would be difficult to estimate by other methods. The disappearance or appearance of an absorbing species in a chemical reaction can be followed accurately, without in any way disturbing the chemical system, and this finds application in chemical kinetics (p. 430).

13.10
Infra-red spectroscopy

Infra-red spectra arise principally as a result of transitions between the vibrational energy levels of a molecule. The energy gaps between such levels are considerably less than those between the electronic levels involved in visible and ultra-violet spectroscopy; but they are much greater than the spacings between rotational levels (see fig. 13.13).

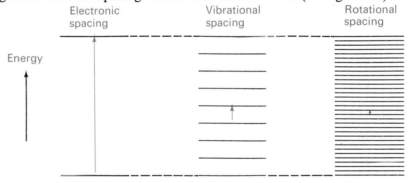

FIG. 13.13. *An indication of the spacing between electronic, vibrational and rotation energy levels*

FIG. 13.14. *Vibration of a diatomic molecule*

v_0 = **fundamental frequency of vibration**
k = **force constant of the bond**
μ = **reduced mass**
$$\left(\frac{m_1 m_2}{m_1 + m_2} \right)$$

Consider a diatomic molecule consisting of two different atoms of mass m_1 and m_2; a very useful model which describes the vibratory motion of such a molecule is one consisting of the two masses joined by a fairly stiff spring. If this system is set into vibration, in the absence of frictional forces, it will execute simple harmonic motion about an equilibrium position, the two extremes of motion being a contracted position and an extended position as shown in fig. 13.14. For such a system it can be shown quite simply that the frequency of vibration about the centre of mass is given by the expression:

$$v_o = \frac{1}{2\pi} \sqrt{\frac{k}{\mu}} \qquad (2)$$

The force constant, k, is a measure of the strength of the bond joining two atoms together, a strong bond having a larger force constant than a weaker bond. The relationship given in equation (2) proves to be a good description of the vibratory motion of a diatomic molecule provided, that its use is restricted to the lower vibrational energy levels (at higher energies the vibratory motion tends to deviate from simple harmonic).

Unlike the classical system comprising two vibrating masses which can take up energy continuously, the vibrating atoms in a molecule can only take up energy in a quantised manner, as given by the expression:

$$E_{\text{vib}} = (v + \tfrac{1}{2})h v_{\text{o}} \qquad (3)$$

v = vibrational quantum number 0, 1, 2 etc.

v_0 = fundamental frequency of vibration

h = Planck's constant

(The number $\tfrac{1}{2}$ appears since quantum mechanics stipulates that a molecule must always have some residual vibrational energy, even at absolute zero. This so-called zero-point energy is $\tfrac{1}{2}h v_{\text{o}}$). If follows that the energy absorbed in raising a molecule from its lowest energy level to the next highest, i.e. from v_{o} to v_1 is given by

$$\Delta E = \tfrac{3}{2}h v_{\text{o}} - \tfrac{1}{2}h v_{\text{o}} = h v_{\text{o}} \qquad (4)$$

Combining equations (2) and (4), the energy difference between two adjacent vibrational levels for a diatomic molecule is given by the expression:

$$\Delta E = \frac{h}{2\pi}\sqrt{\frac{k}{\mu}} \qquad (5)$$

For a molecule to absorb in the infra-red there must be a change of dipole moment (p. 87) when it vibrates. Such a fluctuating dipole moment is able to interact with the fluctuating electric component of the incident radiation of the same frequency, and hence absorb energy from it. The hydrogen chloride molecule H—Cl, which is polar, absorbs in the infra-red whereas the non-polar hydrogen molecule H—H does not.

In general most of the molecules studied are more complex than the simple diatomic molecules we have been discussing so far. Thus we have already seen that there are a total of $3N$-5 modes of vibration for a linear molecule (p. 130) and of this total N-1 are bond-stretching motions and

FIG. 13.15. *The vibrational modes of carbon dioxide*

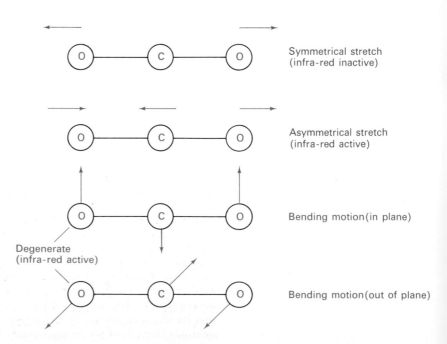

Symmetrical stretch (infra-red inactive)

Asymmetrical stretch (infra-red active)

Bending motion (in plane)

Degenerate (infra-red active)

Bending motion (out of plane)

186

2*N*-4 are bending motions. For carbon dioxide, therefore, there will be a total of 4 vibrational modes of which 2 are **bond-stretching** motions and 2 are **bending** motions (see fig. 13.15).

It will be seen from fig. 13.15 that the two bending motions for carbon dioxide are degenerate (have the same energy) and this means that only one characteristic absorption occurs for this vibration. Of the two bond-stretching motions the one labelled **symmetrical stretch** does not involve a dipole moment change and is said to be **infra-red inactive**, whereas the **asymmetric stretch** does involve a dipole moment change and is **infra-red active**. The infra-red absorption spectrum of carbon dioxide does show the two absorptions at frequencies of 667 cm^{-1} (bending motion) and 2349 cm^{-1} (asymmetric stretch) but, in addition, there are other absorptions due to interaction between the principal vibratory modes. This interaction increases with the complexity of the molecule.

In the case of a non-linear molecule, such as water the total number of vibratory modes is 3*N*-6 of which *N*-1 are bond-stretching motions and 2*N*-5 are bending motions. For benzene, C_6H_6, there will be a total of 30 modes of vibration but not all of these will be active since some of the vibrations will not involve a dipole moment change. The spectrum, which is shown in fig. 13.16 is fairly complex because absorptions due to the molecule as a whole occur in addition to those characteristic of particular bonds. Before proceeding further it should be noted that infra-red spectra are normally plots of percentage transmittance and not absorbance (the two are related inversely); in addition, the wavenumber, $\bar{\nu}$, which is the reciprocal of the wavelength (in cm) is employed and this quantity is related directly to the energy of the transition.

By careful analysis it is possible to assign the various peaks in the spectrum to certain vibrations in the molecule, and the major ones have been marked on fig. 13.16.

wave number = $\bar{\nu}$

where $\bar{\nu} = \dfrac{1}{\lambda} = \dfrac{E}{hc}$

FIG. 13.16. *Infra-red spectrum of benzene, C_6H_6 (liquid film)*

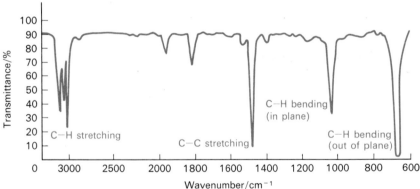

By recording the infra-red spectra of a large number of compounds it has been possible to compile extensive data of characteristic infra-red absorptions, typical of most functional groups in both inorganic and organic compounds. The most useful part of the spectrum for identification purposes is the region around 700–1500 cm^{-1}, known as the 'fingerprint' region, since compounds give a unique set of absorption peaks in this area. The region 1500–3100 cm^{-1} usually includes peaks arising from the stretching vibrations of many common groups in organic compounds (such as C=C, C—H, C=O and C≡C), whereas the 'fingerprint' region also includes the many bending vibrations which arise as a result of the many possible substituents which can be included in the structure. As we have noted previously the force constant, *k*, is a measure of the strength of the bond joining two atoms together (p. 185). Hence for the bonds C—C, C=C and C≡C there is a progressive increase in force constant and a consequent increase in energy (and wavenumber) associated with the stretching vibrations of these groups. Fig. 13.17 summarises the main areas where some functional groups absorb. As a final illustration, the infra-red spectrum of ethyl ethanoate, $CH_3{-}C{-}O{-}CH_2{-}CH_3$, is shown in fig. 13.18 with three characteris-

tic peaks marked.

FIG. 13.17. *Regions in the infra-red where functional groups absorb*

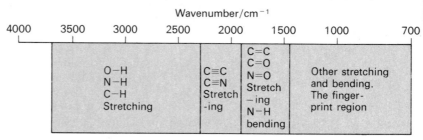

FIG. 13.18. *Infra-red spectrum of ethyl ethanoate, $CH_3CO_2CH_2CH_3$ (liquid film)*

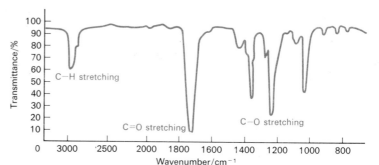

FIG. 13.19. *Rotation of a diatomic molecule A–B about three mutually perpendicular axes*

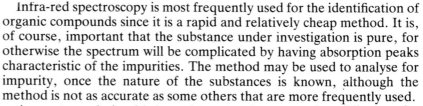

Infra-red spectroscopy is most frequently used for the identification of organic compounds since it is a rapid and relatively cheap method. It is, of course, important that the substance under investigation is pure, for otherwise the spectrum will be complicated by having absorption peaks characteristic of the impurities. The method may be used to analyse for impurity, once the nature of the substances is known, although the method is not as accurate as some others that are more frequently used.

Accurate analysis of infra-red spectra shows a series of lines under high resolution which arise through the absorption of electromagnetic radiation by rotational modes in a molecule. Although much useful information concerning the structure of the molecule can be gained from a detailed study of the **rotational fine structure**, the method is a complex one. The absorption of radiation by rotation within a molecule is generally studied at a higher wavelength in the microwave region and this is described in the next section.

13.11 Microwave spectroscopy

Molecules in the gas phase undergo unhindered rotation at low pressure which can be resolved about three mutually perpendicular axes.* For a diatomic molecule, and this is the only sort we shall be considering, this rotation is illustrated in fig. 13.19. The molecule *A-B* has the same numerical value for the moments of inertia about the *y*- and *z*-axes but

* These axes pass through the centre of gravity of the molecule.

its moment of inertia about the x-axis is negligible. A heteronuclear diatomic molecule, i.e. one with a dipole moment, is able to interact with radiation in the microwave region of the electromagnetic spectrum and, as a result, is raised to a higher rotational energy level. Only certain rotational energy levels are permitted, since they are quantised, and are given by the formula:

$$E_{\text{rot}} = \frac{h^2 J(J + 1)}{8\pi^2 I} \tag{6}$$

By the application of simple mechanics, it can be shown quite easily that the moment of inertia for a diatomic molecule of masses m_1 and m_2 about an axis passing through the centre of gravity is given by the formula

$$I = \frac{m_1 m_2}{(m_1 + m_2)} r_{\text{o}}^2 \text{ (see fig. 13.20)}$$

J = rotational quantum number 0, 1, 2, 3 etc.
h = Planck's constant
I = moment of inertia

FIG. 13.20. *Rotation of a diatomic molecule about an axis passing through its centre of gravity*

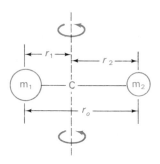

The energy levels available to a rotating molecule can be determined by allowing J to take integral values 0, 1, 2, 3 etc. and substituting these values into equation (6) above. These energy levels are shown in fig. 13.21, together with the allowed transitions and the corresponding energies of these transitions.

FIG. 13.21. *Rotational energy levels for a diatomic molecule, showing the allowed transitions and the energies of transitions*

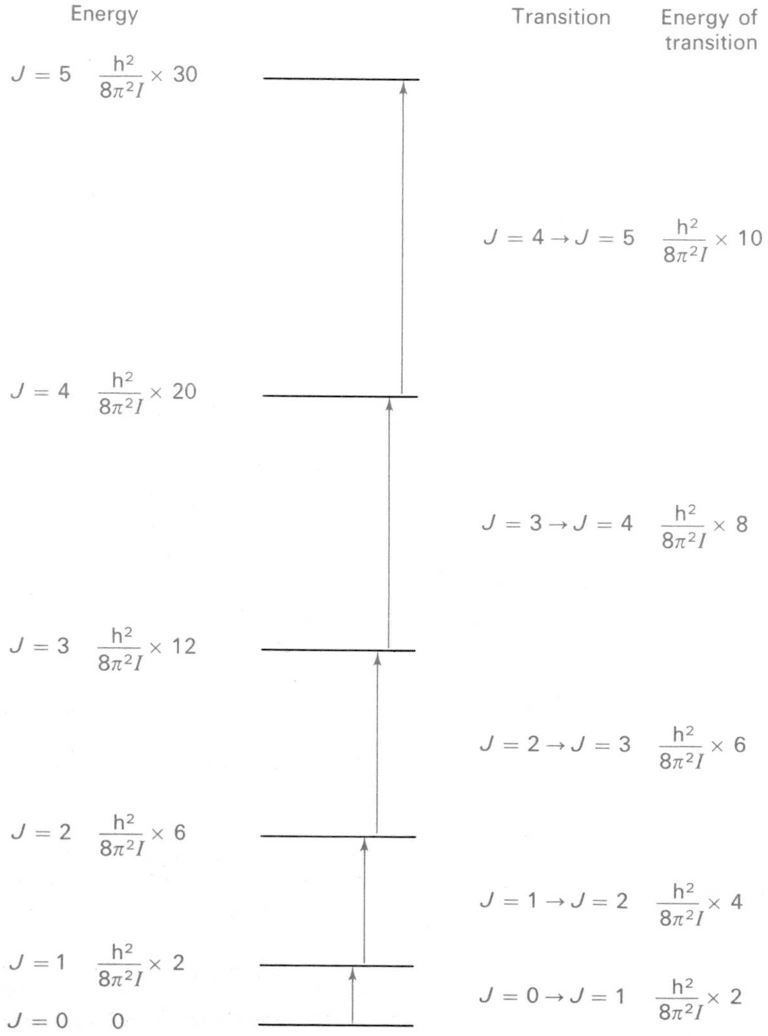

Energy		Transition	Energy of transition
$J = 5$	$\frac{h^2}{8\pi^2 I} \times 30$		
		$J = 4 \rightarrow J = 5$	$\frac{h^2}{8\pi^2 I} \times 10$
$J = 4$	$\frac{h^2}{8\pi^2 I} \times 20$		
		$J = 3 \rightarrow J = 4$	$\frac{h^2}{8\pi^2 I} \times 8$
$J = 3$	$\frac{h^2}{8\pi^2 I} \times 12$		
		$J = 2 \rightarrow J = 3$	$\frac{h^2}{8\pi^2 I} \times 6$
$J = 2$	$\frac{h^2}{8\pi^2 I} \times 6$		
		$J = 1 \rightarrow J = 2$	$\frac{h^2}{8\pi^2 I} \times 4$
$J = 1$	$\frac{h^2}{8\pi^2 I} \times 2$		
		$J = 0 \rightarrow J = 1$	$\frac{h^2}{8\pi^2 I} \times 2$
$J = 0$	0		

As can be seen in fig. 13.21 (under the column headed 'energy of transition') the energy changes, ΔE, increase uniformly and this means that the rotational spectrum of a diatomic molecule should consist of a number of equally spaced absorption lines. From measurements on the spacings between the rotational absorption lines, the moment of inertia and hence the internuclear distance can be determined for a diatomic molecule as the following example shows.

In practice the lines get slightly closer together as the energy of rotation increases. This is because the bond stretches, consequently increasing the moment of inertia and hence decreasing the energy of rotation, E_{rot} (see equation (6)).

Example

Hydrogen chloride, $H^{35}Cl$, has rotational absorption lines at the following positions (R. L. Hausler and R. A. Oetjen, J. Chem. Phys. 21, 1340 (1953): 83·32, 104·13, 124·73, 145·37, 165·89, 186·23, 206·60, 226·86 cm^{-1}. Calculate the moment of inertia and the bond length of the molecule.

There is, with minor exceptions, a slight decrease in the distance between successive lines. The average separation distance from the above figures is about 20·5 cm^{-1}. Clearly the first line given, at 83·32 cm^{-1}, corresponds to a transition from $J = 3$ to $J = 4$. How do we know this? Let us calculate the moment of inertia and bond length for the separation distance $104·13 - 83·32 = 20·81$ cm^{-1}. Since wavenumber, $\bar{\nu}$, is equal to $1/\lambda$, $\nu = c/\lambda$, and $\Delta E = h\nu$, we have

$$\Delta E = hc\bar{\nu}$$

or

$$\bar{\nu} = \frac{\Delta E}{hc} = \frac{2h^2}{8\pi^2 Ihc} = \frac{h}{4\pi^2 Ic} \tag{7}$$

From equation (7) it follows that:

$$I = \frac{h}{4\pi^2 c\bar{\nu}} = \frac{6·6256 \times 10^{-34}}{4\pi^2 \times 2·998 \times 10^{10} \times 20·81}$$
$$= 2·69 \times 10^{-47} \text{ kg m}^2$$

(Note: the speed of light, c, must be in cm s^{-1} since wave number, $\bar{\nu}$, is in cm^{-1}).

$$I = \frac{m_1 m_2}{(m_1 + m_2)} r_o^2 \tag{8}$$

The masses of the hydrogen and chlorine atoms must be in kg and these are determined by dividing the respective relative atomic masses by the Avogadro constant, $6·02 \times 10^{23}$ and by 10^3 (to convert from g to kg). Thus substituting in equation (8) we have:

$$2·69 \times 10^{-47} = \frac{1 \times 35 \times 10^{-6}}{6·02 \times 10^{23} \times 36 \times 10^{-3}} r_o^2$$

hence

$$r_o^2 = \frac{2·69 \times 6·02 \times 36 \times 10^{-21}}{35} = 0·01666 \times 10^{-18}$$

thus

$$r_o = 0·129 \times 10^{-9} \text{ m} = 0·129 \text{ nm}$$

The moment of inertia, I, is $2·69 \times 10^{-47}$ kg m^2 and the bond length, r_o, is 0·129 nm.

The moment of inertia and hence the bond length of a diatomic molecule will vary slightly depending upon which particular spacing value is employed. It is, however, possible to make allowances for the stretching of the bond at higher rotational levels and thus obtain some accurate 'average' bond lengths.

Bond distances in simple diatomic molecules such as H_2, N_2, Cl_2 and O_2 cannot be determined by microwave spectroscopy, since these molecules do not have a dipole moment.

The essential features of a microwave spectrometer are:

(a) A klystron valve emitting virtually monochromatic radiation over a narrow range of frequency and which can be tuned electronically.

(b) A 'wave-guide', which is generally a hollow copper tube, to direct the radiation on to the gas sample. Since the beam is subject to atmospheric absorption the system must be evacuated before the gas sample, at a pressure in the region of $1\ N\ m^{-2}$, is admitted.

(c) A crystal detector linked to an oscilloscope or chart recorder to display the signals.

13.12
Nuclear magnetic resonance spectroscopy

It is an experimental fact that any nucleus consisting of either an odd number of protons or an odd number of neutrons (or both) has the property of nuclear spin (p. 61). Nuclei which possess spin include ^1H, ^{13}C and ^{19}F but not ^{12}C and ^{16}O. This section will be concerned almost exclusively with the nuclear spin of the proton ^1H.

A spinning proton will generate a circulating electric current which in turn will produce a magnetic field. It is therefore helpful to think of a spinning proton as behaving rather like a small magnet. When a proton is placed in a magnetic field it can align itself in one of two possible orientations with respect to this field. The low energy alignment of the nucleus, in which the magnetic field of the nucleus is in the same direction as the applied magnetic field, and the high-energy position in which the two magnetic fields are opposed are shown in fig. 13.22. A transition from the low-energy state of the nucleus to the high-energy state can be achieved by supplying an amount of energy equal to the difference in energy between the two states, ΔE. This energy may be supplied by electromagnetic radiation, and since the energies themselves are dependent on the magnitude of the applied field, it is possible to choose an external field to give absorption of electromagnetic radiation in a suitable part of the spectrum. In practice a field of approximately $2\cdot3\ T$ (Tesla) (23 000 Gauss) results in absorption at radio frequencies of 100 MHz, and this is the region in which so-called nuclear magnetic resonance (NMR) is normally observed.

Part of the difficulty in setting up the arrangement for the observation of the absorption of radiation by a proton is that of obtaining a magnet of sufficiently high field strength and also uniformity of field (1 part in 10^8 is required). In contrast to most other absorption methods the frequency of the radiation is kept constant and the magnitude of the applied magnetic field is varied until absorption of radiation by the proton occurs. This slight variation in the magnetic field is obtained by passing current through wire coils wrapped round the pole-pieces of the magnet. Absorption of radiation is detected on a radio frequency bridge (a type of Wheatstone bridge arrangement) which goes out of balance when absorption occurs and the resulting signal is then amplified and recorded. This is schematically illustrated in fig. 13.23.

So far we have restricted ourselves to discussing the resonance condition for the isolated proton. In a chemical compound protons appear in different chemical environments and these protons absorb radiation at slightly different field strengths. The reason for this is that such protons are surrounded by electron clouds, and in a magnetic field electrons circulate around the lines of force and produce a local field, B_{local}, in opposition to the applied field, B_o. A proton surrounded by electron clouds therefore experiences a field slightly less than that applied, and this field must consequently be increased for absorption of energy to occur. Now the extent to which a proton is shielded by electron clouds—

FIG. 13.22. *The two alignments of the magnetic moment of a proton in a magnetic field*

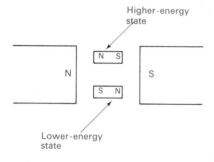

Higher-energy state

Lower-energy state

FIG. 13.23. *Basic arrangement for NMR spectroscopy*

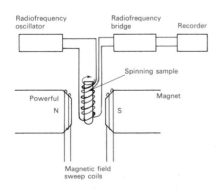

Radiofrequency oscillator

Radiofrequency bridge

Recorder

Spinning sample

Powerful

Magnet

Magnetic field sweep coils

as this condition is called—is determined by the electron-attracting properties of neighbouring atoms and groups. For example, the proton in O—H is not so well shielded as in C—H (since oxygen is more electronegative than carbon) and consequently resonance occurs for the O—H proton at a lower field strength than for the C—H proton. In practice these differences are very slight (the order of parts per million), but the whole technique of NMR is based upon measuring these slight variations in applied field with precision.

In practice, field differences are not measured in actual magnetic field units but by reference to the resonance condition for an arbitrary standard. The standard adopted is the compound tetramethylsilane, $Si(CH_3)_4$, abbreviated to TMS. This compound has twelve equivalent protons and thus has a single absorption peak. The compound under investigation, either the pure liquid or a solution in a non-protonic solvent, e.g. tetrachloromethane, is mixed with a small amount of TMS and the spectrum recorded. It will show an absorption for TMS at a magnetic flux density $B_{standard}$ and for a particular proton in the compound at say B_{sample}. Then the chemical shift for this type of proton is given by

$$\delta = \frac{\Delta B}{B_o} \times 10^6 \quad \text{where } \Delta B = B_{standard} - B_{sample}$$

B_o is the magnetic flux density of the magnet employed. The factor 10^6 is included to make the numerical values convenient to handle and they are subsequently quoted as parts per million. Chemical shifts are also quoted in terms of τ units (Tau units) in which TMS is given the arbitrary value 10 and other values related to it by the expression:

$$\tau = 10 - \delta$$

The NMR spectrum of ethanol, C_2H_5OH, is shown in fig. 13.24 with the chemical shifts for the various types of proton given in τ units. The three absorption peaks mean that there are three different types of proton in ethanol, i.e. CH_3 protons, CH_2 protons and an OH proton.

FIG. 13.24. *Low resolution NMR spectrum of ethanol, C_2H_5OH*

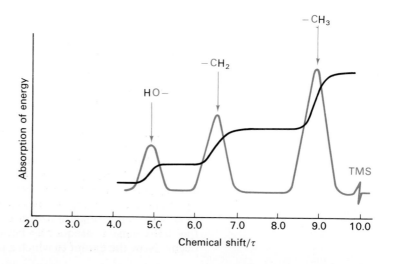

The area under each absorption peak is proportional to the number of protons in any given environment, and this is shown by the integrator trace which is incorporated into the instrument. From the step heights it can be seen that these protons are in the ratio of $3:2:1$, with the OH proton resonance occurring at the lowest field strength and the CH_3 protons at the highest field strength. This is exactly as expected in view of the electron-attracting ability of the oxygen atom, i.e. the less the shielding of a particular proton the lower the field strength (lower τ value) at which resonance occurs.

Table 13.4 Chemical shifts for a variety of differently located protons in organic compounds

Type of proton	Chemical shift/τ
CH_3—C—	9·1
CH_3—C$=$C	8·4
C—CH_2—C	8·6
$CH_2$$=$C	5·3
CH$\equiv$C	8·0
H—⬡	2·7
CH_3—⬡	7·8
CH_3—C— $\overset{\|}{O}$	7·9
CH_3—O—	6·7
CH_3—O—C— $\overset{\|}{O}$	6·3
—$\overset{\|}{\underset{\|}{C}}$—$CH_2$—Br	6·5

Table 13.4 shows some values of chemical shifts for a variety of differently located protons in organic compounds. The values are generally within $\pm 0\cdot 1\ \tau$, although larger variations are possible.

At this stage it is a useful exercise to attempt to predict the NMR spectra of some simple organic compounds. We shall consider two compounds, 1,3-dibromopropane, $BrCH_2CH_2CH_2Br$, and 1,4-dimethylbenzene, $4\text{-}CH_3C_6H_4CH_3$.

(a) 1,3-dibromopropane, $BrCH_2CH_2CH_2Br$
In this compound there are two different types of proton in the ratio $4:2$ or $2:1$, thus the NMR spectrum should show two absorption peaks with areas in the ratio of $2:1$. The larger absorption peak should be centred at $6\cdot 5\ \tau$ and the smaller one at $8\cdot 6\ \tau$ (see Table 13.4). The actual spectrum does show two peaks with areas in the ratio of $2:1$, but the smaller one is centred at $7\cdot 7\ \tau$ and not $8\cdot 6\ \tau$, a consequence of the inductive effect of the two bromine atoms being felt at the central position in the molecule.

193

(b) 1,4-dimethylbenzene, 4-$CH_3C_6H_4CH_3$

Once again there are two different types of proton in the molecule in the ratio 6:4 or 3:2, thus there should be two peaks in the NMR spectrum with relative areas of 3:2. The larger one should be centred at 7·8 τ and the smaller one at 2·7 τ (see Table 13.4). The actual spectrum is almost identical with that predicted, with peaks of relative areas in the ratio 3:2 located at 7·8 τ and 3·1 τ.

High resolution NMR spectroscopy

There is a further feature of NMR which increases the diagnostic value of the method considerably. When a very precise homogeneous magnetic field is applied to the substance under investigation, it is found that the spectrum does not always consist of single peaks at different values of τ, but some of the peaks are split (or resolved) into several parts. For example the high resolution spectrum of ethoxyethane, $CH_3CH_2OCH_2CH_3$, has two main peaks centred at 6·6 τ and 8·85 τ respectively. The one at 6·6 τ is split into four components (a quartet) and the other at 8·85 τ into three components (a triplet). (See fig. 13.25).

This fine splitting of the peaks in the NMR spectrum arises from the interaction between protons on adjacent atoms (those on the same atom do not interact with one another and are said to be equivalent, while those which are on non-adjacent atoms are too far apart to interact with each other). We have already seen that it is the electronic environment of each proton which distinguishes it from all others in the molecule; it seems that the magnetic field associated with a given proton can be affected by the magnetic field of an adjacent proton. In ethoxyethane

$$\overset{1}{C}H_3-\overset{2}{C}H_2-O-CH_2-CH_3$$

the magnetic field experienced by the protons on carbon atom 1 will depend slightly on the way in which the magnetic moments of the protons on carbon atom 2 are aligned with respect to the applied field. There are three possible alignments for these two protons:

(a) Both protons can have their magnetic moments arranged so as to reinforce the applied field.

(b) Both protons can have their magnetic moments arranged so as to oppose the applied field.

(c) One proton can have its magnetic moment aligned to reinforce the applied field while the other proton has its magnetic moment aligned to oppose the applied field. There are two ways of achieving this arrangement, hence this condition is twice as probable as either (a) or (b). These three different arrangements can be represented thus:

$$\uparrow\uparrow$$
$$\Updownarrow \qquad \Uparrow\Downarrow$$
$$\downarrow\downarrow$$

The three equivalent protons on carbon atom 1 can experience three slightly different field strengths. In a large number of ethoxyethane molecules one of these fields will be twice as probable as either of the other two and the result will be a triplet absorption pattern with relative intensities of 1:2:1. If we now consider the way in which the three protons on carbon atom 1 influence the magnetic field experienced by the protons on carbon atom 2 we get the following arrangement of magnetic moments:

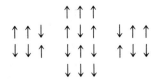

In this case the protons on carbon atom 2 can experience four slightly different field strengths and, in a collection of molecules, two arrangements are three times as probable as either of the other two. The result is a quartet absorption pattern with relative intensities of 1:3:3:1. Examination of fig. 13.25 will show the splitting pattern we have been discussing. In general, the splitting pattern can readily be obtained from Pascal's triangle as below

Number of protons causing splitting				Splitting pattern (relative intensities)					
1						1	1			
2					1		2		1	
3				1		3		3		1
4			1		4		6		4		1
5		1		5		10		10		5	1

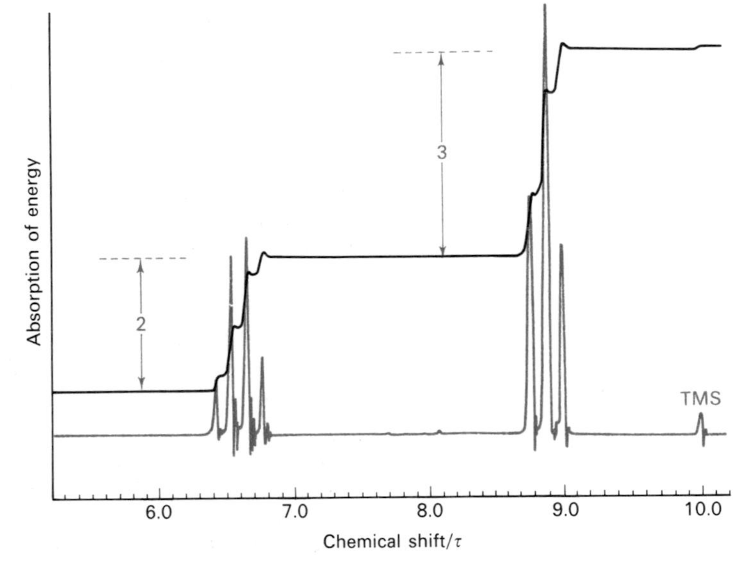

FIG. 13.25. *High resolution NMR spectrum of ethoxyethane,* $CH_3CH_2OCH_2CH_3$

We have previously discussed the general features of the spectra of 1,3-dibromopropane, $BrCH_2CH_2CH_2Br$, and 1,4-dimethylbenzene, 4-$CH_3C_6H_4CH_3$, and we can now take the analysis a little further. In the case of the former compound

$$Br-\overset{1}{C}H_2-\overset{2}{C}H_2-\overset{3}{C}H_2-Br$$

the protons on carbon atom 2 are surrounded by four adjacent protons, so a quintet absorption pattern with relative intensities of 1:4:6:4:1 should appear for the —CH_2— proton resonance. Similarly the two

195

sets of equivalent protons on carbon atoms 1 and 3 will give rise to an absorption peak, split into a triplet of intensities 1:2:1 by the two protons on the central carbon atom. These features are in fact present in the spectrum of the compound. There are no splittings for the compound 1,4-dimethylbenzene, since there are no protons adjacent to the six equivalent methyl protons nor to the four equivalent aromatic ring protons.

A detailed analysis of the splitting patterns of more complex molecules is beyond the scope of the book, but it is possible to discover extremely fine detail by the application of this technique. Thus the patterns can tell us where a substituent is placed in an aromatic ring and whether two groups in an alkene are cis- or trans- to one another.

FIG. 13.26. *The structures of (a) BrF$_5$ and (b) SF$_4$*

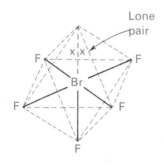

(a)

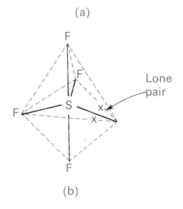

(b)

Summary of information obtainable from high resolution NMR spectroscopy

(a) Integration trace gives the relative numbers of protons of different types.
(b) Chemical shifts gives the type of structure associated with particular protons.
(c) The splitting patterns give the number of protons on atoms adjacent to the group which contains proton(s) whose resonance is being measured.

Other applications of nuclear magnetic resonance spectroscopy

Although the major uses of this technique are in organic chemistry, it is employed in the determination of inorganic structures. For example, ^{19}F has a nuclear spin and thus compounds containing this atom exhibit nuclear magnetic resonance effects. Bromine pentafluoride, BrF$_5$, has a spectrum consisting of two main peaks of relative intensities 4:1, with the more intense peak split into a doublet (1:1) and the other peak split into a quintet (1:4:6:4:1). The molecule thus contains one unique fluorine atom and a group of four equivalent ones. Similarly the spectrum of sulphur tetrafluoride, SF$_4$, consists of two triplets, i.e. the molecule contains two pairs of non-equivalent fluorine atoms. The spectra of these two compounds are consistent with the structures shown in fig. 13.26 and are what one would predict on the basis of the electron-pair repulsion theory (p. 84).

13.13
Mass spectrometry

If a substance can be vaporized without decomposition, the vapour can be ionised by electron bombardment and the positively charged ions can be accelerated electrostatically in a mass separator and detected electronically. Hence the positive ions are separated and detected according to their mass/charge ratio (m/e ratio). This method of separation of ions leads to a very useful technique for structural determination, known as mass spectrometry. It is a complementary process to the spectroscopic methods described in the preceeding sections of this chapter, and is based on a quite different principle.

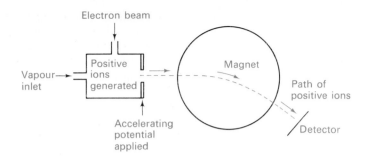

FIG. 13.27. *Schematic representation of a simple mass spectrometer*

The basic features of a mass spectrometer

A schematic diagram of a mass spectrometer is given in fig. 13.27. Gases and volatile liquids can be introduced directly into the ionisation chamber at very low pressure (in the region of $10^{-4} \, \text{N m}^{-2}$), while less volatile liquids and solids are introduced via a heated inlet system which generates a sufficiently high vapour pressure for the sample to be studied.

The sample vapour is bombarded in the ionisation chamber with a high energy electron beam and molecular ions are formed; in many cases chemical bonds are broken in the molecule and a range of fragment ions, in addition to the parent molecular ion, are often produced. The ions are accelerated electrostatically and concentrated into a narrow beam before passing through a magnetic field. Here ions of the same m/e move along the same circular arc, and for ions of the same charge (in practice most carry one single positive charge) the lightest ions are deflected the most. The separated streams of ions are now focussed successively on to a detecting device, which is generally an electron multiplier, by varying the magnetic field. The final signal is either displayed on an oscilloscope screen or recorded photographically.

The theoretical principles

For any gaseous molecule, positive ions are produced by electron bombardment:

$$M(g) + e^- \longrightarrow M^+(g) + 2e^-$$
$$M(g) + e^- \longrightarrow M^{2+}(g) + 3e^-$$

The ions are accelerated by a potential V, in the ionisation chamber, when they attain a velocity of v given by

$$\tfrac{1}{2}mv^2 = eV \tag{9}$$

When the positive ions pass through the magnetic field of flux density B, they are constrained to move along a circular arc of radius r, and

$$Bev = \frac{mv^2}{r} \tag{10}$$

Combining equations (9) and (10) we have

$$\frac{m}{e} = \frac{B^2 r^2}{2V} \tag{11}$$

As equation (11) indicates the extent of the deflection of any ion is related to m/e. By keeping V constant and varying B, ion beams of different m/e can be brought into focus on the detector.

The double-focussing mass spectrometer

In practice positive ions are produced in the ionisation chamber which possess a fairly narrow spread of kinetic energies. In order to achieve greater resolution it is necessary to reduce this range of energies before the ions pass through the magnetic field. This is achieved by incorporating an electrostatic analyser into the instrument, which deflects the ions in relation to their kinetic energies and allows just one extremely narrow range of energies to pass through the magnetic field for detection. While the mass spectrometer which relies simply on magnetic separation of the lines of different m/e values (single-focussing) can achieve resolution of ions which differ by one mass unit, the double-focussing instrument can determine masses to an accuracy of about 1 part in 10^6. Many modern instruments incorporate both facilities, since it is not always essential to work at high resolutions. The basic features of a double-focussing mass spectrometer are shown in fig. 13.28.

FIG. 13.28. *Basic features of the double-focussing mass spectrometer*

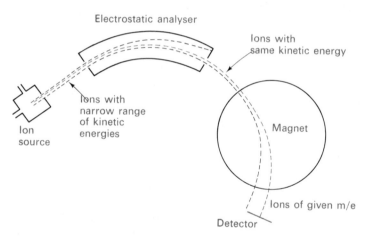

The appearance of a mass spectrum is illustrated in Plate 9, and is a plot of signal from the recorder against the integral values of m/e. The spectrum obtained from the instrument is usually in the form of sharp peaks, but it is common practice to convert the spectrum into a 'stick diagram' (fig. 13.29) in which the major peaks only are drawn and the

FIG. 13.29. *Stick diagram of the mass spectrum of ethanol*

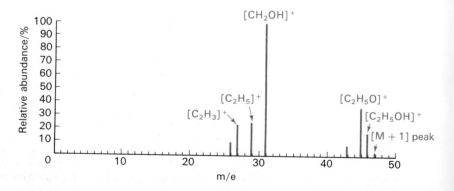

198

Plate 9. Mass spectrum of ethanol, CH₃CH₂OH. The traces (a) to (c) are recorded simultaneously with different degrees of amplification

height of the line represents a percentage of the height of the highest peak present. The information contained in a mass spectrum may be valuable to the chemist in a variety of ways, and some of these are discussed below.

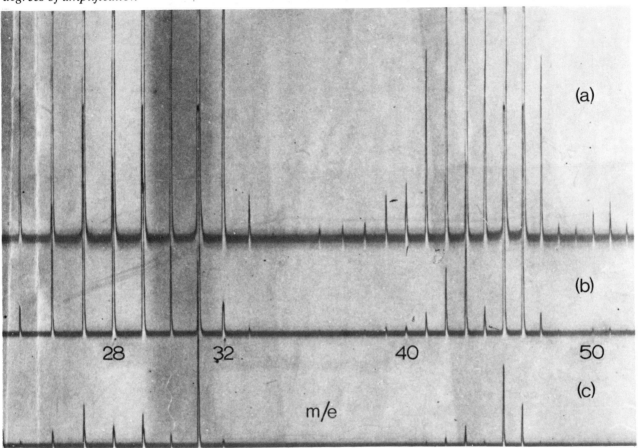

Applications of mass spectrometry

An obvious application of the mass spectrometer is for the measurement of relative atomic masses, and for the determination of the isotopic abundance of a sample of an element. The mass spectrum will lead directly to the accurate relative atomic mass of each isotope present, and the intensity of the lines will enable the percentage distribution of each isotope to be calculated.

In the same way, the mass spectrometer can be used to determine relative molecular masses of compounds, since the line with the highest m/e value in the spectrum is likely to be the single-positively charged ion of the complete molecule, M^+, known as the **molecular ion**. This is not always the case, however, and in a few instances the molecular ion is completely fragmented in the ionisation chamber and does not appear in the mass spectrum. The spectrum of ethanol (fig. 13.29) shows a molecular ion at m/e 46, but note that it is not the most intense peak present.

There is a very small peak at (M + 1) for organic compounds and this is due to the presence of molecules containing the ^{13}C isotope, which is present with a natural abundance of 1·1% of ^{12}C. In the mass spectrum of ethanol this (M + 1) peak is about 2·2% of the height of the molecular ion peak (M peak) and this indicates that there are two carbon atoms in the molecule (for an organic compound containing N carbon atoms the height of the (M + 1) peak is approximately $N\%$ of the height of the M peak).

Because the high resolution mass spectrometer enables very accurate measurements of masses to be made, it is very often possible to distinguish between ions which may have the same integral mass on the ^{12}C scale. Consider four ions which commonly occur in mass spectra at an integral m/e value of 28, i.e. N_2^+, CO^+, H_2CN^+ and $C_2H_4^+$; their accurate masses are respectively 28·0061, 27·9949, 28·0187 and 28·0313, and it is possible to identify the atom or group of atoms which constitutes the particular ion in the spectrum. High resolution mass spectrometry enables the relative molecular mass of a compound to be determined with great precision (if the molecular ion appears in the mass spectrum) and hence allows the molecular formula to be determined.

FIG. 13.30. *Mass spectrum of problem worked through in the text*

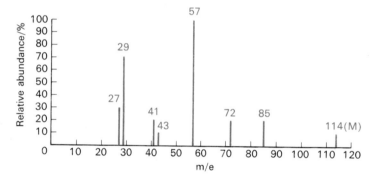

FIG. 13.30. *Mass spectrum of problem worked through in the text*

However, since we are more concerned with the determination of structure, rather than with the establishment of relative atomic and molecular masses, we shall turn our attention to the major use of the mass spectrometer in enabling structural formulae of compounds to be determined. We have already seen that the impact of electrons on the gaseous molecule of a compound can lead to a weakening of the bonds present in the molecule and thus to sequential fragmentation of the substance. It is the fragments of the parent molecule which give rise to the majority of lines in the mass spectrum, and the major ones for ethanol are marked in fig. 13.29. A careful study of the fragmentation pattern can lead to the structure of the parent compound and we illustrate this for the mass spectrum shown in fig. 13.30. If we assume that this spectrum includes the molecular ion, then the relative molecular mass of the compound is 114. There are prominent lines at m/e values of 29, 43, 57, and 85 and these would be consistent with the parent molecule being an isomer of octane, C_8H_{18}, ($M_r = 114$), giving rise to fragments $C_2H_5^+$ (m/e 29), $C_3H_7^+$ (m/e 43), $C_4H_9^+$ (m/e 57) and $C_6H_{13}^+$ (m/e 85), all of which are common ions from straight-chain hydrocarbon molecules. However, the existence of a strong line at m/e 72 would rule out the possibility of the parent molecule being C_8H_{18}, and we must look for other combinations of atoms which fit the complete picture. Other possibilities include $C_7H_2N_2$ ($M_r = 114$) and $C_7H_{14}O$ ($M_r = 114$), but the former is not likely to be the compound owing to the lack of hydrogen atoms. The latter might be an aldehyde, a ketone or even a cyclic compound. A possible structure would be the ketone $CH_3CH_2CCH_2CH_2CH_2CH_3$ with a breakdown pattern:

$$\overset{\|}{\underset{O}{}}$$

$$CH_3CH_2CO^+ + CH_3CH_2CH_2CH_2\cdot$$
$$m/e\ 57$$

$$[CH_3CH_2COCH_2CH_2CH_2CH_3]^{\cdot+}$$

$$CH_3CH_2CH_2CH_2CO^+ + CH_3CH_2\cdot$$
$$m/e\ 85$$

By loss of carbon monoxide the ion $CH_3CH_2CH_2CH_2CO^+$ would give rise to $CH_3CH_2CH_2CH_2^+$ (m/e 57) and by further fragmentation $CH_3CH_2CH_2^+$ (m/e 43) and $CH_3CH_2^+$ (m/e 29) could be generated. The line at m/e 41 is probably due to the loss of a hydrogen atom from the ion $C_3H_7^+$; a similar process would account for the strong line at m/e 27 by loss of hydrogen from $C_2H_5^+$:

$$C_3H_7^+ \longrightarrow C_3H_5^+ + H_2$$
$$m/e\ 41$$
$$C_2H_5^+ \longrightarrow C_2H_3^+ + H_2$$
$$m/e\ 27$$

An interpretation of the mysterious line at m/e 72 might be that the molecule undergoes some form of chemical reaction (or rearrangement) in the gas phase during its time in the ionisation chamber, and in the case of this particular compound the reaction

$$C_7H_{14}O(g) \longrightarrow C_4H_8O(g) + C_3H_6(g)$$

is found to occur. Since the compound C_4H_8O is formed in the ionisation chamber the ion $C_4H_8O^+$ is also formed and is responsible for the line at m/e 72 in the spectrum.

The above example is intended to show that it is possible to use the evidence from mass spectra, in association with complementary evidence from other techniques, to determine chemical structure. Clearly it would be necessary to produce confirmatory evidence in the above example that the compound C_4H_8O is a decomposition product of the parent molecule at that temperature, and this is easily established. The more complex the molecule the greater the task, but the coupling of a computer to the signal output of the mass spectrometer means that data can now be assimilated, decoded and presented in tabular form in the space of a few minutes.

13.14 Magnetic susceptibility

Magnetic measurements enable the presence of unpaired electrons in atoms and molecules to be detected. In the context of transition metal chemistry, in particular, such measurements have proved of great value in understanding the bonding and the stereochemistry of complexes.

Background

Any material placed in a magnetic field will experience some effect due to the field, although in many cases the effect will be so small as to appear non-existent. There are, however, two main effects which may be distinguished from each other quite clearly. Imagine that two different materials are suspended between the poles of a strong electromagnet so that only part of the specimen is in the region of the magnetic field when the magnet is switch on. When the magnetic field is subsequently established specimen 1 moves out of the space between the poles of the magnet, from the stronger to the weaker part of the field. Specimen 2, however, moves in the opposite direction, from the weaker to the stronger part of the field. Those substances which behave like specimen 1, which move out of the field, are said to be **diamagnetic**; and those like specimen 2, which move into the field, are known as **paramagnetic** substances. There is a further type of effect, which is an extension of paramagnetism, known as ferromagnetism which is a property of a few substances only, principally iron, cobalt and nickel. Ferromagnetism is an abnormal effect in that its magnitude is of the order of one million times greater than the corresponding paramagnetic effect. Ferromagnetic substances also acquire a permanent magnetic moment. Since it only occurs in a few cases, ferromagnetism will not be discussed futher in this book—the chemical applications being relatively unimportant compared with those of paramagnetism.

Diamagnetism is a property possessed by all materials and which is capable of a fairly simple interpretation. In all substances there are electrons revolving around the nuclei of the atoms composing them; when an external magnetic field is imposed on the moving electrons their speeds will change, inducing a magnetic field which is opposed to the applied field—this being an example of Lenz's law. A parallel situation is that of a current-carrying conductor placed in a magnetic field. Insofar as all substances are composed of atoms, and therefore electrons moving around nuclei, all substances exhibit a diamagnetic effect to a greater or lesser degree.

Paramagnetism, on the other hand, is associated with the presence in the substance of unpaired electrons, which cause the substance to behave like a collection of small permanent magnets, each with an associated magnetic moment. In an applied magnetic field these little magnets tend to align themselves parallel with the field and this ordering effect is resisted by the thermal agitation of the particles (ions, atoms or molecules) of which the substance is composed. As the temperature increases the thermal energy of the particles increases and the paramagnetic effect is consequently reduced; this is in direct contrast to the diamagnetic effect which is not temperature dependent. In general any effect due to paramagnetism in a molecule is about one hundred times as great as the effect due to diamagnetism, so that although the diamagnetism cannot be ignored its magnitude is much smaller than that of paramagnetism.

Experimental procedure

From a knowledge of the electronic structures of the transition elements (p. 32), it is clear that paramagnetism can be expected for such species and this can be demonstrated quite easily. The apparatus used originally was devised by Gouy, and consists essentially of an analytical balance and a powerful magnet (fig. 13.31). A cylindrical glass tube containing the sample under investigation is weighed in the absence of any magnetic field. It is then lowered into a strong uniform magnetic field of about 0·5 T (5 000 Gauss) in such a manner that the bottom of the tube is in the region of maximum field and the level of the top of the sample is completely out of the field. It is now necessary to increase the mass on the compensating arm to restore the balance, showing that the sample is being drawn into the magnetic field. If quantitative

FIG. 13.31. *Essential features of
the method for determining
magnetic susceptibilities*

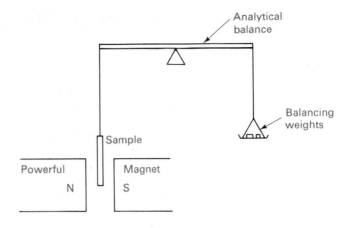

measurements are required in order to determine the so-called magnetic susceptibility
of the specimen (from which the magnetic moment can be calculated) the repulsion
due to diamagnetism of the empty sample tube is determined. The apparatus is also
calibrated with a compound of known magnetic susceptibility, for example iron (II)
ammonium sulphate, $FeSO_4(NH_4)_2SO_4.6H_2O$.

From such measurements a value for the magnetic moment of the substance may be
calculated.

For elements of the first transition series there is a simple relationship
between the magnetic moment and the number of unpaired electrons in
the atom or ion concerned. This is obtained by expressing the magnetic
moment in units known as the Bohr magneton, defined as follows:

e = electronic charge
m = mass of the electron
c = speed of light
h = Planck's constant

$$1 \text{ Bohr magneton} = \frac{he}{4\pi mc}$$

The value of the Bohr magneton is $9 \cdot 27 \times 10^{-24}$ A m^2. For a complex
ion of the first transition series, the magnetic moment μ_B is related to
the number of unpaired electrons, n, by the formula:

$$\mu_B = \sqrt{n(n+2)} \text{ B.M. (Bohr magnetons)} \qquad (12)$$

**Table 13.5 Comparison of calculated and observed magnetic moments
for some transition metal ions**

Ion	Number of unpaired electrons	Calculated magnetic moment (B.M.)	Observed magnetic moment (B.M.)
Ti^{3+}	1	1·73	1·73
V^{3+}	2	2·84	2·75–2·85
V^{2+}	3	3·87	3·80–3·90
Cr^{2+}	4	4·90	4·75–4·90
Mn^{2+}	5	5·92	5·65–6·10
Fe^{2+}	4	4·90	5·10–5·70
Co^{2+}	3	3·87	4·30–5·20
Ni^{2+}	2	2·84	2·80–3·50
Cu^{2+}	1	1·73	1·70–2·20

For example, an ion <u>containing</u> two unpaired electrons will have a magnetic moment of $\sqrt{2(2 + 2)}$ or 2·84 B.M. From measurements of the magnetic susceptibilities of complexes it is a relatively simple matter to calculate the magnetic moment of the particular species and hence the number of unpaired electrons present. Table 13.5 shows some values of calculated and experimental magnetic moments for a number of transition metal ions. The agreement, in most cases, is quite close.

There is much greater disagreement between the calculated and observed magnetic moments for ions of the second and third transition series if the calculated values are obtained from application of equation (12). The reason for this is that the paramagnetic moment is made up of two parts, a contribution from the electron as it revolves about the nucleus (orbital momentum) and a contribution from the electron as it spins about its own axis (spin momentum). In the first transition series the orbital contribution is quite small compared with the spin contribution and so equation (12) is a good approximation to the truth. In the case of the second and third transition series, however, the orbital contribution is appreciable and equation (12) is no longer valid.

The information given in Table 13.5, which refers to ions of the first transition series, is displayed in an alternative manner in fig. 13.32. The magnetic moments of Sc^{3+} and Zn^{2+} are given for completion of the picture. There are two features which are immediately obvious from a study of fig. 13.32, namely the symmetrical nature of the graph and the peak which occurs at the d^5 electron configuration. Both of these features are a consequence of the way in which the d orbitals are progressively filled by electrons, as seen in chapter 3. In the d^5 configuration all the electrons are unpaired and occupy the five d orbitals singly, the magnetic moment corresponding to the five unpaired electrons. However, the d^6 configuration has the sixth electron paired with one of the other five, leaving four unpaired electrons. This will, in turn, result in the d^6 configuration having the same magnetic moment as the d^4 configuration. Similarly the pairs d^3 and d^7, d^2 and d^8, d^1 and d^9 will have the same magnetic moments as will d^0 and d^{10} (in this case zero since there are no unpaired electrons present).

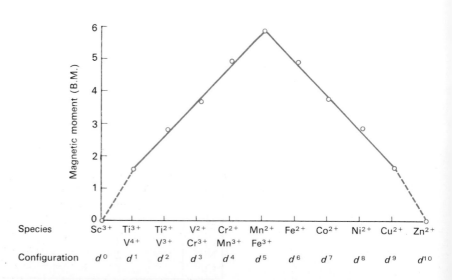

FIG. 13.32. *Magnetic moments for ions from Sc to Zn in the Periodic Table*

Applications of the method

Two examples are given to illustrate the use of magnetic measurements in inorganic chemistry.

(a) Compounds containing the hexacyanoferrate(II) ion, $Fe(CN)_6^{4-}$, are diamagnetic (no unpaired electrons) whereas compounds containing the hexacyanoferrate(III) ion, $Fe(CN)_6^{3-}$, have a paramagnetic moment of about 1·73 (one unpaired electron). At first sight there seems to be no simple way of reconciling these two results with configurations d^6 and d^5 for Fe(II) and Fe(III) compounds respectively. However, in an octahedral environment (in this case provided by the six cyanide ions) the d orbitals, which are normally degenerate, interact with the six surrounding groups, and are split into a lower energy group of three and a higher energy group of two as shown in fig. 13.33. The energy difference between the two sets of d levels is determined by the nature of the six groups which surround the ion. In the case of both $Fe(CN)_6^{4-}$ and $Fe(CN)_6^{3-}$ it is energetically more favourable for the electrons (six and five respectively) to pair in the lower energy level. The complex ions thus contain 0 and 1 unpaired electron respectively in accordance with the experimental results.

FIG. 13.33. *The splitting of the 3d orbitals in an octahedral environment (note that the splitting is greater for Fe(III) than for Fe(II))*

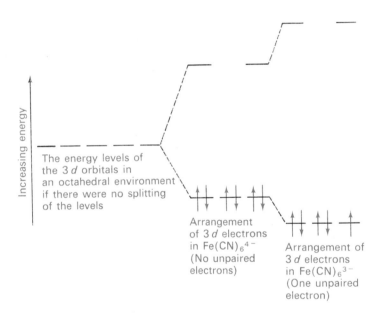

FIG. 13.34 *The arrangement of the 3d electrons in the FeF$_6^{3-}$ and the Fe(CN)$_6^{3-}$ complex ions*

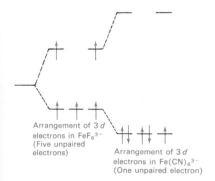

(b) The paramagnetic moment associated with the FeF_6^{3-} ion is consistent with the ion having five unpaired electrons whereas, as we have seen, the ion $Fe(CN)_6^{3-}$ has one unpaired electron. These experimental results can be explained if it is assumed that the six fluoride ions do not cause as great a splitting of the d orbitals as do the six cyanide ions. In the case of FeF_6^{3-} the increase of energy resulting from pairing the electrons in the lower energy d orbitals is greater than that which results when the d levels are singly occupied; consequently the latter arrangement is preferred. For the ion $Fe(CN)_6^{3-}$ the reverse situation is more energetically favourable. These two arrangements are shown in fig. 13.34. Spectroscopic measurements confirm that the cyanide ion does indeed cause a greater splitting than the fluoride ion.

1 Describe any ONE spectrometric method for the determination of molecular structure (possible methods include infra-red spectroscopy or mass spectrometry but NOT diffraction methods). For the method of your choice you should deal with the practical procedure AND with the interpretation of the results. N

2 (a) Carbon consists of 99% of the isotope ^{12}C and 1% of the isotope ^{13}C. One way of estimating the number of carbon atoms in the molecule of a compound is to examine the mass spectrometer peaks which the molecule gives. The peak corresponding to the second highest mass is caused by the molecule ion having only ^{12}C in it but the highest mass peak is due to a molecule ion with one ^{13}C atom in it.

A hydrocarbon X gives a peak at mass M and a smaller peak at mass $M + 1$. There are no significant peaks at higher mass than this. The peak at mass M is 12·5 times as intense as the peak at mass $M + 1$.

When a quantity of X was completely oxidised, 0·11 g of CO_2 and 0·023 g of water were formed. X decolourizes bromine water and 1 mole of X reacts with 1 mole of bromine molecules.

What is the molecular formula and probable structure of X?

(b) Bromine consists of a mixture of isotopes ^{79}Br and ^{81}Br. Assuming no bonds are broken in the mass spectrometer and that hydrogen has only one significant isotope, how many peaks will be given by the compound CH_2Br_2 and what will each peak be due to? N (S)

3 State what you understand by **each** of the following terms: *atomic number, isotope, atomic mass, mole.*

Draw a diagram of a mass spectrometer, labelling the parts. Explain how the instrument may be used to determine (a) molecular mass, (b) isotopic ratio.

What limitations are there in the determination of molecular mass by mass spectrometry? W

4 Figure 1 is the low resolution mass spectrum of chlorobenzene, C_6H_5Cl.

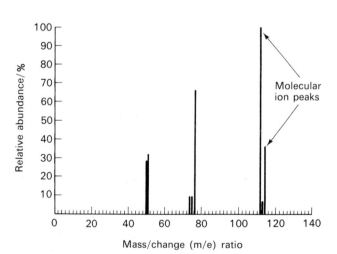

FIG. 1

(i) Explain the presence of two molecular ion peaks.
(ii) What fragment corresponds to the m/e ratio of 77?

5 Figures 1 and 2 show the mass spectrometer traces obtained from two hydrocarbons A and B whose molecular masses are given.

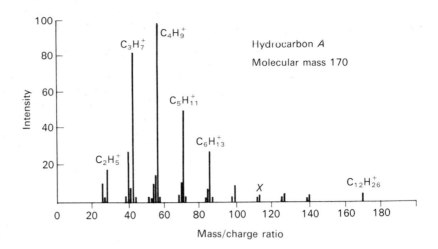

FIG. 1

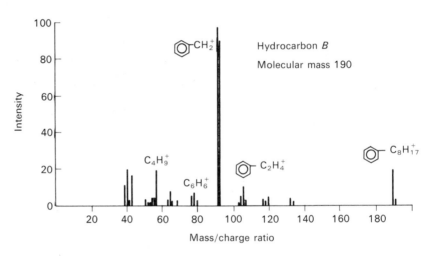

FIG. 2

    (i) Explain how in these examples a single substance gives traces of several different species.

    (ii) What is the connection between the species of highest mass/charge ratio and the substance being analysed?

    (iii) Explain why each peak is accompanied by several smaller peaks.

    (iv) Write the structural formula for the hydrocarbon A, and explain how it is deduced from the mass spectrogram.

    (v) Draw two structures which could give the mass spectrogram shown for B.

    (vi) What is the probable formula for the species giving the trace marked X in the spectrogram for A?

    (vii) State TWO ways in which samples of A and B could be distinguished experimentally.   N

6 Compound A reacts with 2,4-dinitrophenylhydrazine to give a solid derivative but does not react with Fehling's solution. Its empirical formula is C_4H_8O, and the IR spectrum of the compound indicates the presence of a $>C=O$ group.

The NMR spectrum of the compound, shown in Figure 1, shows three 'peaks' centred at 7·6, 7·9 and 9·0 τ.

    (i) In how many different environments are hydrogen atoms found in this compound?

    (ii) From the integration, what is the ratio of hydrogen atoms in each environment?

207

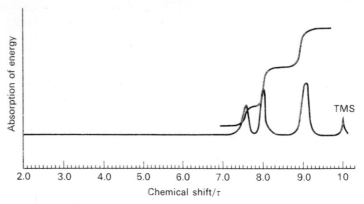

FIG. 1

(iii) Using the answers to (i) and (ii) in conjunction with the other information given, determine the structure of compound A.

Confirm your answer by reference to Appendix II on page 460.

7 Quantitative analysis of compound B shows it to contains 22·2% carbon, 4·6%
hydrogen and 73·2% bromine.

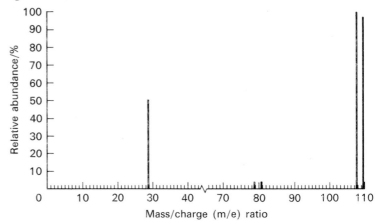

FIG. 1

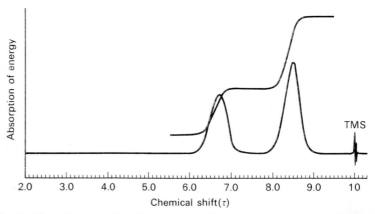

FIG. 2

The mass spectrum and NMR spectrum of Compound B are given in Figures 1
and 2 respectively.

(i) From the quantitative analysis data calculate the empirical formula of
compound B.

(ii) From the NMR spectrum determine the ratio of hydrogen atoms in each
enviroment in the molecule.

(iii) Using the answers to (i) and (ii) in conjunction with Appendix II on page

460 propose a structural formula for compound B.

    (iv) Is the mass spectrum consistent with the structural formula proposed? Explain your reasoning.

8 ·Compound C was found to have an empirical formula of C_2H_3Cl. On reduction with hydrogen, it formed compound D which has the NMR spectrum shown in Figure 1.

    (i) From the NMR spectrum determine the ratio of hydrogen atoms in each environment in a molecule of compound D.

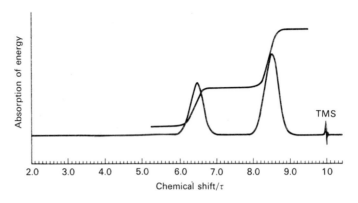

FIG. 1

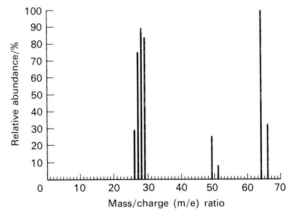

FIG. 2

    (ii) Using the answer to (i) in conjunction with Appendix II on page 460, propose a structural formula for compound D.
        The mass spectrum of compound D is given in Figure 2.

    (iii) Is the mass spectrum consistent with the structural formula proposed for compound D? Explain your reasoning.

    (iv) From the answers above, propose a structural formula for compound C. Explain your reasoning.

9 Compound D is found to have a relative molecular mass of 60. Its IR and mass spectra are shown in Figures 1 and 2 respectively.

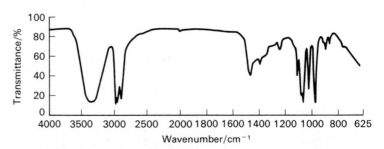

FIG. 1

209

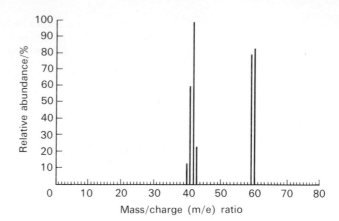

FIG. 2

Oxidation of the substance yielded a compound whose IR spectrum showed absorptions at:

$$3\ 000\ cm^{-1} \text{ and } 1\ 740\ cm^{-1}$$

(i) Considering only those absorptions listed, deduce what you can about the structure of the compound using Appendix I on page 459.

Frequency/cm^{-1}

2 880–2 960
1 380
1 050–1 100

(ii) Suggest a molecular formula and structure for the initial substance.

(iii) Identify the oxidation product from the data given.

(iv) Is the mass spectrum (Figure 2) consistent with the proposed structure? Justify your answer.

(v) What chemical tests would you use to verify your suggestions in (ii) and (iii)?
State what you would observe in each case, giving equations wherever possible.

10 The Bragg law for crystal diffraction is given by the following equation:

$$2d \sin \theta = n\lambda$$

(i) Explain clearly and concisely the meaning of diffraction.

(ii) To what do the symbols d, θ, n and λ refer in the Bragg equation? Show how this equation is derived.

(iii) Explain how the equation can be used to give information about the structure of crystals such as sodium chloride.   AEB (Physical Science)

11 X-ray diffraction studies show that potassium chloride is a 6:6 co-ordinated compound and has a cubic close packed structure for which the unit cell is a cube of side 0·628 nm containing four potassium ions and four chloride ions. Calculate:

(i) the volume of the unit cell;

(ii) the Avogadro constant, given that the volume of one mole of potassium chloride is 37·3 cm^3.

12 Describe the use of X-rays to determine the structure of solids. Illustrate and discuss the different types of X-ray diffraction pattern that can be obtained and mention any limitations of the method.

X-ray studies of lithium chloride using X-rays of wavelength 0·0585 nm produced a strong diffraction at an angle of diffraction, θ, of 6·3° and another diffraction at 8·8°. There were also two related weaker diffractions at 5·4° and 10·9°. Use the Bragg diffraction equation to calculate the separation of the crystal planes in lithium chloride.

Draw suitable diagrams to show how the three sets of crystal planes are related to the unit cell of lithium chloride (face-centred cubic).   N

Energetics

Thermochemistry

14.1 Introduction

Chemical reactions are invariably associated with a transfer of energy; if the amount of energy transferred can be measured—and fortunately it often can—then it is possible to obtain vital information about the interactions which occur during a chemical reaction. Most frequently, energy transfer in chemical reactions takes place in the form of heat. Many reactions are exothermic and transfer heat to the surroundings; the oxidation of aluminium provides a good example:

$$4Al(s) + 3O_2(g) \longrightarrow 2Al_2O_3(s)$$

In the above reaction 1676 kJ of heat are evolved to the surroundings for every mole of aluminium oxide formed at 298 K (strictly speaking at constant pressure). The progress of this reaction might be shown diagramatically as in fig. 14.1. Clearly we should have to put energy into the system to return aluminium oxide to its constituent elements.

FIG. 14.1. *Energy diagram for the reaction*
$4Al(s) + 3O_2(g) \rightarrow 2Al_2O_3(s)$

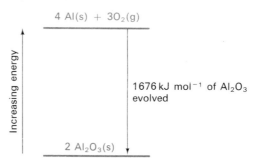

Many reactions, however, are endothermic and take place with a transfer of heat from the surroundings to the chemical system. An example is provided by the formation of nitrogen oxide, NO, a reaction important in atmospheric pollution:

$$N_2(g) + O_2(g) \longrightarrow 2NO(g)$$

In this reaction 90·4 kJ of heat are absorbed by the chemical system for every mole of nitrogen oxide formed at 298 K. This could be represented by fig. 14.2 and the direction of energy transfer tells us that the

FIG. 14.2. *Energy diagram for the reaction*
$N_2(g) + O_2(g) \rightarrow 2NO(g)$

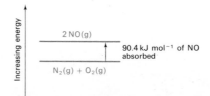

product is less stable than the elements from which it was formed. Just why the reaction should have occurred at all under these circumstances is a question that is postponed until chapter 15.

Before we can interpret energy transfers, or energy changes as they are commonly called, in terms of molecular interactions taking place during a chemical reaction (and more specifically in terms of the breaking and forming of chemical bonds) we must look closely at what it is we are measuring and at the conditions under which we are taking the measurements.

14.2
Changes in internal energy

It is useful to think of a chemical system as possessing an internal energy, U. It is not possible to measure the absolute value of the internal energy of a system but it is possible to measure changes in internal energies, ΔU. The internal energy of a system is the sum of the internal energies of its constituent molecules and these are comprised of several contributions, some kinetic and some potential, as shown below.

(a) Translational energy associated with the translational motion of a molecule. This is significant only for atoms and molecules in gases and liquids.

(b) Rotational energy associated with the rotation of a molecule about its centre of gravity. As we have seen previously (p. 127) there is no rotational contribution for a monatomic gas.

(c) Vibrational energy associated with the vibrational motions of the atoms within the molecule. This is possessed by all polyatomic molecules in the solid, liquid and gaseous states.

(d) Electronic energy associated with the electrostatic interactions between the various charged particles that make up the molecule.

The first two contributions are entirely kinetic, the third is partly kinetic and partly potential, while the fourth is completely potential. Since we are going to be mainly interested in the interactions between the charged particles within a molecule, and the way in which these interactions change when the molecule reacts and bonds are broken and formed, we shall need to focus attention on the changes in electronic potential energy. This can be done by eliminating, so far as is possible, the kinetic energy changes by ensuring that the products of the reaction return to the original temperature of the reactants. In this way the transfer of kinetic energy is minimised and the measured value of ΔU can be attributed, in the main, to changes in electronic energy:

$$\Delta U = U_{\text{final}} - U_{\text{initial}}$$

Conventionally values of ΔU are quoted at 298 K, even though many of them cannot be measured directly at this temperature. They refer to **constant volume** conditions. A further important convention attaches to the sign given to ΔU. If the system transfers heat to the surroundings (that is, the reaction is exothermic) then ΔU is negative. Conversely, for endothermic reactions the system gains heat from the surroundings and ΔU is positive.

14.3
Changes in enthalpy

Consider the energy change associated with the reaction between zinc and dilute hydrochloric acid:

$$Zn(s) + 2H^+(aq) \longrightarrow Zn^{2+}(aq) + H_2(g)$$

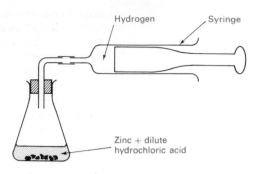

If the reaction is carried out at constant pressure, e.g. in the apparatus shown in fig. 14.3, then work is done by the hydrogen gas in expanding against the atmospheric pressure. The performance of work involves the transfer of energy and the value of ΔU which we measure will not be a true measure of the change in internal energy. However, a simple correction can be made. If the hydrogen evolved in the reaction produces a change in volume ΔV at a constant pressure p, then the work done by the system in expanding the gas will be $p\Delta V$ and this will be equal to the energy transferred as work. We now define a new term, ΔH, the **enthalpy change** (or the heat of reaction as it is commonly known), which takes into account the work done when a system expands or contracts during a reaction at constant pressure:

$$\Delta H = \Delta U + p\Delta V$$

If work is done **by** the system, $p\Delta V$ is positive. If work is done **on** the system, $p\Delta V$ is negative, for example in the following reaction:

$$N_2(g) + 3H_2(g) \longrightarrow 2NH_3(g)$$

Example 1
Calculate ΔU at 298 K for the reaction

$$\tfrac{1}{2}N_2(g) + \tfrac{3}{2}H_2(g) \longrightarrow NH_3(g)$$

given that ΔH is $-46\cdot0$ kJ mol^{-1} of ammonia formed.

Applying the ideal gas law $pV = nRT$ it follows that

$$p\Delta V = \Delta nRT$$

In this particular case one mole of ammonia is formed and 0·5 mole of nitrogen and 1·5 moles of hydrogen are consumed, i.e. there is an overall decrease of 1 mole of gas. Since

$$\Delta H = \Delta U + p\Delta V = \Delta U + \Delta nRT \qquad (1)$$

and $\Delta n = -1$, it follows that $\Delta nRT = -1 \times 8\cdot314 \times 298$ or -2478 J (approx. $-2\cdot5$ kJ). Substituting in equation (1) we get:

$$-46\cdot0 = \Delta U - 2\cdot5$$

or

$$\Delta U = -43\cdot5 \text{ kJ mol}^{-1} \text{ of ammonia formed}$$

Except in reactions involving changes in gas volumes, the difference between ΔU and ΔH is unimportant. However, to avoid any confusion, all energy changes quoted from this point in the chapter will be enthalpy changes, ΔH, (the heat change that occurs when the reaction in question is carried out at constant pressure, as most reactions commonly are). The sign convention applies to ΔH as it does to ΔU, so that a negative sign implies an exothermic reaction.

14.4 Measuring heat energy changes

The heat energy change in a chemical reaction is measured by transferring it to a known amount of substance whose 'heat absorbing capacity' (heat capacity) is known, and recording its temperature rise.

This is the basis of calorimetry and the sophistication of the technique is determined by the degree of accuracy that is required. However, two chief problems have to be solved in any practical measurement of ΔH.

(a) The heat will probably be transferred to several materials of different heat capacities, e.g. the calorimeter vessel, the water contained in it and the thermometer recording the temperature change. Approximate measurements can be obtained in calorimeters of negligible heat capacity (for example, polystyrene beakers) ignoring the heat absorbed by the thermometer. Accurate determinations of ΔH require that the heat capacity of the whole apparatus be determined, either by previously carrying out a reaction whose value of ΔH is known or by transferring a known amount of electrical energy to the calorimeter and contents and measuring the temperature rise.

(b) Precautions must be taken to ensure either that no heat is transferred to the surroundings, or else that the amount that is transferred can be estimated. The former condition is achieved in an **adiabatic** calorimeter which encloses the reaction vessel in an outer jacket, whose temperature is automatically maintained as nearly as possible equal to that of the calorimeter.

FIG. 14.4. *The bomb calorimeter*

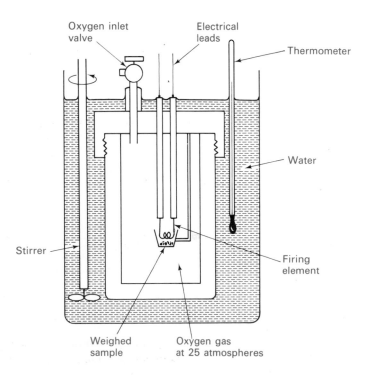

Measurements taken with the bomb calorimeter (see later) enable a value of ΔU to be obtained, since the reaction is carried out at constant volume. This can be used to calculate ΔH by the method outlined in the example (p. 214), using the equation:

$$\Delta H = \Delta U + \Delta nRT$$

Values of ΔU and hence ΔH accurate to 1 part in 10^4 can be obtained using a **bomb calorimeter**. The cross section of a typical bomb calorimeter is shown in fig. 14.4 in which a known mass of the material is ignited electrically in a sealed vessel containing pure oxygen at a known pressure. The heat evolved is transferred to the water in the surrounding jacket whose temperature rise is measured. The heat capacity of the bomb calorimeter may be determined by combustion of a standard substance (usually benzoic acid) or, more commonly, by electrical heating. For accurate work, account must be taken of the energy transferred through the wire used to start the combustion process.

The values obtained refer to one particular type of reaction, namely combustion in oxygen, for example:

$$C_6H_{12}O_6(s) + 6O_2(g) \longrightarrow 6CO_2(g) + 6H_2O(1)$$

However, as we shall see later (p. 225), these values may be used to calculate enthalpy changes for many other types of reaction.

14.5
Conventions, symbols and standard states

As we have seen in section 14.2, the magnitude of the heat energy transfer in a chemical reaction depends upon the changes that occur in the different factors which contribute to the internal energy of the system. If possible, we want to eliminate all of these changes except the one involving the electronic potential energy contribution. Only then will it be possible to make valid comparisons between the ΔH values for different reactions. This means that certain standard conditions must operate and certain conventions must be obeyed. The most important of these are summarised below.

Amount of substance

Clearly, the amount of energy transferred will depend upon the amount of substances which react. Any ambiguity is removed by quoting the complete equation, together with the appropriate value of ΔH. Thus the statement

$$H_2(g) + Cl_2(g) \longrightarrow 2HCl(g) \qquad \Delta H = -184 \cdot 6 \text{ kJ mol}^{-1}$$

signifies that when one mole of hydrogen gas reacts with one mole of chlorine gas to produce two moles of hydrogen chloride gas completely, the change is accompanied by the output of 184·6 kJ of heat energy to the surroundings.

Temperature

For the purposes of valid comparison, values of ΔH are usually quoted at a standard temperature of 298 K (although the choice of this temperature is quite arbitrary).

In practice, many enthalpy changes cannot be measured at 298 K so we need to know the manner in which ΔH varies with temperature so that a conversion can be made. Fortunately the relationship is a simple one, first derived by Kirchhoff in 1858:

$$\frac{d(\Delta H)}{dT} = \Delta C_p$$

where ΔC_p is the difference between the heat capacities of the products and reactants at constant pressure and d/dT is the usual calculus notation for rate of change. On integration this becomes:

$$\Delta H_2 - \Delta H_1 = \int_{T_1}^{T_2} \Delta C_p dT \qquad (2)$$

and since the way in which C_p varies with temperature can be expressed in the form

$$C_p = a + bT + cT^2 + \cdots$$

where a, b, and c are constants for a particular substance, the integration can be carried out with little difficulty (for a temperature range of about 1000 K the first three terms only in the above equation are required).

In practice it is often sufficient to use the average heat capacity values in the temperature range of interest and, where this is valid, equation (2) reduces to:

$$\Delta H_2 - \Delta H_1 = \Delta C_p(T_2 - T_1)$$

where ΔC_p is the difference between the average heat capacities of the products and reactants at temperatures of T_2 and T_1. For very small temperature changes the differences between the enthalpy changes for a particular reaction are usually quite small. For example:

$$H_2(g) + \tfrac{1}{2}O_2(g) \longrightarrow H_2O(g) \quad \Delta H(298 \text{ K}) = -241 \cdot 81 \text{ kJ mol}^{-1}$$
$$\Delta H(291 \text{ K}) = -241 \cdot 75 \text{ kJ mol}^{-1}$$

Pressure

ΔH values are quoted for a pressure of 1 atmosphere (standard atmospheric pressure).

Standard enthalpy changes

Enthalpy changes that have been measured under the conditions of temperature and pressure stated above are known as standard enthalpy changes and the symbol is written thus:

$$\Delta H^\ominus$$

The temperature should be specified in a bracket, i.e. $\Delta H^\ominus(298 \text{ K})$, but if it is not, it may be taken as 298 K.

State symbols

Since an appreciable energy change is involved in changing a substance from one state of matter to another, it is important that a state symbol is attached to all the formulae given in an equation, e.g.

$$H_2(g) + \tfrac{1}{2}O_2(g) \longrightarrow H_2O(l) \quad \Delta H^\ominus(298 \text{ K}) = -285 \cdot 9 \text{ kJ mol}^{-1}$$
$$H_2(g) + \tfrac{1}{2}O_2(g) \longrightarrow H_2O(g) \quad \Delta H^\ominus(298 \text{ K}) = -241 \cdot 8 \text{ kJ mol}^{-1}$$

The difference between these two values arises since in the first case **liquid** water is formed and in the second case water **vapour** is formed;

217

the difference of $44 \cdot 1$ kJ mol^{-1} represents the molar enthalpy of vaporization of water.

If a substance involved in a reaction can exist in more than one allotropic form, then this should be specified, e.g.

$$C(graphite) + O_2(g) \longrightarrow CO_2(g) \quad \Delta H^{\ominus}(298 \text{ K}) = -393 \cdot 5 \text{ kJ mol}^{-1}$$
$$C(diamond) + O_2(g) \longrightarrow CO_2(g) \quad \Delta H^{\ominus}(298 \text{ K}) = -395 \cdot 4 \text{ kJ mol}^{-1}$$

Substances in aqueous solution are denoted by the symbol (aq).

14.6
Standard enthalpies of formation

Up until now we have been considering the enthalpy changes that accompany any general chemical reaction. In this section we look at one particular kind of reaction—the reaction in which a compound is formed from its elements. For example, the formation of ethene:

$$2C(graphite) + 2H_2(g) \longrightarrow C_2H_4(g) \quad \Delta H^{\ominus}(298 \text{ K}) = +52 \cdot 3 \text{ kJ mol}^{-1}$$

The enthalpy change which occurs when one mole of a compound is formed, under standard conditions, from its elements in their standard states is known as the standard enthalpy (or heat) of formation of the compound, and is written $\Delta H_f^{\ominus}$. Some standard enthalpies of formation are listed in Table 14.1.

Not all compounds can be made by the direct combination of their constituent elements but, in such cases, $\Delta H_f^{\ominus}$ values can be calculated indirectly (see later, p. 225).

Table 14.1 Standard enthalpies of formation of some common compounds, $\Delta H_f^{\ominus}$ (298 K)/kJ mol^{-1}

Compound	$\Delta H_f^{\ominus}$(298 K)/kJ mol^{-1}	Compound	$\Delta H_f^{\ominus}$(298 K)/kJ mol^{-1}
Al_2O_3	-1676	$AlCl_3$	-704
CaO	-636	CCl_4	-136
CO	-111	PCl_3	-320
CO_2	-393	KCl	-436
$H_2O(g)$	-242	NaCl	-411
MgO	-602	$SiCl_4$	-640
N_2O	$+82 \cdot 0$		
NO	$+90 \cdot 4$		
HF	$-271 \cdot 1$	CH_4	$-74 \cdot 8$
HCl	$-92 \cdot 3$	C_2H_6	$-84 \cdot 6$
HBr	$-36 \cdot 2$	C_2H_4	$+52 \cdot 3$
HI	$+26 \cdot 5$	C_6H_6	$+49 \cdot 0$
NH_3	$-46 \cdot 0$	C_2H_5OH	-278
PH_3	$+5 \cdot 4$	CH_3CHO	-192
AsH_3	$+66 \cdot 4$		

Elements have, by definition, zero enthalpies of formation in their standard states at 298 K; thus the $\Delta H_f^{\ominus}$ of a compound represents the energy transferred to or from the surroundings when chemical bonds in the elements are broken and new bonds are formed in the compound. It is therefore a measure of the stability of the compound relative to its constituent elements. For example, if the compound is less stable than its constituent elements then energy must be absorbed from the surroundings. Conversely, if energy is evolved to the surroundings then the compound will be more stable than its constituent elements.

Before proceeding further it is worth stressing several factors at this stage.

(a) Since we are taking elements as standards, it is not very helpful to compare $\Delta H_f^\ominus$ values of, say, CO with $SiCl_4$ for neither compound contains an element common to both. On the other hand, the comparison

Compound	HF(g)	HCl(g)	HBr(g)	HI(g)
$\Delta H_f^\ominus(298\text{ K})/\text{kJ mol}^{-1}$	$-271\cdot1$	$-92\cdot3$	$-36\cdot2$	$+26\cdot5$

is helpful since it indicates the effect of the increasing size of the halogen atom and the remarkable stability of the fluoride.

(b) The word 'stability' is used in the context of energetic stability and not kinetic stability. For example, the fact that $\Delta H_f^\ominus$ of hydrogen peroxide is equal to $-187\cdot6$ kJ mol^{-1} does not mean that the reaction

$$H_2(g) + O_2(g) \longrightarrow H_2O_2(l)$$

will necessary take place at an observable rate (see chapters 24 and 25). Furthermore, hydrogen peroxide is itself energetically unstable with respect to water and oxygen:

$$H_2O_2(l) \longrightarrow H_2O(l) + \tfrac{1}{2}O_2(g) \quad \Delta H^\ominus(298\text{ K}) = -98\cdot3 \text{ kJ mol}^{-1}$$

which underlines the fact that $\Delta H_f^\ominus$ values refer only to the elements as standards.

(c) As is made clear in chapter 15, enthalpy change is only an approximate guide to energetic stability but, as a rule of thumb, it is adequate for our present purposes except at very high temperatures.

The standard enthalpies of formation of the oxides of the short period from sodium to chlorine are shown in Table 14.2. In order to be able to make valid comparisons, enthalpies of formation per mole of oxygen consumed are quoted in the third row.

Table 14.2 Enthalpies of formation of the oxides of elements from sodium to chlorine

Compound	Na_2O	MgO	Al_2O_3	SiO_2	P_4O_{10}	SO_3	Cl_2O_7
$\Delta H_f^\ominus(298\text{ K, mole of compound})/\text{kJ mol}^{-1}$	-416	-601	-1676	-911	-2984	-395	$+265$
$\Delta H_f^\ominus(298\text{ K, mole of oxygen})/\text{kJ mol}^{-1}$	-832	-1202	-1117	-911	-597	-264	$+76$

The variation in stability is clearly indicated, reflecting the powerful reducing properties of magnesium and aluminium and the explosive instability of dichlorine heptoxide, Cl_2O_7.

Another interesting comparison between standard enthalpies of formation of the Group IA oxides and those of the Group 4B oxides is shown in Table 14.3.

219

Table 14.3 Enthalpies of formation of Group 1A and Group 4B oxides

Group 1A oxides $\Delta H_f^\ominus(298\ K)/kJ\ mol^{-1}$	Li_2O −596	Na_2O −416	K_2O −362	Rb_2O −330	Cs_2O −318
Group 4B oxides $\Delta H_f^\ominus(298\ K)/kJ\ mol^{-1}$	CO_2 −393	SiO_2 −911	GeO_2 −551	SnO_2 −581	PbO_2 −277

The Group IA oxides all have ionic structures and the decrease in stability is principally caused by the increasing size of the metal ion which reduces the electrostatic force of attraction between ions of opposite charge. The Group 4B oxides have different structures; carbon dioxide is a gas, silicon dioxide and germanium(IV) oxide have giant covalent structures, and tin(IV) oxide and lead(IV) oxide are predominantly ionic in character. These structural differences account for the random variation in the $\Delta H_f^\ominus$ values.

14.7 Particular enthalpies of reaction

The standard enthalpy changes associated with particular types of chemical reaction have acquired special names and are usually referred to by these names. The most important of these are discussed below.

Enthalpies of combustion

We have already seen that the most accurate calorimetric measurements are carried out in a bomb calorimeter (p. 215). The values obtained are known as standard enthalpies of combustion, defined as: the enthalpy change that takes place when one mole of the substance is completely burned in oxygen, converted to standard conditions. A knowledge of enthalpies of combustion is often of commercial importance, particularly for internal combustion engineers. For example, the hydrocarbon 3,3-dimethylpentane is an important component of high grade octane petrol:

$$C_7H_{16}(l) + 11O_2(g) \longrightarrow 7CO_2(g) + 8H_2O(l)$$
$$\Delta H^\ominus(298\ K) = -4802{\cdot}8\ kJ\ mol^{-1}$$

Table 14.4 Some standard enthalpies of combustion, $\Delta H^\ominus(298\ K)/kJ\ mol^{-1}$

Compound	Formula	$\Delta H^\ominus(298\ K)/kJ\ mol^{-1}$
Methane	CH_4	− 890
Ethane	C_2H_6	−1560
Propane	C_3H_8	−2220
Butane	C_4H_{10}	−2877
Methanol	CH_3OH	− 726
Ethanol	C_2H_5OH	−1367
Propan-1-ol	C_3H_7OH	−2017
Butan-1-ol	C_4H_9OH	−2675

Table 14.4 lists the standard enthalpies of combustion of some common organic compounds at 298 K. It will be seen that the numerical values increase as an homologous series of compounds is ascended.

Enthalpy of neutralization

The enthalpy of neutralization of an acid with an alkali is the enthalpy change which takes place when an amount of acid or alkali is neutralized to form one mole of water. The reaction is carried out in dilute aqueous solution.

Strong acids and alkalis are, by definition, virtually completely ionised, so the reaction between any strong acid and any strong alkali is effectively the reaction between aqueous hydrogen and hydroxyl ions:

$$H^+(aq) + OH^-(aq) \longrightarrow H_2O(l)$$

This is confirmed by the fact that the enthalpy of neutralization of any strong acid by any strong alkali is approximately constant at $\Delta H^{\ominus}(298 \text{ K}) = -57 \cdot 3 \text{ kJ mol}^{-1}$. If a weak acid or alkali is used, or if both are weak, then the enthalpy of neutralization differs significantly from the value given above. This is because weak acids and alkalis are only slightly ionised in aqueous solution, and an enthalpy of ionization term is involved, together with enthalpies of hydration. Although the numerical value is generally less than $-57 \cdot 3$ kJ , in some cases it can be greater:

$$CH_3COOH(aq) + NaOH(aq) \longrightarrow CH_3COONa(aq) + H_2O(l)$$
$$\Delta H^{\ominus}(298 \text{ K}) = -55 \cdot 2 \text{ kJ mol}^{-1}$$
$$HF(aq) + NaOH(aq) \longrightarrow NaF(aq) + H_2O(l)$$
$$\Delta H^{\ominus}(298 \text{ K}) = -68 \cdot 6 \text{ kJ mol}^{-1}$$
$$CH_3COOH(aq) + NH_3(aq) \longrightarrow CH_3COONH_4(aq) + H_2O(l)$$
$$\Delta H^{\ominus}(298 \text{ K}) = -50 \cdot 4 \text{ kJ mol}^{-1}$$

Enthalpy of solution

The dissolving of a solute in a solvent is an interesting case of an enthalpy change accompanying a chemical process. On the one hand the structure of the solute is broken down (for example, the sodium chloride lattice into isolated ions) while on the other hand new bonds are formed between the solute particles and the solvent.

The first of these processes requires an input of energy to overcome the forces holding the solute structure together. Energy known as lattice energy (p. 69) will be absorbed from the surroundings. The second process will involve the release of energy to the surroundings. The enthalpy of solution will measure the difference between these two processes, and is defined as: the enthalpy change which takes place when one mole of solute is completely dissolved in enough solvent so that no further heat change takes place on adding more solvent. Under such conditions the solution is said to be at infinite dilution. When the symbol (aq) is used in a thermochemical equation it can be assumed that the solution is at infinite dilution.

The standard enthalpies of solution of the alkali metal chlorides are listed in Table 14.5. With the exception of lithium fluoride, the dissolving of the other chlorides is an endothermic process. The fact that they

221

nevertheless dissolve spontaneously is a problem that will be tackled in chapter 15.

Table 14.5 Standard enthalpies of solution of the alkali metal chlorides, $\Delta H^\ominus(298\text{ K})/\text{kJ mol}^{-1}$

Compound	Formula	$\Delta H^\ominus(298\text{ K})/\text{kJ mol}^{-1}$
Lithium chloride	LiCl	$-37\cdot2$
Sodium chloride	NaCl	$+ 3\cdot9$
Potassium chloride	KCl	$+17\cdot2$
Rubidium chloride	RbCl	$+16\cdot7$
Caesium chloride	CsCl	$+17\cdot9$

The mechanism of solubility is treated in greater detail in chapter 21, and the solvation energy mentioned above is fully explained there.

Enthalpy of atomisation

The enthalpy change that takes place when a substance decomposes to form one mole of atoms in the gas phase under standard conditions is called the enthalpy of atomisation, and some values are given in Table 14.6.

In the case of liquids and solids the enthalpy of atomisation includes the enthalpy of vaporization, and the enthalpies of fusion and vaporization respectively.

Table 14.6 Some standard enthalpies of atomisation, $\Delta H^\ominus$ (298 K)/ kJ mol^{-1}

Process	$\Delta H^\ominus(298\text{ K})/\text{kJ mol}^{-1}$
$\frac{1}{2}F_2(g) \longrightarrow F(g)$	$+79\cdot1$
$\frac{1}{2}Cl_2(g) \longrightarrow Cl(g)$	$+121\cdot1$
$\frac{1}{2}Br_2(l) \longrightarrow Br(g)$	$+112\cdot0$
$\frac{1}{2}I_2(s) \longrightarrow I(g)$	$+106\cdot6$
$\frac{1}{2}H_2(g) \longrightarrow H(g)$	$+218\cdot0$
$\frac{1}{2}O_2(g) \longrightarrow O(g)$	$+249\cdot2$
$\frac{1}{2}N_2(g) \longrightarrow N(g)$	$+472\cdot8$

Note the relatively low value for fluorine and the very high value for nitrogen which we discuss later (p. 232).

14.8
The first law of thermodynamics

This law, despite its forbidding title, states one of the simplest generalisations of science, namely that energy cannot be created nor destroyed. Energy can be transformed from one kind to another: electrical to heat, kinetic to electrical, potential to kinetic and so on. Energy can never be 'lost' or 'made'. A more useful statement of the law for the chemist summarises the problem we tackled earlier (p. 214), namely that a change in internal energy, ΔU, may not be manifested entirely in the form of heat; it may also be shown by the performance of work **on** or **by** the system. We can express this as follows:

$$\Delta U = q + w$$

ΔU is independent of the path by which the change is carried out and is accordingly said to be a function of state, i.e. a change in U depends only on the initial and final states of the system. It does not matter what route is taken; the same is also true for ΔH.

It is not difficult to see why the value of ΔH for a reaction must be independent of the way in which the reaction is brought about. Consider the following simple reaction sequence in which 1 mole of hydrogen gas and 1 mole of oxygen gas are converted, in two different ways, into 1 mole of water and 0·5 mole of oxygen gas:

Route A $H_2(g) + O_2(g) \longrightarrow H_2O(l) + \frac{1}{2}O_2(g)$
Route B $H_2(g) + O_2(g) \longrightarrow H_2O_2(l) \longrightarrow H_2O(l) + \frac{1}{2}O_2(g)$

In each case the initial and final states are the same. If the enthalpy change was different in one of the routes it would be possible to carry out a cyclic process shown in fig. 14.5 which would create or destroy energy. Experience tells us that this is not possible, and it is no surprise to find the overall standard enthalpy changes by Routes A or B are identical ($\Delta H^{\ominus}(298 \text{ K}) = -286 \text{ kJ mol}^{-1}$). This consequence of the First law of Thermodynamics is summarised in **Hess's Law** (1840) which states:

The change in enthalpy accompanying a chemical reaction is independent of the pathway between the initial and final states.

The importance of this law is considerable as the following examples illustrate.

Energy level diagrams

Hess's law enables us to break down a reaction into several intermediate steps and assign to each step an individual enthalpy change. The sum of the individual changes must, of course, equal the overall enthalpy change provided the initial and final states are the same in each case.

FIG. 14.5. *Energy cannot be created or destroyed*

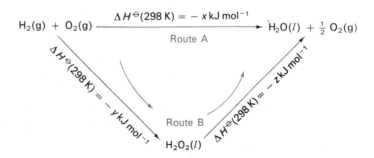

Suppose $x > (y + z)$, then the reaction could proceed by Route A and some of the heat evolved could be absorbed by the products of this *reaction*, $(y + z)$ kJ, and the initial reactants obtained by reversal of Route B. We are back where we started, and have created $x - (y + z)$ kJ of energy from nowhere. This is impossible. If $x < (y + z)$ we could proceed by Route B and reverse Route A. Again we create energy from nowhere. It therefore follows that $x = (y + z)$

This subdivision is best illustrated in the form of an energy level or enthalpy diagram on which elements in their standard states are assigned to a zero level while the positive and negative areas of the diagram indicate decreasing and increasing energetic stability with respect to these elements. An energy level diagram for the reaction cycle described in fig. 14.5 is shown in fig. 14.6. A more involved example, showing the formation of hydrogen chloride from its elements and the particular factors which contribute to the overall enthalpy change, is illustrated in fig. 14.7.

FIG. 14.6. *Energy level diagram to illustrate Hess's law*

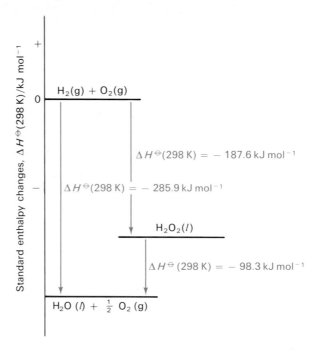

FIG. 14.7. *The standard enthalpy change for the formation of two moles of hydrogen chloride*

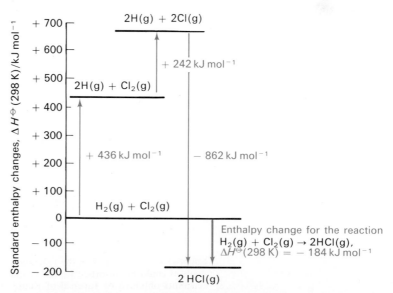

The last example is important in that it allows us to isolate particular factors which contribute to the overall enthalpy change and to make comparisons from one compound to another. Secondly, it allows us to calculate enthalpy changes which cannot be measured directly. The following two examples illustrate an indirect determination of enthalpy changes.

Calculating standard enthalpies of formation

Very few enthalpies of formation can be measured directly by experiment. However, Hess's law enables us to calculate enthalpies of formation from other data, and especially from enthalpies of combustion which can be measured very accurately (p. 220). In order to calculate the enthalpy of formation of a compound we need to know its enthalpy of combustion and the enthalpies of combustion of its constituent elements.

Example 2
Calculate the standard enthalpy of formation of methane from the following standard enthalpies of combustion:

$$C(graphite) + O_2(g) \longrightarrow CO_2(g) \qquad \Delta H_1^{\ominus}(298 \text{ K}) = -393 \text{ kJ mol}^{-1}$$
$$H_2(g) + \tfrac{1}{2}O_2(g) \longrightarrow H_2O(l) \qquad \Delta H_2^{\ominus}(298 \text{ K}) = -286 \text{ kJ mol}^{-1}$$
$$CH_4(g) + 2O_2(g) \longrightarrow CO_2(g) + 2H_2O(l) \quad \Delta H_3^{\ominus}(298 \text{ K}) = -890 \text{ kJ mol}^{-1}$$

We require to calculate the standard enthalpy change associated with the reaction:

$$C(graphite) + 2H_2(g) \longrightarrow CH_4(g) \quad \Delta H_4^{\ominus}(298 \text{ K}) = x \text{ kJ mol}^{-1}$$

FIG. 14.8. *An enthalpy cycle to determine the standard enthalpy of formation of methane*

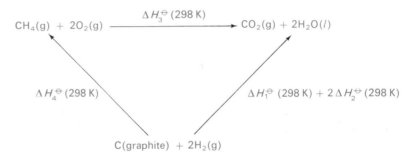

From the enthalpy cycle shown in fig. 14.8 it can be seen that

$$\Delta H_3^{\ominus} + \Delta H_4^{\ominus} = \Delta H_1^{\ominus} + 2\Delta H_2^{\ominus}$$

Thus the standard enthalpy of formation of methane, $\Delta H_4^{\ominus}$, is given by:

$$\Delta H_4^{\ominus} = -393 - 572 + 890 = -75 \text{ kJ mol}^{-1}$$

Calculating enthalpies of reaction

Consider the general reaction expressed as:

$$A + B \longrightarrow C + D$$

In order to calculate the overall enthalpy change for the reaction it is necessary to know only the standard enthalpies of formation of the substances A, B, C and D. As in the previous example, Hess's law is applied.

225

Example 3
Calculate the enthalpy change for the following reaction:

$$2CO(g) + O_2(g) \longrightarrow 2CO_2(g)$$

The enthalpy of formation of carbon dioxide is the enthalpy change for the reaction:

$$C(graphite) + O_2(g) \longrightarrow CO_2(g) \quad \Delta H_1^{\ominus}(298\ K) = -393\ kJ\ mol^{-1}$$

Similarly for carbon monoxide:

$$C(graphite) + \tfrac{1}{2}O_2(g) \longrightarrow CO(g) \quad \Delta H_2^{\ominus}(298\ K) = -111\ kJ\ mol^{-1}$$

FIG. 14.9. *An enthalpy cycle to determine the standard enthalpy change for the reaction:*
$2CO(g) + O_2(g) \rightarrow 2CO_2(g)$

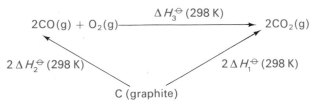

The various enthalpy changes are shown in fig. 14.9 where it can be seen that:

$$2\Delta H_2^{\ominus} + \Delta H_3^{\ominus} = 2\Delta H_1^{\ominus}$$

Thus the enthalpy change for the reaction

$$2CO(g) + O_2(g) \longrightarrow 2CO_2(g)$$

is given by:

$$\Delta H_3^{\ominus} = -786 + 222 = -564\ kJ\ mol^{-1}$$

14.9
The energetics of crystals

The lattice energy of an ionic crystal (p. 69) is the enthalpy change* for the general process:

$$nA^{m+}(g) + mB^{n-}(g) \longrightarrow A_nB_m(s)$$

It represents the energy released to the surroundings when 1 mole of the ionic compound is formed from its isolated ions in the gaseous state under standard conditions. Lattice energies cannot be obtained by direct experimental methods but, as we have already seen (p. 71), it is possible to calculate lattice energies from an electrostatic model of an ionic solid. We are now in a position to determine lattice energies by another route using Hess's law.

The overall process described in the above equation can be broken down into several hypothetical stages. This is not to imply that these steps actually represent the mechanism of the process, but simply that we can equate the overall enthalpy change with the sum of the changes for each step, since the initial and final states are the same.

The method is best illustrated on an energy diagram, and fig. 14.10 illustrates the process for sodium chloride, Na^+Cl^-. As usual we start with the elements in their standard states on the zero energy level.

Applying Hess's law to the energy level diagram, showing the numbered enthalpy changes, we have:

$$\Delta H_1^{\ominus} + \Delta H_2^{\ominus} + \Delta H_3^{\ominus} + \Delta H_4^{\ominus} + \Delta H_5^{\ominus} = \Delta H_6^{\ominus} \qquad (3)$$
$$\text{(where } \Delta H_5^{\ominus} \text{ is the lattice energy)}$$

---

* Lattice energies are not the same as lattice enthalpies but the difference is small and often ignored, but see the footnote on p. 227.

FIG. 14.10. *The Born-Haber cycle for sodium chloride*

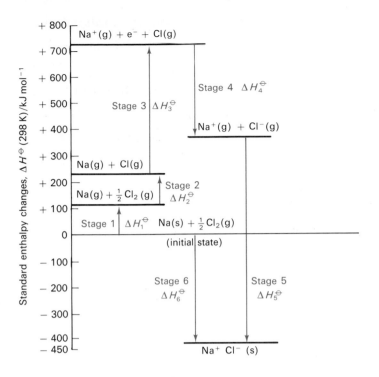

The enthalpy changes can be identified as follows:

Stage 1 The enthalpy of atomisation of sodium:
  $Na(s) \longrightarrow Na(g)$ $\Delta H_1^\ominus = +108 \cdot 4 \text{ kJ mol}^{-1}$

Stage 2 The enthalpy of atomisation of chlorine:
  $\frac{1}{2}Cl_2(g) \longrightarrow Cl(g)$ $\Delta H_2^\ominus = +121 \cdot 1 \text{ kJ mol}^{-1}$

Stage 3 The ionisation energy of sodium:
  $Na(g) \longrightarrow Na^+(g) + e^-$ $\Delta H_3^\ominus = +496 \text{ kJ mol}^{-1}$

Stage 4 The electron affinity of chlorine:
  $Cl(g) + e^- \longrightarrow Cl^-(g)$ $\Delta H_4^\ominus = -348 \text{ kJ mol}^{-1}$

Stage 5 The lattice energy which we are determining:
  $Na^+(g) + Cl^-(g) \longrightarrow Na^+Cl^-(s)$ $\Delta H_5^\ominus$

Stage 6 The standard enthalpy of formation of sodium chloride:
  $Na(s) + \frac{1}{2}Cl_2(g) \longrightarrow Na^+Cl^-(s)$ $\Delta H_6^\ominus = -411 \text{ kJ mol}^{-1}$

If we substitute these enthalpy values into equation (3) above, we get:

$$+108 \cdot 4 + 121 \cdot 1 + 496 - 348 + \Delta H_5^\ominus = -411$$
$$+377 \cdot 5 + \Delta H_5^\ominus = -411$$
$$\Delta H_5^\ominus = -411 - 377 \cdot 5$$
$$= -788 \cdot 5 \text{ kJ mol}^{-1*}$$

The splitting-up of the overall energy terms into several stages is known, in the case of ionic crystals, as the Born-Haber cycle.

---

* Strictly speaking this value should be corrected by adding $2RT$, i.e. 5 kJ mol^{-1}, since lattice energy is an internal energy change.

Another application of the Born-Haber cycle

We are now in a position to discuss the stoichiometry of ionic compounds in some detail. For example, we shall enquire into the reasons why calcium chloride has the empirical formula $Ca^{2+}(Cl^-)_2$ and not, say, Ca^+Cl^- nor $Ca^{3+}(Cl^-)_3$. To do this, we need to know the lattice energies for the three alternative structures and these are calculated by the method as discussed for sodium chloride in chapter 6 (p. 71). For the hypothetical structures, i.e. Ca^+Cl^- and $Ca^{3+}(Cl^-)_3$, these have been estimated from the calculated lattice energies of K^+Cl^- and $Sc^{3+}(Cl^-)_3$ respectively (potassium and scandium are adjacent to calcium in the Periodic Table). The various enthalpy terms are set out in Table 14.7 for the three cases.

Table 14.7 The Born–Haber cycle for Ca^+Cl^-, $Ca^{2+}(Cl^-)_2$ and $Ca^{3+}(Cl^-)_3$

Enthalpy change in the process, $\Delta H^{\ominus}$(298 K)/kJ mol^{-1}		Ca^+Cl^-	$Ca^{2+}(Cl^-)_2$	$Ca^{3+}(Cl^-)_3$
Ca(s)	$\longrightarrow$ Ca(g)	+193	+193	+193
Ca(g)	$\longrightarrow$ Ca$^+$(g) + e$^-$	+590	+590	+590
Ca$^+$(g)	$\longrightarrow$ Ca^{2+}(g) + e$^-$		+1145	+1145
Ca^{2+}(g)	$\longrightarrow$ Ca^{3+}(g) + e$^-$			+4912
$\frac{1}{2}$Cl$_2$(g)	$\longrightarrow$ Cl(g)	+121		
Cl$_2$(g)	$\longrightarrow$ 2Cl(g)		+242	
$\frac{3}{2}$Cl$_2$(g)	$\longrightarrow$ 3Cl(g)			+363
Cl(g) + e$^-$	$\longrightarrow$ Cl$^-$(g)	-348		
2Cl(g) + 2e$^-$	$\longrightarrow$ 2Cl$^-$(g)		-696	
3Cl(g) + 3e$^-$	$\longrightarrow$ 3Cl$^-$(g)			-1044
Ca$^+$(g) + Cl$^-$(g)	$\longrightarrow$ Ca$^+$Cl$^-$(s)	-711		
Ca^{2+}(g) + 2Cl$^-$(g)	$\longrightarrow$ Ca^{2+}(Cl$^-$)$_2$(s)		-2237	
Ca^{3+}(g) + 3Cl$^-$(g)	$\longrightarrow$ Ca^{3+}(Cl$^-$)$_3$(s)			-4803

By summing the entries in the three columns of Table 14.7 we may obtain the enthalpy changes for the following three processes (enthalpy changes as shown):

$$Ca(s) + \tfrac{1}{2}Cl_2(g) \longrightarrow Ca^+Cl^-(s) \quad \Delta H^{\ominus}(298\ K) = -155\ kJ\ mol^{-1}$$
$$Ca(s) + Cl_2(g) \longrightarrow Ca^{2+}(Cl^-)_2(s) \quad \Delta H^{\ominus}(298\ K) = -763\ kJ\ mol^{-1}$$
$$Ca(s) + \tfrac{3}{2}Cl_2(g) \longrightarrow Ca^{3+}(Cl^-)_3(s) \quad \Delta H^{\ominus}(298\ K) = +1356\ kJ\ mol^{-1}$$

As can be seen, the process that 'actually' occurs is the most exothermic of the three, and this is the reason why calcium chloride has the empirical formula $Ca^{2+}(Cl^-)_2$. The two most important energy terms are ionisation energy and lattice energy; lattice energy increases from Ca^+Cl^- to $Ca^{3+}(Cl^-)_3$ but so too does ionisation energy (in the opposite sense). The 'actual' structure in this case is, by and large, dictated by the relative values of these two terms in the Born-Haber cycle. Similar arguments can be used to discuss the stability, or lack of it, of other ionic structures.

14.10
Bond energies

There is strong evidence to suggest that a particular chemical bond is associated with a bond enthalpy or, to use a more familiar term, bond energy. When a particular bond is broken or formed, a particular amount of energy is absorbed from or released to the surroundings respectively.

For example, the bond energies of the Cl—Cl, H—H and H—Cl molecules are respectively 242, 436 and 431 kJ mol^{-1} at 298 K. From these values we can calculate the standard enthalpy of formation of hydrogen chloride as follows:

$$H_2(g) + Cl_2(g) \longrightarrow 2HCl(g)$$

Energy required to break bonds (H—H and Cl—Cl)

$$= +436 + 242 = +678 \text{ kJ mol}^{-1}$$

Energy gained on forming bonds ($2 \times$ H—Cl)

$$= 2 \times (-431) = -862 \text{ kJ mol}^{-1}$$

Overall enthalpy change, $\Delta H^{\ominus}(298 \text{ K}) = +678 + (-862) = -184 \text{ kJ mol}^{-1}$. This agrees well with the accepted value of $-184 \cdot 6$ kJ mol^{-1} (for 2 moles of hydrogen chloride), but we must be careful not to get into a circular argument because it is likely that the bond energies themselves were derived from the standard enthalpy of formation of hydrogen chloride. However, there are a number of independent measurements which support these particular values.

The most striking evidence in support of the principle of assigning an enthalpy value to a particular bond comes from the enthalpies of combustion of an homologous series of organic compounds. The standard enthalpies of combustion of the straight chain saturated alcohols from CH_3OH to $C_8H_{17}OH$ and the straight chain alkanes from CH_4 to C_6H_{14} are plotted against number of carbon atoms in the molecule (fig. 14.11).

FIG. 14.11. *Standard enthalpies of combustion of (a) some straight chain alkanes (b) some straight chain alcohols*

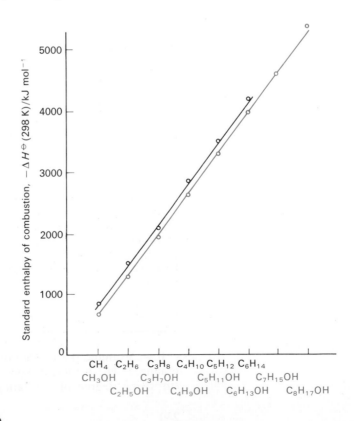

In each case there is a roughly constant increment in enthalpy associated with the addition of each successive group, which suggests that an enthalpy value can be attached to the bonds in this —CH_2— group which are broken in the process of combustion.

Bond dissociation energies

In a molecule like methane it is possible to break successively each of the four bonds and the energy involved, which will differ for each bond, is known as the bond dissociation energy.

$$CH_4(g) \longrightarrow CH_3(g) + H(g) \qquad \Delta H^\circ(298 \text{ K}) = +425 \text{ kJ mol}^{-1}$$
$$CH_3(g) \longrightarrow CH_2(g) + H(g) \qquad \Delta H^\circ(298 \text{ K}) = +470 \text{ kJ mol}^{-1}$$
$$CH_2(g) \longrightarrow CH(g) \; + H(g) \qquad \Delta H^\circ(298 \text{ K}) = +416 \text{ kJ mol}^{-1}$$
$$CH(g) \longrightarrow C(g) \quad + H(g) \qquad \Delta H^\circ(298 \text{ K}) = +335 \text{ kJ mol}^{-1}$$

When quoting bond dissociation energies, the bond to which reference is being made should be specified. Other examples are given in Table 14.8.

Table 14.8 Some bond dissociation energies

Molecule	Bond	$\Delta H^\circ(298 \text{ K})/\text{kJ mol}^{-1}$
H_2O	HO—H	+494
	H—O	+430
CO_2	OC=O	+531
	O=C	+1075

Average bond energies

The enthalpy change for the total process

$$CH_4(g) \longrightarrow C(g) + 4H(g)$$

is the sum of the individual bond dissociation energies, 1646 kJ mol^{-1} This figure divided by 4 gives a value for the average bond energy of the C—H bond of 412 kJ mol^{-1}.

Table 14.9 Some average bond energies

Bond	$\Delta H^\circ(298 \text{ K})/\text{kJ mol}^{-1}$	Bond	$\Delta H^\circ(298 \text{ K})/\text{kJ mol}^{-1}$
H—H	+436	C—H	+413
C—C	+346	N—H	+388
O—O	+146	O—H	+463
F—F	+158	F—H	+562
Cl—Cl	+242	Cl—H	+431
Br—Br	+193	Br—H	+366
I—I	+151	I—H	+299

If we examine the C—H bond in the context of other molecules, e.g. ethane, C_2H_6, we find a slightly different value, but in general there is a surprising degree of constancy from one molecule to another. For a number of different types of molecule the average bond energy of the C—H bond is taken as 413 kJ mol^{-1}. Table 14.9 lists some values of bond energies, and from this point on we shall be concerned with average bond energies (henceforth called bond energies) rather than with bond dissociation energies, although in the case of a homonuclear diatomic molecule, e.g. Cl—Cl, the two values are identical.

Determination of bond energies

Bond energies are determined in three ways:
(a) By spectroscopic methods
(b) By electron impact methods
(c) By thermochemical methods
The last of these three methods is illustrated in the following example, which refers to the average C=O bond energy in carbon dioxide. We require to know the standard enthalpy change for the reaction:

$$CO_2(g) \longrightarrow C(g) + 2O(g) \qquad \Delta H^\ominus(298\ K) = x\ kJ\ mol^{-1}$$

FIG. 14.12. *The standard enthalpy change for the reaction:* $CO_2(g) \rightarrow C(g) + 2O(g)$

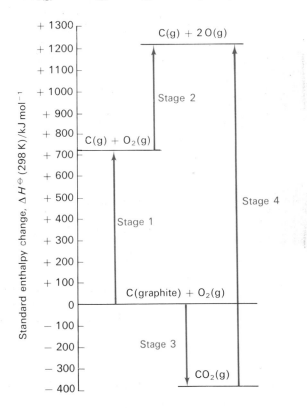

At this point it is worth stressing that in determining bond energies all the species formed must be in the gas phase. We can construct the energy level diagram as shown in fig. 14.12, starting as usual with the elements in their standard states. The enthalpy changes at each stage can be identified as follows:

Stage 1   The enthalpy of atomisation of graphite. This is the key value in the determination of the energy of any bond containing carbon:

$$C(graphite) \longrightarrow C(g) \qquad \Delta H^\ominus(298\ K) = +715\ kJ\ mol^{-1}$$

Stage 2   The enthalpy of atomisation of oxygen, determined spectroscopically, multiplied by 2:

$$O_2(g) \longrightarrow 2O(g) \qquad \Delta H^\ominus(298\ K) = +498\ kJ\ mol^{-1}$$

Stage 3 The enthalpy of formation of carbon dioxide:

$$C(\text{graphite}) + O_2(g) \longrightarrow CO_2(g) \qquad \Delta H^{\ominus}(298 \text{ K}) = -393 \text{ kJ mol}^{-1}$$

Stage 4 The enthalpy of atomisation of carbon dioxide—the value we wish to measure:

$$CO_2(g) \longrightarrow C(g) + 2O(g) \qquad \Delta H^{\ominus}(298 \text{ K}) = x \text{ kJ mol}^{-1}$$

By reference to fig. 14.12, applying Hess's law, we have:

$$+715 + 498 = -393 + x$$
$$x = +1606 \text{ kJ mol}^{-1}$$

Since this is the amount of energy needed to break two C=O bonds, the average bond energy of the C=O bond in carbon dioxide is $+1606/2$ or $+803$ kJ mol^{-1}.

Factors affecting average bond energies

The bond energy is a measure of the force operating between two bonded atoms. This in turn is dependent upon the electronic arrangement within the bond. For example, the I—I bond (bond energy = $+151$ kJ mol^{-1}) is clearly weaker than the H—H bond (bond energy = $+436$ kJ mol^{-1}).

(a) Effect of bond length
The larger the atoms joined by a particular bond, the longer the bond length. Large atoms tend to have more electrons than smaller ones and this results in an increased screening of the nucleus, and an increase in electron cloud repulsion. Both these effects contribute to a weakening of the bond, as is seen for the halogens in Table 14.10.

From the bond length values given in Table 14.10 one would predict a value for the F—F bond energy in the region of about 300 kJ mol^{-1}. In fact the measured value is surprisingly low as can be seen, and this has been attributed to the high degree of lone-pair repulsion in a bond of such short length. The same explanation is given for the low bond energy of O—O ($+146$ kJ mol^{-1}).

Table 14.10 The halogen-halogen bond energies

Bond	Bond length/nm	$\Delta H^{\ominus}(298 \text{ K})/\text{kJ mol}^{-1}$
F—F	0·142	+158
Cl—Cl	0·199	+242
Br—Br	0·228	+193
I—I	0·266	+151

(b) Effect of number of bonding electrons
The more electrons that constitute a bond, the greater the strength of the bond. Triple bonds would be expected to be stronger than double bonds which, in turn, should be stronger than single bonds. This trend is true as the values in Table 14.11 show, but we must be careful only to compare 'like with like', e.g. C—C with C=C and not with C=O.

Table 14.11 Bond energies of some single, double and triple bonds

Bond	$\Delta H^{\ominus}(298 \text{ K})/\text{kJ mol}^{-1}$
C—C	+346
C=C	+610
C≡C	+837
C—O	+358
C=O (in ketones)	+745
N—N	+163
N=N	+410
N≡N	+945

The fact that only 264 kJ mol^{-1} are required to break the second bond in C=C suggests that it is not the same kind of bond as the first, where 346 kJ mol^{-1} is required. A C=C bond consists of a σ-bond and a π-bond (p. 109).

(c) Effect of bond polarity
A bond is said to become more polar as the difference in electronegativity between the two bonded atoms increases. This increases the ionic character of the bond and, in turn, increases the bond strength. One scale of electronegativity values that is widely used (the Pauling scale) is derived from the variation of bond energies with the polarity of the bond (p. 89).

Table 14.12 lists the bond energies for three situations where there is a steady increase in electronegativity (the bond lengths are comparable).

Table 14.12 Bond energies of some polar bonds

Bond	Electronegativity difference between the atoms	ΔH°(298 K)/kJ mol^{-1}
N—H	0·9	+388
O—H	1·4	+463
F—H	1·9	+562

It is interesting to note that a bond energy value decreases if the polarity is shared out between a greater number of bonds. For example, the P—Cl bond energy is +319 kJ mol^{-1} in PCl$_3$ but only +258 kJ mol^{-1} in PCl$_5$. Similarly the Ti—Cl bond energy in TiCl$_2$, TiCl$_3$ and TiCl$_4$ is respectively 502, 456 and 427 kJ mol^{-1}

14.11
Bond energies and structure

We have seen how to calculate the enthalpy change for a chemical reaction knowing the bond energies of the various bonds that are broken and formed during the reaction (p. 229). Sometimes the calculated value for the enthalpy change is in serious disagreement with the experimentally determined value. This discrepancy between the two values often provides some useful information about the structures of the molecules concerned. Two examples are discussed below.

Benzene

The discrepancy between the calculated and measured enthalpy of formation of benzene, C$_6$H$_6$, provides the classic example of delocalisation energy (or resonance energy or stabilisation energy). If we assume the structure of benzene to be

the calculated enthalpy of formation is +221 kJ mol^{-1}. The measured value (applying Hess's law) is +49 kJ mol^{-1}, which suggests an extra stability of 172 kJ mol^{-1} owing to electron delocalisation over the entire molecule (p. 112). The reader is recommended to verify the 'calculated' value using the bond energy values given in this chapter. (Note that the enthalpy of vaporization of benzene, which is 30·8 kJ mol^{-1}, is needed in the calculation).

Cyclopropane

The enthalpy of formation of cyclopropane is $+55 \cdot 2$ kJ mol^{-1}. However, the application of the 'bond energy method' to the equation

$$3C(\text{graphite}) + 3H_2(g) \longrightarrow C_3H_6(g)$$

gives a value of -63 kJ mol^{-1}. This means that the molecule is apparently $118 \cdot 2$ kJ mol^{-1} less stable than the bond energies would predict. This is attributed to the strain in bending the molecule into its triangular shape, since the angles between the bonds forming the ring are $60°$ whereas sp^3 orbitals are at angles of $109°28'$ (p. 108). As a consequence, the overlap between the pairs of orbitals is not so complete as in a straight-chain alkane.

14.12
Questions on chapter 14

1 (a) (i) Define enthalpy of formation ΔH_f of a compound.
   (ii) What extra conditions must be imposed to specify the standard enthalpy of formation $\Delta H_f^{\ominus}$ of a compound?
   (b) When ethanol burns in oxygen, carbon dioxide and water are formed.
   (i) Write the equation which describes this reaction.
   (ii) Using the data

   $\Delta H_f^{\ominus}$ for ethanol(l) $= -277 \cdot 0$ kJ mol^{-1}
   $\Delta H_f^{\ominus}$ for carbon dioxide(g) $= -393 \cdot 7$ kJ mol^{-1}
   $\Delta H_f^{\ominus}$ for water(l) $= -285 \cdot 9$ kJ mol^{-1}

   calculate the value of $\Delta H^{\ominus}$ for the combustion of ethanol.

   JMB (Syllabus B)

2 State *Hess's law*.
   Define (*a*) *heat (enthalpy) of combustion*, and (*b*) *heat (enthalpy) of formation*.
   Calculate the heat (enthalpy) of hydrogenation of (i) cyclohexene, (ii) benzene, using the following heat of combustion data.

Substance	$\Delta H^{\ominus}$ (combustion)/ kJ mol^{-1}
Benzene	-3268
Cyclohexane	-3920
Cyclohexene	-3754
Hydrogen	-286

What conclusions can you draw from these heats of hydrogenation about the structure of benzene compared with that of cyclohexene?
   The heats (enthalpies) of neutralisation of 1 mol of sodium hydroxide in aqueous solution by various acids are as follows:
   hydrochloric acid, $-57 \cdot 3$ kJ mol^{-1}; nitric acid, $-57 \cdot 3$ kJ mol^{-1}; ethanoic (acetic) acid, $-55 \cdot 2$ kJ mol^{-1}.
   Comment on the significance of these data and explain them as far as you can.

   C (Overseas)

3 (a) Define (i) *enthalpy change of neutralization*, (ii) *enthalpy change of formation*, (iii) *endothermic compound*.
   (b) State Hess's Law of Constant Heat Summation.
   (c) Comment on the statement that the enthalpy changes of neutralization of many acids are approximately the same. Account for any exceptions.
   (d) A natural gas may be assumed to be a mixture of methane and ethane only. On complete combustion of 10 litres (measured at s.t.p.) of this gas, the evolution of heat was $474 \cdot 6$ kJ.
   Assuming ΔH combustion $[CH_4(g)] = -894$ kJ mol^{-1}

ΔH combustion $[C_2H_6(g)] = -1560$ kJ mol^{-1}

calculate the percentage by volume of each gas in the mixture. AEB

4 State Hess's Law of Constant Heat Summation.
   Define
   (a) *enthalpy of formation* (*heat of formation*),
   (b) *heat of neutralization*.
   Describe how you would determine the heat of neutralization of nitric acid by sodium hydroxide solution.
   Explain why approximately the same numerical result would be obtained by using hydrochloric acid, but not hydrofluoric acid, in place of nitric acid.
   Using the following data collected at 25°C and standard atmospheric pressure, in which the negative sign indicates heat evolved, calculate the enthalpy of formation of rubidium sulphate, $Rb_2SO_4(s)$

ΔH(298 K)/kJ mol^{-1}

(i) $H^+(aq) + OH^-(aq)$	$= H_2O(l)$	$-57\cdot3$
(ii) $RbOH(s)$	$= Rb^+(aq) + OH^-(aq)$	$-62\cdot8$
(iii) $Rb_2SO_4(s)$	$= 2Rb^+(aq) + SO_4^{2-}(aq)$	$+24\cdot3$
(iv) $Rb(s) + \frac{1}{2}O_2(g) + \frac{1}{2}H_2(g)$	$= RbOH(s)$	$-414\cdot0$
(v) $H_2(g) + S(s) + 2O_2(g)$	$= 2H^+(aq) + SO_4^{2-}(aq)$	$-907\cdot5$
(vi) $H_2(g) + \frac{1}{2}O_2(g)$	$= H_2O(l)$	$-285\cdot0$

   Assume that $H_2SO_4(aq)$ is fully ionized in solution. S

5 Define the *heat of solution* and *heat of combustion* of a compound. What additional thermochemical evidence is necessary to determine the enthalpy of formation of ethane from its enthalpy of combustion?
   From the standard enthalpies of combustion given below (in kJ mol^{-1}), and the standard enthalpy of formation of acetylene (ethyne) find the standard enthalpy change when acetylene (ethyne) is hydrogenated to ethane.
   Carbon -394, Hydrogen -286, Acetylene (ethyne) -1300, Ethane -1560.

Camb. Entrance

6 (a) State the first law of thermodynamics.
   (b) If ΔH and ΔU are the heat changes in a given process at constant pressure and constant volume respectively, write an equation which relates ΔH and ΔU.
   (c) Which of the two quantities ΔH or ΔU is the more useful when studying chemical processes? Give a reason for your answer.
   (d) The following are the enthalpies of hydrogenation of ethene and of benzene to ethane and to cyclohexane respectively:
   $C_2H_4(g) + H_2(g) \longrightarrow C_2H_6(g)$ $\Delta H = -132$ kJ mol^{-1}
   $C_6H_6(g) + 3H_2(g) \longrightarrow C_6H_{12}(g)$ $\Delta H = -208$ kJ mol^{-1}
   (i) Use the data above to deduce the relative stabilities of ethene and benzene.
   (ii) Explain your answer to (d) (i) in terms of the electronic structures of ethene and benzene. JMB (Syllabus B)

7 What do you understand by the expression 'bond energy term'? Explain how bond energy terms are determined by referring to suitable examples. Using a Book of Data, show how the bond energy terms of the bonds formed by an element are related to its position in the Periodic Table. N

8 (a) Explain what is meant by the terms *electron affinity* and *lattice energy*, and define them both.
   (b) Use the data given below to calculate the electron affinity of chlorine.
   Data

Standard enthalpy of formation of sodium chloride	-411 kJ mol^{-1}
Lattice energy of sodium chloride	-781 kJ mol^{-1}
First ionisation energy of sodium	$+500$ kJ mol^{-1}
Enthalpy of atomisation of sodium	$+108$ kJ mol^{-1}
Bond dissociation enthalpy of molecular chlorine	$+242$ kJ mol^{-1}

   (c) Comment on the following:
   (i) The energy change on the addition of one electron to a chlorine atom has a large, negative value, but the energy change on the addition of a second electron has a large, positive value.

(ii) Despite the large numerical value for the lattice energy of sodium chloride the enthalpy of solution for sodium chloride in water is almost zero $(+7 \text{ kJ mol}^{-1})$. O (S)

9 (a) Discuss the various experimental values of energy changes needed to calculate the lattice energy of a typical ionic salt.

(b) Comment on the values shown below:

	Theoretical lattice energy ($kJ\ mol^{-1}$)	Experimental lattice energy ($kJ\ mol^{-1}$)
KI	−630·9	−631·8
AgI	−735·9	−865·4

(c) Suggest how energetic considerations can help in the understanding of the stoichiometry of some compounds.

Oxford Schol. and Entrance (Physical Science)

10 State the First Law of Thermodynamics.

Define the terms *heat of formation* or *enthalpy of formation* and *heat of solution* or *enthalpy of solution*.

Determine from the following data the enthalpy of solvation of $Mg^{2+}_{(g)}$ ions by water.

Enthalpy of atomisation of $Mg_{(s)}$	$167\cdot2 \text{ kJ mol}^{-1}$
Enthalpy of dissociation of $Cl_{2(g)}$	$241\cdot6 \text{ kJ mol}^{-1}$
Enthalpy of formation of $MgCl_{2(s)}$	$-639\cdot5 \text{ kJ mol}^{-1}$
Enthalpy of solution of $MgCl_{2(s)}$	$-150\cdot5 \text{ kJ mol}^{-1}$
Enthalpy of solvation of $Cl^-_{(g)}$	$-383\cdot7 \text{ kJ mol}^{-1}$
Electron affinity of $Cl_{(g)}$	$-3\cdot78 \text{ eV}$
First ionisation energy (or ionisation potential) of $Mg_{(g)}$	$7\cdot65 \text{ eV}$
Second ionisation energy of $Mg_{(s)}$	$15\cdot03 \text{ eV}$

$[1 \text{ eV} \equiv 96\cdot48 \text{ kJ mol}^{-1}$. The electron affinity of $Cl_{(g)}$ is defined as the enthalpy change for $Cl_{(g)} + e^- \rightarrow Cl^-_{(g)}$.] Camb. Entrance

11 Show how Hess's law may be used (i.e. the Born-Haber cycle), in calculations of lattice energies of alkali halides. State clearly what quantities are used in the calculations.

Using the following enthalpies of formation, (in $kJ\ mol^{-1}$), of crystal lattices, discuss the factors governing the magnitude of lattice energies:

LiF	−612	LiI	−271
NaF	−569	NaI	−288
CaO	−621	CaF$_2$	−1203

W (S)

Chemical thermódynamics

15.1 Introduction

The question 'why do chemical reactions occur?' is one of the fundamental questions of chemistry and has occupied the minds of scientists since the times of Newton. For example, why does the reaction

$$Zn(s) + Cu^{2+}(aq) \longrightarrow Zn^{2+}(aq) + Cu(s)$$

apparently occur spontaneously (and by spontaneous is meant without the assistance of any external agency) while the reaction

$$Cu(s) + Zn^{2+}(aq) \longrightarrow Cu^{2+}(aq) + Zn(s)$$

simply does not occur at all?

Why do some reactions become spontaneous when the temperature is raised? Ice, for example, spontaneously melts as soon as the temperature rises above 273 K.

This chapter will be concerned with the answers to these questions as we try to define the driving force behind chemical reactions. We are still concerned with the energetics and not with the rates of reactions. The reaction between hydrogen and oxygen gas is energetically spontaneous, but at room temperature it takes place at a very slow rate indeed. However, the addition of a catalyst of finely-divided platinum will speed up the rate to explosive proportions, but it will in no way affect the overall energy change in the reaction. In short, a reaction which is energetically feasible at a particular temperature may not take place at an observable rate. A reaction which is not energetically feasible will simply not take place at all. This chapter is concerned with the energetic feasibility of chemical reactions.

15.2 Potential energy decrease

A sensible starting point in our search is the examination of the enthalpy changes that accompany spontaneous and non-spontaneous processes. We have previously noted (p. 216) that the enthalpy change is a measure (at constant pressure) of the change in the potential energy of a system.

Common sense tells us that spontaneity in everyday life is associated with a decrease in potential energy. Water always flows downhill; we never observe stones spontaneously rolling uphill. Elastic bands when stretched spontaneously contract when the tension is released. All of these processes involve a decrease in the potential energy of the system as, in these examples, it is transferred into kinetic energy.

An important point to note here is that work can be derived from all of these spontaneous processes, e.g. the water-wheel. Indeed a spontaneous process can be reversed and the ball made to roll uphill, for example, only if work is done on the system. We shall return to the importance of work and spontaneity later in the chapter.

Working by analogy with the examples given above we could suggest that at the atomic and molecular levels spontaneity is also attended by a decrease in potential energy. This seems to make sense because such a decrease in energy leads to products of greater stability than the reactants. Weaker bonds are broken and stronger bonds formed, the energy released being transferred from the system to the surroundings in an exothermic reaction.

The suggestion, then, is quite simply that spontaneous reactions will be exothermic reactions; indeed in 1878 Berthelot wrote "Every chemical change accomplished without the intervention of an external agency tends towards the body or the system of bodies that sets free the most heat". There are many reactions which lend support to this hypothesis. For example:

$$Zn(s) + Cu^{2+}(aq) \rightarrow Zn^{2+}(aq) + Cu(s) \quad \Delta H^{\ominus}(298\,K) = -217\,kJ\,mol^-$$
$$H_2(g) + \tfrac{1}{2}O_2(g) \rightarrow H_2O(l) \quad \Delta H^{\ominus}(298\,K) = -286\,kJ\,mol^-$$
$$C(graphite) + O_2(g) \rightarrow CO_2(g) \quad \Delta H^{\ominus}(298\,K) = -393\,kJ\,mol^-$$

However, we do not have to look far before we find examples of endothermic reactions which are nevertheless spontaneous. One of the commonest examples is the dissolving of many salts in water. For example

$$NH_4^+NO_3^-(s) + aq \rightarrow NH_4^+(aq) + NO_3^-(aq)$$
$$\Delta H^{\ominus}(298\ K) = +25 \cdot 8\ kJ\ mol^{-1}$$

$$K^+Cl^-(s) + aq \rightarrow K^+(aq) + Cl^-(aq)$$
$$\Delta H^{\ominus}(298\ K) = +17 \cdot 2\ kJ\ mol^{-1}$$

Although the above two changes involve the breaking and making of chemical bonds they are often thought of as purely physical processes; but there are 'genuine' chemical reactions which are both spontaneous and endothermic. The reaction between solid hydrated cobalt chloride and sulphur dichloride oxide (thionyl chloride) is probably one of the most spectacular but should be carried out in a fume cupboard:

$$CoCl_2.6H_2O(s) + 6SOCl_2(l) \rightarrow CoCl_2(s) + 6SO_2(g) + 12HCl(g)$$

Clearly our initial suggestion must be rejected.

The clue to our next line of enquiry lies in the state symbols shown in the above equation. One mole of solid and six moles of liquid produce one mole of solid and eighteen moles of gas. In this example there has been an increase in the number of molecules during the reaction, particularly in the gaseous phase, and in a crude sort of way we can say that there has been an increase in disorder.

15.3
Changes occurring on mixing

We now turn to another form of spontaneous change which involves no energy transfer at all. Indeed, we eliminate the possibility of transfer of energy taking place by carrying out the process in an isolated system, i.e. one that is suitably lagged to prevent energy entering or leaving it. In this way we ensure that the internal energy remains constant (fig. 15.1(a)). In flask A of our apparatus we have only 'blue molecules' of gas and in flask B we have only 'black molecules' of gas. If the two flasks are connected the gases spontaneously mix to produce the situation depicted in fig. 15.1(b). We know from experience that the process cannot be reversed spontaneously—the gases do not unmix. What is the driving force behind this process? Clearly it has nothing to do with changes in potential energy. The answer lies in the study of the probability of each possible arrangement of the system.

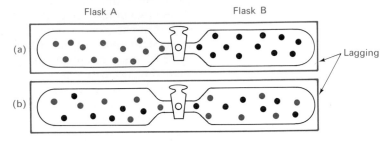

FIG. 15.1. *The mixing of 'blue molecules' and 'black molecules' of gas (a) before mixing (b) after mixing*

In order to pursue this further, let us simplify our system by letting flask A initially contain just 4 'blue molecules' and flask B initially be empty. The different possible arrangements of the system on opening the tap are shown in fig. 15.2. If the behaviour of the molecules is

FIG. 15.2. *The different possible arrangements for four 'blue molecules' distributed between flasks A and B of equal volume*

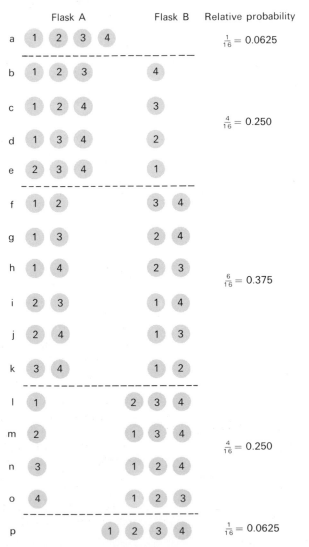

completely random, the probability of the system adopting one of the mixed states *b* to *o* is higher than the probability of it adopting its initial state because there are more of the mixed states available. The relative probabilities of the various kinds of distribution are shown in fig. 15.2. It can be seen that the relative probability of two molecules being in flask A and two in flask B, i.e. an even distribution, is greater than any other arrangement. As the number of molecules increases the probability of the system unmixing gets smaller and smaller. If we are considering one mole of gas, i.e. approximately 6×10^{23} molecules, the probability of unmixing occurring is vanishingly small and can be completely discounted.

The different arrangements of the system are known as microstates and, other things being equal, the system will tend to adopt the arrangement that is characterised by the maximum number of equivalent microstates. In the example we have been considering, there is a total of 16 microstates and there are 6 equivalent microstates with an even distribution of molecules between the two flasks. This then is the most likely arrangement adopted. This is what we meant by 'an increase in disorder'.

15.4
Distribution of energy

Our probability model has explained the spontaneous change that occurs when gases mix or diffuse, but we must now apply the same kind of argument to the way in which energy is distributed in a chemical system.

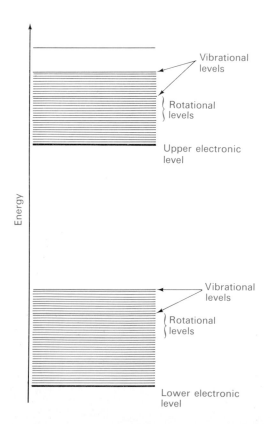

FIG. 15.3. *The vibrational and rotational energy levels associated with two electronic energy levels (not to scale). The translational quanta are too small, by comparison, to include*

We discussed in chapter 9 (p. 133) the way in which changes in energy of a chemical system of gas molecules can be attributed to contributions from electronic, vibrational, rotational and translational energies. We noted too that the transfer of these energy contributions is subject to quantum restrictions, such that the transfer can occur only in definite integral numbers of quanta; and finally we noted that the size of these quanta are in the order:

$$\text{Electronic} > \text{vibrational} > \text{rotational} > \text{translational}$$

It is convenient, therefore, to think of the total energy of the system being distributed amongst a number of energy levels which might be represented by fig. 15.3. This is a simplification of the actual state of affairs but its acceptibility as a model is supported by evidence from the spectra of molecules.

The way in which the total energy of the system is distributed between these levels is again determined by the laws of probability. To see how this works let us consider a simpler system. Fig. 15.4 shows how a total of 3 energy units can be distributed amongst three molecules in energy levels 1 unit apart. While there is only one way of achieving state (c), there are three ways (assuming each molecule to be distinguishable) of achieving state (a) and six ways of achieving state (b). If we assume that each individual microstate is equally possible (and there are 10 of them in total), then the relative probabilities of the states (a), (b) and (c) will be:

$$\text{State (a)} : \text{State (b)} : \text{State (c)}$$
$$\tfrac{3}{10} = 0 \cdot 3 \ : \ \tfrac{6}{10} = 0 \cdot 6 \ : \ \tfrac{1}{10} = 0 \cdot 1$$

FIG. 15.4. *Possible ways in which three units of energy can be distributed between three molecules*

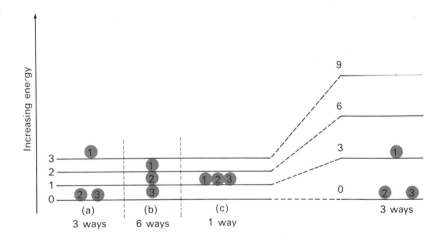

Let us suppose that the gap between the energy levels is suddenly increased, so that the size of the energy quanta becomes larger and each new level represents 3 of the old levels. The number of arrangements is at once drastically reduced, in fact to three, and this is shown as an extension to fig. 15.4.

We have seen how the quantisation of energy leads to a definite number of ways in which a chemical system can organise its energy. The system will adopt the arrangement which offers the maximum statistical probability. The smaller the size of the energy quanta, the greater the number of arrangements or microstates that the system can adopt.

In a true chemical system which contains large numbers of molecules the distribution of energy is described by the Boltzmann Distribution Law. If there are N atoms or molecules in the system which are distributed amongst energy levels 0, 1, 2, 3, etc. such that N_o occupy level 0, N_1 occupy level 1, N_2 occupy level 2 and so on, then

$$\frac{N_1}{N_o} = \exp -(E_1 - E_o)/kT \qquad \text{etc.}$$

where $E_1 - E_o$ is the energy difference between energy levels represented by 0 and 1, and k is the Boltzmann constant.

15.5 Entropy

A chemical system will adopt the most probable distribution of energy, which is the one with the maximum number of arrangements or microstates. If a system is not in its most probable state then it will change until it is; that is, a spontaneous change will occur in an isolated system if it results in an increase in the number of microstates available to the system.

The entropy of a system, S, is a measure of the number of microstates W and the two quantities are related through the expression:

$$S = k \ln W \qquad (1)$$

It is a logarithmic function since entropies are additive and probabilities are multiplicative.

The **Second Law of Thermodynamics** is often stated in the form '**the total entropy of a system and its surroundings increase during all spontaneous processes.**' At absolute zero, where all molecular and atomic motion is 'frozen', there is just one single arrangement of the system with the lowest vibrational energy level occupied. All systems possess this zero-point energy ($E = \frac{1}{2}h\nu_o$) and therefore

$$S = k \ln 1 = 0$$

The result, expressed in the above equation, is known as the **Third Law of Thermodynamics**, often stated in the form: '**at absolute zero, perfect crystals have zero entropy**'.

As the temperature rises more energy levels become accessible, and when a solid melts there will be a significant increase in the number of microstates, W, as the particles in the system gain translational as well as rotational energy. On boiling, the system becomes gaseous and the number of microstates increases dramatically.

The way in which entropy varies with temperature is shown in fig. 15.5, although no attempt has been made to draw the graph to scale. Entropy values are measured in J K^{-1} mol^{-1} and Table 15.1 lists some values of the standard molar entropies at 298 K of some common substances. As can be seen, the entropy of a gas is generally greater than that of a liquid which in turn is generally greater than that of a solid. There are, however, some exceptions; thus solid iodine has a higher entropy than water.

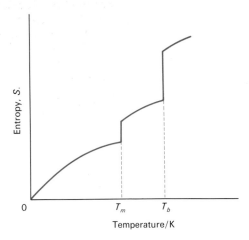

FIG. 15.5. *Variation of entropy with temperature. At* T_m *the solid melts and at* T_b *the liquid boils*

Table 15.1 Standard molar entropies, $S^{\ominus}(298 \text{ K})/\text{J K}^{-1} \text{ mol}^{-1}$, of some common substances

Cu(s)	33·3	$H_2O(l)$	69·9	$Cl_2(g)$	233·0
$I_2(s)$	116·8	$CHCl_3(l)$	201·8	$CO_2(g)$	213·6
$SiO_2(s)$	41·8	$Br_2(l)$	151·6	$UF_6(g)$	379·7

Two important facts emerge from Table 15.1 in addition to the general observations already made above:

(a) The more complicated the molecule, the more energy levels there are available and the larger the value of W. Note the high entropy value of solid iodine compared with that of copper, and the high value for UF_6 compared with that for the much simpler chlorine molecule.

(b) The structure of solids and liquids will affect the value of the entropy. For example, the values for copper and silicon dioxide, both giant structures, are low; so too is the value for water which has a fairly ordered hydrogen-bonded structure.

Equation (1) (p. 242) is of little value when it comes to measuring entropy changes. The number of distinguishable microstates cannot be calculated for any but the very simplest of chemical systems. The concept of entropy was initially developed through a study of heat engines carried out in the nineteenth century by the French physicist Sadi Carnot. For the chemist, this approach is rather less illuminating than the statistical approach which we have already considered. However, it does produce one result of fundamental importance which we cannot ignore, namely the relationship for the entropy change that occurs when a system absorbs heat from its surroundings:

dq = heat absorbed
T = absolute temperature

$$dS = \frac{dq}{T} \qquad (2)$$

Calculus notation is used in equation (2) because it applies only to an infinitesimally small change. For a finite entropy change we have

243

$$\Delta S = S_2 - S_1 = \int_{T_1}^{T_2} \frac{dq}{T} \tag{3}$$

Equation (3) holds only for a reversible change, that is to say one in which the system is at equilibrium with its surroundings at all times throughout the change.

If we consider the entropy change, at constant pressure, for one mole of substance as the temperature is raised from T_1 to T_2 we can write

$$dq = C_p.dT$$

hence

$$dS = \frac{dq}{T} = \frac{C_p.dT}{T}$$

which on integrating gives

$$\Delta S = S_2 - S_1 = \int_{T_1}^{T_2} \frac{C_p.dT}{T} \tag{4}$$

and the integration may be carried out graphically by plotting C_p/T against T as shown in fig. 15.6.

FIG. 15.6. *Entropy changes on raising the temperature from* T_1 *to* T_2 *at constant pressure (it is assumed that no phase change occurs)*

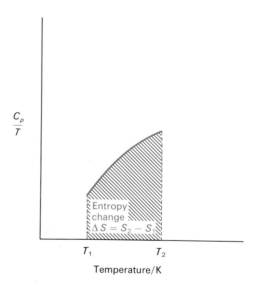

Note that equation (4) measures only entropy changes. Absolute entropies can be determined by reference to the entropy at absolute zero.

Example 1
Calculate the entropy change when one mole of water at 373 K is converted into steam.
Clearly the phase change occurs at constant temperature and we can write:

ΔH_v = molar enthalpy of vaporization
T_b = boiling point (K)

$$\Delta S = \frac{\Delta H_v}{T_b} = \frac{40 \cdot 7 \times 10^3}{373} = 109 \text{ J K}^{-1} \text{ mol}^{-1}$$

244

Entropy changes in chemical reactions

Absolute values of entropies enable us to calculate the entropy change that occurs during a chemical reaction. Three examples are discussed below, the entropies being listed under each substance.

Example 2
Calculate the standard entropy change for the following reaction:

$$CaCO_3(s) \longrightarrow CaO(s) + CO_2(g)$$
$$S^{\ominus}(298 \text{ K})/\text{J K}^{-1} \text{ mol}^{-1} \quad +92 \cdot 9 \qquad +39 \cdot 7 \qquad +213 \cdot 6$$
$$\Delta S^{\ominus}(298 \text{ K}) = (39 \cdot 7 + 213 \cdot 6) - 92 \cdot 9$$
$$= +160 \cdot 4 \text{ J K}^{-1} \text{ mol}^{-1}$$

Note that entropy changes can be calculated by the same method used for enthalpy changes.

In the above example we would anticipate a marked increase in entropy, since one mole of gas is being formed from one mole of solid.

Example 3
Calculate the standard entropy change for the formation of one mole of water vapour from its elements.

The reaction is:

$$H_2(g) + \tfrac{1}{2}O_2(g) \longrightarrow H_2O(g)$$
$$S^{\ominus}(298 \text{ K})/\text{J K}^{-1} \text{ mol}^{-1} \quad +130 \cdot 6 \quad \tfrac{1}{2} \times (+204 \cdot 9) \qquad +188 \cdot 7$$
$$\Delta S^{\ominus}(298 \text{ K}) = 188 \cdot 7 - 102 \cdot 4 - 130 \cdot 6$$
$$= -44.3 \text{ J K}^{-1} \text{ mol}^{-1}$$

One and a half moles of gas react to produce one mole of vapour, so a moderate decrease in entropy is expected. For the formation of liquid water from its elements

$$H_2(g) + \tfrac{1}{2}O_2(g) \longrightarrow H_2O(l)$$
$$\Delta S^{\ominus}(298 \text{ K}) = -163 \cdot 1 \text{ J K}^{-1} \text{ mol}^{-1}$$

which reflects the much lower entropy of **liquid** water compared with water vapour.

Example 4
Calculate the standard entropy change for the following reaction:

$$H^+(aq) + OH^-(aq) \longrightarrow H_2O(l)$$
$$S^{\ominus}(298 \text{ K})/\text{J K}^{-1} \text{ mol}^{-1} \quad 0 \qquad -10 \cdot 7 \qquad +70$$
$$\Delta S^{\ominus}(298 \text{ K}) = 70 - (-10 \cdot 7)$$
$$= +80 \cdot 7 \text{ J K}^{-1} \text{ mol}^{-1}$$

(Note that the standard entropy of $H^+(aq)$ is arbitrarily assigned a value of zero at 298 K and the entropies of other hydrated ions are related to this value)

The entropy change in the above example might appear surprising at first sight, since one mole of hydrated hydrogen ions and one mole of hydrated hydroxyl ions produce one mole of water. However, the neutralisation reaction releases water molecules which were strongly bound to the ions (to the H^+ ion in particular) and this results in an increase in 'disorder':

$$H^+ . xH_2O + OH^- . yH_2O \longrightarrow (x + y + 1)H_2O$$

15.6
Free energy changes

A reaction is spontaneous in the thermodynamic sense if there is an overall increase in entropy. Consider the reaction:

$$H_2(g) + \tfrac{1}{2}O_2(g) \longrightarrow H_2O(l) \quad \Delta S^{\ominus}(298 \text{ K}) = -163 \cdot 1 \text{ J K}^{-1} \text{ mol}^{-1}$$
$$\Delta H^{\ominus}(298 \text{ K}) = -286 \text{ kJ mol}^{-1}$$

The formation of water results in a decrease in entropy but the reaction is exothermic and releases heat to the surroundings. The overall entropy change is positive—and must be if it is to be spontaneous—since the entropy of the surroundings increases more than the entropy of the chemical system decreases.

It is usually simpler to consider only changes occurring in the chemical system itself, so we need some criterion for spontaneity which allows us to ignore the entropy changes in the surroundings as such (but they are, of course, allowed for in the new function). We define a new function which incorporates both the enthalpy term and the entropy term for the chemical system. This is called the free energy of the system, G, and the change in standard free energy during a reaction, $\Delta G^\ominus$, is related to the change in standard enthalpy, $\Delta H^\ominus$, and the change in standard entropy, $\Delta S^\ominus$, and the absolute temperature by the expression:

$$\Delta G^\ominus = \Delta H^\ominus - T\Delta S^\ominus \qquad (5)$$

From the earlier part of the chapter, we might expect a reaction to be spontaneous if $\Delta H^\ominus$ is negative (heat released to the surroundings and an increase in entropy of the surroundings) and if $\Delta S^\ominus$ is positive (an increase in entropy of the chemical system). This would suggest that a decrease in free energy ($\Delta G^\ominus$ is negative) is the essential condition for a spontaneous change.

In order to see how this works out in practice, we give some values for the various terms in equation (5) for several reactions in Table 15.2.*

Table 15.2 The values of $\Delta G^\ominus$, $\Delta H^\ominus$, and $T\Delta S^\ominus$ at 298 K for some reactions

Reaction	$\Delta G^\ominus$/kJ mol^{-1}	$\Delta H^\ominus$/kJ mol^{-1}	$T\Delta S^\ominus$/kJ mol^{-1}	Spontaneous?
$CaCO_3(s) \longrightarrow CaO(s) + CO_2(g)$	+130	+178	+48	No
$N_2(g) + O_2(g) \longrightarrow 2NO(g)$	+173	+181	+8	No
$H_2(g) + \frac{1}{2}O_2(g) \longrightarrow H_2O(l)$	−237	−286	−49	Yes
$Ca^{2+}(aq) + CO_3^{2-}(aq) \longrightarrow CaCO_3(s)$	−48	+13	+61	Yes

The results in Table 15.2 show that spontaneous reactions are associated with a decrease in free energy, while reactions which do not occur spontaneously involve an increase in free energy. One must be careful in designating a reaction 'non-spontaneous' because it may be kinetically controlled and taking place at too slow a rate to observe. However, the existence of vast deposits of calcium carbonate on earth over a period of billions of years, together with an atmosphere of nitrogen and oxygen, suggests that these two reactions do not occur spontaneously—at least at temperatures around 298 K. The importance of kinetic control is shown by the thermal stability of nitrogen oxide, NO, once it has been formed; decomposition into nitrogen and oxygen occurs only at temperatures above about 1300 K.

The importance of temperature in determining spontaneity is apparent from equation (5) above. Values of $\Delta H^\ominus$ are quoted in kJ mol^{-1} while values of $\Delta S^\ominus$ are quoted in J K^{-1} mol^{-1} since they are generally much smaller. Thus the $T\Delta S^\ominus$ term is not the dominant one generally at room temperature (the first three reactions in Table 15.2) but there are, of course, exceptions (the fourth reaction in Table 15.2). However, at

* Entropy changes are in J K^{-1} mol^{-1} and must be divided by 1000 before insertion into equation (5)

high temperatures the $T\Delta S^\ominus$ term clearly increases in magnitude and it can become the all-important term in equation (5); many reactions become spontaneous at high enough temperatures, e.g. the decomposition of calcium carbonate will occur spontaneously at about 1100 K.

In conclusion we can say that the condition for a reaction to occur spontaneously is that there should be a decrease in the free energy of the chemical system. This, in effect, is an alternative statement of the Second Law of Thermodynamics, since it is equivalent to saying that a spontaneous change is always accompanied by an increase in entropy of the chemical system and surroundings—one may decrease but the other must increase by a greater amount to compensate. The $\Delta H^\ominus$ term is the one which controls the heat transfer to or from the surroundings, and hence the entropy of the surroundings.

Example 5
Assuming $\Delta H^\ominus$ and $\Delta S^\ominus$ to be independent of temperature, at what temperature will the reaction given below become spontaneous?

$$N_2(g) + O_2(g) \longrightarrow 2NO(g)$$

We first write down the standard entropy values for each species:

$$
\begin{array}{cccc}
 & N_2(g) & + O_2(g) & \longrightarrow 2NO(g) \\
S^\ominus/\text{J K}^{-1}\text{ mol}^{-1} & +191\cdot4 & +204\cdot9 & 2 \times (+210\cdot5)
\end{array}
$$
$$\Delta S^\ominus = (2 \times 210\cdot5) - 191\cdot4 - 204\cdot9$$
$$= +24\cdot7 \text{ J K}^{-1}\text{ mol}^{-1}$$

$\Delta H^\ominus$ is, of course, simply twice the standard enthalpy of formation of nitrogen oxide:

$$= 2 \times (+90\cdot4) = +180\cdot8 \text{ kJ mol}^{-1}$$

Thus
$$\Delta G^\ominus = 180\cdot8 - (T \times 24\cdot7 \times 10^{-3}) \text{ kJ mol}^{-1}$$

For spontaneity, $\Delta G^\ominus < O$ and this occurs when

$$T > \frac{180\cdot8 \times 10^3}{24\cdot7} = 7320 \text{ K}$$

The reaction becomes spontaneous above a temperature of 7320 K.

15.7
Free energy and work

A spontaneous change can be made to do work (p. 237); likewise, a spontaneous chemical reaction can be carried out in a way that produces work, e.g. in a cell which produces electrical work. The work that a given cell reaction can yield increases as the current is reduced, since this results in less and less heat being generated in the connecting wires, and becomes a maximum when the current is zero. Of course we cannot extract any work when the current is zero, nevertheless we may get as close to the maximum work as we like by ensuring that only a tiny current flows. The electrical work is obtained by multiplying the charge (in coulombs) by the potential difference (in volts) through which this charge is transferred. For the maximum work yield, the potential difference will be the e.m.f. of the cell which can be measured with a potentiometer.

E = e.m.f. of cell
n = number of moles of electrons transferred in the reaction
F = Faraday
1 faraday = 96 500 coulombs

$$\text{Maximum work} = -nFE \tag{6}$$

The details of the method of measurement need not concern us here since cell reactions are discussed in detail in chapter 18. The negative sign in equation (6) means that work is obtained **from** the reaction.

Let us consider the reaction:

$$2Ag^+(aq) + Zn(s) \longrightarrow Zn^{2+}(aq) + 2Ag(s)$$

The standard enthalpy change for this reaction can be determined calorimetrically. The accepted value is given by

$$\Delta H^\ominus(298 \text{ K}) = -363 \cdot 5 \text{ kJ mol}^{-1}$$

The e.m.f. of the cell in which the above reaction can take place (see chapter 18 for a discussion of cell reactions) is 1·56 V. From equation (6) we thus have:

$$\text{Maximum work} = -2 \times 96\,500 \times 1 \cdot 56 \times 10^{-3} \text{ kJ mol}^{-1}$$
$$= -301 \cdot 1 \text{ kJ mol}^{-1}$$

Thus the available work is appreciably less than the available heat energy and this is the crux of the whole matter. The maximum amount of available work is also the free energy change of the reaction, i.e. the energy that is free to do work. What about the (363·5−301·1) or 62·4 kJ mol^{-1} which is evidently not 'free'? This energy is used in changing the system from Zn(s) and Ag$^+$(aq) into Zn^{2+}(aq) and Ag(s) and is thus 'unavailable' energy. The system at the end is more ordered that it was at the beginning, and 62·4 kJ mol^{-1} of heat are transferred to the surroundings in order to achieve this condition.

A spontaneous reaction, then, is not necessarily one in which there is an evolution of heat to the surroundings, e.g. reaction four in Table 15.2 where $\Delta H^\ominus = +13$ kJ mol^{-1} but $\Delta G^\ominus = -48$ kJ mol^{-1}. Rather it is one in which there is a decrease in the capacity of the system to do work and these two phenomena are not the same.

It is easy to fall into the trap of thinking that the available work (as measured by the free energy change) is always less than the available heat energy. In the case of a reaction which occurs with an increase in entropy the reverse will be true because there will be a gain from the 'disordering' of the chemical system.

15.8
Free energy and stability

Free energy changes, like enthalpy changes, are measured in kJ mol^{-1} and when measured under standard conditions are known as standard free energy changes and written $\Delta G^\ominus$. Hess's law applies to free energy changes—that is to say, the overall free energy change in a reaction is independent of the route by which the reaction is carried out. We can thus construct free energy diagrams in an analogous fashion to enthalpy diagrams (p. 223).

Compounds are assigned a standard free energy of formation which can be used to calculate the overall free energy change in any particular reaction. $\Delta G^\ominus$ values for elements in their standard states at 298 K are zero.

Previously in chapter 14 we have taken the enthalpy of formation of a compound as a measure of the energetic stability of that compound with respect to the elements from which it was formed (p. 219). For example, silicon dioxide, SiO$_2$, ($\Delta H^\ominus(298 \text{ K}) = -911$ kJ mol^{-1}) is more stable than the elements silicon and oxygen, whereas silicon hydride, SiH$_4$, ($\Delta H^\ominus(298 \text{ K}) = +34 \cdot 3$ kJ mol^{-1}) is less stable than silicon and hydrogen, under standard conditions.

248

To a first approximation, the standard enthalpy of formation is usually a good enough indication of the energetic stability of a compound. However, the true indication of stability is the free energy of formation, $\Delta G^{\ominus}$, since this represents the work that must be done on the compound to get it back again to its original state (the constituent elements). For many compounds there is little difference between $\Delta H^{\ominus}$ and $\Delta G^{\ominus}$. Table 15.3 compares the values for the oxides of the short period sodium to chlorine.

Table 15.3 Comparison of $\Delta H^{\ominus}(298\ K)/kJ\ mol^{-1}$ and $\Delta G^{\ominus}(298\ K)/kJ\ mol^{-1}$ for the oxides of the elements of the second short period

Oxide	Na_2O	MgO	Al_2O_3	SiO_2	P_4O_{10}	SO_3	Cl_2O_7
$\Delta H^{\ominus}(298\ K)/kJ\ mol^{-1}$	-416	-601	-1676	-911	-2984	-395	$+265$
$\Delta G^{\ominus}(298\ K)/kJ\ mol^{-1}$	-377	-569	-1582	-857	-2698	-370	—

Clearly the trend in values is the same in each case, although the numerical value of $\Delta G^{\ominus}$ is lower than for $\Delta H^{\ominus}$ and is due to the fact that a decrease in entropy occurs when most oxides are formed from their elements. The difference between $\Delta G^{\ominus}$ and $\Delta H^{\ominus}$ is due to the $T\Delta S^{\ominus}$ term in equation (5) (p. 246), but at 298 K this is generally small unless the entropy change is very large. An example of a large entropy change is provided by the following reaction:

$$N_2(g) + 2O_2(g) \longrightarrow N_2O_4(g)$$

In this case $\Delta S^{\ominus}(298\ K)$ is $-298\ J\ K^{-1}\ mol^{-1}$ and this means that 1 mole of N_2O_4 is energetically only slightly more stable than 2 moles of NO_2 with respect to their constituent elements at 298 K, although their $\Delta H^{\ominus}$ values would indicate a much higher energetic stability (Table 15.4).

Table 15.4 Comparison of $\Delta H^{\ominus}(298\ K)/kJ\ mol^{-1}$ and $\Delta G^{\ominus}(298\ K)/kJ\ mol^{-1}$ for $NO_2(g)$ and $N_2O_4(g)$

Oxide	$NO_2(g)$	$N_2O_4(g)$
$\Delta H^{\ominus}(298\ K)/kJ\ mol^{-1}$	$+33{\cdot}9$	$+9{\cdot}7$
$\Delta G^{\ominus}(298\ K)/kJ\ mol^{-1}$	$+51{\cdot}8$	$+98{\cdot}3$

Standard free energies of formation are concerned only with energetic stability; they do not indicate the **rate** at which a reaction will take place. It is possible that a particular product will be favoured for kinetic reasons even though it may be unfavoured in energetic terms. For example, carbon and oxygen ($\Delta G_f^{\ominus}(298\ K) = 0$ for each) are energetically less stable than carbon dioxide ($\Delta G_f^{\ominus}(298\ K) = -393\ kJ\ mol^{-1}$) but coke does not spontaneously catch fire in air. Diamond ($\Delta G_f^{\ominus}(298\ K) = +2{\cdot}8\ kJ\ mol^{-1}$) is less stable than graphite ($\Delta G_f^{\ominus}(298\ K) = 0$) but at 298 K the rate of change of diamond to graphite is, happily, infinitesimally small.

Relative energetic stability is of crucial importance when it comes to the extraction of metals from their ores. The discussion which follows is limited to the reduction of oxides but the same principles have been applied to many different ores, in particular sulphides and chlorides.

The reduction of a metallic oxide, MO, by a reducing agent R can be represented by the general equation:

$$MO + R \longrightarrow RO + M$$

Since R and M are elements ($\Delta G_f^{\ominus}(298 \text{ K}) = 0$) the overall free energy change for the reaction will depend on the values of $\Delta G_f^{\ominus}(298 \text{ K})$ for the two oxides RO and MO.

Consider the reduction of chromium(III) oxide by aluminium (the Thermit process):

$$\begin{array}{cccc} & Cr_2O_3(s) + 2Al(s) \longrightarrow & Al_2O_3(s) + & 2Cr(s) \\ \Delta G_f^{\ominus}(298 \text{ K})/\text{kJ mol}^{-1} & -1047 \qquad 0 & -1582 & 0 \end{array}$$

hence $\Delta G^{\ominus}(298 \text{ K})$ for the reaction $= -1582 - (-1047)$
$$= -535 \text{ kJ mol}^{-1}$$

The reaction is energetically feasible—even at 298 K—but for kinetic reasons it must be initiated by heating. This simple example serves to illustrate the general point that an element will reduce an oxide if its own oxide is more stable than the one being reduced. All we need do then is to consult tables of $\Delta G^{\ominus}$ values.

Example 6
Determine whether or not it is possible for sodium to reduce aluminium oxide to aluminium at 298 K.
The reaction is

$$\begin{array}{ccc} & Al_2O_3(s) + 6Na(s) \longrightarrow & 3Na_2O(s) + 2Al(s) \\ \Delta G_f^{\ominus}(298 \text{ K})/\text{kJ mol}^{-1} & -1582 \qquad 0 & 3 \times (-377) \quad 0 \end{array}$$

hence $\Delta G^{\ominus}(298 \text{ K})$ for the reaction $= -1131 + 1582$
$$= +451 \text{ kJ mol}^{-1}$$

Evidently the reaction cannot occur since $\Delta G^{\ominus}(298 \text{ K})$ is positive. Yet sodium was used to produce aluminium before the Hall electrolytic process was developed in 1886. The problem is resolved when we remember that values are measured at 298 K. We need to know how the free energy changes vary with temperature so that we can determine their values at the temperature at which the reaction is carried out.

Assuming that $\Delta H^{\ominus}$ and $\Delta S^{\ominus}$ do not vary with temperature (and this approximation is justifiable for our purposes) we may differentiate equation (5) (p. 246) with respect to temperature and obtain:

$$\frac{d(\Delta G^{\ominus})}{dT} = -\Delta S^{\ominus}$$

Thus for a reaction which involves a decrease in entropy ($\Delta S^{\ominus}$ is negative) the free energy change becomes less negative.

Consider the reaction below:

$$\begin{array}{cccc} & 2\text{Mg(s)} & + \text{O}_2(\text{g}) \longrightarrow & 2\text{MgO(s)} \\ S^{\ominus}(298 \text{ K})/\text{J K}^{-1} \text{ mol}^{-1} & 2 \times (+32 \cdot 7) & +205 & 2 \times (+26 \cdot 8) \end{array}$$

hence $\Delta S^{\ominus}(298 \text{ K}) = +53 \cdot 6 - (+65 \cdot 4) - (+205)$
$$= -216 \cdot 8 \text{ J K}^{-1} \text{ mol}^{-1}$$

The variation of $\Delta G^{\ominus}$ for this reaction with temperature is shown in fig. 15.7 (blue line). The discontinuities occur at the melting- and boiling-points of magnesium when $\Delta S^{\ominus}$ becomes even more negative.

FIG. 15.7. *Variation of free energies with temperature*

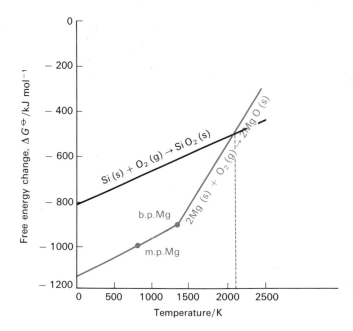

We can now superimpose upon this curve the analogous plot for the reaction between silicon and oxygen (fig. 15.7 black line):

$$\text{Si(s)} + \text{O}_2(\text{g}) \longrightarrow \text{SiO}_2(\text{s})$$

It is clear from fig. 15.7 that below a temperature of about 2000 K magnesium oxide is more stable than silicon dioxide, and magnesium will reduce silicon dioxide to silicon. Above about 2000 K, however, the reverse is true and silicon will reduce magnesium oxide to magnesium.

Free energy diagrams of this kind are known as Ellingham diagrams after their originator. A more comprehensive diagram is shown in fig. 15.8. A few interesting features of the diagram are noted below.

(a) Gold(III) oxide, Au_2O_3, has a positive free energy of formation. Gold is therefore found in the native state.

(b) Carbon has two oxides, and carbon monoxide is the only oxide which becomes more stable as the temperature increases. This is of great commercial importance, since it means that almost any metallic oxide can be reduced with coke at a high enough temperature. But cheapness of coke must be balanced against the high cost of maintaining plant at high temperatures and in practice most reactive metals are produced electrolytically.

(c) Hydrogen gas is not a particularly good reducing agent and its 'line' runs roughly parallel with those of most metallic oxides. Moreover, it

251

forms interstitial hydrides which affect the properties of the metal. Hydrogen is not used in large-scale production of metals.

FIG. 15.8. *Typical Ellingham diagrams*

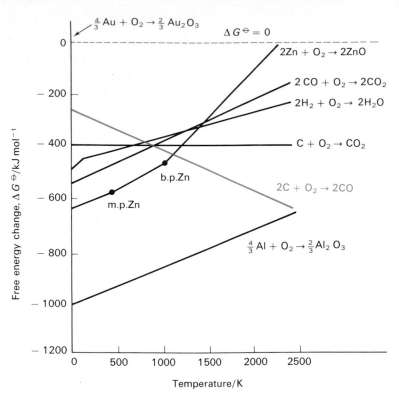

(d) The reduction of zinc oxide by carbon is carried out at a temperature above the boiling point of zinc. This causes problems on cooling, since the reaction

$$ZnO(s) + C(s) \longrightarrow Zn(g) + CO(g)$$

reverses and the zinc becomes coated with a film of zinc oxide. A plant is now operating at an increased pressure, thus raising the boiling point of the metal above the temperature required to achieve reduction of the oxide with carbon.

Many metals are found as sulphides rather than oxides, e.g. zinc sulphide, lead sulphide and copper(II) sulphide, but the reaction

$$2MS + C \longrightarrow 2M + CS_2$$

is very unfavourable energetically because of the instability of carbon disulphide ($\Delta G_f^{\ominus}(298 \text{ K}) = +63\cdot6 \text{ kJ mol}^{-1}$). For this reason, roasting of the sulphide ore in air to convert it into the oxide is generally the first stage in the extraction of metals from their sulphide ores.

15.10
Questions on chapter 15

1 (a) Using the same axes, plot the free energy changes, ΔG, against temperature for the following reactions:
   (i) $2C(s) + O_2(g) \rightarrow 2CO(g)$;
   (ii) $2Zn(s) + O_2(g) \rightarrow 2ZnO(s)$;

Temperature in K	400	600	720	800	1000	1120	1200	1400	1600	1800
$-\Delta G$ in kJ mol^{-1}										
Reaction (i)	300	335	350	370	405	425	440	475	505	540
Reaction (ii)	615	595	580	555	495	465	425	340	260	165

(b) From the graphs deduce whether carbon will reduce zinc oxide at (i) 1000 K, (ii) 1200 K. Give an explanation.

(c) Calculate the free energy change of the following reaction
$$ZnO + C \longrightarrow Zn + CO$$
at (i) 1000 K, (ii) 1200 K.

State whether your results agree with the predictions you made in (a). Give an explanation.

(d) State **two** advantages and **two** disadvantages of producing zinc by reduction of its oxide with carbon.   AEB (Syllabus II)

2 Discuss the reduction of metal oxides in the light of the information in the diagram where the Gibbs free energy of formation of oxides per mole of oxygen is plotted against temperature.

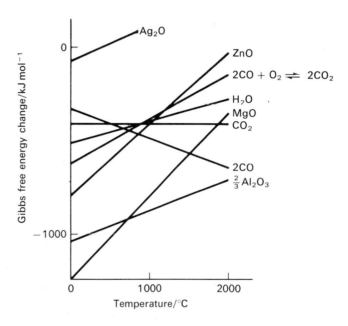

Oxford Schol. and Entrance

3 If chemical change always proceeds in the direction of greater stability, why is it that endothermic reactions occur at all? Illustrate your answer with some specific examples.   Oxford Schol. and Entrance

4 What do you understand by the 'standard molar entropy of a substance at 298 K'? What patterns are there in the entropy of substances and what is a possible interpretation of the patterns?   N (S)

5 Calculate the standard free energy change for the following reaction at 1 atm and 298 K from the following data.

	$H_2(g)$ +	$Cl_2(g) \rightarrow$	$2HCl(g)$
$\Delta H^{\ominus}$(298 K)	0	0	$-92 \cdot 3 \times 2$ kJ mol^{-1}
$S^{\ominus}$(298 K)	131	223	187×2 J K^{-1} mol^{-1}

6 The free energy change for the reaction
$$C(s) + H_2O(g) \rightleftharpoons CO(g) + H_2(g)$$
at 1 atm is given by the equation:
$$\Delta G = \Delta H - T\Delta S$$

The enthalpy change, ΔH, for the reaction is very nearly constant at a value of $+126$ kJ mol^{-1}, while $T\Delta S$ increases nearly linearly from a value of 0 at 0 K to $+502$ kJ mol^{-1} at 3500 K. By graphical plotting determine the approximate temperature at which reaction can occur. If the reaction did not occur at this temperature what would be the likely reason?

7 Would you expect the standard entropy change, $\Delta S^{\ominus}(298$ K) to increase, decrease or remain approximately constant in the following reactions?

    (a) $2NH_3(g) \rightarrow N_2(g) + 3H_2(g)$
    (b) $N_2(g) + O_2(g) \rightarrow 2NO(g)$
    (c) $2Mg(s) + O_2(g) \rightarrow 2MgO(s)$
    (d) $CH_4(g) + 2O_2(g) \rightarrow CO_2(g) + 2H_2O(l)$

Now use a Data Book and determine the actual values of the standard entropy changes for these reactions.

Equilibrium

Chapter 16

The nature of equilibrium

16.1
Introduction

One of the earliest experiments encountered in a study of chemistry is that between an aqueous solution of copper(II) sulphate and an excess of iron filings. If this mixture is shaken in a test tube, the blue colour due to the presence of $Cu^{2+}(aq)$ ions is replaced by a light green solution containing $Fe^{2+}(aq)$ ions. A reddish-brown deposit of copper is formed:

$$Cu^{2+}(aq) + Fe(s) \longrightarrow Fe^{2+}(aq) + Cu(s)$$

We have written the above equation with an arrow pointing in the direction from left to right, since there is no evidence to suggest that the reverse reaction takes place to any significant extent.

On the other hand the reaction between aqueous solutions of silver sulphate and iron(II) sulphate does not proceed to completion (see chapter 18, p. 316). When the reaction *apparently* ceases we say that a state of **equilibrium** has been achieved, and the mixture contains all four species represented in the equation below. Such a situation is denoted by writing the symbol $\rightleftharpoons$.

$$Ag^+(aq) + Fe^{2+}(aq) \rightleftharpoons Fe^{3+}(aq) + Ag(s)$$

We have been careful to avoid concluding that the reaction ceases when a state of equilibrium is reached by adding the word 'apparently'. Let us use a mechanical analogy: Two well-matched children sitting on opposite sides of a see-saw are in a state of **static** equilibrium (fig. 16.1). However, an athlete training on a moving conveyor belt is in a state of **dynamic** equilibrium if his speed is exactly matched by the speed of the

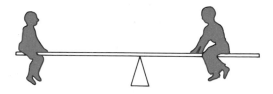

FIG. 16.1. *Static equilibrium. A state of balance is achieved and nothing is happening*

FIG. 16.2. *Dynamic equilibrium. The athlete is running at the same speed as the conveyor belt but in an opposite direction. A state of balance is achieved but something is definitely happening*

256

conveyor belt in the opposite direction (fig. 16.2). In the example of the see-saw nothing at all is happening. On the other hand the athlete is making no progress so far as a stationary observer is concerned, but it would be a little unkind to suggest that he was not exerting himself.

Chemical reactions which are reversible do not cease when equilibrium is attained, i.e. they are analogous to the athlete and the conveyor belt, and a state of **dynamic equilibrium** is achieved. The forward and backward reactions are proceeding at the same rates at equilibrium and no macroscopic change is observed. Is it possible to prove the dynamic nature of chemical equilibria? The answer to this question is yes it can be done by using radioactive isotopes. For example, if a little radioactive silver is added to the equilibrium mixture depicted in fig. 16.3, and this will in no way affect the equilibrium (see section 16.6, p. 270), then after a little while some of the radioactivity will be present in the aqueous solution. The result proves that radioactive $Ag^+(aq)$ is formed, i.e. at equilibrium the reaction is taking place in opposite directions at the same rates.

FIG. 16.3. *An example of a reversible reaction*

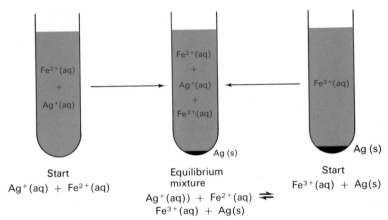

Start
$Ag^+(aq) + Fe^{2+}(aq)$

Equilibrium mixture
$Ag^+(aq)) + Fe^{2+}(aq) \rightleftharpoons$
$Fe^{3+}(aq) + Ag(s)$

Start
$Fe^{3+}(aq) + Ag(s)$

16.2
Some features of chemical equilibrium

Consider a general chemical reaction such as the one below carried out in a closed system, i.e. one which does not allow matter to enter or leave but does allow the free transfer of energy:

$$aA + bB \rightleftharpoons cC + dD$$

The letters A, B, C and D represent the chemical substances and a, b, c and d are the stoichiometric coefficients.

Suppose known concentrations of A and B are mixed and the reaction allowed to proceed at a fixed temperature in a thermostat. As particles of A and B collide they will react to form C and D which will collide with increasing frequency to reform A and B. Eventually the rate at which the reaction is proceeding from left to right is exactly matched by the rate of the reverse reaction, and a state of dynamic equilibrium is achieved. The same condition is attained if known concentrations of C and D are allowed to react to form A and B.

If the concentrations of the various substances at equilibrium are determined experimentally, then the extent to which the reaction proceeds from left to right is given by a simple expression as follows:

$$K_c = \frac{[C]^c_{eqm}[D]^d_{eqm}}{[A]^a_{eqm}[B]^b_{eqm}} \qquad (1)$$

$$\Delta n = (c + d - a - b)$$

where $[A]_{eqm}$ is the concentration of substance A at equilibrium (mol dm^{-3}), similarly for substances B, C and D. K_c is called the equilibrium constant for the reaction (the subscript c indicates that the units of concentration are mol dm^{-3}) and has units of (mol dm^{-3})$^{\Delta n}$. For a particular reaction K_c is independent of the initial concentrations, i.e. any combination of concentrations of A, B, C and D can be reacted together, yet the same value of K_c is obtained, within experimental error, when the respective equilibrium concentrations are inserted into equation (1).

A catalyst has no effect on the value of an equilibrium constant and thus may be used to study reactions which are otherwise very slow; its function is simply to speed up the rate at which equilibrium is achieved. In fact equilibrium constants are affected only by a change of temperature, hence the need to carry out reactions being studied in a thermostat and to quote the temperature at which an equilibrium constant is measured.

As can be seen from equation (1), a large value of K_c means that the reaction proceeds extensively from left to right—it may be so large in a particular example that the reaction can be considered to go to completion, e.g. the reaction between an aqueous solution of copper(II) sulphate and iron filings. Conversely a small value of K_c means that very little reaction occurs from left to right. However, as we shall see later in connection with the Haber process (p. 267), a surprisingly low value for the equilibrium constant still allows a reasonable yield of ammonia to be obtained.

16.3 Two classical experiments to determine equilibrium constants

So far the discussion of chemical equilibria has been qualitative. It is now time to examine some experimental results, and the two systems described below are classical examples of equilibrium reactions.

The equilibrium between ethanoic acid, ethanol, ethyl ethanoate and water

This was one of the earliest studies of chemical equilibrium and can be represented thus:

$$CH_3COOH(l) + C_2H_5OH(l) \rightleftharpoons CH_3COOC_2H_5(l) + H_2O(l)$$

Table 16.1 Data for the reaction
$CH_3COOH(l) + C_2H_5OH(l) \rightleftharpoons CH_3COOC_2H_5(l) + H_2O(l)$ at 373 K

Initial moles ethanoic acid	Initial moles ethanol	Equilibrium moles ethyl ethanoate	K_c
1·00	0·18	0·171	3·9
1·00	0·33	0·293	3·3
1·00	0·50	0·414	3·4
1·00	1·00	0·667	4·0
1·00	2·00	0·858	4·6
1·00	8·00	0·966	3·9

258

The experiment was originally carried out by sealing known amounts of the acid and alcohol in a number of glass tubes which were then heated to a temperature of 373 K (100°C) until equilibrium was attained. The tubes were then rapidly cooled to 'freeze' the equilibrium and the amount of acid left determined by titration with a standard solution of alkali. The assumption is made that the equilibrium does not shift position significantly during the analysis. From the amount of acid left, it is a simple matter to calculate the amount used and hence the amounts of ester and water produced and the amount of alcohol remaining. Some typical results are shown in Table 16.1.

Although the values of K_c are more variable than one might expect, an error of only 3% in determining the equilibrium concentration of ethyl ethanoate is sufficient to cause a discrepancy of about 17% in the final answer.

We can illustrate the method used to determine K_c for this reaction by using the last set of results in Table 16.1. Since 0·966 mole of ethyl ethanoate is formed, then (1·00 − 0·966) or 0·034 mole of ethanoic acid is left. Similarly (8·00 − 0·966) or 7·034 moles of ethanol remain and 0·966 mole of water is formed. If V is the total volume of the liquid mixture in dm^3 we have:

$$CH_3COOH(l) + C_2H_5OH(l) \rightleftharpoons CH_3COOC_2H_5(l) + H_2O(l)$$

Initial conc./ mol dm^{-3}	$\dfrac{1\cdot00}{V}$	$\dfrac{8\cdot00}{V}$	$\dfrac{0}{V}$	$\dfrac{0}{V}$
Equilibrium conc./ mol dm^{-3}	$\dfrac{0\cdot034}{V}$	$\dfrac{7\cdot034}{V}$	$\dfrac{0\cdot966}{V}$	$\dfrac{0\cdot966}{V}$

$$K_c = \frac{[CH_3COOC_2H_5]_{eqm}[H_2O]_{eqm}}{[CH_3COOH]_{eqm}[C_2H_5OH]_{eqm}}$$

$$= \frac{\left(\dfrac{0\cdot966}{V}\right)\left(\dfrac{0\cdot966}{V}\right)}{\left(\dfrac{0\cdot034}{V}\right)\left(\dfrac{7\cdot034}{V}\right)}$$

$$= \frac{0\cdot966 \times 0\cdot966}{0\cdot034 \times 7\cdot034} = 3\cdot90$$

Notice that the equilibrium constant for this reaction has no units. This is always the case when there is no change in the number of moles.

If we take the value of K_c to be 4·0, then for an equimolar ratio of ethanoic acid and ethanol there is approximately 67% conversion into ethyl ethanoate and water (see the fourth set of results in Table 16.1). However, for a given quantity of ethanoic acid (in this case one mole) a large excess of ethanol increases the amounts of ethyl ethanoate and water produced (approaching 97% conversion of ethanoic acid in the last set of results). This illustrates an important point, namely that the **position of equilibrium** can be altered by varying the relative concentrations of the participating substances, but the equilibrium constant itself remains constant unless the temperature is altered.

The equilibrium between hydrogen, iodine and hydrogen iodide

The equilibrium reaction

$$H_2(g) + I_2(g) \rightleftharpoons 2HI(g)$$

was first studied by Bodenstein towards the end of the nineteenth century and subsequently investigated very carefully by Taylor and Crist. The results of the latter two workers are given in Table 16.2.

The reaction was carried out in the gaseous phase by sealing known amounts of hydrogen and iodine in several quartz vessels and heating them to a temperature above 673 K (400°C). After equilibrium had been attained the vessels were rapidly cooled to 'freeze' the equilibrium and analysed for iodine by titration with standard sodium thiosulphate solution. From the amount of iodine remaining it is a relatively simple matter to determine the equilibrium concentrations of hydrogen, iodine and hydrogen iodide in each case. A series of results was also obtained by sealing known amounts of hydrogen iodide in quartz vessels, heating to a temperature in excess of 673 K followed by analysis for iodine. The equilibrium was therefore approached from both sides.

Table 16.2 Data for the reaction $H_2(g) + I_2(g) \rightleftharpoons 2HI(g)$ at a temperature of 730·6 K (457·6°C)

(Initial conc. $\times 10^3$)/ mol dm^{-3}			(Equilibrium conc. $\times 10^3$)/ mol dm^{-3}			K_c
H_2	I_2	HI	H_2	I_2	HI	
11·967	6·9436	0	5·617	0·5936	12·70	48·37
12·281	9·964	0	3·841	1·524	16·87	48·61
12·010	8·403	0	4·580	0·9733	14·86	49·54
0	0	3·573	1·696	1·696	1·181	48·48
0	0	3·866	1·433	1·433	1·000	48·71
0	0	11·369	4·213	4·213	2·943	48·81

As can be seen from the final column of Table 16.2 the value of the equilibrium constant, K_c, is remarkably constant.

Let us now examine this reaction in a little more detail. Suppose that a moles of hydrogen, b moles of iodine and 0 mole of hydrogen iodide are used in an experiment. If x moles of hydrogen are used up in the reaction, then clearly x moles of iodine are also consumed and $2x$ moles of hydrogen iodide are produced (see the stoichiometric equation below). At equilibrium, the number of moles of the three species will be respectively $(a - x)$, $(b - x)$ and $2x$. If the total volume of the vessel is V dm^3, then the initial and equilibrium concentrations are as given below:

$$H_2(g) \quad + I_2(g) \quad \rightleftharpoons 2HI(g)$$

Initial conc./mol dm^{-3} $\quad \dfrac{a}{V} \qquad \dfrac{b}{V} \qquad \dfrac{0}{V}$

Equilibrium conc./mol dm^{-3} $\quad \dfrac{(a - x)}{V} \quad \dfrac{(b - x)}{V} \quad \dfrac{2x}{V}$

$$K_c = \frac{[HI]^2_{eqm}}{[H_2]_{eqm}[I_2]_{eqm}}$$

$$= \frac{\left(\dfrac{2x}{V}\right)^2}{\left(\dfrac{a-x}{V}\right)\left(\dfrac{b-x}{V}\right)}$$

$$= \frac{(2x)^2}{(a-x)(b-x)} \text{ or } \frac{4x^2}{(a-x)(b-x)}$$

Substituting the first set of results (Table 16.2) into the above equation we have:

$$K_c = \frac{(12 \cdot 70 \times 10^{-3})^2}{(5 \cdot 617 \times 10^{-3})(0 \cdot 5936 \times 10^{-3})}$$
$$= 48 \cdot 37$$

Once again, the equilibrium constant has no units.

16.4 Homogeneous equilibria

Chemical equilibrium may conveniently be classified as homogeneous and heterogeneous. A homogeneous equilibrium has all the participating substances present in one phase only, e.g. reactions in liquid solution and in the gaseous phase. Both of the reactions discussed in the last section are examples of homogeneous equilibria. A heterogeneous equilibrium has the participating substances present in more than one phase, e.g. the reaction between a solid and a gas. This section will be concerned with homogeneous equilibria, and the features of heterogeneous equilibria will be discussed at a later stage.

Neither of the two examples of homogeneous equilibria described in the preceeding section involves a change in the number of moles when reaction proceeds from left to right or vice-versa (it is for this reason that the volume, V, does not appear in the final expression for the equilibrium constant). In many reactions there is a change in the number of moles and we now examine such reactions, including two of major industrial importance.

The equilibrium between dinitrogen tetroxide, N_2O_4, and nitrogen dioxide, NO_2, in the gaseous phase

This reaction involves an increase in the number of moles as it proceeds from left to right as indicated below:

$$\underset{\text{(colourless)}}{N_2O_4(g)} \rightleftharpoons \underset{\text{(dark brown)}}{2NO_2(g)}$$

(a) The equilibrium constant, K_c, in terms of concentration units (mol dm^{-3})
A convenient way of determining the equilibrium constant for this reaction is to determine the relative molecular mass of the 'mixture' at a fixed temperature by the method discussed in chapter 10 (p. 149). The relative molecular mass for this mixture at 323 K and 1 atm pressure is 65·7, whereas the relative molecular mass of dinitrogen tetroxide, N_2O_4, is 92. Consider the equilibrium in which we take 1 mole of N_2O_4 initially:

$$N_2O_4(g) \rightleftharpoons 2NO_2(g)$$

	$N_2O_4(g)$	$2NO_2(g)$
Initial moles	1	0
Equilibrium moles	$(1 - x)$	$2x$
Equilibrium conc./mol dm^{-3}	$\dfrac{(1 - x)}{V}$	$\dfrac{2x}{V}$

The volume V is the equilibrium volume and not the initial volume since it changes as reaction proceeds.

$$K_c = \frac{[NO_2]_{eqm}^2}{[N_2O_4]_{eqm}} = \frac{\left(\dfrac{2x}{V}\right)^2}{\dfrac{(1 - x)}{V}}$$

$$= \frac{4x^2}{(1 - x)V}$$

In order to calculate K_c we need to know the values of x and V. The value of x can be determined as follows.

From 1 mole of N_2O_4 a total of $(1 - x) + 2x$ or $(1 + x)$ moles of the equilibrium mixture is obtained. The ratio $1/(1 + x)$ is given by the expression

R.M.M. = Relative molecular mass

$$\frac{1}{(1 + x)} = \frac{\text{Observed R.M.M. (for equilibrium mixture)}}{\text{Calculated R.M.M. (for } N_2O_4)}$$

hence

$$\frac{1}{(1 + x)} = \frac{65 \cdot 7}{92} \quad \text{and} \quad 92 = 65 \cdot 7 + 65 \cdot 7x$$

$$65 \cdot 7x = (92 - 65 \cdot 7) = 26 \cdot 3$$

$$x = \frac{26 \cdot 3}{65 \cdot 7} = 0 \cdot 4$$

Since 1 mole of gas at 273 K and 1 atm pressure has a volume of $22 \cdot 4$ dm^3 (this is an accurate enough value for our purpose), then $(1 + x)$ or $1 \cdot 4$ moles of the equilibrium mixture of gases should have a volume of $1 \cdot 4 \times 22 \cdot 4$ dm^3 at 1 atm and 273 K. But the temperature of the equilibrium mixture is 323 K, thus the volume will be given by:

$$V = \frac{1 \cdot 4 \times 22 \cdot 4 \times 323}{273} = 37 \cdot 1 \text{ dm}^3$$

$$K_c = \frac{4x^2}{(1 - x)V} = \frac{4 \times 0 \cdot 4^2}{(1 - 0 \cdot 4)37 \cdot 1}$$

$$= 0 \cdot 029 \text{ mol dm}^{-3}$$

Note that this equilibrium constant has units of mol dm^{-3} since $2x$ mole of NO_2 is formed when x mole of N_2O_4 is consumed.

Suppose we now increase the pressure so that the volume of the equilibrium mixture is halved; what effect will this have on the composition of the equilibrium mixture? Since the temperature remains unaltered the value of K_c does not change, and we can use this to determine the new value of x (denoted by x'). The new volume is $V/2$ or $(37{\cdot}1)/2$ or $18{\cdot}55$ dm^3, thus we have:

$$K_c = 0{\cdot}029 = \frac{4x'^2}{(1 - x')18{\cdot}55}$$

On multiplying out, the above expression can be tidied up to give:

$$4x'^2 + 0{\cdot}532x' - 0{\cdot}532 = 0$$

This equation can be solved by recognising it to be of the form $ax^2 + bx + c = 0$ which has solutions given by

$$x = \frac{-b \pm \sqrt{b^2 - 4ac}}{2a}$$

$$x' = \frac{-0{\cdot}532 \pm \sqrt{(0{\cdot}532)^2 + 16(0{\cdot}532)}}{8}$$

$$= 0{\cdot}30 \text{ and } -0{\cdot}44 \text{ (clearly inadmissible)}$$

The effect of halving the volume is seen to decrease the number of moles of NO_2 from $0{\cdot}80$ to $0{\cdot}60$ (remember that it is $2x$ and not simply x); the position of equilibrium is thus shifted from right to left.

The qualitative effect of a change of volume on this equilibrium reaction is easily seen by considering the expression:

$$K_c = \frac{4x^2}{(1 - x)V}$$

If V is reduced then clearly $4x^2/(1 - x)$ must decrease by the same amount to maintain a constant value of K_c. This can only happen if x decreases. Fig. 16.4 shows the effect of volume changes on this system.

For equilibrium reactions in which there is no change in the number of moles, the position of equilibrium is independent of the volume of the system.

FIG. 16.4. *The effect of volume (or pressure) changes on the equilibrium $N_2O_4(g) \rightleftharpoons 2NO_2(g)$*

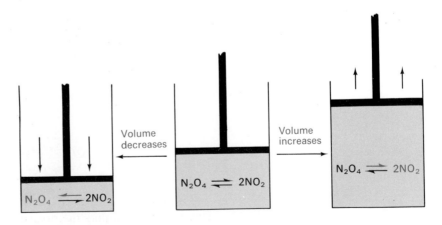

(b) *The equilibrium constant, K_p, in terms of pressure units*
When reactions are carried out in the gaseous phase it is more appropri-

263

ate to use pressure units as a measure of concentration (rather than mol dm^{-3}). In such cases the equilibrium constant is designated K_p; everything we have said about K_c applies equally to K_p. At the risk of repetition, temperature alone alters the value of the equilibrium constant for a particular reaction.

Let us examine the N_2O_4/NO_2 equilibrium to see how this works out in practice.

$$N_2O_4(g) \rightleftharpoons 2NO_2(g)$$

Initial moles	1	0
Equilibrium moles	$(1 - x)$	$2x$
Partial pressures	$\dfrac{(1 - x)}{(1 + x)} P$	$\dfrac{2x}{(1 + x)} P$

The partial pressures of the two gases at equilibrium are determined as follows. There is a total of $(1 + x)$ moles of gas and, of this, $(1 - x)$ is contributed by the N_2O_4 and $2x$ by the NO_2; each gas therefore contributes to the total pressure P in the ratio of $(1 - x)/(1 + x)$ to $2x/(1 + x)$, giving partial pressures as shown above, P being the total pressure. As a check, the sum of the partial pressures should equal the total pressure (Dalton's law of partial pressures (p. 143)).

If p_{NO_2} and $p_{N_2O_4}$ are the equilibrium partial pressures of NO_2 and N_2O_4 respectively, then K_p is given by the expression

$$K_p = \frac{p_{NO_2}^2}{p_{N_2O_4}} = \frac{\left(\dfrac{2x}{1 + x}\right)^2 P^2}{\left(\dfrac{(1 - x)}{(1 + x)}\right) P}$$

which simplifies to

$$K_p = \frac{4x^2 P}{(1 - x^2)}$$

Substituting the value $x = 0.4$ and $P = 1$ atm into this equation we obtain

$$K_p = \frac{4(0.4)^2 \times 1}{(1 - 0.16)} = 0.762 \text{ atm}$$

Notice that the units of K_p in this example are atmospheres.

Suppose the pressure is now doubled; since the temperature is unchanged there is no effect on the value of K_p, but x will change to a new value, say y. We thus have

$$K_p = 0.762 = \frac{4y^2 \times 2}{(1 - y^2)}$$

$$0.762(1 - y^2) = 8y^2$$
$$8.762 \, y^2 = 0.762$$
$$y = 0.29$$

The effect of doubling the pressure is to decrease the number of moles of NO_2 from 0.80 to 0.58. This is as expected, since increasing the pressure amounts to decreasing the volume, which we have previously seen results in the equilibrium being shifted to the left.

Why do the two results not give the same value for x' and y when (a) the equilibrium volume is halved ($x' = 0.30$), (b) the equilibrium pressure is doubled ($y = 0.29$)? In other words, why does halving the equilibrium volume not quite amount to the same thing as doubling the equilibrium pressure?

The Haber Process—the equilibrium between nitrogen, hydrogen and ammonia

This process involves reacting nitrogen and hydrogen together at a high pressure (at least 200 atm) in the presence of an iron catalyst and at a temperature in the region of 700 K. Under these conditions about 15%[*] of the equilibrium mixture is ammonia. The gases are cooled, while still under pressure, and the ammonia removed as a liquid; the unreacted nitrogen and hydrogen are recycled for further conversion into ammonia.

$$N_2(g) + 3H_2(g) \rightleftharpoons 2NH_3(g) \qquad \Delta H^{\ominus}(298\text{ K}) = -92 \text{ kJ mol}^{-1}$$

The process is of enormous industrial importance and accounts for practically all the ammonia used. Plate 10 shows the original small scale experimental apparatus used by Fritz Haber in 1911.

Ammonia, as such, is used in the manufacture of nitric acid, carbamide (urea) and nitrogenous fertilisers, e.g. ammonium nitrate.

As can be seen from the above equation, the reaction proceeds from left to right with a decrease in the number of moles, so clearly an

Plate 10. Haber's experimental apparatus for the synthesis of ammonia (photo: The Deutsches Museum, Munich)

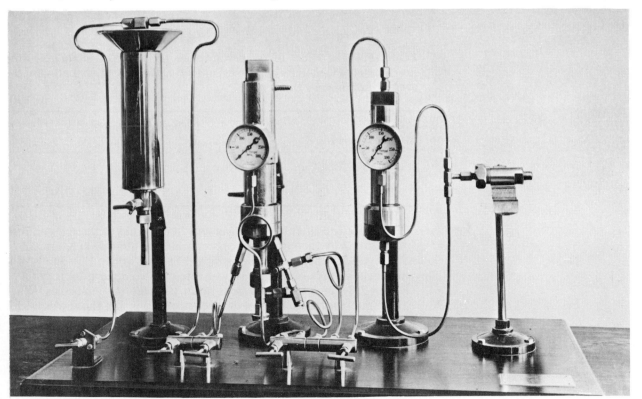

---

[*] Using stoichiometric amounts of nitrogen and hydrogen.

265

increase in pressure will move the equilibrium over to the right and thus increase the percentage yield of ammonia. Table 16.3 gives some data relating to percentage yield of ammonia at three different pressures.

Table 16.3 The effect of pressure on the yield of ammonia using stoichiometric amounts of nitrogen and hydrogen at 723 K in the presence of an iron catalyst

Pressure/atm	Percentage molar yield of ammonia
1	Negligible
100	7
1000	41

In addition to increasing the yield, the reaction reaches equilibrium faster at higher and higher pressures, since the number of gaseous collisions increases. However, with increasing pressure the cost of industrial plant becomes more expensive, e.g. pipes have to be made thicker to withstand the pressure, and a compromise is reached between high yield at high cost and lower yield at lower cost.

The synthesis of ammonia is an exothermic reaction and approximately 46 kJ of heat are evolved for every mole of ammonia synthesised; the reverse reaction is endothermic to the same extent. In chapter 15 we noted that more and more endothermic reactions take place at higher and higher temperatures, so we should expect to obtain a lower yield of ammonia if the temperature is increased. Some data are given in Table 16.4.

Table 16.4 The effect of temperature on the yield of ammonia using stoichiometric amounts of nitrogen and hydrogen at 200 atm in the presence of an iron catalyst

Temperature/K	Percentage molar yield of ammonia
1273	Negligible
723	15
473	88

There is a considerable improvement in the yield as the temperature is reduced. Why then is the process operated at about 700 K instead of at, say, 473 K where the yield is 88%? The answer to this question is that reactions take place at a slower rate if the temperature is reduced (chapter 25), since fewer collisions occur between the molecules. What really matters is the number of energetic collisions, i.e. those which lead to chemical reaction, and these are drastically reduced if the temperature is lowered. At a temperature of 473 K the high yield of ammonia takes a long time to achieve; it is more economical to accept a lower yield of 15% at a higher temperature since equilibrium is established much more rapidly. In practice a 15% yield is better than it sounds, since liquid ammonia is drained off from the system and the unreacted nitrogen and hydrogen recycled. In effect this means that all the reactant gases are eventually converted into ammonia.

If we assume that nitrogen and hydrogen are reacted in the molar ratio of $1:3$ at 700 K and at a pressure of 200 atm, and that a 15% yield of ammonia is achieved, we can calculate a value of K_p for the reaction as follows.

The percentage of nitrogen and hydrogen in the equilibrium mixture will be $(100 - 15)$ or 85%, and of this total $85 \times \frac{1}{4}$ will be nitrogen and $85 \times \frac{3}{4}$ will be hydrogen. The percentage of each gas will therefore be:

$$\text{ammonia} = 15\%, \text{nitrogen} = 21\cdot25\%, \text{hydrogen} = 63\cdot75\%$$

If P is the equilibrium pressure, then the partial pressures of the three gases will be given by:

$$p_{NH_3} = 15P/100 = 0\cdot15P$$
$$p_{N_2} = 21\cdot25P/100 = 0\cdot2125P$$
$$p_{H_2} = 63.75P/100 = 0.6375P$$

For this reaction, the equilibrium constant is given by the expression:

$$K_p = \frac{p_{NH_3}^2}{p_{N_2} \, p_{H_2}^3} = \frac{(0\cdot15)^2 P^2}{(0\cdot2125)P(0\cdot6375)^3 P^3}$$

Substituting $P = 200$, we get

$$K_p = \frac{(0\cdot15)^2}{(0\cdot2125)(0\cdot6375)^3(200)^2} = 1\cdot02 \times 10^{-5} \text{ atm}^{-2}$$

Since this calculation assumes that gases behave ideally at a pressure of 200 atm, which certainly will be far from true (p. 138), the order of magnitude only of this result will be correct. A surprisingly low value of K_p gives a reasonable yield of ammonia, and this must always be borne in mind when equilibrium expressions contain concentration terms raised to different powers.

The Contact Process—the equilibrium between sulphur dioxide, oxygen and sulphur trioxide

This process, which is the basis of the industrial production of sulphuric acid, involves the reaction of highly purified sulphur dioxide and oxygen in the presence of a vanadium(V) oxide catalyst at a temperature of about 723 K (450°C). Under these conditions, using an SO_2/O_2 molar ratio of approximately $1:1$, about 98% of the sulphur dioxide is converted into sulphur trioxide using a pressure only slightly higher than atmospheric.

$$2SO_2(g) + O_2(g) \rightleftharpoons 2SO_3(g) \qquad \Delta H^{\ominus}(298 \text{ K}) = -189 \text{ kJ mol}^{-1}$$

The reaction mixture is cooled and the sulphur trioxide dissolved in 98% sulphuric acid to give either 100% sulphuric acid or oleum, $H_2S_2O_7$, (see a textbook of Inorganic Chemistry for details).

As in the Haber process there is a decrease in the number of moles as the reaction proceeds from left to right, hence an increase in pressure will result in a larger yield of sulphur trioxide. In practice a good yield of sulphur trioxide is obtained at a pressure only slightly in excess of atmospheric (which is necessary to push the reactants through the catalyst chamber), so the additional cost involved in working at higher pressures is not justified. The reaction is exothermic, so the yield of sulphur trioxide will drop as the temperature is increased. The actual temperature employed, about 723 K, is a compromise arrived at by balancing the conflicting factors of high yield and rapid rate of attainment of equilibrium.

The following example is based on this reaction.

Example 1
The oxidation of sulphur dioxide is a reversible process:

$$2SO_2(g) + O_2(g) \rightleftharpoons 2SO_3(g)$$

Calculate the value of the equilibrium constant, K_p, at 1000 K from the following equilibrium partial pressures.

<table>
<tr><td></td><td>**Partial pressures/atm**</td><td></td></tr>
<tr><td>p_{SO_2}</td><td>p_{O_2}</td><td>p_{SO_3}</td></tr>
<tr><td>0·273</td><td>0·402</td><td>0·325</td></tr>
</table>

If the above equilibrium mixture was obtained by starting with a mixture of sulphur dioxide and oxygen in a sealed vessel at 1000 K, what were the initial partial pressures of these two gases?
For this reaction, K_p, is given by the expression:

$$K_p = \frac{p_{SO_3}{}^2}{p_{SO_2}{}^2 \, p_{O_2}} = \frac{(0·325)^2}{(0·273)^2(0·402)}$$
$$= 3·53 \text{ atm}^{-1}$$

Initially there is no SO_3 whereas at equilibrium there is 0·325 atm of this gas. This must be produced by the disappearance of 0·325 atm of SO_2 and 0·1625 atm of O_2 (see the stoichiometric coefficients in the equation). The initial pressures of SO_2 and O_2 are therefore respectively (0·273 + 0·325) or 0·598 atm and (0·402 + 0·1625) or 0·565 atm.

16.5
Factors which affect the position of equilibrium, the equilibrium constant and the rate at which equilibrium is achieved

In the discussion so far we have seen that there are three factors of importance when considering chemical equilibria, namely
(a) the position of equilibrium,
(b) the magnitude of the equilibrium constant,
(c) the rate at which equilibrium is achieved.
The effect of changing conditions on the first two can be summed up in terms of **Le Chatelier's Principle** which states:
If any constraint is applied to a system in equilibrium, then the system will change in such a manner as to counteract this constraint as far as is possible.

For example, in the Haber process the effect of increasing the pressure is to move the position of equilibrium further over to the right, since this results in a decrease in volume and consequently a decrease in applied pressure (the constraint applied). Similarly the injection of more nitrogen into this system, in such a manner that the equilibrium pressure remains unchanged, will result in more ammonia being formed. In this case the constraint (more nitrogen) is counteracted to some extent by further combination with hydrogen to give additional ammonia.

Le Chatelier's principle also allows one to predict which way an equilibrium constant will be affected qualitatively by changes of temperature. For exothermic reactions an increase in temperature (the constraint applied) moves the equilibrium in the direction which absorbs heat, i.e. in the reverse direction, and the magnitude of the equilibrium constant decreases. Conversely for an endothermic reaction the magnitude of the equilibrium constant is increased.

The N_2O_4/NO_2 equilibrium is endothermic as written from left to right:

$$N_2O_4 \rightleftharpoons 2NO_2$$
$$\text{(colourless)} \quad \text{(dark brown)}$$

If a bulb containing the equilibrium mixture at room temperature is plunged into a beaker of boiling water, the colour darkens as more NO_2 is produced (equilibrium constant increases). Conversely if the bulb is placed in a mixture of water and ice the colour lightens (equilibrium constant decreases). The two situations are shown in fig. 16.5.

FIG. 16.5. *The effect of temperature changes on the equilibrium* $N_2O_4(g) \rightleftharpoons 2NO_2(g)$

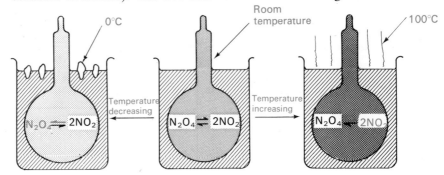

The third factor, the rate at which equilibrium is achieved, is a problem of kinetics (chapter 24). A catalyst increases the rate at which equilibrium is achieved by speeding up both the forward and backward reactions to the same extent, and has no effect whatsoever on the equilibrium constant or on the position of equilibrium. An increase of temperature increases the rate at which equilibrium is achieved but it will affect the magnitude of the equilibrium constant for an exothermic reaction adversely.

Table 16.5 summarises the effects on a system at equilibrium of varying the factors discussed above.

Table 16.5 The effect of varying conditions on the equilibrium $aA + bB \rightleftharpoons cC + dD$

Changes imposed on the system in equilibrium	Equilibrium position moves	Equilibrium constant	Any other points
Conc. of A and/or B increased	To right	No change	—
Conc. of C and/or D increased	To left	No change	—
Pressure increased	To right if $(c + d) < (a + b)$ To left if $(a + b) < (c + d)$	No change	Very little effect, if any, on reactions in liquid solution
Temperature increased	To left if the reaction is exothermic To right if the reaction is endothermic	Value decreased Value increased	Equilibrium achieved faster
Addition of a catalyst	No change	No change	Equilibrium achieved faster

269

16.6
Heterogeneous equilibria

In heterogeneous equilibria the participating substances are present in more than one phase, a typical example being the dissociation of calcium carbonate:

$$CaCO_3(s) \rightleftharpoons CaO(s) + CO_2(g)$$

We can write down an expression for the equilibrium constant of this reaction:

$$K_c = \frac{[CaO]_{eqm}[CO_2]_{eqm}}{[CaCO_3]_{eqm}}$$

The concentrations of calcium oxide and calcium carbonate are constant (if their masses are altered, their volumes will change correspondingly). The concentrations $[CaO]_{eqm}$ and $[CaCO_3]_{eqm}$ are incorporated into a new equilibrium constant, and we may write:

$$K_p = p_{CO_2}$$

This final expression means that if we heat calcium carbonate in a closed vessel and measure its dissociation pressure at a fixed temperature, then this pressure is independent of the quantity of calcium carbonate. If more calcium carbonate and/or calcium oxide is added to the equilibrium mixture it will not have any affect on the value of the dissociation pressure. If a larger pressure of carbon dioxide is injected into this system, then this extra carbon dioxide will combine with some of the calcium oxide to reform calcium carbonate until the original dissociation pressure of carbon dioxide is re-established.

The dissociation of calcium carbonate is an endothermic process, so the value of K_p (which has units of atm) will increase with a rise in temperature, i.e. the dissociation pressure of carbon dioxide will increase.

Another example of a heterogeneous equilibrium

Calculations involving heterogeneous equilibria can be carried out in exactly the same manner as for homogeneous equilibria as the following example shows.

Example 2
The equilibrium pressure in a system originally consisting of a solid sample of ammonium hydrogen sulphide, NH_4HS, at 298 K is 0·66 atm. Calculate the value of the equilibrium constant K_p in terms of partial pressures. What would be the partial pressure of ammonia in the mixture if 0·1 atm of this gas were added to the system?

$$NH_4HS(s) \rightleftharpoons NH_3(g) + H_2S(g)$$
$$K_p = p_{NH_3} \, p_{H_2S}$$

The final pressure 0.66 atm is composed of 0·33 atm of ammonia and 0·33 atm of hydrogen sulphide, therefore the equilibrium constant is given by:

$$K_p = (0·33)(0·33) = 0·33^2 = 0·109 \text{ atm}^2$$

Addition of more ammonia will alter the pressures of the two gases. If the partial pressure of hydrogen sulphide is now y atm, the partial pressure of ammonia will be $(y + 0·1)$ atm. Using the value of K_p determined above we have

$$K_p = 0·109 = y(y + 0·1)$$

hence
$$y^2 + 0·1y - 0·109 = 0$$

On solving this quadratic equation by the method used previously (p. 263) we have:

$$y = 0.28 \text{ or } -0.38 \text{ (clearly inadmissible)}$$

The partial pressure of ammonia is therefore $(0.28 + 0.1)$ or 0.38 atm. Once again, the equilibrium is shifted over to the left.

16.7
Some more points about equilibrium constants

It is possible to calculate equilibrium constants from tabulated values of standard free energies of formation (p. 248). The equation which links $\Delta G^\ominus$ and K_p for a reaction in the gaseous phase can be derived by rather involved thermodynamic arguments and is given below without proof:

$$\Delta G^\ominus = -RT \ln K_p \qquad (2)$$

Since standard free energy changes are defined in terms of a standard pressure of 1 atm, equilibrium constants derived for gaseous reactions from standard free energy data are therefore K_p (and not K_c) values. The actual numerical value of K_p relates to equilibrium partial pressures expressed in atm (and not any other pressure unit such as N m^{-2}). Of course conversion of an actual K_p value to another expressed, say, in N m^{-2} can be carried out but it is an unnecessary extra step.

To return to the main discussion of this section, we now show how to calculate the equilibrium constant for the N_2O_4/NO_2 equilibrium.

Calculation of K_p for the equilibrium $N_2O_4(g) \rightleftharpoons 2NO_2(g)$ at 298 K

$$
\begin{array}{ccc}
 & N_2O_4(g) & \rightleftharpoons & 2NO_2(g) \\
\Delta G_f^\ominus (298\ K)/\text{kJ mol}^{-1} & +98.3 & & 2 \times 51.8
\end{array}
$$

hence
$$
\begin{aligned}
\Delta G^\ominus(298\ K) \text{ for this reaction} &= 103.6 - (+98.3) \\
&= +5.3 \text{ kJ mol}^{-1}
\end{aligned}
$$

Substituting this value of $\Delta G^\ominus(298\ K)$ into equation (2) above we have

R = 8·314 J K^{-1} mol^{-1}
The expression on the right-hand side is divided by 1000 to convert to kJ

$$\Delta G^\ominus(298\ K) = +5.3 = -\frac{8.314 \times 298}{1000} \ln K_p$$

hence
$$\ln K_p = -\frac{5.3 \times 1000}{8.314 \times 298} = -2.139$$

and
$$K_p = 0.118 \text{ atm at 298 K}$$

Variation of K_p with temperature

It is possible to derive an expression showing how K_p varies with temperature from equation (2) and the expression:

$$\Delta G^\ominus = \Delta H^\ominus - T\Delta S^\ominus \qquad (\text{p. 246})$$

Thus
$$\Delta G^\ominus = -RT \ln K_p = \Delta H^\ominus - T\Delta S^\ominus$$

and
$$\ln K_p = -\frac{\Delta H^\ominus}{RT} + \frac{\Delta S^\ominus}{R}$$

If we assume that both $\Delta H^\ominus$ and $\Delta S^\ominus$ are independent of temperature (and for small temperature changes this is generally a good approximation) we may write:

$$\ln K_p = -\frac{\Delta H^\ominus}{RT} + \text{Constant} \qquad (3)$$

Let us now denote the values of K_p at two different temperatures T_1 and T_2 by K_1 and K_2, then

$$\ln K_2 = -\frac{\Delta H^\ominus}{RT_2} + \text{Constant} \qquad (4)$$

$$\ln K_1 = -\frac{\Delta H^\ominus}{RT_1} + \text{Constant} \qquad (5)$$

Subtracting equation (5) from equation (4) we obtain

$$\ln \frac{K_2}{K_1} = \frac{\Delta H^\ominus}{R}\left[\frac{1}{T_1} - \frac{1}{T_2}\right]$$

or

$$\ln \frac{K_2}{K_1} = \frac{\Delta H^\ominus}{R}\left[\frac{T_2 - T_1}{T_2 T_1}\right] \qquad (6)$$

Equation (6) enables K_p to be calculated at a specific temperature, if its value at another one is known.

Before working through a typical calculation, let us see what equation (6) tells us qualitatively. Consider an endothermic reaction in which the temperature is increased from T_1 to T_2; since $\Delta H^\ominus$ is positive the right hand side of equation (6) is positive, thus $\ln(K_2/K_1)$ is greater than unity and $K_2 > K_1$. Conversely for an exothermic reaction the right-hand side of equation (6) is negative, so $\ln(K_2/K_1)$ is negative and $K_2 < K_1$. These conclusions were arrived at previously by applying Le Chatelier's principle (p. 268).

Calculation of K_p for the equilibrium $N_2O_4(g) \rightleftharpoons 2NO_2(g)$ at 323 K

The standard enthalpy change for this reaction at 298 K is given by twice the standard enthalpy of formation of NO_2 minus the standard enthalpy of formation of N_2O_4, i.e.

$$2 \times (+33\cdot9) - (+9\cdot7) \text{ or } 58\cdot1 \text{ kJ mol}^{-1} \text{ (58 100 J mol}^{-1}\text{)}.$$

Substituting this value and also $T_2 = 323$ and $T_1 = 298$ into equation (6) we have

$$\ln \frac{K_2}{K_1} = \frac{58\ 100(323 - 298)}{8\cdot314 \times 323 \times 298} = 1\cdot815$$

thus

$$\frac{K_2}{K_1} = 6\cdot14$$

Using the value $K_1 = 0\cdot118$ atm found previously (p. 271), the value of K_p at 323 K is:

$$K_2 = 6 \cdot 141 \times K_1 = 6 \cdot 141 \times 0 \cdot 118$$
$$= 0 \cdot 725 \text{ atm}$$

This value compares with an experimentally determined one of 0·762 atm (p. 264), so the agreement is quite good.

Calculation of equilibrium constants for reactions in solution

These can be calculated in the same way as for gaseous reactions, but with the substitution of K_c for K_p since the standard state is defined in terms of unit concentration (1 mol dm^{-3}). In fact, some other corrections have to be applied but they need not concern us here. Thus:

$$\Delta G^{\ominus} = -RT \ln K_c$$

Once a value of K_c has been determined at a particular temperature, this can then be used to calculate the value of the equilibrium constant at another one by application of equation (6) (p. 272).

16.8
Questions on chapter 16

1 Write a *short account* of the factors affecting the position of equilibrium of a balanced reaction, the rate at which equilibrium is attained and the value of the equilibrium constant.
(a) Using partial pressures, show that for gaseous reactions of the type
$$XY(g) \rightleftharpoons X(g) + Y(g)$$
at a given temperature, the pressure at which XY is exactly *one-third* dissociated is numerically equal to *eight* times the equilibrium constant at that temperature.
(b) When one mole of ethanoic acid (acetic acid) is maintained at 25°C with 1 mole of ethanol, one-third of the ethanoic acid remains when equilibrium is attained. How much would have remained if one-half of a mole of *ethanol* had been used instead of one mole at the same temperature?   S

2 The following equilibrium is involved in the industrial preparation of sulphuric acid.
$$2SO_2(g) + O_2(g) \rightleftharpoons 2SO_3(g); \quad \Delta H = -188 \text{ kJ mol}^{-1}.$$
(a) Write the expression for the equilibrium constant K_p for the above equilibrium.
(b) State the effect upon the equilibrium of, and give an explanation for, each of the following changes.
  (i) Increase of pressure at constant temperature
  (ii) Increase of temperature at constant pressure
(c) Give approximate values of the temperature and pressure which are used in the industrial preparation of sulphur(VI) oxide.   JMB (Syllabus B)

3 (a) Discuss those aspects of chemical kinetics and chemical equilibria which help to determine the best conditions for the industrial conversion of nitrogen and hydrogen to ammonia.
(b) Deduce the increase in yield of ammonia, per unit volume of nitrogen, when the operating pressure is doubled. You can assume that the extra amount of ammonia formed does not change the partial pressures of the other gases.
(c) Discuss the advantages of a large plant compared to a small plant in terms of
  (i) fixed charges
  (ii) operating charges, per tonne of ammonia.
  Mention two disadvantages of a large plant.   JMB (Syllabus A)

4 Explain the effect of temperature on rate of reaction.
   Methanol is manufactured by reacting carbon monoxide and hydrogen in the

presence of a catalyst. The following equilibrium is attained

$$CO(g) + 2H_2(g) \rightleftharpoons CH_3OH(g); \quad \Delta H = -65 \text{ kJ mol}^{-1}$$

(a) What is the function of the catalyst?

(b) Deduce the optimum conditions of temperature and pressure, explaining your reasoning.

(c) Deduce an expression for K_P in terms of the total pressure of the system, P, and the number of moles of carbon monoxide, a, of hydrogen, b, and of methanol, c, present at equilibrium.

(d) In practice, if carbon monoxide and hydrogen in the molar ratio of $1:2$ are used and at equilibrium 15% of the carbon monoxide has been converted, calculate the number of moles of carbon monoxide, of hydrogen and of methanol present at equilibrium. If the equilibrium constant K_p is $4 \times 10^{-10}/\text{kPa}^2$, calculate the pressure in the system. (101 kPa = 1 atm.) AEB

5 (a) Explain what you understand by the term *equilibrium constant*.

(b) Write a brief account of the influence of change of pressure and temperature on the position of equilibrium of a balanced homogeneous gaseous reaction, the rate at which equilibrium is attained and the value of the equilibrium constant.

(c) Carbon monoxide will react with steam under appropriate conditions according to the following reversible reaction

$$CO(g) + H_2O(g) \rightleftharpoons CO_2(g) + H_2(g); \quad \Delta H = -40 \text{ kJ mol}^{-1}$$

(i) Calculate the number of moles of hydrogen in the equilibrium mixture when three moles of carbon monoxide and three moles of steam are placed in a reaction vessel of constant volume and maintained at a temperature at which the equilibrium constant has the numerical value of 4·00.

(ii) Calculate the mole fractions of reactants and products.

(iii) Sketch a graph to show how the amount of carbon monoxide changes during the course of the reaction.

(iv) Sketch similar graphs on the same scale to illustrate what happens when the reaction is repeated exactly as before except that in the first the temperature is raised, in the second the pressure is lowered and in the third an industrial catalyst is introduced, the other factors remaining constant. Add a brief note to clarify what you have shown. S

6 (a) Write an equation to show the relationship between the equilibrium constant, K_p, and the partial pressures, $P_{N_2O_4}$ and P_{NO_2} of the reactants in the following gaseous equilibrium:

$$N_2O_4 \rightleftharpoons 2NO_2; \quad \Delta H(298 \text{ K}) = 54 \text{ kJ mol}^{-1}$$

(b) State the effect, if any, on the above equilibrium of (i) increasing the pressure, (ii) raising the temperature. Give reasons for your answers.

(c) It was found that one dm³ of the gaseous mixture weighed 2·777 g at 50·0°C and under a pressure of $1·01 \times 10^5 \text{ Nm}^{-2}$ (=1 atmosphere). Calculate:

(i) the fraction of the N_2O_4 that is dissociated;

(ii) the percentage of NO_2 molecules in the mixture;

(iii) the value of K_p.

[One mole of a gas occupies 22·4 dm³ at s.t.p.] O

7 What do you understand by the terms *partial pressure* and *concentration* as applied to gases? Deduce the relationship between these two quantities for an ideal gas.

Write down expressions for the equilibrium constants K_p and K_c for the reaction

$$SO_2Cl_2(g) \rightleftharpoons SO_2(g) + Cl_2(g)$$

What is the effect (at constant temperature) on the position of this equilibrium and on K_p of:

(a) adding a catalyst,

(b) compressing the system?

At a temperature of 375°C and an overall pressure of 101 325 N m⁻² a sample of SO_2Cl_2 in the gas phase was found to be 84% dissociated. What is the value of K_p for the above reaction under these conditions? O and C

8 (a) State *Le Chatelier's principle*.

(b) For the gaseous equilibrium represented by the equation
$$H_2(g) + I_2(g) \underset{k_1}{\overset{k_{-1}}{\rightleftharpoons}} 2HI(g); \quad \Delta H \text{ positive,}$$
state and explain how the equilibrium constant K_c and the rate constants k_1 and k_{-1} change, if at all,

(i) with increase in temperature,

(ii) in the presence of a catalyst.

(c) How would you expect the composition of the equilibrium mixture of the reaction in (b) to be affected, if at all, by diluting the system with argon?

(d) Write an expression for the equilibrium constant K_c for the reaction in (b) and outline an experimental method which would enable you to determine K_c in the laboratory. C

9 Write a brief account of the influence of change of pressure and temperature on the position of equilibrium of a balanced reaction, the rate at which equilibrium is attained and the value of the equilibrium constant.

Write down equilibrium constants, K_p or K_c as appropriate, indicating their units (if any), for the following equilibrium
$$P + Q \rightleftharpoons 2R + S$$
where, under the conditions of the experiments,

(i) P, Q, R and S are all gases,

(ii) P, Q, R and S are all liquids,

(iii) P and S are gases, Q and R, solids.

The vapour of dinitrogen tetroxide is partially dissociated
$$N_2O_4 \rightleftharpoons 2NO_2$$
4·80 g of dinitrogen tetroxide occupies a volume of 1·50 dm^3 at normal atmospheric pressure, $1·00 \times 10^5$ Pa (1 atmosphere) and 27°C.

Calculate

(i) the degree of dissociation,

(ii) the equilibrium constant, K_p, at this temperature.

N = 14, O = 16

$R = 8·314$ J mol^{-1} K^{-1}

J = kg m^2 s^{-2} = N m

Pa (pascal, N m^{-2}) is the SI unit of pressure. S

10 The table below gives information about the values at different temperatures of the equilibrium constant for the reaction
$$N_2(g) + O_2(g) \rightleftharpoons 2NO(g)$$

		Partial pressure of NO/atm $\times 10^2$	
Temperature/K	$10^4 K_p$	$p_{N_2} = 0·8$ atm $p_{O_2} = 0·2$ atm	$p_{N_2} = 0·8$ atm $p_{O_2} = 0·05$ atm
1800	1·21	0·44	0·22
2000	4·08	0·81	0·40
2200	11·00	1·33	0·67
2400	25·10	2·00	1·00
2600	50·30	2·84	1·42

It also gives the partial pressures of NO in equilibrium with two different mixtures of nitrogen and oxygen at the given temperatures.

(a) Write an expression for the equilibrium constant K_p for the given reaction.

(b) Use the expression you have written to explain why the values in the fourth column are half those in the third.

(c) What conditions of temperature and pressure should be used to obtain the best yield of NO? Justify your answer.

(d) Is the reaction exothermic or endothermic? Justify your answer.

11 What do you understand by the term *dynamic equilibrium*? Describe a simple experiment which would demonstrate whether a state of dynamic equilibrium exists in a selected system.

The reaction

$$ZnO(s) + CO(g) \longrightarrow Zn(g) + CO_2(g)$$

has an equilibrium constant of $1 \cdot 00$ atm at 1500 K. Calculate the equilibrium partial pressure of zinc vapour in a reaction vessel if an equimolar mixture of CO and CO_2 is brought into contact with solid zinc oxide at 1500 K and at a total pressure of $1 \cdot 0$ atm.

This reaction is highly endothermic. Zinc is manufactured by bubbling the effluent gas from such a reaction vessel through molten lead at 700 K. Suggest reasons why this is done.

O and C (S)

12 Discuss the factors that affect the position of equilibrium and its rate of attainment in a chemical reaction.

In the industrial production of ammonia a mixture of nitrogen and hydrogen flows over a catalyst where the two gases react with evolution of heat. The diagram below shows the percentage conversion of the stoichiometric mixture to ammonia as a function of temperature for a particular catalyst. The flow rate for the lower curve is four times that for the upper curve. Explain the main features of the diagram and discuss the optimum conditions for the manufacture of ammonia.

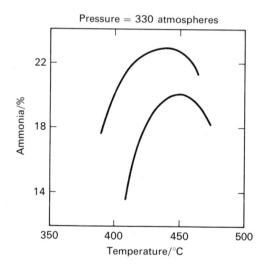

Pressure = 330 atmospheres

Oxford Schol. and Entrance

276

Chapter 17

Acid-base equilibria

17.1
Acids and bases

The definitions of an acid and a base are inseparable. The most useful definition (Brønsted-Lowry) states that an acid is a substance which can donate a proton, H^+, to a base; a base is thus defined as a substance which can accept a proton from an acid. The two go hand in hand, and acid-base reactions are proton transfer reactions:

$$HA + B^- \rightleftharpoons HB + A^-$$
$$\text{Acid} \quad \text{Base}$$

A base is usually a negative ion or a neutral molecule with a lone-pair of electrons and it forms a co-ordinate bond (a dative bond) with the proton.

When an acid loses its proton to a base, it forms a base itself; similarly the base in accepting a proton forms an acid. Every acid has its conjugate base, and every base its conjugate acid. An acid-base reaction is a dynamic equilibrium between two conjugate acid-base pairs:

$$\text{Acid 1} + \text{Base 2} \rightleftharpoons \text{Base 1} + \text{Acid 2}$$

For example:

$$HCl(aq) + H_2O(l) \rightleftharpoons Cl^-(aq) + H_3O^+(aq)$$
$$\text{Acid 1} \quad \text{Base 2} \quad \text{Base 1} \quad \text{Acid 2}$$

$$HCl(aq) + NH_3(aq) \rightleftharpoons Cl^-(aq) + NH_4^+(aq)$$
$$\text{Acid 1} \quad \text{Base 2} \quad \text{Base 1} \quad \text{Acid 2}$$

The solvent in acid-base equilibria

Water is a particularly versatile solvent in that it can act both as an acid and as a base:

$$HCl(aq) + H_2O(l) \rightleftharpoons Cl^-(aq) + H_3O^+(aq)$$
$$\text{Acid 1} \quad \text{Base 2} \quad \text{Base 1} \quad \text{Acid 2}$$

$$H_2O(l) + NH_3(aq) \rightleftharpoons OH^-(aq) + NH_4^+(aq)$$
$$\text{Acid 1} \quad \text{Base 2} \quad \text{Base 1} \quad \text{Acid 2}$$

A substance that can act both as an acid and as a base is said to be amphoteric. The hydrogen sulphate ion, HSO_4^-, provides another example of amphoteric behaviour:

$$H_3O^+(aq) + HSO_4^-(aq) \rightleftharpoons H_2O(l) + H_2SO_4(aq)$$
$$\text{Acid 1} \quad \text{Base 2} \quad \text{Base 1} \quad \text{Acid 2}$$

$$HSO_4^-(aq) + OH^-(aq) \rightleftharpoons SO_4^{2-}(aq) + H_2O(l)$$
$$\text{Acid 1} \quad \text{Base 2} \quad \text{Base 1} \quad \text{Acid 2}$$

Plate 11. Justus von Liebig, 1803–73 (The Mansell Collection) *He proposed that the replacement of hydrogen by metals was a characteristic of acids.*

A change of solvent can have a dramatic effect on the behaviour of acids and bases. For example, in aqueous solution hydrogen chloride behaves as an acid by giving up its proton to water:

$$HCl(aq) + H_2O(l) \rightleftharpoons H_3O^+(aq) + Cl^-(aq)$$

If magnesium is added to the solution, hydrogen is evolved:

$$Mg(s) + 2H_3O^+(aq) \rightarrow Mg^{2+}(aq) + 2H_2O(l) + H_2(g)$$

In dry methylbenzene (toluene), however, anhydrous hydrogen chloride does not react with magnesium since hydrogen chloride, HCl, is unable to donate a proton under these conditions and hence cannot function as an acid.

Liquid ammonia has been used extensively as a solvent and, despite its unpleasant nature, it is fairly easily handled. In liquid ammonia, ammonium chloride, normally regarded as a salt, will react with magnesium to liberate hydrogen:

$$NH_4Cl(solv) + NH_3(l) \rightleftharpoons NH_4^+(solv) + Cl^-(solv)$$
$$Mg(s) + 2NH_4^+(solv) \rightarrow Mg^{2+}(solv) + 2NH_3(l) + H_2(g)$$

These two equations are analogous to the two above showing the action of HCl(aq) on magnesium. In this case the species are solvated with liquid ammonia, e.g. $NH_4^+(solv)$ is analogous to $H_3O^+(aq)$ in the water system.

17.2
Aqueous equilibria

Since we are predominantly concerned with acid-base equilibria in water, it is important to look more closely at the equilibrium that already exists in water. Before doing this, however, it is convenient to discuss the acidic species $H_3O^+(aq)$, which we have been writing in equations.

The hydrated proton

The proton has a diameter of about 3×10^{-15} m and it would be impossible for a particle of such a high charge density to exist as H^+ in a solvent as polar as water. Like other positive ions it is hydrated in aqueous solution, and it is thought that one molecule of water is closely bound to the proton to form the hydroxonium ion H_3O^+ (also called the hydronium ion or the oxonium ion). This ion is analogous to the ammonium ion NH_4^+. Evidence for its existence comes from the spectroscopic studies of hydrated chloric(VII) acid, $HClO_4 \cdot H_2O$. These show that the three hydrogen atoms are equivalent and all bonded to oxygen; a type of intra-molecular acid-base reaction has occurred and the formula should really be written as $H_3O^+ClO_4^-$.

It has been suggested that the predominant form of the proton in water is $H_9O_4^+$, i.e. $H^+(H_2O)_4$, in which the hydroxonium ion is hydrogen-bonded to three water molecules. However, it is unnecessarily complicated to use this formula in writing acid-base equilibria and the most commonly used forms are $H_3O^+(aq)$, H_3O^+, or just simply $H^+(aq)$—the hydrated proton. We shall generally use the latter in this book, although we shall use $H_3O^+(aq)$ where this is felt to be more appropriate.

FIG. 17.1. *Structures of H_3O^+ and $H_9O_4^+$ ions*

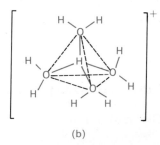

(a)

(b)

278

The ionic product, K_w, for water

Water is ionised to a very small extent:

$$H_2O(l) \rightleftharpoons H^+(aq) + OH^-(aq)$$

and this equilibrium is characterised by an equilibrium constant, K, at a given temperature:

$$K = \frac{[H^+(aq)]_{eqm}[OH^-(aq)]_{eqm}}{[H_2O(l)]_{eqm}}$$

Since the degree of ionisation is extremely small, the concentration of unionised water molecules is virtually constant and the equation can be rewritten as

$$K_w = [H^+(aq)]_{eqm}[OH^-(aq)]_{eqm} \qquad (1)$$

where $\qquad K_w = K[H_2O(l)]_{eqm}.$

Temperature/K	K_w/mol^2 dm^{-6}
273	0.11×10^{-14}
283	0.30×10^{-14}
293	0.68×10^{-14}
298	1.00×10^{-14}
323	5.47×10^{-14}
373	51.3×10^{-14}

The equilibrium constant, K_w, is called the **ionic product for water** and has a value of 10^{-14} mol^2 dm^{-6} at 298 K, provided that concentrations are expressed in the customary units of mol dm^{-3} (see chapter 23, p. 419 for the method of determining this constant). This constant determines absolutely the relative concentrations of $H^+(aq)$ and $OH^-(aq)$ in water at this temperature. If the solution is neutral, the concentrations of $H^+(aq)$ and $OH^-(aq)$ must be the same, and

$$[H^+(aq)]_{eqm} = [OH^-(aq)]_{eqm} = 10^{-7} \text{ mol dm}^{-3}$$

(this figure means that approximately 1 in every 5×10^8 molecules of water is split up into ions at 298 K. There are approximately 6×10^{16} $H^+(aq)$ ions in every dm^3 of water).

If the solution is acidic, then $[H^+(aq)]_{eqm} > [OH^-(aq)]_{eqm}$ but the ionic product, K_w, remains constant at 10^{-14} mol^2 dm^{-6} at 298 K. For example, if $[H^+(aq)]_{eqm} = 10^{-3}$ mol dm^{-3}, then

$$[OH^-(aq)]_{eqm} \times 10^{-3} = 10^{-14}$$

hence

$$[OH^-(aq)]_{eqm} = 10^{-11} \text{ mol dm}^{-3}$$

Plate 12. Potentilla, a plant that flourishes naturally in alkaline soils, pH > 7 (Harry Smith Horticultural Photographic Collection)

The pH scale

In order to avoid the inconvenience of negative indices and to accommodate the very wide range of $H^+(aq)$ and $OH^-(aq)$ concentrations which are commonly encountered in acid-base reactions, Sørensen devised the logarithmic pH scale in 1909:

$$pH = -lg[H^+(aq)] \qquad (2)$$

Thus for a neutral solution at 298 K, $[H^+(aq)] = 10^{-7}$ mol dm^{-3} and so

$$pH = -lg(10^{-7})$$
$$= 7$$

279

It follows that acidic aqueous solutions have a pH of less than 7; alkaline solutions have a pH greater than 7 (see fig. 17.2).

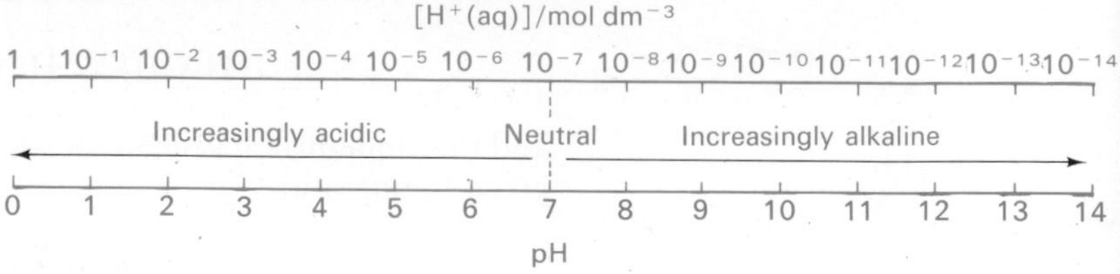

FIG. 17.2. *The pH scale of acidity/alkalinity*

$[H^+(aq)]/mol\ dm^{-3}$

1   10^{-1} 10^{-2} 10^{-3} 10^{-4} 10^{-5} 10^{-6} 10^{-7} 10^{-8} 10^{-9} 10^{-10} 10^{-11} 10^{-12} 10^{-13} 10^{-14}

Increasingly acidic   Neutral   Increasingly alkaline

0   1   2   3   4   5   6   7   8   9   10   11   12   13   14

pH

A strongly acidic solution can, in theory, have a pH less than zero, e.g. a 10M HCl solution has a theoretical pH $= -1$; similarly, a strongly alkaline solution may have a pH value greater than 14. However, pH values generally lie within the range 0 to 14.

It is also possible to define pOH as follows:

$$pOH = -lg[OH^-(aq)]$$

and therefore at 298 K it is is possible to convert equation (1) into the relationship

$$pH + pOH = 14 \qquad (3)$$

Calculations involving equation (2) often cause confusion, and some worked examples are given below.

Example 1
If a solution contains a concentration of $H^+(aq)$ equal to 8.6×10^{-9} mol dm^{-3}, what is the pH of the solution?

$$pH = -lg[H^+(aq)] = -lg(8.6 \times 10^{-9})$$
$$= -(0.9345 - 9)$$
$$= 8.06(55) = 8.07$$

Example 2
If a solution contains an $OH^-(aq)$ concentration equal to 4.7×10^{-3} mol dm^{-3}, what is the pH of the solution?

$$pOH = -lg(4.7 \times 10^{-3})$$
$$= -(0.6721 - 3)$$
$$= 2.3279$$

From equation (3) pH + pOH = 14
so
$$pH = 14 - 2.3279$$
$$= 11.67(21) = 11.67$$

Example 3
If the pH of a solution is 2.73, what is the concentration of $H^+(aq)$ ions?

$$pH = -lg[H^+(aq)] = 2.73$$
$$lg[H^+(aq)] = -2.73$$
$$= \bar{3}.27$$
$$[H^+(aq)] = antilg\ \bar{3}.27$$
$$= 0.001862 = 1.86 \times 10^{-3}\ mol\ dm^{-3}$$

Plate 13. Erica (heather), a plant that flourishes naturally in acidic soils, pH < 7 (Harry Smith Horticultural Photographic Collection)

The measurement of pH values based on buffer solutions and indicators is described later in the chapter. The most convenient method which employs a pH meter is discussed in chapter 18. A third method which is based upon electrolytic conductivity measurements is described in chapter 23.

Salt hydrolysis

In pure water the concentrations of $H^+(aq)$ ions and $OH^-(aq)$ ions are the same. Certain salts, however, which are not classified as acids or bases can disturb this balance when they are dissolved in water; the phenomenon is known as salt hydrolysis.

The highly charged small cations of metals, e.g. those of transition metals, are strongly hydrated in aqueous solution. Furthermore, the small, highly charged cations exert a considerable attraction on the oxygen atoms of the water molecules, thereby polarising the bonds between the hydrogen and oxygen atoms. Under these conditions, solvent water molecules are able to act as a base and thus give rise to an acidic solution. Aqueous solutions of iron(III) chloride, nickel(II) sulphate and copper(II) nitrate, for instance, show acid reactions:

$$\underset{\text{Acid 1}}{[Fe(H_2O)_6]^{3+}} + \underset{\text{Base 2}}{H_2O(l)} \rightleftharpoons \underset{\text{Base 1}}{[Fe(H_2O)_5(OH)]^{2+}} + \underset{\text{Acid 2}}{H^+(aq)}$$

There is little tendency for the chloride, sulphate and nitrate anions to detach hydrogen ions from water molecules, and thereby counteract this acidity, because they are poor proton acceptors. The hydrated sodium and potassium ions do not give rise to acidity, since the cations are only singly charged and they have ionic radii which are large by comparison with the cations of transition metals. The chloride, sulphate and nitrate of these metals consequently show a neutral reaction in aqueous solution.

The carbonates and ethanoates of sodium and potassium, for example, show an alkaline reaction in aqueous solution. This is due to the fact that both the carbonate ion, $CO_3^{2-}(aq)$, and the ethanoate ion, $CH_3COO(aq)^-$, function as bases in aqueous solution; in other words their conjugate acids, carbonic and ethanoic acid respectively, are weak:

$$\underset{\text{Base 1}}{CO_3^{2-}(aq)} + \underset{\text{Acid 2}}{H_2O(l)} \rightleftharpoons \underset{\text{Acid 1}}{HCO_3^-(aq)} + \underset{\text{Base 2}}{OH^-(aq)}$$

followed to some extent by

$$\underset{\text{Base 1}}{HCO_3^-(aq)} + \underset{\text{Acid 2}}{H_2O(l)} \rightleftharpoons \underset{\text{Acid 1}}{H_2CO_3(aq)} + \underset{\text{Base 2}}{OH^-(aq)}$$

$$\underset{\text{Base 1}}{CH_3COO^-(aq)} + \underset{\text{Acid 2}}{H_2O(l)} \rightleftharpoons \underset{\text{Acid 1}}{CH_3COOH(aq)} + \underset{\text{Base 2}}{OH^-(aq)}$$

In the case of salts composed of cations and anions which act respectively as acids and bases in aqueous solution, the resulting solutions may well be neutral owing to the balancing out of the two tendencies. Ammonium ethanoate, for example, is practically neutral in aqueous solution.

17.3
The strengths of acids and bases

When an acid HA dissolves in water, the following equilibrium is established,

$$HA(aq) + H_2O(l) \rightleftharpoons H^+(aq) + A^-(aq)$$

and the position of equilibrium is an indication of the strength of the acid in question. For a strong acid, such as hydrochloric acid, the reaction goes virtually to completion:

$$HCl(aq) + H_2O(l) \rightleftharpoons H^+(aq) + Cl^-(aq)$$

while in the case of a weak acid, such as ethanoic acid, the position of equilibrium favours the reactants:

$$CH_3COOH(aq) + H_2O(l) \rightleftharpoons CH_3COO^-(aq) + H^+(aq)$$

A precise indication of the position of this equilibrium is provided by an equilibrium constant:

$$K = \frac{[H^+(aq)]_{eqm}[A^-(aq)]_{eqm}}{[HA(aq)]_{eqm}[H_2O(l)]_{eqm}}$$

Since $[H_2O(l)]_{eqm}$ is virtually constant we can define a new equilibrium constant, $K_a = K[H_2O(l)]_{eqm}$ which gives the expression

$$K_a = \frac{[H^+(aq)]_{eqm}[A^-(aq)]_{eqm}}{[HA(aq)]_{eqm}} \tag{4}$$

where K_a is called the **acid dissociation constant**, and has units of mol dm^{-3}. As in the case of pH, and for the same reasons, it is often convenient to refer to pK_a where

$$pK_a = -\lg K_a$$

and values of pK_a are generally quoted at 298 K.

The more positive the value of pK_a, the smaller the value of K_a and the weaker is the acid. Table 17.2 lists some pK_a values for common acids.

Table 17.2 The pK_a values for some common acids at 298 K

	Acid(aq)	pK_a	Conjugate base(aq)	
increasing acidic strength	HCl	−7	Cl$^-$	increasing basic strength
	H$_3$PO$_4$	2·1	H$_2$PO$_4^-$	
	CH$_3$COOH	4·8	CH$_3$COO$^-$	
	HCN	9·3	CN$^-$	
	H$_2$O	15·7	OH$^-$	

In the case of acids which can donate more than one proton (polyprotic acids) there is more than one equilibrium to consider and thus more than one pK_a value. These are designated pK_{a1}, pK_{a2} etc. and Table 17.3 lists the values for phosphoric acid, H$_3$PO$_4$.

Table 17.3 Successive pK_a values for phosphoric acid in aqueous solution at 298 K

Acid	pK_a	Equilibrium
H_3PO_4	2·1	$H_3PO_4(aq) \rightleftharpoons H_2PO_4^-(aq) + H^+(aq)$
$H_2PO_4^-$	7·2	$H_2PO_4^-(aq) \rightleftharpoons HPO_4^{2-}(aq) + H^+(aq)$
HPO_4^{2-}	12·4	$HPO_4^{2-}(aq) \rightleftharpoons PO_4^{3-}(aq) + H^+(aq)$

In a completely analogous way, we can express the equilibrium for the reaction of a base with water:

$$B(aq) + H_2O(l) \rightleftharpoons BH^+(aq) + OH^-(aq)$$

and so

$$K_b = \frac{[BH^+(aq)]_{eqm}[OH^-(aq)]_{eqm}}{[B(aq)]_{eqm}}$$

and K_b is called the **base dissociation constant**. In a similar manner we define a pK_b scale where

$$pK_b = -\lg K_b$$

If we now consider the case of a particular acid, ethanoic acid, CH_3COOH, whose conjugate base is the ethanoate ion, CH_3COO^-, we can write

$$K_a = \frac{[CH_3COO^-(aq)]_{eqm}[H^+(aq)]_{eqm}}{[CH_3COOH(aq)]_{eqm}}$$

$$K_b = \frac{[CH_3COOH(aq)]_{eqm}[OH^-(aq)]_{eqm}}{[CH_3COO^-(aq)]_{eqm}}$$

Multiplying these two expressions together we have

$$K_a \times K_b = [H^+(aq)]_{eqm}[OH^-(aq)]_{eqm} = K_w$$

and

$$pK_w = pK_a + pK_b = 14 \text{ (at 298 K)} \tag{5}$$

Provided the acid dissociation constant for a particular acid is known, the base dissociation constant of its conjugate base can be determined by using equation (5).

Strong acids

Strong acids such as nitric and hydrochloric acids may be considered to be totally ionised in aqueous solution. The evidence to support this comes from pH measurements. For example, the pH of a 0·1M solution of hydrochloric acid is found to be 1 and as the acid is diluted with water successively to give 0·01M, 0·001M etc., the pH of the solution rises to 2, 3, etc. Since pH measures the concentration of $H^+(aq)$ ions, these results can only mean that the acid has reacted completely with the water:

$$HCl(aq) + H_2O(l) \longrightarrow H^+(aq) + Cl^-(aq)$$

It should be noted that successive 10-fold increases in concentration of 0·1M hydrochloric acid do not cause an exactly uniform decrease in pH from 1 to 0, -1, -2 etc. This is because, at high concentrations, association between ions of opposite charge becomes appreciable; positive and negative ions no longer act independently.

Since water itself functions as a base, all strong acids in aqueous solution donate their protons completely to it and give the appearance of being equal in strength. This is the same as saying that $H^+(aq)$ is the strongest acid that can exist in aqueous solution. Water is said to exert a levelling effect on strong acids and the same is true with strong bases, since $OH^-(aq)$ is the strongest base that can exist in water.

Evidence that strong acids and bases in aqueous solution can be regarded as $H^+(aq)$ and $OH^-(aq)$ respectively comes from the enthalpy of neutralisation (chapter 14, p. 221) which is constant for all strong acid-strong base combinations, implying the common reaction:

$$H^+(aq) + OH^-(aq) \longrightarrow H_2O(l)$$

Strong acids will only show an order of acid strength in the presence of a weaker base than water; pure ethanoic acid provides a suitable medium, hydrogen chloride reacting as follows:

$$HCl + CH_3COOH \rightleftharpoons CH_3COOH_2^+ + Cl^-$$

The position of equilibrium for this system does not lie completely over on the right. In pure ethanoic acid as solvent, the common 'strong acids' show the following order of decreasing acid strengths:

$$HClO_4 > H_2SO_4 > HCl > HNO_3$$

Weak acids

The acid dissociation constant for a weak acid, HA, is written according to equation (4):

$$K_a = \frac{[H^+(aq)]_{eqm}[A^-(aq)]_{eqm}}{[HA(aq)]_{eqm}}$$

Suppose 1 mole of the acid HA is dissolved in water to produce V dm^3 of an aqueous solution, and that at equilibrium a fraction α mole has reacted with the water. Then the initial and equilibrium concentrations of each species in the above equation will be as follows (water ignored):

$$HA(aq) + H_2O(l) \rightleftharpoons H^+(aq) + A^-(aq)$$

Initial conc./mol dm^{-3} $\quad\quad \dfrac{1}{V} \quad\quad\quad\quad \dfrac{0}{V} \quad\quad \dfrac{0}{V}$

Equilibrium conc./mol dm^{-3} $\dfrac{(1-\alpha)}{V} \quad\quad \dfrac{\alpha}{V} \quad\quad \dfrac{\alpha}{V}$

We can now substitute these equilibrium concentrations into the equilibrium expression and obtain

$$K_a = \frac{\dfrac{\alpha}{V} \cdot \dfrac{\alpha}{V}}{\dfrac{(1-\alpha)}{V}} = \frac{\alpha^2}{(1-\alpha)V} \tag{6}$$

The above relationship is known as **Ostwald's Dilution Law**.

For a weak acid which undergoes relatively little ionisation in solution, α is small and so $(1 - \alpha)$ approximates to 1, and equation (6) may be written in the approximate form:

$$K_a = \frac{\alpha^2}{V} \qquad (7)$$

Calculations, which we carry out below, involving equations (6) and (7) show that for weak acids such as ethanoic acid ($K_a = 1\cdot8 \times 10^{-5}$ mol dm^{-3} at 298 K) the approximate form of the Dilution Law introduces errors no greater than about 1–2% for aqueous solutions of dilution up to 0·01 M.

Example 4
Calculate the degree of ionisation of ethanoic acid of concentration 0·1M using (a) Ostwald's Dilution Law, (b) the approximate form of the Dilution Law.
Since the solution is 0·1 M there is 1 mole in 10 dm^3, and so $V = 10$. Substituting in equation (6) we have

$$1\cdot8 \times 10^{-5} = \frac{\alpha^2}{(1 - \alpha)10}$$
$$\alpha^2 + 1\cdot8 \times 10^{-4}\alpha - 1\cdot8 \times 10^{-4} = 0$$

On solving this quadratic equation, by the method shown in chapter 16, p. 263, we get $\alpha = 0\cdot0133$.

Substituting into equation (7) we get

$$1\cdot8 \times 10^{-5} = \frac{\alpha^2}{10} \qquad \alpha^2 = 1\cdot8 \times 10^{-4}$$

hence

$$\alpha = (1\cdot8 \times 10^{-4})^{1/2} = 0\cdot0134$$

The difference between these two values is less than 1% so we are justified in using the approximate expression.

It would be useful for the reader to work through the calculation again for an acid with $K_a = 10^{-3}$ mol dm^{-3}, to see whether or not the approximate expression is accurate enough in this particular case.

Continuing to use the approximate expression, we can see that from equation (7)

$$\alpha = (K_a V)^{1/2}$$

and since $[H^+(aq)] = \alpha/V$, then $[H^+(aq)] = (K_a/V)^{1/2}$

c = 1/V, i.e., conc. in mol dm^{-3}

and
$$pH = -\tfrac{1}{2}\lg K_a + \tfrac{1}{2}\lg V$$
$$= -\tfrac{1}{2}\lg K_a - \tfrac{1}{2}\lg c$$
hence
$$pH = \tfrac{1}{2}(pK_a - \lg c) \qquad (8)$$

This means that for a weak acid, since pK_a is constant, a ten-fold increase in concentration will decrease the pH by about 0·5 of a unit (compared with a decrease of 1 unit for a strong acid, as we have already seen (p. 283)). Experimental observations confirm this relationship.

Strong bases

The commonest water-soluble bases are sodium and potassium hydroxides and they are termed alkalis. Both these compounds are classified as strong bases as are the hydroxides of the other Group 1A and 2A metals, with the exception of beryllium hydroxide which is amphoteric

(p. 288). In the solid state these hydroxides are predominantly ionic, and the process of solution in water simply amounts to the separation of the constituent ions of the solid as hydrated species. As in the case of strong acids, evidence to support the claim that they are strong bases comes from pH measurements. For example, the pH of a 0·1M solution of sodium hydroxide is found to be 13 and as the alkali is diluted with water successively to give 0·01M, 0·001M etc., the pH of the solution falls to 12, 11, etc. These pH values are explained as follows.

0·1M NaOH(aq) contains 0·1 mole of OH^-(aq) if it is a strong alkali. The pOH of the solution is given by

$$pOH = -lg[OH^-(aq)] = -lg\ 0·1 = 1$$

Since

$$pH + pOH = 14 \text{ (at 298K)}$$
$$pH = 14 - 1 = 13$$

The pH values 12, 11, etc. are found in a similar manner.

Successive dilution of an aqueous solution of an alkali cannot take the pH below 7 (the pH of pure water); the OH^-(aq) concentration from water, which can normally be disregarded by comparison with the OH^-(aq) concentration from the alkali, is now the major contribution, i.e. the pOH of 10^{-8}M NaOH(aq) is not 8 but some value lower than 7, giving a pH value somewhat greater than 7. It is for the same kind of reason that the pH of an aqueous solution of an acid can never be greater than 7 however much it is diluted.

Weak bases

Ammonia is a typically weak base and reacts with water in the following way:

$$NH_3(aq) + H_2O(l) \rightleftharpoons NH_4^+(aq) + OH^-(aq)$$

If K_b is the base dissociation constant we can write

$$K_b = \frac{[NH_4^+(aq)]_{eqm}[OH^-(aq)]_{eqm}}{[NH_3(aq)]_{eqm}}$$

Ostwald's law applies equally to weak bases as it does to weak acids in aqueous solution and we shall apply the approximate expression (equation (7)) in the example below.

Example 5
Calculate the pOH and hence the pH of a 0·1M solution of aqueous ammonia, given that $K_b = 1·8 \times 10^{-5}$ mol dm^{-3} at 298 K.
  In this case

$$\alpha = (K_b V)^{1/2}$$

and since $V = 10$ (there is 1 mole of ammonia in 10 dm^3 of solution) we have

$$\alpha = (1·8 \times 10^{-5} \times 10)^{1/2}$$
$$= (1·8 \times 10^{-4})^{1/2} = 0·0134$$

The concentration of OH^-(aq) ions is α/V, so we have

$$[OH^-(aq)] = \frac{0.0134}{10} = 0.00134$$
$$= 1.34 \times 10^{-3} \text{ mol dm}^{-3}$$

$$pOH = -lg(1.34 \times 10^{-3})$$
$$= -(0.1271 - 3) = 2.87(29) = 2.87$$

Since

$$pH + pOH = 14$$
$$pH = 14 - 2.87 = 11.13$$

17.4 Factors affecting acid and base strength

The nature of hydroxides

Oxyacids and alkalis are hydroxides, i.e. they contain an —OH group. In aqueous solution an hydroxide may behave as an acid:

$$EO—H(aq) + H_2O(l) \rightleftharpoons EO^-(aq) + H_3O^+(aq)$$

or as a base:

$$E—OH(aq) + H_2O(l) \rightleftharpoons E—OH_2^+(aq) + OH^-(aq)$$

Behaviour as an acid will be encouraged by an electronegative element E with a tendency to accept electrons and thus facilitate the loss of a proton. In an acid, the element E is therefore generally a non-metal and oxyacids of the type $EO_x(OH)_y$ are particularly common, although they are often written in the form which conceals the presence of the hydroxyl group:

Sulphuric acid	H_2SO_4	or	$SO_2(OH)_2$
Chloric(VII) acid	$HClO_4$	or	$ClO_3(OH)$
Nitric acid	HNO_3	or	$NO_2(OH)$

The larger the number of electronegative oxygen atoms attached to the central non-metal, the easier the release of a proton and the stronger the acid. Thus we observe the following order of acid strength:

$$HClO_4 > HClO_3 > HClO_2 > HClO$$

and

$$HNO_3 > HNO_2$$

If the hydroxyl group is attached to an electropositive element, e.g. metals in Groups 1A and 2A, the hydroxide will behave as a base in aqueous solution; indeed, these hydroxides are predominantly ionic and the OH⁻ ions already exist in the solid compounds.

In the case of elements of intermediate electron-attracting power, there may be occasions when the hydroxide behaves as an alkali and others when it more resembles an oxyacid, and this is so for some hydroxides, e.g. zinc hydroxide and lead(II) hydroxide. Both these hydroxides are only sparingly soluble in water; even so the resulting solutions slowly turn red litmus blue, leaving little doubt that they act as alkalis, under these conditions. However, in the presence of a strong alkali, e.g. sodium hydroxide, sodium salts are produced which can only be interpreted in terms of acidic tendencies. Hydroxides such as these are said to be **amphoteric** and their behaviour is easily explained by assuming an equilibrium to be set up in solution; thus for zinc hydroxide:

$$2OH^-(aq) + Zn^{2+}(aq) \rightleftharpoons Zn(OH)_2(s) + 4H_2O(l) \rightleftharpoons Zn(OH)_4^{2-}(aq) + 2H_3O^+(aq)$$

$$+ \qquad\qquad\qquad\qquad\qquad\qquad\qquad\qquad\qquad\qquad +$$
$$2H_3O^+(aq) \qquad\qquad\qquad\qquad\qquad\qquad\qquad\qquad 2OH^-(aq)$$
$$\updownarrow \qquad\qquad\qquad\qquad\qquad\qquad\qquad\qquad\qquad\qquad \updownarrow$$
$$4H_2O(l) \qquad\qquad\qquad\qquad\qquad\qquad\qquad\qquad\qquad 4H_2O(l)$$

In water the equilibrium lies over to the left, i.e. there is an excess of hydroxyl ions compared with hydroxonium ions and hence an alkaline reaction. The equilibrium can be driven further over to the left by adding $H_3O^+(aq)$ ions (a strong acid), which combine with the $OH^-(aq)$ ions to form water:

$$2OH^-(aq) + 2H_3O^+(aq) \rightleftharpoons 4H_2O(l)$$

The zinc hydroxide consequently dissolves completely and the solution contains $Zn^{2+}(aq)$ ions. If $OH^-(aq)$ ions are added (a strong alkali) the equilibrium is shifted over to the right as hydroxonium ions in the equilibrium mixture are removed to produce water:

$$2H_3O^+(aq) + 2OH^-(aq) \rightleftharpoons 4H_2O(l)$$

The zinc hydroxide again dissolves completely but this time the solution contains zincate ions, $Zn(OH)_4^{2-}(aq)$, sometimes written in the form $ZnO_2^{2-}(aq)$.

Organic acids

The strength of an organic carboxylic acid RCOOH depends again on the ability of the RCOO— group to attract electrons away from the hydrogen atom and thus allow its release as a proton:

$$RCOOH(aq) + H_2O(l) \rightleftharpoons RCOO^-(aq) + H^+(aq)$$

This depends particularly on the nature of the group R—. Table 17.4 gives the pK_a values (at 298 K) for ethanoic acid and the chloroethanoic acids, from which it is clear that the pK_a values decrease (and hence the strengths of the acids increase) with increasing chlorine substitution.

Table 17.4 The pK_a values at 298 K for ethanoic acid and its chlorine-substituted acids

Acid	Formula	R— Group	pK_a
Ethanoic	CH_3COOH	CH_3-	4·76
Chloroethanoic	$CH_2ClCOOH$	$ClCH_2-$	2·86
Dichloroethanoic	$CHCl_2COOH$	Cl_2CH-	1·29
Trichloroethanoic	CCl_3COOH	Cl_3C-	0·65

288

An explanation of the increasing acidity in this series runs as follows: The chlorine atom is powerfully electron-attracting by comparison with the hydrogen atom, and this results in the —OH oxygen atom in the acid being more $\delta+$ than in the parent compound, ethanoic acid. This will tend to facilitate the removal of a proton and hence lead to an increase in acid strength. The net result of introducing three chlorine atoms into the molecule is to make the final acid a moderately strong one. The chlorine atom is said to exert a $-I$ inductive effect by comparison with the hydrogen atom.

When we consider ethanoic acid and other acids in which hydrogen in the CH_3— group is substituted by alkyl groups, we find an increase in pK_a (a decrease in acid strength). The argument now would be as follows: Since alkyl groups are slightly electron-releasing by comparison with the hydrogen atom (they exert a $+I$ inductive effect), they would tend to make the oxygen atom in the —OH group slightly less $\delta+$ than in ethanoic acid and thus the acid more difficult to ionise. The pK_a values of some alkyl-substituted ethanoic acids are shown in Table 17.5.

Table 17.5 The pK_a values at 298 K for ethanoic acid and some alkyl-substituted ethanoic acids

Acid	Formula	R— group	pK_a
Ethanoic	CH_3COOH	CH_3—	4·76
Propanoic	CH_3CH_2COOH	CH_3CH_2—	4·87
Butanoic	$C_2H_5CH_2COOH$	$C_2H_5CH_2$—	4·82
Pentanoic	$C_3H_7CH_2COOH$	$C_3H_7CH_2$—	4·86
2,2-dimethylpropanoic	$(CH_3)_3CCOOH$	$(CH_3)_3C$—	5·05

The factors responsible for determining the strength of an acid in aqueous solution are, in reality, more complex than we have indicated above. For example, any factor which stabilised $RCOO^-$ with respect to $RCOOH$ would be expected to lead to an increase in acid strength. Electron-withdrawing substituents, such as chlorine atoms, would be expected to lead to extra stability by delocalising the negative charge on the anion (p. 111); but the nature of the solvent is also important and we should be comparing the difference in the stabilities of $RCOO^-(aq)$ and $RCOOH(aq)$, and how this is affected by the substituents in the acid. This in turn depends on free energy changes (p. 273) so both enthalpy and entropy factors should be taken into consideration. Factors such as these are considered in relation to the hydrogen halides in the next section.

17.5
The energetics of acid-base equilibria

The hydrogen halides provide an interesting illustration of the way in which the principles of energetics may be applied to acid-base equilibria. The order of decreasing acid strength of the hydrogen halides is:

$$HI > HBr > HCl > HF$$

This order contradicts the principle used in the preceding section, namely that the more electronegative the atom in association with the 'acidic' hydrogen atom the stronger the acid. Despite the fluorine atom being the most electronegative of the halogens, hydrogen fluoride, HF, is by far the weakest acid ($pK_a = 3\cdot25$) of these four. In aqueous solution HI, HBr and HCl all behave as strong acids, but in a more acidic medium than water such as methanoic acid, (HCOOH), which consequently functions as a base, the above order of decreasing acid strength is revealed. A study of the energy changes associated with each equilibrium is able to explain this unexpected order.

The acid dissociation constant, K_a, is related to the standard free energy change for the process

$$HX(aq) + H_2O(l) \rightleftharpoons H^+(aq) + X^-(aq)$$

by the expression:

$$\Delta G^\ominus = -RT\ln K_a$$

This means that, for a given temperature, the more negative the free energy change, the larger the dissociation constant and the stronger the acid. The standard free energy change is, in turn, composed of an enthalpy term and an entropy term:

$$\Delta G^{\ominus} = \Delta H^{\ominus} - T\Delta S^{\ominus}$$

The values of $\Delta H^{\ominus}$(298 K) for each equilibrium are:

	HF	HCl	HBr	HI
Standard enthalpy of dissociation, $\Delta H^{\ominus}$/kJ mol^{-1}	-13	-59	-63	-57

The standard enthalpy change for the dissociation of HF in aqueous solution is considerably less exothermic than for the other three acids. In order to trace the reason for this, it is necessary to write down a Born-Haber type cycle for each dissociation and compare terms. When this is done it turns out that the high enthalpy of solution of unionised HF molecules (due to hydrogen-bonding) and the high H—F bond energy are the principal reasons for this.

The values of $T\Delta S^{\ominus}$(298 K) for each equilibrium are:

	HF	HCl	HBr	HI
$T\Delta S^{\ominus}$/kJ mol^{-1}	-29	-13	-4	$+4$

Here the trend is fairly uniform, showing that the entropy change becomes more favourable towards greater acid strength in preceding from HF to HI.

The combined enthalpy and entropy factors are shown below, together with the calculated values of K_a, which confirm the observed order of acid strength and the anomalously low strength of HF:

	HF	HCl	HBr	HI
$\Delta G^{\ominus}$/kJ mol^{-1}	$+16$	-46	-59	-61
K_a (calc)/mol dm^{-3}	10^{-3}	10^{8}	10^{10}	10^{11}

17.6
Acid-base titrations

Neutralisation occurs when the appropriate stoichiometric quantity of acid is added to a given amount of base, or vice-versa. For example, complete neutralisation of one mole of sulphuric acid requires a volume of aqueous sodium hydroxide which contains exactly two moles of base:

$$H_2SO_4(aq) + 2NaOH(aq) \longrightarrow Na_2SO_4(aq) + 2H_2O(l)$$

Neutralisation is usually carried out by titration, i.e. by adding controlled quantities of one reactant (usually the acid) from a burette to a known quantity of the other (usually the base) contained in a flask. The point of neutralisation, also known as the end-point, may be determined by means of a visual indicator (p. 294) or by some physical technique, for example electrolytic conductivity (conductimetric titration, p. 421) or enthalpy changes (thermometric titration).

Titration curves

The graphs obtained by plotting pH against the volume of alkali added to 25 cm^3 of acid are shown in fig. 17.3. The four curves refer in turn to each acid-base combination of strong and weak acids (hydrochloric and ethanoic acids respectively) and strong and weak bases (sodium hydroxide and aqueous ammonia respectively) each of 0·1M concentration in aqueous solution.

In the case of reactions between monobasic acids and monoacid bases (one mole of protons transferred per mole of acid and base), neutralisation in the stoichiometric sense occurs when equimolar portions of acid and alkali have reacted. This corresponds to the vertical part of each graph. However, it is apparent that the pH at this point is not always 7, and the solution, therefore, does not necessarily contain equal concentrations of $H^+(aq)$ and $OH^-(aq)$ ions.

FIG. 17.3. *The four titration curves showing the pH changes that occur when alkali is added to acid*

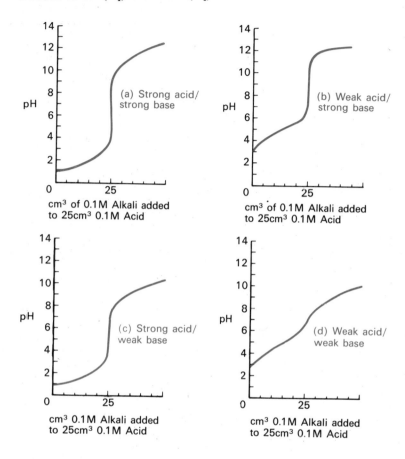

(a) Strong acid/strong base combination

Since strong acids and bases are completely ionised in aqueous solution, the process of neutralisation may be regarded simply as:

$$H^+(aq) + OH^-(aq) \longrightarrow H_2O(l)$$

At the end-point, equal concentrations of $H^+(aq)$ and $OH^-(aq)$ will be present from the slight self-ionisation of water so the solution will have a pH value of 7.

It is a simple matter to calculate the pH of the solution after the addition of any quantity of alkali to the acid, and this is done in the example below. There is close agreement between the 'calculated' and the 'observed' titration.

Example 6
Calculate the pH values of the solutions obtained by the addition of varying amounts of $0\cdot 1M$ NaOH(aq) to 25 cm³ of $0\cdot 1M$ HCl(aq).

Initially $[H^+(aq)]$ is $0\cdot 1$ or 10^{-1} mol dm⁻³ and the pH = 1. After the addition of x cm³ of $0\cdot 1M$ NaOH(aq), where x is less than 25, there will be $(25 - x)$ cm³ of un-neutralised acid but the volume has increased to $(25 + x)$ cm³. The concentration

291

of $H^+(aq)$ is therefore given by the expression:

$$[H^+(aq)] = \frac{10^{-1}(25 - x)}{(25 + x)}$$

If varying values of x are substituted into the above expression, $[H^+(aq)]$ and hence the pH of the resulting solutions may be calculated. Let us do the calculations for $x = 10$. Substitution into the expression gives

$$[H^+(aq)] = \frac{10^{-1}(25 - 10)}{(25 + 10)} = 4\cdot29 \times 10^{-2} \text{ mol dm}^{-3}$$

$$pH = -lg[H^+(aq)] = -lg(4\cdot29 \times 10^{-2})$$
$$= -(0\cdot6325 - 2)$$
$$= 1\cdot36(75) = 1\cdot37$$

The results for various values of x are shown below. For values of x greater than 25 we use the expression

$$[OH^-(aq)] = \frac{10^{-1}(x - 25)}{(25 + x)}$$

and then calculate pOH. This may then be converted to pH by using the expression:

$$pH + pOH = 14$$

x/cm^3	5	10	15	20	24	24·5	24·9
$[H^+(aq)]/\text{mol dm}^{-3}$	0·067	0·0429	0·025	0·0111	0·00204	0·00101	0·00020
pH	1·10	1·37	1·60	1·95	2·69	3·00	3·70

x/cm^3	24·99	25·00					
$[H^+(aq)]/\text{mol dm}^{-3}$	0·000020	10^{-7}					
pH	4·70	7					

x/cm^3	25·01	25·1	25·5	26	30		
$[OH^-(aq)]/\text{mol dm}^{-3}$	0·000020	0·00020	0·00099	0·00197	0·0091		
pH	9·30	10·30	11·00	11·30	11·96		

(b) Weak acid/strong base combination

In aqueous solution a weak acid is only slightly ionised. For example:

$$CH_3COOH(aq) + H_2O(l) \rightleftharpoons CH_3COO^-(aq) + H^+(aq)$$

from which we derive the acid dissociation constant:

$$K_a = \frac{[CH_3COO^-(aq)]_{eqm}[H^+(aq)]_{eqm}}{[CH_3COOH(aq)]_{eqm}} \tag{9}$$

When a strong alkali, such as sodium hydroxide solution, is added, the ethanoic acid is converted to its salt sodium ethanoate which may be taken to be fully ionised. The concentration of ethanoate ions, $CH_3COO^-(aq)$, produced in this way greatly exceeds that which is due to the ionisation of the weak acid itself. Moreover, we can ignore the slight ionisation of the weak acid when determining the concentration of undissociated acid at equilibrium. The errors are negligible if we rewrite equation (9) thus:

$$K_a = \frac{[\text{salt}]_{\text{eqm}}[\text{H}^+(\text{aq})]_{\text{eqm}}}{[\text{acid}]_{\text{eqm}}}$$

This expression may now be written in the form:

$$[\text{H}^+(\text{aq})]_{\text{eqm}} = \frac{K_a[\text{acid}]_{\text{eqm}}}{[\text{salt}]_{\text{eqm}}}$$

and so

$$\text{pH} = -\lg[\text{H}^+(\text{aq})]_{\text{eqm}} = -\lg K_a + \lg \frac{[\text{salt}]_{\text{eqm}}}{[\text{acid}]_{\text{eqm}}}$$

hence

$$\text{pH} = pK_a + \lg \frac{[\text{salt}]_{\text{eqm}}}{[\text{acid}]_{\text{eqm}}} \qquad (10)$$

This relationship enables the pH of the solution to be calculated during the titration.

Example 7
Calculate the pH of the solution after 15 cm³ of 0·1M sodium hydroxide solution have been added to 25 cm³ of 0·1M ethanoic acid ($pK_a = 4.76$).
Initially $[\text{CH}_3\text{COOH}(\text{aq})]$ is 0·1 or 10^{-1} mol dm⁻³. After the addition of 15 cm³ of 0·1M NaOH(aq) there are $(25 - 15)$ cm³ of un-neutralised acid but the volume of solution has increased to $(25 + 15)$ cm³. Thus

$$[\text{acid}]_{\text{eqm}} = \frac{10^{-1}(25 - 15)}{(25 + 15)} = \frac{10^{-1} \times 10}{40}$$

15 cm³ of NaOH(aq) have been converted into $\text{CH}_3\text{COO}^-(\text{aq})$ ions (salt) but, because of the dilution of the solution, the $[\text{salt}]_{\text{eqm}}$ will not be 0·1 or 10^{-1} mol dm⁻³ but will be given by

$$[\text{salt}]_{\text{eqm}} = \frac{10^{-1} \times 15}{(15 + 25)} = \frac{10^{-1} \times 15}{40}$$

Substituting into equation (10) we have

$$\text{pH} = 4.76 + \lg \left(\frac{10^{-1} \times 15}{40}\right)\bigg/\left(\frac{10^{-1} \times 10}{40}\right)$$

$$= 4.76 + \lg 1.5$$

$$= 4.94$$

(c) Strong acid/weak base combination
In aqueous solution a weak base is only slightly ionised, in contrast with a strong acid which is completely ionised. For example, the following equilibrium exists for aqueous ammonia:

$$\text{NH}_3(\text{aq}) + \text{H}_2\text{O}(\text{l}) \rightleftharpoons \text{NH}_4^+(\text{aq}) + \text{OH}^-(\text{aq})$$

from which we derive the base dissociation constant:

$$K_b = \frac{[\text{NH}_4^+(\text{aq})]_{\text{eqm}}[\text{OH}^-(\text{aq})]_{\text{eqm}}}{[\text{NH}_3(\text{aq})]_{\text{eqm}}} \qquad (11)$$

As a strong acid, such as hydrochloric acid, is added to aqueous ammonia, ammonium chloride is formed which may be assumed to be fully ionised. The concentration of ammonium ions, $NH_4^+(aq)$, produced in this way greatly exceeds that which is due to the ionisation of aqueous ammonia itself. In addition, we may ignore the slight ionisation of the aqueous ammonia when determining the concentration of undissociated base at equilibrium. The errors are negligible if we rewrite equation (11) thus:

$$K_b = \frac{[\text{salt}]_{\text{eqm}}[OH^-(aq)]_{\text{eqm}}}{[\text{base}]_{\text{eqm}}}$$

This expression may now be written in the form:

$$[OH^-(aq)]_{\text{eqm}} = \frac{K_b[\text{base}]_{\text{eqm}}}{[\text{salt}]_{\text{eqm}}}$$

from which the pOH and hence the pH of the solution may be calculated at any stage during the titration (the process is very similar to that in example 7, p. 293).

(d) Weak acid/weak base combination

A typical combination is an aqueous solution of ethanoic acid and aqueous ammonia. At the end-point of this titration the salt produced, ammonium ethanoate, may be taken to be fully ionised, and the following equilibria will exist in solution:

$$CH_3COO^-(aq) + H_2O(l) \rightleftharpoons CH_3COOH(aq) + OH^-(aq)$$
$$NH_4^+(aq) + H_2O(l) \rightleftharpoons NH_3(aq) + H^+(aq)$$

The pH of the resulting solution will be determined by the relative magnitudes of the equilibrium constants for the above two equilibria. If pK_b for the base ($CH_3COO^-(aq)$) is greater than pK_a for the acid ($NH_4^+(aq)$), then the solution will have a pH value less than 7. Conversely if pK_b is less than pK_a the solution will have a pH greater than 7. However, since both the acid and base are only partially ionised, the value of the pH at the end-point would not be expected to vary greatly from 7, and this is generally true. The shape of the pH curve is less steep at the end-point and consequently there is no satisfactory indicator for determining it.

17.7 Indicators

Acid-base titrations are carried out in order to establish the equivalent (stoichiometric) amounts of acid and base which are required to neutralise each other. The most convenient way of locating the end-point is by the use of an indicator, which changes colour at the end-point of the titration. It is important to emphasise that the indicator must identify the equivalence point of the titration and not necessarily the point at which the solution has a pH of 7. It is evident from fig. 17.3 that, except in the case of a strong acid/strong base titration, these two points are not necessarily the same. If we only wished to identify the point in the titration at which the pH is 7, then one indicator could be used for all titrations. This is obviously not the case, and a whole range of indicators is used according to the exact nature of the acid and base which are reacting together.

An indicator is a weak acid and is conveniently represented by HIn, a molecule which has a different colour in the unionised form, HIn, than in the ionised form, In^-.

$$HIn(aq) + H_2O(l) \rightleftharpoons H^+(aq) + In^-(aq)$$
$$\text{(colour I)} \qquad\qquad\qquad\qquad \text{(colour II)}$$

In a strongly acid solution the unionised form, HIn, will predominate, since the equilibrium will be well over to the left, but in a strongly alkaline solution there will be a preponderance of $In^-(aq)$. Thus the colour will change according to the pH of the solution. At very high or very low pH values the colour of the indicator will be predictable, but for intermediate pH values we shall need to apply the equation

$$pH = pK_{in} + lg\frac{[In^-(aq)]_{eqm}}{[HIn(aq)]_{eqm}}$$
$$\text{(see equation (10), p. 293)}$$

where pK_{in} is the pK of the indicator. When $[In^-(aq)]_{eqm}$ is equal to $[HIn(aq)]_{eqm}$ we may write

$$pH = pK_{in} + lg\ 1$$

and so

$$pH = pK_{in}$$

It should be appreciated that the very small amount of indicator added during a titration does not affect the pH of the solution, and it is rather the case that the pH of the solution determines the equilibrium point of the indicator. Hence, whatever the indicator colours may be in strongly acid or strongly alkaline solution, at the pH approximately equal to pK_{in} the solution will pass through a range of colours.

It is not until one form of the indicator, i.e. HIn(aq) or $In^-(aq)$, is in approximately ten-fold excess that one colour will be clearly distinguished by the human eye. Suppose $[HIn(aq)]_{eqm}$ is equal to $10[In^-(aq)]_{eqm}$, then the pH is given by:

$$pH = pK_{in} + lg(\tfrac{1}{10})$$
$$= pK_{in} - 1$$

Similarly if $[In^-(aq)]_{eqm}$ is equal to $10[HIn(aq)]_{eqm}$ then

$$pH = pK_{in} + lg\ 10$$
$$= pK_{in} + 1$$

Thus between the points at which the human eye can detect the unionised form in excess and the anion form in excess there will be a change in pH of approximately 2 units. This range of pH is known as the 'theoretical' working range of the indicator.

Some 'experimental' working ranges of indicators are given in Table 17.6.

In choosing an indicator for a particular acid/base titration it is clear that two considerations should be kept in mind. First of all the value of pK_{in} should be close to the value of the pH of the solution at the end-point, so that the colour change occurs as closely as possible to the equivalence point. Secondly, the indicator working range should lie on the vertical part of the pH curve (as shown in fig. 17.3), otherwise the indicator colour will change either before the equivalence point or at some value after the equivalence point. For example, the titration of

0·1M ethanoic acid against 0·1M sodium hydroxide reaches an equivalence point at a pH value of about 8·7 (p. 291), which means that if bromophenol blue is used as indicator (working range 2·8–4·6) it would change colour too early in the titration. Clearly, phenolphthalein (working range 8·2–10·0) would be more suitable.

Table 17.6 Experimental working range of some indicators

Indicator	pK_{in}	pH range	Colour change	
			Colour in acid	Colour in alkali
Thymol blue	1·7	1·2–2·8	Red	Yellow
Methyl orange	3·7	3·2–4·2	Red	Yellow
Bromophenol blue	4·0	2·8–4·6	Yellow	Blue
Bromocresol green	4·7	3·8–5·4	Yellow	Blue
Methyl red	5·1	4·2–6·3	Red	Yellow
Bromothymol blue	7·0	6·0–7·6	Yellow	Blue
Phenol red	7·9	6·8–8·4	Yellow	Red
Thymol blue (2nd range)	8·9	8·0–9·6	Yellow	Blue
Phenolphthalein	9·3	8·2–10·0	Colourless	Red
Thymolphthalein	9·7	8·3–10·5	Colourless	Blue

Table 17.6 shows that the indicator thymol blue has two working ranges. This happens because the indicator is a weak dibasic acid, and there are therefore two ionised forms in addition to the unionised form of the indicator:

$$H_2In(aq) + H_2O(l) \rightleftharpoons HIn^-(aq) + H^+(aq)$$
$$HIn^-(aq) + H_2O(l) \rightleftharpoons In^{2-}(aq) + H^+(aq)$$

There are three colours corresponding to $H_2In(aq)$, $HIn^-(aq)$ and $In^{2-}(aq)$. Since the two pK_{in} values occur at widely different values (1·7 and 8·9), the indicator may be used over two quite different ranges.

17.8
Buffer solutions

The progress of many chemical reactions is affected critically by the pH of the solution in which they take place, and it is therefore often necessary to control the pH in a specific manner. The pH of aqueous solutions is usually sensitive to addition of even small amounts of acids and alkalis. Consider, for example, the addition of one drop (say 0·01 cm^3) of M NaOH(aq) to 1 dm^3 of 10^{-5}M HCl(aq). The pH of the acid solution is 5, but on the addition of the drop of sodium hydroxide solution 10^{-5} mole of the OH$^-$(aq) will neutralise the 10^{-5} mole of the H$^+$(aq) present from the acid. The concentration of H$^+$(aq) will consequently fall to 10^{-7} mol dm^{-3} and the pH of the solution will be 7. Such a change of two units of pH could have a critical effect on the course of any reaction being carried out in the solution.

A buffer solution is able to counteract the effect of relatively small quantities of acid or alkali on the pH of the solution. Such solutions have important applications in many fields of chemistry and biochemistry; for example, the human blood is a natural buffer system which maintains a pH of 7·4 which is necessary for the efficient transport of oxygen from the lungs. A variation of 0·4 units could prove fatal, despite the fact that some body cells contain fluid with a pH value as low as 5. The buffering action of the blood is able to counteract the pH changes which the release of these fluids might make, and hence the essential constant pH of the blood is maintained.

Buffer solutions are important for two reasons. First of all, as we have seen above, they are able to absorb the addition of small quantities of acid and alkali without a significant change in the pH of the solution, and secondly because they can be designed to provide a solution of a particular pH. An important point to realise is that buffer solutions are mixtures; they can either be a mixture of a weak acid and its conjugate base (for pH values below 7) or a weak base and its conjugate acid (for pH values greater than 7). For example, a mixture of ethanoic acid and sodium ethanoate (providing the conjugate base, ethanoate) in aqueous solution is a common buffer which provides a stable pH around 5. The buffering action of such a mixture can be seen in the titration curve (fig. 17.3(b)). After an initial rise of pH the buffering action begins to take effect as more ethanoate ions are produced in the course of the titration, and only just before the equivalence point does the pH change appreciably. The initial rise in pH occurs because, at the beginning of the titration, only a very small quantity of ethanoate ions have been produced, and the solution is almost unionised ethanoic acid. The principal equilibrium in the system is

$$CH_3COOH(aq) + H_2O(l) \rightleftharpoons CH_3COO^-(aq) + H^+(aq)$$

and according to equation (10), p. 293, we may write

$$pH = pK_a + \lg \frac{[CH_3COO^-(aq)]_{eqm}}{[CH_3COOH(aq)]_{eqm}}$$

This equation, coupled with the equilibrium expression, provides the key to the explanation of buffer action. If a quantity of alkali is added to the buffer mixture, it converts some of the ethanoic acid to ethanoate ions; if a quantity of acid is added it converts ethanoate ions into ethanoic acid molecules. In each case the concentrations of $OH^-(aq)$ and $H^+(aq)$ respectively are drastically reduced from what they would otherwise be:

$$CH_3COOH(aq) + OH^-(aq) \rightleftharpoons CH_3COO^-(aq) + H_2O(l)$$
$$CH_3COO^-(aq) + H^+(aq) \rightleftharpoons CH_3COOH(aq) + H_2O(l)$$

The ratio $[CH_3COO^-(aq)]_{eqm}/[CH_3COOH(aq)]_{eqm}$ will change but the above equation involves the logarithm of this ratio which will change even less, e.g. the ratio of the concentrations of ethanoate ions to ethanoic acid molecules must change by a factor of ten to produce a pH change of one unit. In the majority of cases the effect on the pH of the solution of adding small quantities of acid or alkali will be extremely slight. The most efficient buffering action occurs when the pH of the solution is equal to the pK_a value of the acid or the pK_b value of the base. In the above example the most effective pH for an ethanoic acid/ethanoate buffer will be 4·76.

The larger the concentrations of acid and conjugate base (or base and conjugate acid) the greater the 'reservoir' of the mixture, and the greater the buffering capacity of the solution.

The example below illustrates the action of a buffer solution.

Example 8

Calculate the change in pH when 1 drop (0·01 cm³) of M NaOH(aq) is added to 1 dm³ of buffer solution containing 0·1M ethanoic acid and 0·1M sodium ethanoate.

The pH of the buffer solution is given by the equation

$$pH = pK_a + lg \frac{[CH_3COO^-(aq)]_{eqm}}{[CH_3COOH(aq)]_{eqm}}$$
$$= 4·76 + lg 1$$
$$= 4·76$$

On the addition of 0·01 cm³ of M NaOH(aq), 10^{-5} mole of OH⁻(aq) is added to the acid which produces a further 10^{-5} mole of ethanoate ions by the reaction:

$$CH_3COOH(aq) + OH^-(aq) \rightleftharpoons CH_3COO^-(aq) + H_2O(l)$$

The total concentration of ethanoate ions now becomes $(0·1 + 0·00001)$ mol dm⁻³ or 0·10001M, and the concentration of ethanoic acid becomes $(0·1 - 0·00001)$ mol dm⁻³ or 0·09999M. The new pH is then given by

$$pH = 4·76 + lg \left(\frac{0·10001}{0·09999}\right)$$
$$= 4·76 \text{ (using 4-figure logarithms)}$$

The effect of adding this small amount of alkali to the buffer solution is negligible.

This result contrasts strongly with that obtained at the beginning of this section for the change in pH when 1 drop of M NaOH(aq) is added to an unbuffered solution of hydrochloric acid (p. 296).

The mode of action of a weak base/conjugate acid buffer is similar. The reader may care to verify that the relevant equation is

$$pH = pK_a + lg \frac{[\text{weak base}]_{eqm}}{[\text{conjugate acid}]_{eqm}}$$

where pK_a refers to the conjugate acid.

For example, in the case of a buffer made by mixing aqueous ammonia with ammonium chloride solution, the pH is given by:

$$pH = pK_a + lg \frac{[NH_3(aq)]_{eqm}}{[NH_4^+(aq)]_{eqm}}$$

The following example shows how to calculate quantities when making a buffer solution of a certain pH.

Example 9

Calculate the volume of 0·1M HCOONa(aq) that must be added to 1 dm³ of 0·1M methanoic acid solution to give a buffer solution of pH = 3·50. The pK_a for methanoic acid is 3·75.

Let x dm³ be the volume added. The solution will become more dilute and the concentrations of HCOO⁻(aq) and HCOOH(aq) will be given by

$$[HCOO^-(aq)] = \frac{x}{(1 + x)} 0·1 \quad [HCOOH(aq)] = \frac{1}{(1 + x)} 0·1$$

$$pH = pK_a + lg \frac{[HCOO^-(aq)]_{eqm}}{[HCOOH(aq)]_{eqm}}$$

$$3·50 = 3·75 + lg \frac{x}{1}$$

hence

$$lg x = -0·25$$
$$x = antilg -0·25 = antilg \bar{1}·75$$
$$= 0·562 \text{ dm}^3 \text{ or } 562 \text{ cm}^3$$

Determination of the pH of a solution using buffer solutions

The pH of a solution can be determined to ± 1 unit by adding a few drops of 'Universal' indicator solution and comparing the colour with that of the colour chart supplied with the indicator. Having determined the approximate pH, a set of buffer solutions can be made up to cover this range. Equal volumes of these buffer solutions are placed in several identical tubes and two or three drops of an indicator, whose working range corresponds to this range of pH values, is placed in each tube. The same volume of the unknown solution is placed in an identical tube and an equal number of drops of indicator added as before. On looking down the length of this tube against a white background a 'colour match' is obtained with one of the buffer solutions. By this means the pH of a solution may be determined to $0 \cdot 1$ of a unit and in some cases even more accurately.

Another method is to add a few drops of a suitable indicator to the solution of unknown pH and then determine the $[\text{In}^-(\text{aq})]_{\text{eqm}}/[\text{HIn}(\text{aq})]_{\text{eqm}}$ ratio with a colorimeter. Knowing the pK_{in} value of the indicator used, the pH is then calculated using the equation

$$\text{pH} = pK_{in} + \lg \frac{[\text{In}^-(\text{aq})]_{\text{eqm}}}{[\text{HIn}(\text{aq})]_{\text{eqm}}}$$

17.9 The determination of acid dissociation constants

The acid dissociation constant, K_a, plays a central role in the study of acid-base equilibria and we end this discussion with a summary of the methods used to determine this constant.

For any weak acid, HA, the following equilibrium is established in aqueous solution:

$$\text{HA}(\text{aq}) + \text{H}_2\text{O}(\text{l}) \rightleftharpoons \text{H}^+(\text{aq}) + \text{A}^-(\text{aq})$$

If the original concentration of the aqueous acid solution is known, then the equilibrium constant for the dissociation may be calculated by determining either $[\text{H}^+(\text{aq})]$ or $[\text{A}^-(\text{aq})]$. In practice the concentration of $\text{H}^+(\text{aq})$ is much easier to measure than the concentration of $\text{A}^-(\text{aq})$, and methods of determining K_a are therefore essentially methods of determining the pH of a solution of acid of known concentration. Some of these methods are listed below.

(a) Potentiometric Titrations

The $\text{H}^+(\text{aq})$ concentration of an aqueous solution of an acid can be determined using a reversible cell arrangement, such as that described in chapter 18 (p. 320). A hydrogen electrode or the more convenient glass electrode is employed. If the concentration of the acid solution is c mol dm^{-3}, then according to equation (8), p. 285, we have

$$\text{pH} = \tfrac{1}{2}(pK_a - \lg c)$$

and

$$pK_a = 2\text{pH} + \lg c$$

(b) Indicator method

This has been discussed in connection with buffer solutions at the end of section 17.8, above.

299

(c) *Electrolytic conductivity*

This method, which is discussed in chapter 23, can be employed to determine the degree of dissociation, α, of a weak acid. Since $[H^+(aq)] = \alpha c$, the hydrogen ion concentration and hence K_a can be determined.

(d) *Titration curves*

Consideration of the equation

$$pH = pK_a + \lg \frac{[\text{salt}]_{\text{eqm}}}{[\text{acid}]_{\text{eqm}}}$$

shows that the point in the titration when the concentration of the salt is equal to the concentration of the acid is given by

$$pH = pK_a + \lg 1 = pK_a$$

This point will be reached at the half-way stage of the titration, and it is therefore a matter of determining the pH (with a pH meter) at each stage of the titration, and reading off a graph the pH at this half-way stage. Alternatively, the determination may be made by titrating the acid with standard base to locate the equivalence point and then, in a separate experiment, adding half the base required in the titration. The pH of the half-neutralised solution may then be measured directly by one of the several methods already described.

(e) If the standard free energy change ΔG° for the dissociation

$$HA(aq) + H_2O(l) \rightleftharpoons H^+(aq) + A^-(aq)$$

can be determined, the K_a may be calculated from the equation:

$$\Delta G^\circ = -RT\ln K_c$$

More often ΔG° is calculated from readily available thermodynamic data (ΔH° and ΔS°) which have been determined experimentally. It is possible to calculate the magnitude of K_a for strong acids from this relationship (p. 289).

17.10 The Lewis theory of acids and bases

According to this theory an acid is 'a substance which can accept a pair of electrons (an electron pair)' whereas a base is a 'substance which can donate a pair of electrons'. The scope of acid-base reactions is thus considerably extended to embrace reactions in which protons are not involved. The reaction between boron trifluoride and ammonia is an example:

$$\begin{array}{ccccc}
\text{F} & & \text{H} & & \text{F}\ \ \text{H} \\
| & & | & & |\ \ \ | \\
\text{F—B} & + & :\text{N—H} & \longrightarrow & \text{F—B}{\leftarrow}\text{N—H} \\
| & & | & & |\ \ \ | \\
\text{F} & & \text{H} & & \text{F}\ \ \text{H}
\end{array}$$

Lewis acids, e.g. boron trifluoride, have important parts to play in catalysing organic reactions.

Other examples of Lewis acid-base behaviour include the formation of complex ions, e.g.

$$Cu^{2+} + 4NH_3 \longrightarrow [Cu(NH_3)_4]^{2+}$$
$$\text{(Acid)} \quad \text{(Base)}$$

and the reaction of sulphur trioxide with metallic oxides, e.g.

$$Ca^{2+}O^{2-}(s) + SO_3(s) \longrightarrow Ca^{2+}SO_4^{2-}(s)$$

or

$$O^{2-}(s) + SO_3(s) \longrightarrow SO_4^{2-}(s)$$
$$\text{(Base)} \quad \text{(Acid)}$$

17.11
Questions on chapter 17

1 Discuss the ways in which the concepts of 'acidity' and 'alkalinity' have developed since 1777 when Lavoisier generalized that all acids contained oxygen.
   You may like to illustrate your answer by reference to some or all of the following species, but equal credit will be given for using examples of your own choice.
   HCl, H_2SO_4, HSO_4^-, NH_3, H_2O, H_3O^+, NH_4^+
                                                                          L

2 (a) Define the terms *Brönsted acid* and *Brönsted base*.
  (b) Explain whether **each** of the following species normally acts as an acid or a base or both in (i) aqueous solution and (ii) liquid ammonia: NH_4^+, HSO_4^-, NH_2^-, NH_3.
  (c) Discuss the factors which have to be taken into account in determining the equivalence point (end point) in the titration of an aqueous solution of ethanoic acid (acetic acid) with sodium hydroxide solution.   W

3 The following statements have been made at various times about acids:
  (1) An acid is a substance containing hydrogen which can be replaced by a metal.
  (2) An acid is a compound containing hydrogen which dissolves in water to give an excess of hydrogen ions over hydroxide ions.
  (3) An acid is a substance capable of donating a proton to a base.
  Briefly discuss to what extent you consider each of these statements to provide a satisfactory definition of the term 'acid'.
     Consider the application of the statements to the following species in aqueous solution: HCl, OH^-, NH_4^+ and HCO_3^-, and to the following species in concentrated sulphuric acid: H_2O and HNO_3.   L

4 (a) (i) What is the Brønsted-Lowry (proton transfer) theory of acids and bases?
      (ii) Illustrate the application of the theory to solvents other than water.
     (iii) What is the difference between strength and concentration as applied to acids?
      (iv) How may the strengths of acids be compared? (Practical detail is NOT required.)
  (b) What volume of 0.001 M HCl would you need to add to 10 cm^3 of 0.001 M NaOH in order to change its pH by 1 unit?   N

5 (a) Explain what you understand by an *acid* and by a *base*. Discuss briefly the acid-base properties of (i) the water molecule, (ii) the hydrogen carbonate ion.
  (b) What does pH represent and why is it used?
  (c) Explain carefully what is meant by the pK_a of a weak monobasic acid, HA.
     What would be the approximate pH of a solution of methanoic (formic) acid, $HCOOH$, containing one mole of the acid in 10 dm^3 of solution, if the value of its pK_a is 3.75?   O

6 What do you understand by the terms *acid* and *base*?
   Explain, using the equilibria involved, why *neither* sodium carbonate *nor* iron(III) sulphate gives a neutral aqueous solution.

Discuss the relation between structure and acid strength of some of the inorganic **or** organic acids with which you are familiar. W (S)

7 Write an account of buffer solutions. In your answer you should consider, giving examples, (a) their composition, (b) how they function, (c) their uses, and (d) their importance in nature. L

8 Define (a) pH, and (b) the acid dissociation constant (K_a) of a weak acid.

The pH of water at 25°C is 7. What is the corresponding value of the acid dissociation constant of water?

The acid dissociation constant of a monobasic acid is 10^{-5} mol litre^{-1}. Calculate

(i) the pH of a 0·1 M solution of the acid,

(ii) the pH of the solution formed when 100 cm^3 of a 0·1 M solution of the sodium salt of the acid is added to 100 cm^3 of a 0·1 M solution of the acid. State clearly any approximation that your calculation requires.

C (Overseas)

9 (a) Sodium hydroxide solution was added to 25 cm^3 of an aqueous solution of ethanoic acid (acetic acid) of concentration 0·1 mol/l, and the pH was measured at intervals giving the following results:

cm^3 of sodium hydroxide solution	0	4·0	8·0	12·0	16·0	18·0	20·0	22·0	22·5	23·0	23·5	24·0	28·0
pH	2·8	3·5	4·0	4·5	5·1	5·5	5·8	7·0	9·0	10·5	11·0	11·4	12·3

Plot these results on graph paper.

Use this graph to determine the following:

(i) the pH at the end-point; account for this value;

(ii) the concentration in mol/l of the sodium hydroxide solution;

(iii) the dissociation constant, K_a, of ethanoic acid.

(b) Name an indicator suitable for this titration, and give its colour in acid and alkaline solutions.

(c) Explain why ethanoic acid is called a *weak* acid. AEB

10 (a) Define pH.

Calculate the pH of 0·10 mol dm^{-3} aqueous sodium hydroxide.

(b) What is *Ostwald's dilution law*? Illustrate your answer by reference to propanoic acid.

The pH of 0·10 mol dm^{-3} aqueous propanoic acid at room temperature is 2·94. Calculate

(i) the degree of dissociation α of the acid,

(ii) the dissociation constant K_a of the acid at room temperature.

State clearly any approximations you make in your calculation.

Comment briefly on the usefulness of the degree of dissociation and the dissociation constant as an indication of the strength of an acid.

C (Overseas)

11 (a) The concentration of a dilute aqueous solution of hydrochloric acid (approximate concentration 0·1 mol/dm^3) is determined by titrating 25·00 cm^3 of the acid with dilute aqueous sodium hydroxide solution of concentration 0·100 mol/dm^3 using a pH meter.

Describe how you would carry out this experiment, and sketch the graph obtained by plotting the volume in cm^3 of dilute sodium hydroxide solution against the pH. Explain how you would use the graph to determine the concentration of the dilute hydrochloric acid.

(b) An aqueous solution of an acid HX has a concentration of 0·015 mol/dm^3. Given that the ionization (dissociation) constant of the acid, K_a, is $4·5 \times 10^{-4}$ mol/dm^3, calculate the pH of the solution. State any assumptions you make in your calculation AEB (Syllabus II)

12 (a) Draw diagrams to show the approximate change in pH when

(i) aqueous sodium hydroxide solution of concentration 0·1 mol/l is added, in portions, to 25 cm^3 of an aqueous solution of ethanoic acid (acetic acid) of concentration 0·1 mol/l,

(ii) aqueous ammonia solution of concentration 0·1 mol/l is added, in portions, to 25 cm^3 of an aqueous solution of hydrochloric acid of concentration 0·1 mol/l.

(b) By reference to these diagrams, explain the principle which underlies the choice of indicators in acid/alkali titrations. Give the name of an indicator suitable for each of the titrations in (a) and give its colours in acid and in alkaline solutions.

(c) Write equations for the reactions which take place when (i) ethanoic acid, (ii) aminoethane are separately dissolved in water. Explain why the first is an acid, and the second is a base.

(d) Write an equation for the reaction which takes place when amino ethanoic acid, $H_2N—CH_2—COOH$, dissolves in water, and suggest why the solution is nearly neutral. AEB

13 Describe how the pH varies when a solution of phosphoric(v) acid (ortho phosphoric acid) is titrated with a strong alkali. Explain how such titration curves can be used to find the dissociation constant (K_a) values for the acid.

The oxyacids of the p-block elements show considerable variation in K_a (mol dm^{-3}) values as shown below.

	H_3PO_4	H_2SO_4	$HClO_4$	$HClO_3$	$HClO_2$	$HClO$
K_1	7.5×10^{-3}	Very large	Strongest known	Large	1.1×10^{-2}	2.9×10^{-8}
K_2	6.2×10^{-8}	1.2×10^{-2}				
K_3	1.0×10^{-12}					

For 1M aqueous solutions of **each** of the six acids state which species predominates. Explain the variation in the K_a values given above.

Calculate the pH of a buffer solution obtained by mixing equal volumes of 0·1M NaH_2PO_4 (sodium dihydrogen phosphate) and 0·2M Na_2HPO_4 (disodium hydrogen phosphate). W (S)

14 An indicator in aqueous solution ionizes according to the equilibrium

$$HR \rightleftharpoons H^+ + R^-$$
$$(1 - \alpha) \quad \alpha \quad \quad \alpha$$

The ion R^- has an intense red colour whereas the unionized form is colourless. Given that the ionization constant

$$K_a = \frac{[H^+][R^-]}{[HR]} = 10^{-9} \text{ mol dm}^{-3}$$

derive an expression for the fraction α in the coloured form, and use the result to plot a graph of α against pH. Comment on the range over which this indicator would be expected to give an accurate measure of pH.

Oxford Schol. and Entrance

15 Explain what is meant by pH and pK_a. What are the meanings of the terms 'weak' and 'strong' acids? What is the pH of:

(i) 0·001M sulphuric acid;
(ii) 0·001M ethanoic acid ($K_a = 1.8 \times 10^{-5}$);
(iii) 0·001M trichloroethanoic acid ($K_a = 2.3 \times 10^{-1}$)?

What is (a) the pK_a of trichloroethanoic acid and (b) the pK_b of the trichloroethanoate ion? At what pH will the concentration of trichloroethanoic acid and the trichloroethanoate be equal? Oxford Schol. and Entrance

Equilibrium in redox systems

18.1
Introduction

A redox reaction is one in which a transfer of electrons takes place between the reactants. This requires—as the name 'redox' implies—the simultaneous involvement of a reducing agent and an oxidising agent. The reducing agent is the species which is able to donate electrons (for example, metals or metal ions in low valence states) and the oxidising agent is the species able to accept electrons (for example, non-metals or metal ions in high valence states).

Since a redox reaction must involve both oxidation and reduction components, it is possible to divide the total reaction into two half-reactions showing the donation and acceptance of electrons. For example, the reaction between chlorine and aqueous iodide ions may be written thus:

$$Cl_2(g) + 2I^-(aq) \longrightarrow I_2(aq) + 2Cl^-(aq)$$

This equation can be written as two half-reactions to show chlorine as the oxidising agent:

$$Cl_2(g) + 2e^- \longrightarrow 2Cl^-(aq)$$

and iodide ions as the reducing agent:

$$2I^-(aq) \longrightarrow I_2(aq) + 2e^-$$

This method of writing a redox reaction as two separate half-reactions gives flexibility to the concept of redox reactions, since chlorine in the presence of a reducing agent will behave according to the half-reaction shown above. Thus it will oxidise iron(II) to iron(III), for which the half-reaction is

$$Fe^{2+}(aq) \longrightarrow Fe^{3+}(aq) + e^-$$

according to the complete equation:

$$Cl_2(g) + 2Fe^{2+}(aq) \longrightarrow 2Fe^{3+}(aq) + 2Cl^-(aq)$$

This complete equation is derived from the two half-equations simply by balancing out the electrons.

Oxidising or reducing power is a relative term (like acid or base power) and it would clearly be useful if we could develop a scale to express it in more quantitative terms. This is done later in the chapter.

Oxidation numbers

The concept of oxidation number, although somewhat artificial, is very useful in the study of redox reactions. In assigning oxidation numbers to elements in a compound it is assumed that all compounds are ionic and the oxidation number is the charge on the atom in question.

In the case of a truly ionic compound this presents no problem, e.g. potassium chloride is K^+Cl^- and the oxidation number of the potassium atom is $+1$ while that of the chlorine atom is -1. For a covalent molecule such as CCl_4, the bonding electrons must be considered to belong to the more electronegative atom constituting the bond, in this case chlorine. The oxidation numbers in this example are $+4$ for the carbon atom and -1 for the chlorine atom.

Sometimes ionic and covalent bonding will exist in the same species and in this case a combination of both methods is needed. Potassium permanganate, for example, is $K^+MnO_4^-$, and the oxidation number of the potassium atom is $+1$. In the covalently bonded MnO_4^- ion, oxygen is more electronegative than manganese hence the oxidation number of the oxygen atom is -2; since there are four oxygen atoms and the overall charge on the ion is -1, the oxidation number of the manganese atom must be $+7$.

In the discussion so far, we have anticipated some of the rules which are used in deciding oxidation numbers, and these we set out in full below.

The oxidation number of:

(a) the atoms of uncombined elements is 0;

(b) a monatomic ion is equal to the charge on the ion;

(c) fluorine is equal to -1;

(d) the other halogens is -1 except when bonded to a more electronegative atom (for example in ICl where the oxidation number of iodine is $+1$);

(e) oxygen is -2, except in peroxides when it is -1 and in combination with fluorine;

(f) hydrogen is $+1$, except when bonded to a metal (for example in LiH where it is -1).

The sum of the oxidation numbers of all the elements in a species must equal the charge on the species, for example in the ion MoO_2^+ the oxidation numbers respectively of Mo and O are $+5$ and -2, giving the charge on the ion a value of $+5 + (2 \times -2)$ which is $+1$ as it should be.

An element may show two different oxidation numbers within the same compound. An example of this is iron in Fe_3O_4 which is more

Table 18.1 Oxidation numbers exhibited by some elements

Oxidation number	Chromium	Manganese	Iron	Nitrogen	Sulphur
+7		MnO_4^-			
+6	CrO_4^{2-}	MnO_4^{2-}	FeO_4^{2-}		SF_6
+5				N_2O_5	
+4		MnO_2		NO_2	SO_2
+3	$CrCl_3$	$Mn(OH)_3$	Fe_2O_3	NCl_3	
+2	$CrCl_2$	$MnCl_2$	$FeCl_2$	NO	
+1				N_2O	
0	Cr	Mn	Fe	N_2	S
-1					
-2					H_2S
-3				NH_3	

accurately written as $Fe(III)_2Fe(II)O_4$. Many elements, the transition elements being the best examples, exhibit a variety of oxidation numbers and these may be conveniently shown in a diagram such as Table 18.1.

Oxidation numbers enable us to interpret redox reactions more easily since oxidation involves an increase in oxidation number of the element; likewise, reduction involves a decrease in the oxidation number.

Example 1
Which is the oxidising and which is the reducing agent in the following reaction?

$$2H_2S(g) + SO_2(g) \longrightarrow 3S(s) + 2H_2O(l)$$

According to the oxidation number rules given above, the oxidation number of hydrogen is +1 and that of oxygen −2. Sulphur changes its oxidation number during the reaction from −2 in H_2S to 0 in sulphur itself, and from +4 in SO_2 to 0 in sulphur. Hydrogen sulphide is consequently oxidised (increase in oxidation number) while the sulphur dioxide is reduced (decrease in oxidation number). Hydrogen sulphide is the reducing agent and sulphur dioxide the oxidising agent.

18.2
Electrochemical cells

An electrochemical cell is simply a device which allows a redox reaction to proceed in such a way that a flow of electrons is produced. This electron flow can be used to do useful work and so these cells have considerable practical application.

The redox reaction

$$2Ag^+(aq) + Zn(s) \longrightarrow 2Ag(s) + Zn^{2+}(aq)$$

can be separated into two half-reactions:

$$Zn(s) \longrightarrow Zn^{2+}(aq) + 2e^- \qquad \text{(oxidation)}$$
$$2Ag^+(aq) + 2e^- \longrightarrow 2Ag(s) \qquad \text{(reduction)}$$

In the cell shown in fig. 18.1 these two half-reactions proceed at the negative and positive poles respectively. The electrons flow from the zinc to the silver electrode through the external circuit, and this flow can be detected with a sensitive galvanometer. The reaction cannot proceed at all if the two cells are totally isolated and therefore a salt-bridge (containing an ionic solution which does not interfere with the reaction being studied) or a porous partition between the two solutions completes the circuit, and allows the reaction to proceed while maintaining electrical neutrality.

FIG. 18.1. *A typical electrochemical cell*

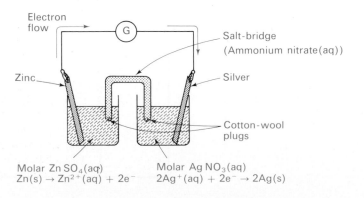

Electron flow

Zinc

Salt-bridge (Ammonium nitrate(aq))

Silver

Cotton-wool plugs

Molar $Zn\,SO_4(aq)$
$Zn(s) \rightarrow Zn^{2+}(aq) + 2e^-$

Molar $Ag\,NO_3(aq)$
$2Ag^+(aq) + 2e^- \rightarrow 2Ag(s)$

Effectively, a cell is thus a way of carrying out a redox reaction in such a way that it produces the maximum amount of work with the minimum amount of heat exchange with the surroundings.

Electrode equilibria

If we focus attention on the left-hand side of the cell in fig. 18.1 we are studying the equilibrium:

$$Zn(s) \rightleftharpoons Zn^{2+}(aq) + 2e^-$$

Zinc is tending to pass into solution as hydrated zinc ions to leave a surplus of electrons on the electrode (see fig. 18.2(a)). We have at this stage no direct evidence for this equilibrium which, at first sight, seems unlikely. However, if we chose a similar system with a different metal, silver, we can postulate a similar equilibrium:

$$Ag(s) \rightleftharpoons Ag^+(aq) + e^-$$

However, the tendency of silver to form hydrated ions is very much lower than that of zinc and there is a deficit of electrons on the silver electrode (see fig. 18.2(b)). An electron pressure difference, i.e. a potential difference, is established between the two electrodes and, when the circuit is completed with a salt-bridge, a cell is formed and a current flows.

FIG. 18.2. *A potential difference is set up when a metal is placed in contact with a solution of its ions*

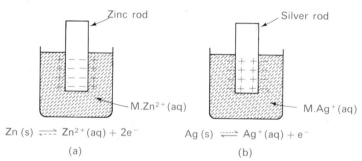

$$Zn(s) \rightleftharpoons Zn^{2+}(aq) + 2e^- \qquad Ag(s) \rightleftharpoons Ag^+(aq) + e^-$$
(a)   (b)

It does not matter if the two half-reactions appear to involve the release of electrons. For example, we might choose:

$$Zn(s) \rightleftharpoons Zn^{2+}(aq) + 2e^-$$
$$Ni(s) \rightleftharpoons Ni^{2+}(aq) + 2e^-$$

Provided the two systems have different tendencies to form ions in solution, a difference in electron density will be established on the electrodes and a potential difference set up.

Clearly the greater the difference in ion-forming tendencies of the two half-reactions, the larger the potential difference. A large potential difference implies an appreciable difference in the stabilities of the two types of ion, which is measured in terms of their standard free energies of formation. Standard free energies of formation of ions are generally expressed relative to $\Delta G^{\ominus}(298 \text{ K}) = 0$ for $H^+(aq)$.

The combination of zinc and silver ($\Delta G^{\ominus}(298 \text{ K}) = -147 \cdot 1$ kJ mol^{-1} for $Zn^{2+}(aq)$ and $\Delta G^{\ominus}(298 \text{ K}) = +77 \cdot 1$ kJ mol^{-1} for $Ag^+(aq)$) might be expected to produce a larger potential difference than the zinc and nickel combinations ($\Delta G^{\ominus}(298 \text{ K}) = -46 \cdot 3$ kJ mol^{-1} for $Ni^{2+}(aq)$).

307

So far we have considered equilibria between a metal and its ions in solution. Another important class of redox reactions involves the half-reaction between ions of the same metal in two different oxidation states. For example, the complete reaction

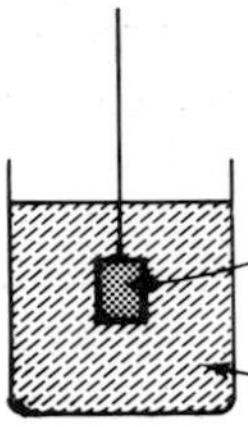

$$Ag^+(aq) + Fe^{2+}(aq) \rightleftharpoons Ag(s) + Fe^{3+}(aq)$$

can be separated into the two half-reactions

$$Ag^+(aq) + e^- \rightleftharpoons Ag(s)$$
$$Fe^{2+}(aq) \rightleftharpoons Fe^{3+}(aq) + e^-$$

and the second of these involves two different types of ion but no metal.

In the half-cell in which this reaction is carried out an inert metal, usually platinum, is used as the electrode (see fig. 18.3). Many redox half-reactions are of the ion-ion type. Two common examples are given below:

$$Sn^{4+}(aq) + 2e^- \rightleftharpoons Sn^{2+}(aq)$$
$$Ce^{4+}(aq) + 2e^- \rightleftharpoons Ce^{2+}(aq)$$

FIG. 18.3. *A half-cell containing aqueous ions of two different oxidation states derived from the same metal*

Platinum

Solution containing $Fe^{2+}(aq)$ and $Fe^{3+}(aq)$

18.3
E.m.f.s of cells

Suppose the cell shown in fig. 18.1 is set up with molar solutions of aqueous silver ions and zinc ions. The moment the salt-bridge connection is made a current begins to flow, the reaction

$$2Ag^+(aq) + Zn(s) \longrightarrow Zn^{2+}(aq) + 2Ag(s)$$

starts to take place, and the free energy change will decrease until at equilibrium $\Delta G = 0$ and the cell is run down. The reaction proceeds slowly but irreversibly. This means that the reaction is not continually in equilibrium with the surroundings, and the measured potential difference between the electrodes, while the current is flowing, is always less than the e.m.f. of the cell. If no current is taken from the cell the potential difference between the electrodes will be a maximum and will be equal to the e.m.f. of the cell. Under these conditions, i.e. a temperature of 298 K, and unit effective concentrations for the solutions, the e.m.f. is called the standard e.m.f. of the cell and is given the symbol $E^{\ominus}$.

FIG. 18.4. *A bridge circuit for determining the e.m.f. of an electrochemical cell*

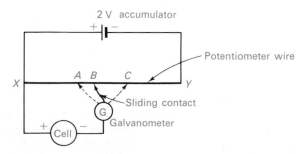

Conditions of zero current flow can be achieved if the potential difference is measured on a bridge circuit (fig. 18.4). This is essentially a method of comparing potential differences by counteracting the potential drop across a uniform wire XY by the potential difference of the cell, such that no current is detected by the galvanometer in the circuit. The ratio of the lengths XY/XB is the ratio of the e.m.f. of a fully charged accumulator (2 V) to the e.m.f. of the cell in question:

$$XY/XB = 2/E^\ominus \quad \text{or} \quad E^\ominus - 2XB/XY$$

The conditions of the bridge circuit allow the cell to function reversibly, since a change in length XB to XA allows current to be drawn from the cell while a change from XB to XC allows the cell reaction to be reversed. An infinitessimal change in the balance point can change the direction of the reaction and the system is maintained in continual equilibrium with its surroundings. Under these conditions the maximum potential difference between the cell electrodes is measured and this is the e.m.f. of the cell.

Conventions

The conventions for symbolising cells and for assigning a sign to the e.m.f. of a cell are those recommended by the International Union of Pure and Applied Chemistry (IUPAC). To see how these conventions work in practice, consider the Daniell cell in which the following chemical reaction occurs:

$$\text{Zn(s)} + \text{Cu}^{2+}\text{(aq)} \longrightarrow \text{Zn}^{2+}\text{(aq)} + \text{Cu(s)}$$

FIG. 18.5. *The Daniell cell*

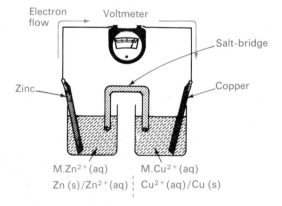

The cell is set up in the form shown in fig. 18.5. Electrodes are represented by a solid line /, the salt-bridge by a ⋮ dotted line. If an inert electrode is used in a cell it will be shown thus

$$\text{Pt/Fe}^{2+}\text{(aq), Fe}^{3+}\text{(aq)}$$

for a platinum electrode immersed in a solution of iron(II) and iron(III) ions.

The Daniell cell reaction would be represented as

$$\text{Zn(s)/Zn}^{2+}\text{(aq)} \vdots \text{Cu}^{2+}\text{(aq)/Cu(s)}$$

and the e.m.f., E, then represents the limiting value for the reaction which proceeds spontaneously from left to right. For reasons which will become clear later, p. 311, the e.m.f. is assigned a positive value. Thus for the cell represented above, $E^\ominus = +1\cdot10$ V if all the conditions correspond to standard ones. The electrode potential of an individual

309

electrode (or half-cell) is the e.m.f. of a cell in which one half-cell is the standard hydrogen electrode (p. 312).

In assigning a positive value to the e.m.f. of the cell above, we are implying that the reactions

$$Zn(s) \longrightarrow Zn^{2+}(aq) + 2e^-$$
$$Cu^{2+}(aq) + 2e^- \longrightarrow Cu(s)$$

occur spontaneously. Thus the conventional representation indicates that in the cell reaction, electrons flow from the species on the left to the species on the right.

Example 2
The e.m.f. of the silver-zinc cell shown in fig. 18.1 has a value of 1·56 V. Write down the conventional representation of this cell.
Zinc, being more electropositive than silver, will react as follows:

$$Zn(s) \longrightarrow Zn^{2+}(aq) + 2e^-$$

Electrons will therefore flow from zinc to silver through the external circuit and the zinc half-cell is consequently placed on the left:

$$Zn(s)/Zn^{2+}(aq) \vdots Ag^+(aq)/Ag(s) \qquad E^{\ominus} = +1·56 \text{ V}$$

Unless it is stated otherwise it is assumed that a cell is operating under standard conditions, i.e. elements in their standard states, solutions at unit effective concentration, a temperature of 298 K and gases at 1 atmosphere pressure. For a half-cell which comprises two different types of ion derived from the same metal, e.g. $Fe^{2+}(aq)$ and $Fe^{3+}(aq)$, the solution must be at unit effective concentration with respect to both ions. In general, the state symbols for all the participating species should be written in to avoid any ambiguity.

The connection between cell e.m.f. and free energy change

The e.m.f. of a cell, measured under reversible conditions, represents the maximum amount of work that is obtainable from the cell. If we take the extreme case of irreversible conditions, we can see that the available work would be zero. This would occur if the cell were short-circuited, in which case the system would get very hot producing an amount of heat equal to that produced by the direct mixing of the two half-cells. None of the energy would be transferred as electrical energy.

However, if the cell is delivering the maximum amount of work, then this will be measured by the product of the charge flowing per mole and the maximum potential difference through which this charge is transferred:

$$\text{Maximum work} = -nFE$$

where n is the number of moles of electrons transferred. If F is measured in coulombs per mole of electrons and E in volts, then the work is calculated in joules. The negative sign appears in the equation since energy is extracted from the cell.

Example 3
The e.m.f. of the Daniell cell operating under standard condittions is 1·10 V. What is the maximum amount of work obtainable from the cell?

310

The reaction is:

$$Zn(s) + Cu^{2+}(aq) \longrightarrow Zn^{2+}(aq) + Cu(s)$$

Two moles of electrons are transferred, so

$$\text{Maximum work} = -nFE^\ominus = -2 \times 96\,500 \times 1{\cdot}10$$
$$= -212\,300 \text{ J}$$
$$= -212{\cdot}3 \text{ kJ}$$

The maximum work that can be derived from a chemical reaction is equal to the free energy change for the reaction.

$$\text{Maximum work} = \Delta G^\ominus \text{ (for standard conditions)}$$

or

$$\Delta G^\ominus = -nFE^\ominus$$

If the free energy change is negative, then the reaction can occur spontaneously. This is why the cell conventions assign a positive value to E when the cell reaction is written down in the direction of spontaneous change.

We have now introduced three expressions from which it is possible to determine standard free energy changes, and these are summarised below:
(a) from the relationship $\Delta G^\ominus = \Delta H^\ominus - T\Delta S^\ominus$ (p. 246).
(b) from the relationship $\Delta G^\ominus = -RT\ln K$ (p. 271). For gaseous reactions K_p is used, while for reactions in solution K_c is the appropriate equilibrium constant.
(c) from the relationship $\Delta G^\ominus = -nFE^\ominus$

The e.m.f.s of three cells are given below for standard conditions:

18.4
Standard electrode (redox) potentials

$$Zn(s)/Zn^{2+}(aq) \vdots Ag^+(aq)/Ag(s) \qquad E^\ominus = +1{\cdot}56 \text{ V}$$

$$Zn(s)/Zn^{2+}(aq) \vdots Cu^{2+}(aq)/Cu(s) \qquad E^\ominus = +1{\cdot}10 \text{ V}$$

$$Cu(s)/Cu^{2+}(aq) \vdots Ag^+(aq)/Ag(s) \qquad E^\ominus = +0{\cdot}46 \text{ V}$$

These values indicate a simple relationship and suggest that each half-cell makes its own definite contribution to the e.m.f. of the whole cell. This is seen more readily if the values are shown on a chart as in fig. 18.6.

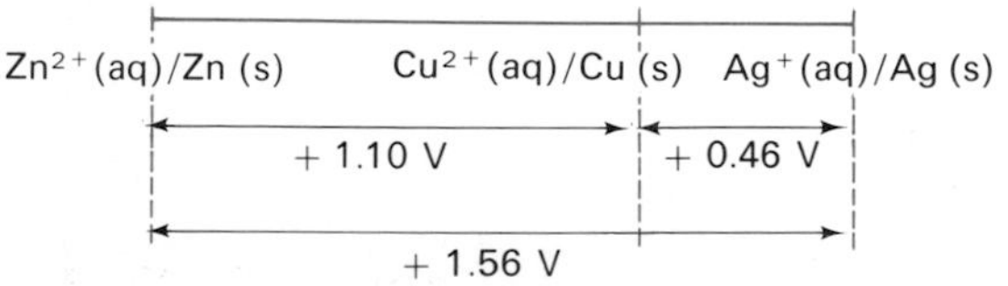

FIG. 18.6. *The e.m.f.s of three cells for standard conditions*

A value can only be assigned to each half-cell if we choose a standard to which all others can be referred. If we choose the half-cell $Zn^{2+}(aq)/Zn(s)$ as our standard and arbitrarily assign a contribution of zero volts, then the other two half-cells will have contributed as follows:

$$Ag^+(aq)/Ag(s) \quad +1{\cdot}56 \text{ V} \qquad Cu^{2+}(aq)/Cu(s) \quad +1{\cdot}10 \text{ V}$$

The standard hydrogen electrode

In practice the half-cell which is chosen as a reference for all other half-cells is not $Zn^{2+}(aq)/Zn(s)$ but $H^+(aq)/\frac{1}{2}H_2(g)$, the hydrogen half-cell. The hydrogen half-cell or standard hydrogen electrode is shown in fig. 18.7. The equilibrium

$$H^+(aq) + e^- \rightleftharpoons \tfrac{1}{2}H_2(g)$$

is established at a freshly-blacked platinum electrode and the half-cell is connected to any other half-cell by a salt-bridge in the normal way.

FIG. 18.7. *The standard hydrogen electrode*

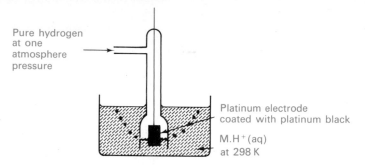

All half-cell potentials or standard electrode (redox) potentials are referred to the standard hydrogen electrode which, by convention, has a standard potential of zero volts (at unit effective concentration, a temperature of 298 K and a hydrogen pressure of one atmosphere).

The standard electrode potential for the system $M^{n+}(aq)/M(s)$ is found by connecting it to a standard hydrogen electrode, via a salt-bridge (aqueous ammonium nitrate or potassium chloride), and reading the potential difference developed, either on a previously calibrated potentiometer or on a high resistance voltmeter, i.e. an electronic voltmeter (so that negligible current flows) (fig. 18.8). The negative pole of the cell is allotted a negative electrode potential. The potential for $Zn^{2+}(aq)/Zn(s)$ is -0.76 V ($E^\ominus = -0.76$ V) and for $Cu^{2+}(aq)/Cu(s)$ is $+0.34$ V ($E^\ominus = +0.34$ V); thus the zinc electrode is the negative and the copper electrode the positive pole when they are both separately connected to a standard hydrogen electrode (fig. 18.9).

FIG. 18.8. *Apparatus for measuring standard electrode potentials*

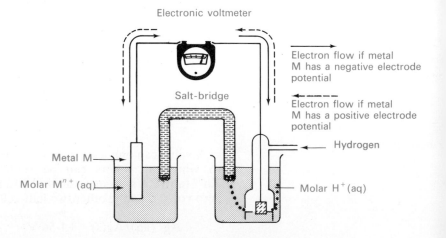

FIG. 18.9. *Zinc forms the negative pole and copper forms the positive pole when they are connected separately to a standard hydrogen electrode*

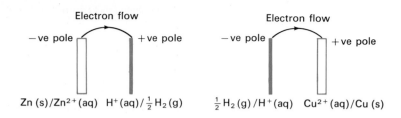

$$Zn\,(s)/Zn^{2+}(aq) \quad H^+(aq)/\tfrac{1}{2}H_2\,(g) \qquad \tfrac{1}{2}H_2\,(g)/H^+(aq) \quad Cu^{2+}(aq)/Cu\,(s)$$

The values of some standard electrode potentials are given in Table 18.2.

Table 18.2 Some standard electrode (redox) potentials

Reaction	$E^{\ominus}/V$
$Li^+(aq) \;+\; e^- \rightleftharpoons Li(s)$	$-3\cdot04$
$K^+(aq) \;+\; e^- \rightleftharpoons K(s)$	$-2\cdot92$
$Ca^{2+}(aq) + 2e^- \rightleftharpoons Ca(s)$	$-2\cdot87$
$Na^+(aq) \;+\; e^- \rightleftharpoons Na(s)$	$-2\cdot71$
$Mg^{2+}(aq) + 2e^- \rightleftharpoons Mg(s)$	$-2\cdot38$
$Al^{3+}(aq) \;+ 3e^- \rightleftharpoons Al(s)$	$-1\cdot66$
$Zn^{2+}(aq) \;+ 2e^- \rightleftharpoons Zn(s)$	$-0\cdot76$
$Pb^{2+}(aq) \;+ 2e^- \rightleftharpoons Pb(s)$	$-0\cdot13$
$H^+(aq) \;+\; e^- \rightleftharpoons \tfrac{1}{2}H_2(g)$	$0\cdot00$
$Cu^{2+}(aq) + 2e^- \rightleftharpoons Cu(s)$	$+0\cdot34$
$\tfrac{1}{2}I_2(s) \;+\; e^- \rightleftharpoons I^-(aq)$	$+0\cdot54$
$Fe^{3+}(aq) \;+\; e^- \rightleftharpoons Fe^{2+}(aq)$	$+0\cdot77$
$Ag^+(aq) \;+\; e^- \rightleftharpoons Ag(s)$	$+0\cdot80$
$\tfrac{1}{2}Br_2(l) \;+\; e^- \rightleftharpoons Br^-(aq)$	$+1\cdot07$
$\tfrac{1}{2}Cl_2(g) \;+\; e^- \rightleftharpoons Cl^-(aq)$	$+1\cdot36$
$\tfrac{1}{2}F_2(g) \;+\; e^- \rightleftharpoons F^-(aq)$	$+2\cdot87$

(left margin: INCREASINGLY POWERFUL OXIDISING AGENTS ↓; right margin: INCREASINGLY POWERFUL REDUCING AGENTS ↑)

Notice that the half-cell reactions given in Table 18.2 are written as reduction processes and it is important to remember this. The following example shows how these electrode potentials may be used.

Example 4
What would be the spontaneous reaction if the two half-cells $Fe^{3+}(aq)/Fe^{2+}(aq)$ and $\tfrac{1}{2}Cl_2(g)/Cl^-(aq)$ were joined together via a salt-bridge?

$$Fe^{3+}(aq) + e^- \rightleftharpoons Fe^{2+}(aq) \qquad E^{\ominus} = +0\cdot77 \text{ V}$$
$$\tfrac{1}{2}Cl_2(g) + e^- \rightleftharpoons Cl^-(aq) \qquad E^{\ominus} = +1\cdot36 \text{ V}$$

Subtracting the first equation from the second we get

$$\tfrac{1}{2}Cl_2(g) - Fe^{3+}(aq) \rightarrow Cl^-(aq) - Fe^{2+}(aq) \quad E^{\theta}_{cell} = +0\cdot59 \text{ V}$$

313

Re-arranging this equation and clearing the fractions we have

$$2Fe^{2+}(aq) + Cl_2(g) \rightarrow 2Fe^{3+} + 2Cl^-(aq) \quad E^{\theta}_{cell} = +0.59 \text{ V}$$

A positive e.m.f. means that the reaction will take place from left to right, i.e. chlorine will oxidise $Fe^{2+}(aq)$ ions to $Fe^{3+}(aq)$ ions. Notice that doubling the equation has no effect on the cell e.m.f. The e.m.f. of any two half-cell combinations is found simply by taking the algebraic difference of their respective electrode potentials (the signs attached to these values are obviously important).

The calomel electrode

A hydrogen electrode is cumbersome to set up and use in practice and it is more convenient to employ another type of calibrated electrode. The most common of these is the calomel electrode. It consists of a platinum wire dipping into mercury below a saturated solution of mercury(I) chloride (which is not very soluble) in potassium chloride solution of known molarity (fig. 18.10). For M.KCl solution, the electrode potential with respect to the standard hydrogen electrode is +0.280 V at 298 K. This means that 0.280 must be added to any electrode potential measured against a calomel electrode in order to obtain the value relative to the standard hydrogen electrode. For example, when the standard zinc half-cell is connected to a calomel electrode the

FIG. 18.10. *The calomel electrode*

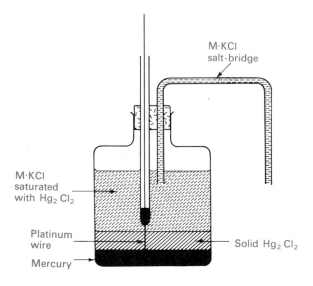

e.m.f. of the cell is 1.040 V. Since zinc constitutes the negative pole of this cell, then the electrode potential of this system with respect to the calomel electrode will be −1.040 V. Addition of 0.280 gives (−1.040 + 0.280) or −0.76 V which is the standard electrode potential of zinc with respect to the standard hydrogen electrode.

The half-reaction for the calomel electrode is:

$$Hg_2Cl_2(aq) + 2e^- \rightleftharpoons 2Hg(l) + 2Cl^-(aq)$$

A closer look at electrode potentials

Standard electrode potentials are of importance in the interpretation of redox reactions in aqueous solution. We shall shortly deal with their interpretation in terms of free energy changes and the equilibrium constant for a particular reaction. But we must first of all note that in the case of the half-cell reaction, a positive value for $E^{\ominus}$ means an oxidising

314

system relative to the hydrogen system, whereas a negative value means a negative value means a reducing system. Consider for example the half-cell $\frac{1}{2}Cl_2(g)/Cl^-(aq)(E^\ominus = +1.36 \text{ V})$. This means that under **standard conditions** chlorine is energetically capable of oxidising all the other half-reactions that lie above it in Table 18.2. However, as we have said before, **a reaction which is energetically feasible may not be observed in practice if it occurs at too slow a rate.**

The $E^\ominus$ values for $Fe^{3+}(aq)/Fe^{2+}(aq)$ which is $+0.77$ V and $Ag^+(aq)/Ag(s)$ which is $+0.80$ V need more careful interpretation. The silver system is the better oxiliser so we should expect the following reaction to occur spontaneously:

$$Ag^+(aq) + Fe^{2+}(aq) \longrightarrow Ag(s) + Fe^{3+}(aq)$$

However, the difference between the two $E^\ominus$ values is small and, as we shall see later (p. 316), the equilibrium constant is only a little greater than unity.

Any system with an $E^\ominus$ value less than zero should reduce $H^+(aq)$ ions to hydrogen gas, e.g. the Group 1A and 2A metals and zinc. There are two metals, however, which merit particular attention, namely aluminium whose standard electrode potential is -1.66 V and lead whose standard electrode potential is -0.13 V. Neither reduces $H^+(aq)$ ions to hydrogen gas at all readily under normal conditions (the metal and dilute acid). The tightly-bound oxide layer on aluminium prevents the equilibrium

$$Al(s) \rightleftharpoons Al^{3+}(aq) + 3e^-$$

from ever becoming established, although if this oxide layer is removed aluminium exhibits the reactivity expected from its large negative electrode potential. In the case of lead we have another example of a reaction which, although energetically feasible, is kinetically too slow to observe.

One final word: $E^\ominus$ values apply only to reactions in aqueous solution under standard conditions. We shall examine what happens when these conditions are changed later on in the chapter.

18.5
Standard electrode potentials and equilibrium constants

If the standard electrode potentials for two half-cells are known, then the e.m.f. of the cell obtained by coupling them together is easily calculated. For example:

$$Cu^{2+}(aq) + 2e^- \rightleftharpoons Cu(s) \qquad E^\ominus = +0.34 \text{ V}$$
$$Zn^{2+}(aq) + 2e^- \rightleftharpoons Zn(s) \qquad E^\ominus = -0.76 \text{ V}$$

Subtracting the second equation from the first, and re-arranging, we have:

$$Zn(s) + Cu^{2+}(aq) \longrightarrow Zn^{2+}(aq) + Cu(s) \qquad E^\ominus_{\text{cell}} = +1.10 \text{ V}$$

The standard free energy change, $\Delta G^\ominus(298 \text{ K})$ for the above reaction is -212.3 kJ mol^{-1} (p. 311).

Bearing in mind that, strictly speaking, the equation should have been written to show a condition of equilibrium, we can calculate the equilibrium constant, K_c, using the equation:

$$\Delta G^\ominus = -RT\ln K_c$$

Substituting values into this equation we have

$$-212\ 300 = -8{\cdot}314 \times 298 \times \ln K_c$$
$$\ln K_c = \frac{212\ 300}{8{\cdot}314 \times 298}$$

hence

$$K_c = 1{\cdot}6 \times 10^{37} \quad \text{(at 298 K)}$$

This exceedingly large value for K_c means that the reaction is effectively complete under standard conditions.

The closer together the $E^\ominus$ values for the separate half-cells become, the smaller the e.m.f. of the cell, the smaller the decrease in free energy and the smaller the equilibrium constant. For example:

$$Ag^+(aq) + e^- \rightleftharpoons Ag(s) \qquad E^\ominus = +0{\cdot}80 \text{ V}$$
$$Fe^{3+}(aq) + e^- \rightleftharpoons Fe^2(aq) \qquad E^\ominus = +0.77 \text{ V}$$

The cell reaction is

$$Ag^+(aq) + Fe^{2+}(aq) \rightleftharpoons Ag(s) + Fe^{3+}(aq) \qquad E^\ominus_{\text{cell}} = +0{\cdot}03 \text{ V}$$

and the standard free energy change is given by:

$$\Delta G^\ominus = -1 \times 96\ 500 \times 0{\cdot}03$$
$$= -2895 \text{ J} = 2{\cdot}895 \text{ kJ mol}^{-1}$$
$$\ln K_c = \frac{2895}{8{\cdot}314 \times 298}$$

hence

$$K_c = 3{\cdot}2 \text{ mol}^{-1} \text{ dm}^3 \text{ (at 298 K)}$$

In this example the equilibrium is more equally balanced between products and reactants.

18.6
Oxidation state diagrams

Many elements can exist in a variety of oxidation states. The transition metals provide the best examples, but the phenomenon is by no means confined to them. Chlorine, for example, can exist in the +7, +5, +3, +1 and −1 oxidation states in the respective ions, chlorate(VII) (ClO_4^-), chlorate(V) (ClO_3^-), chlorate(III) (ClO_2^-), chlorate(I) (ClO^-) and the chloride, Cl^-.

We can use electrode potentials to determine the relative stabilities of the various oxidation states of a particular element with respect to that element. Later we shall see how changes in concentration and temperature influence these stabilities.

Vanadium can exist in the +2, +3, +4 and +5 oxidation states in addition to the zero state of the uncombined metal. The standard electrode potential for the system $V^{2+}(aq)/V(s)$ is −1·2 V, and therefore we may write

$$V^{2+}(aq) + 2e^- \rightleftharpoons V(s) \qquad E^\ominus = -1{\cdot}2 \text{ V}$$

This can be related to the free energy change for the process by the equation:

$$\Delta G^\ominus = -nFE^\ominus$$
$$= -2(-1{\cdot}2)F = +2{\cdot}4F$$

Since F (the Faraday) is a constant, it is convenient to incorporate this on the left-hand side of the equation and use the quantity $\Delta G^{\ominus}/F$ (which has units of volts) as a measure of energetic stability in the present context. The value $+2\cdot4$ above means that V(0) is $2\cdot4$ 'units' less stable than V^{2+}(aq).

For the half cell represented by V^{3+}(aq)/V^{2+}(aq) the standard electrode potential, $E^{\ominus}$, is $-0\cdot26$ V, hence

$$\Delta G^{\ominus}/F = -nE^{\ominus}$$
$$= -1(-0\cdot26) = +0\cdot26 \text{ V}$$

and V^{2+}(aq) is $0\cdot26$ 'units' less stable than V^{3+}(aq).

In this way we can build up a picture of the relative stabilities of the oxidation states of vanadium in aqueous solution. This is most clearly illustrated by drawing a graph of $\Delta G^{\ominus}/F$ against oxidation state, choosing the stability of the zero oxidation state arbitrarily as the zero (fig. 18.11). This method of presentation is known as an oxidation state diagram. It must be remembered that it is not simply $E^{\ominus}$ that we are plotting but $nE^{\ominus}$ and that the stabilities indicated are relative stabilities.

FIG. 18.11. *The oxidation state diagram for vanadium*

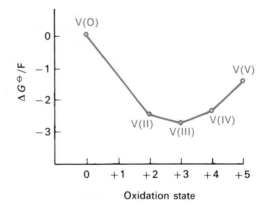

The following rules can be deduced from oxidation state diagrams:
(a) The lower the state on the diagram, the more stable it is. For example, vanadium(IV) is more stable than vanadium(V), but vanadium(III) is the most stable of all, since it is at a minimum point on the diagram:
(b) A maximum point on a diagram or a point on a convex curve will be unstable relative to its two neighbouring states and disproportionation (self oxidation-reduction) will occur. For example, in fig. 18.12 copper(I) will disproportionate to form copper, Cu(0), and copper(II).

FIG. 18.12. *The oxidation state diagram for copper*

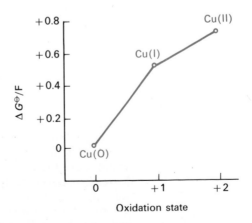

If we apply these rules to the oxidation state diagram of manganese in acid solution (fig. 18.13) we can deduce that:
(a) Manganese(II) is stable relative to all other oxidation states of the metal.
(b) Manganese(VII) is the least stable oxidation state of manganese.

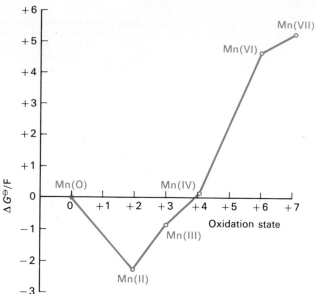

FIG. 18.13. *The oxidation state diagram for manganese in acid solution*

(c) Manganese(VI) should disproportionate into manganese(VII) and manganese(IV)—a fact that is used in the laboratory preparation of potassium permanganate.
(d) Manganese(III) should disproportionate into manganese(IV) and manganese(II).
   In using oxidation state diagrams it should be remembered that they are only applicable to the conditions relating to the particular $E^\ominus$ values. They have nothing to say about the rate at which a particular reaction will take place.

Many redox reactions do not take place under standard conditions, i.e. with solutions of unit effective concentration at 298 K. Moreover some electrode potentials are dependent upon pH, for example the one given below, whose $E^\ominus$ value is for a solution which contains $MnO_4^-(aq)$, $Mn^{2+}(aq)$ and $H^+(aq)$ all at unit effective concentration:

$$MnO_4^-(aq) + 8H^+(aq) + 5e^- \rightleftharpoons Mn^{2+}(aq) + 4H_2O(l)$$

A knowledge of how electrode potentials vary with these conditions is important, since in many cases it is possible to affect the outcome of a reaction by a suitable choice of conditions.

Effect of concentration and temperature

If we consider the half-cell reaction

$$M^{n+}(aq) + ne^- \rightleftharpoons M(s)$$

we should predict that an increase in the concentration of $M^{n+}(aq)$ would shift the position of equilibrium to the right and, in so doing, would make the electrode more positive with respect to the solution. A decrease in concentration of $M^{n+}(aq)$ would clearly be expected to have the opposite effect. In other words, the electrode potential should become more positive as the concentration of the solution is increased and the half-cell should then become a better oxidising agent.
   The exact relationship between electrode potential and concentration can be investigated in a concentration cell, comprising half-cells of similar systems, e.g. $Cu^{2+}(aq)/Cu(s)$, but with different concentrations.

If the two half-cells have initially equal concentrations of $Cu^{2+}(aq)$ the e.m.f. of the cell is zero. If the concentration of one half-cell is progressively diminished to 0·1M, 0·01M, and 0·001M, two effects are observed:

(a) the diluted half-cell electrode becomes negative with respect to the undiluted half-cell electrode;

(b) each ten-fold dilution produces an equal increment in e.m.f. of approximately 30 mV.

Effect (a) confirms the qualitative reasoning given above, but effect (b) suggests that the relationship between E and concentration is logarithmic.

If the copper half-cells are replaced by the system $Ag(s)/Ag^+(aq) : Ag^+(aq)/Ag(s)$ we find a similar relationship, except that each ten-fold dilution now causes an increase in e.m.f. of approximately 60 mV. This suggests that the number of electrons transferred in each cell is important.

These findings are rationalised in the Nernst equation,

$$E = E^\ominus + \frac{RT}{nF} \ln \frac{[\text{oxidised}]}{[\text{reduced}]}$$

where the terms 'oxidised' and 'reduced' refer to the chemical species in a given half-cell reaction, e.g.

$$Cu^{2+}(aq) + 2e^- \rightleftharpoons Cu(s)$$
$$\text{(oxidised)} \qquad\qquad \text{(reduced)}$$

In the case of the $Cu^{2+}(aq)/Cu(s)$ system above this reduces to

$$E = E^\ominus + \frac{RT}{2F} \ln [Cu^{2+}(aq)]$$

since the effective concentration of the metal is unity. At 298 K the Nernst equation for this system becomes

$$E = +0·34 + \frac{0·059}{2} \lg[Cu^{2+}(aq)]$$

(note that we have changed from Napierian logarithms to ordinary logarithms in the last expression, the factor 0·059 being $2·303RT/F$). It is now clear why increments of approximately 30 mV (0·059/2 or 0·0295 V) occur for a ten-fold change in $[Cu^{2+}(aq)]$ where $n = 2$, and approximately 60 mV (0·059 V) in the case of a ten-fold change in $[Ag^+(aq)]$ where $n = 1$.

If both oxidised and reduced forms of the half-cell are aqueous ions then they will both clearly appear in the Nernst equation. The Nernst equation for the system $Fe^{3+}(aq) + e^- \rightleftharpoons Fe^{2+}(aq)$ is thus:

$$E = +0·77 + \frac{RT}{F} \ln \frac{[Fe^{3+}(aq)]}{[Fe^{2+}(aq)]}$$

The effect of pH

If the redox half-equation involves either $H^+(aq)$ or $OH^-(aq)$ ions then its electrode potential will depend upon the pH of the solution. Consider the following half-equation:

$$MnO_4^-(aq) + 8H^+(aq) + 5e^- \rightleftharpoons Mn^{2+}(aq) + 4H_2O(l) \quad E^{\ominus} = +1.52 \text{ V}$$

The Nernst equation for this system at 298 K is

$$E = +1.52 + \frac{0.059}{5} \lg \frac{[MnO_4^-(aq)][H^+(aq)]^8}{[Mn^{2+}(aq)]}$$

Under standard conditions the concentrations of $MnO_4^-(aq)$, $Mn^{2+}(aq)$ and $H^+(aq)$ are unity. Suppose we imagine that the concentration of $H^+(aq)$ is decreased from 1M to 10^{-3}M, i.e. the pH is increased from 0 to 3, without changing the concentrations of either $MnO_4^-(aq)$ or $Mn^{2+}(aq)$. The electrode potential of this new system is given by

$$E = +1.52 + \frac{0.059}{5} \lg(10^{-3})^8$$

$$= +1.24 \text{ V}$$

The value of the electrode potential decreases from $+1.52$ to $+1.24$ V; in other words the system becomes a less powerful oxidising agent. If the pH is increased to 6, the electrode potential falls to $+0.96$ V.

Consider the three systems given below:

$$\frac{1}{2}I_2(s) \quad + e^- \rightleftharpoons I^-(aq) \qquad E^{\ominus} = +0.54 \text{ V}$$
$$\frac{1}{2}Br_2(l) + e^- \rightleftharpoons Br^-(aq) \qquad E^{\ominus} = +1.07 \text{ V}$$
$$\frac{1}{2}Cl_2(g) + e^- \rightleftharpoons Cl^-(aq) \qquad E^{\ominus} = +1.36 \text{ V}$$

The system

$$MnO_4^-(aq) + 8H^+(aq) + 5e^- \rightleftharpoons Mn^{2+}(aq) + 4H_2O(l)$$

will oxidise all the above three systems at a pH of 0, i.e. under standard conditions, but will only oxidise $I^-(aq)$ and $Br^-(aq)$ if the pH is increased to 3. At a pH value of 6 it will only oxidise $I^-(aq)$.

By suitable choice of pH, it is often possible to stabilise certain oxidation states which, under standard conditions, would be too unstable to allow preparation. For example, in acid solution manganese(VI) will disproportionate into manganese(VII) and manganese(IV):

$$MnO_4^{2-}(aq) + 4H^+(aq) + 2e^- \rightleftharpoons MnO_2(s) + 2H_2O(l) \qquad E^{\ominus} = +2.26 \text{ V}$$
$$2MnO_4^-(aq) + 2e^- \qquad\qquad \rightleftharpoons 2MnO_4^{2-}(aq) \qquad E^{\ominus} = +0.56 \text{ V}$$

Subtracting the second equation from the first and re-arranging gives

$$3MnO_4^{2-}(aq) + 4H^+(aq) \longrightarrow 2MnO_4^-(aq) + MnO_2(s) + 2H_2O(l)$$
$$E^{\ominus}_{cell} = +1.70 \text{ V}$$

The positive value for the e.m.f. means that manganese(VI) is unstable with respect to disproportionation into manganese(VII) and manganese(IV) in acid solution. In a sufficiently alkaline solution the electrode potential for the system $MnO_4^{2-}(aq)$, $H^+(aq)/MnO_2(s)$ can be decreased to a value less than $+0.56$ V (the electrode potential of the second half-cell), and consequently manganese(VI) is stabilised under these conditions.

The measurement of pH

A pH meter is simply an electrochemical cell whose e.m.f. is related to the hydrogen ion concentration, or pH, of one half-cell; the meter itself is an electronic voltmeter, a device which draws very little current, and is calibrated in pH units.

Consider the arrangement shown in fig. 18.14 which comprises a standard hydrogen electrode connected, via a potassium chloride salt-bridge, to another hydrogen electrode containing the solution of unknown pH. In both half-cells hydrogen gas at 1 atmosphere pressure is used and the temperature, strictly speaking, is 298 K. The electrode potential of the second half-cell is given by the Nernst equation:

$$E = E^{\ominus} + \frac{RT}{nF} \ln \frac{[\text{H}^+(\text{aq})]}{p_{\text{H}_2}^{1/2}}$$

where $p_{\text{H}_2}^{1/2}$ is 1 since the pressure of hydrogen gas is one atmosphere.

FIG. 18.14. *Apparatus for measuring a solution of unknown pH*

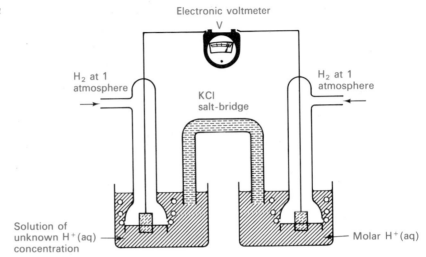

The index $\frac{1}{2}$ appears since the relevant half-equation is:

$$\text{H}^+(\text{aq}) + \text{e}^- \rightleftharpoons \tfrac{1}{2}\text{H}_2(\text{g})$$

Converting to ordinary logarithms and noting that $E^{\ominus} = 0$ we have

$$E = +0{\cdot}059 \lg[\text{H}^+(\text{aq})]$$

but since $\text{pH} = -\lg[\text{H}^+(\text{aq})]$ the equation becomes:

$$E = -0{\cdot}059 \text{ pH}$$

The e.m.f. of the cell is the standard electrode potential minus the electrode potential of the second half-cell, i.e.

$$\text{e.m.f.} = 0{\cdot}059 \text{ pH}$$

In practice it is inconvenient to use a hydrogen electrode and modern pH meters employ other electrode assemblies. One such arrangement comprises a reference electrode consisting of a silver wire, coated with silver chloride, which is in contact with a solution of potassium chloride, the other electrode being the so-called glass electrode. This device consists of a special glass bulb containing M HCl into which a platinum wire is dipping. The glass electrode is immersed in the solution of unknown pH and a potential difference is developed across the glass membrane, whose magnitude depends on the difference in pH between the unknown solution and the M HCl which is contained inside the electrode (fig. 18.15). The e.m.f. of this electrode system is given by an equation of the type:

$$\text{e.m.f.} = \text{Const.} + 0{\cdot}059 \text{ pH}$$

FIG. 18.15. *A typical glass electrode assembly*

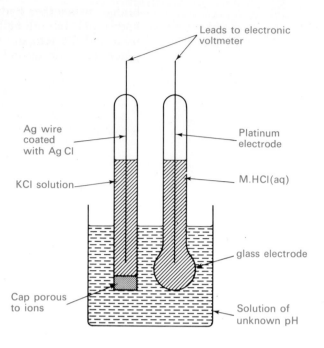

Since the factor 0·059 in the above equation varies with temperature (it is $2\cdot303RT/F$), a control on the instrument can be adjusted to compensate for temperatures which are appreciably different from 298 K.

Before a pH meter can be used to measure an unknown pH it must first be calibrated. This is achieved by dipping the electrode assembly into a buffer solution of known pH, and adjusting the needle of the electronic voltmeter until it records this value.

18.8
Factors which determine the values of standard electrode potentials

The standard electrode potentials for the half-cell systems $Cu^{2+}(aq)/Cu(s)$ and $Zn^{2+}(aq)/Zn(s)$ are very different (+0·34 V and −0·76 V respectively) and yet the two metals are adjacent to each other in the Periodic Table. Moreover the sum of their first two ionisation energies is very similar, i.e. +2705 kJ mol⁻ for copper and +2638 kJ mol⁻¹ for zinc. We have got to look at other energy terms if we are going to explain this wide variation in reactivity between these two metals.

The standard free energy change, $\Delta G^{\ominus}(298\ K)$, for the reaction

$$Zn(s) + Cu^{2+}(aq) \longrightarrow Zn^{2+}(aq) + Cu(s)$$

has been shown to be −212·3 kJ mol⁻¹ (p. 311). The reaction can be broken down into a number of hypothetical stages, for each of which we can write down a free energy change. However, we can simplify the problem since entropy changes are unimportant in this context and we can, without too much error, employ standard enthalpy changes.

The process of zinc passing into solution as its hydrated ions can be considered to involve the following three hypothetical stages:

$$Zn(s) \longrightarrow Zn(g) \qquad \Delta H_1^{\ominus}(298\ K) = +130\ kJ\ mol^{-1}$$
$$Zn(g) \longrightarrow Zn^{2+}(g) + 2e^- \qquad \Delta H_2^{\ominus}(298\ K) = +2638\ kJ\ mol^{-1}$$
$$Zn^{2+}(g) + aq \longrightarrow Zn^{2+}(aq) \qquad \Delta H_3^{\ominus}(298\ K) = -2013\ kJ\ mol^{-1}$$

The overall process is endothermic to the extent of 755 kJ mol⁻¹ i.e.

$$Zn(s) + aq \longrightarrow Zn^{2+}(aq) \qquad \Delta H^{\ominus}(298\ K) = +755\ kJ\ mol^{-1}$$

For copper the corresponding changes are:

$$Cu(s) \longrightarrow Cu(g) \qquad \Delta H_1^\ominus(298\ K) = +339\ kJ\ mol^{-1}$$
$$Cu(g) \longrightarrow Cu^{2+}(g) + 2e^- \qquad \Delta H_2^\ominus(298\ K) = +2705\ kJ\ mol^{-1}$$
$$Cu^{2+}(g) + aq \longrightarrow Cu^{2+}(aq) \qquad \Delta H_3^\ominus(298\ K) = -2069\ kJ\ mol^{-1}$$

The overall process is endothermic to the extent of 975 kJ mol^{-1} i.e.

$$Cu(s) + aq \longrightarrow Cu^{2+}(aq) \qquad \Delta H^\ominus(298\ K) = +975\ kJ\ mol^{-1}$$

The process of zinc passing into solution as its hydrated ions is less endothermic to the extent of 220 kJ mol^{-1} than the similar process for copper. Of course this amount of heat is evolved when 1 mole of zinc displaces 1 mole of copper from copper sulphate, under standard conditions, according to the equation:

$$Zn(s) + Cu^{2+}(aq) \longrightarrow Zn^{2+}(aq) + Cu(s) \qquad \Delta H^\ominus(298\ K) = -220\ kJ\ mol^{-1}$$

It can be seen that the larger ionisation energy of the copper atom ($\Delta H_2^\ominus$) is almost counterbalanced by the larger hydration energy of the copper(II) ion ($\Delta H_3^\ominus$), i.e. the sum of $\Delta H_2^\ominus + \Delta H_3^\ominus$ for both copper and zinc is almost identical. The difference between the two sublimation energies, $\Delta H_1^\ominus(298\ K)$, is 209 kJ mol^{-1}; thus the principal reason why zinc has a negative standard electrode potential as compared with that of copper is because zinc has a lower sublimation energy, i.e. zinc has lower melting- and boiling-points than copper.

This analysis shows that the magnitude of a standard electrode potential is determined by a rather subtle interplay of three energy processes. A metal will clearly have a high negative electrode potential if it has
(a) low melting- and boiling-points;
(b) a low ionisation energy
(c) a high hydration energy.
The fact that lithium has a more negative electrode potential than potassium is due to its higher hydration energy.

18.9
Questions on chapter 18

1 (a) Outline briefly how the concept of oxidation and reduction has developed into that of electron-transfer reactions and say why such reactions are of importance in the study of chemistry.

(b) Certain standard electrode potentials measured at 25°C are listed below. Comment on the selection of examples and on the values given. Explain briefly how the electrode potential arises and with the aid of *one* illustrative example how such an *electromotive force of a half-cell* may be measured.

Zn^{2+}/Zn	$-0.763V$
Fe^{2+}/Fe	$-0.440V$
$H^+/H_2/Pt$	$0V$
Cu^{2+}/Cu	$+0.337V$
$Fe^{2+}, Fe^{3+}/Pt$	$+0.771V$
$Cr^{3+}, Cr_2O_7^{2-}, H^+/Pt$	$+1.33V$

Use these values to decide what happens in the following examples.

(i) Iron filings are placed in a solution containing the following ions in their standard state:
$$Cu^{2+}, Fe^{2+}, Fe^{3+}, H^+, Zn^{2+}$$
Deduce what happens and write equations for the reactions which occur.

(ii) Calculate the standard electromotive force of the cell

$$Zn\ /\ Zn^{2+} \vdots Cu^{2+}\ /\ Cu$$

and compare it with that for the cell

$$Cu\ /\ Cu^{2+} \vdots Zn^{2+}\ /\ Zn.$$

For the first cell write equations for the electrode processes and the cell reactions. Comment on what happens in the cell and explain where the available energy goes when the reaction occurs in a beaker and not in an electrical cell.

(iii) When the following ions (as salts) in their standard states are introduced into a solution which also contains an excess of dilute sulphuric acid, deduce what happens.

$$Fe^{2+}, Fe^{3+}, Cr^{3+}, Cr_2O_7{}^{2-}$$

Write separate ion-electron equations for the reactants and deduce the final equation for the reaction.

If 5·6 g of pure iron wire is dissolved, in the absence of air, in excess of dilute sulphuric acid to which 5·6 g of potassium dichromate is added, which one of the solid reactants would be in excess and by how much?
($Fe = 56$, $K_2Cr_2O_7 = 294$)   S

2   (a)   The following equations represent oxidation/reduction reactions occuring in aqueous solution. By considering each reaction in terms of electron transfer explain which species is *oxidized* and which is *reduced*.
(i) $2FeCl_2 + Cl_2 \rightarrow 2FeCl_3$
(ii) $I_2 + 2Na_2S_2O_3 \rightarrow 2NaI + Na_2S_4O_6$
State, and account for, the colour change observed in each case.

(b)   By considering the ionic half equations:
$$Ce^{4+} + e^- \rightarrow Ce^{3+}$$
$$C_2O_4{}^{2-} \rightarrow 2CO_2 + 2e^-$$
$$Fe^{2+} \rightarrow Fe^{3+} + e^-$$

calculate the volume of cerium(IV) sulphate (of concentration 0·2 mol/l) required to oxidize 25 cm^3 of iron(II) diethanoate (iron(II) oxalate) solution of concentration 0·6 mol/l.

(c)   Given the following standard electrode potentials at 25°C
$$Pb^{2+} + 2e^- \rightleftharpoons Pb \qquad E^{\ominus} = -0·126V$$
$$Zn^{2+} + 2e^- \rightleftharpoons Zn \qquad E^{\ominus} = -0·763V$$
calculate the e.m.f. of the cell:
$$Zn(s) \mid Zn^{2+}(aq, 1M) \vdots Pb^{2+}(aq, 1M) \mid Pb(s) \qquad \text{AEB}$$

3   The following are standard reduction (redox) potentials, $E^{\ominus}$, in volts, in acid solution of concentration 1 mol H$^+$ (aq) ions per dm^3, at 298 K.

		$E^{\ominus}$
$Zn^{2+}(aq) + 2e^- \longrightarrow Zn(s)$		$-0·76$
$Cr^{3+}(aq) + e^- \longrightarrow Cr^{2+}(aq)$		$-0·41$
$SO_4{}^{2-}(aq) + 4H^+(aq) + 2e^- \longrightarrow H_2SO_3(aq) + H_2O$		$+0·17$
$\frac{1}{2}I_2(aq) + e^- \longrightarrow I^-(aq)$		$+0·54$
$Fe^{3+}(aq) + e^- \longrightarrow Fe^{2+}(aq)$		$+0·77$
$\frac{1}{2}Br_2(aq) + e^- \longrightarrow Br^-(aq)$		$+1·07$
$\frac{1}{2}Cr_2O_7{}^{2-}(aq) + 7H^+(aq) + 3e^- \longrightarrow Cr^{3+}(aq) + 3\frac{1}{2}H_2O$		$+1·33$
$MnO_4^-(aq) + 8H^+(aq) + 5e^- \longrightarrow Mn^{2+}(aq) + 4H_2O$		$+1·52$

(a)   Use these data to give explanations of the following:
(i)   the stability in acid solution of Fe^{3+}(aq) ions towards reducing agents, as compared with the stabilities of Cr^{3+}(aq) ions and Zn^{2+}(aq) ions in acid solutions;
(ii) the ease with which iron(II) can be oxidized to iron(III) in aqueous solution compared with the similar process for chromium(II);
(iii) the result of treating an aqueous solution containing chromium(II) with an aqueous solution containing iron(III) compared with the result of treating an aqueous solution of iron(III) with an aqueous solution of bromine.

(b)   State which, if any, of Br$^-$(aq), I$^-$(aq), and H$_2$SO$_3$(aq) will perform any of the following reductions in acid solution, and show your reasoning.
$$Cr_2O_7{}^{2-}(aq) \text{ to } Cr^{3+}(aq), \quad MnO_4^-(aq) \text{ to } Mn^{2+}(aq),$$
$$SO_4{}^{2-}(aq) \text{ to } SO_3{}^{2-}(aq);$$

(c)   From the list above state:
(i)   which combination of electrodes you would use to construct the electrochemical cell having the largest e.m.f.;
(ii) the direction of movement of electrons in the external circuit when the cell in (i) is operating;
(iii) by writing an equation, the overall chemical reaction occurring;
(iv) the e.m.f. of the cell when equilibrium has been attained.
AEB (Syllabus II)

4 This question is about the equation

$$E = E^{\ominus} + 2 \cdot 3 \frac{RT}{nF} \log_{10}[X^{n+}],$$

which relates the electrode potential E, the standard electrode potential $E^{\ominus}$, and the concentration, $[X^{n+}]$, for a metal/metal ion or similar electrode. R, T and F are respectively the gas constant, the temperature and the Faraday constant.

You are given the following special materials and equipment, together with the necessary measuring instruments, etc.:

    Source of hydrogen gas
    Copper and blacked platinum electrodes
    Saturated calomel electrode
    Solutions of HCl, NaOH and $Cu(NO_3)_2$ of convenient, known concentration
    Buffer solutions covering the range pH 2 to pH 11
    Describe how you would use these materials to establish

(a) the relationship between E and the concentration of X^{n+} when X is hydrogen, and the numerical value of $2 \cdot 3 \dfrac{RT}{nF}$,

(b) the way in which the relationship depends on $n+$, the charge on the ion.

          L

5 Describe the usefulness of electrochemical data and electrochemical measurements in chemistry.   W (S)

6 An electrochemical cell

$$\text{Pt}(H_2, 1 \text{ atm})|HNO_3 (m = 1), AgNO_3 (m = 1)|Ag$$

is set up at 298 K. State the e.m.f. of the cell and its polarity on open circuit. What chemical changes begin to occur if it is connected across a high resistance?

The following cell is set up to observe electrolysis at 298 K:

$$\text{Pt}(H_2, 1 \text{ atm})|HNO_3 (m = 1), Cu(NO_3)_2 (m = 1)|Cu.$$

What is the minimum voltage which must be applied, and of what polarity, to cause deposition of copper on the copper electrode? How does the situation differ if:

(a) the left-hand electrode is not as shown but is made of Pt with no supply of hydrogen,

(b) the left-hand electrode is made of silver with no supply of hydrogen,

(c) the electrodes are as shown but the electrolyte is diluted by a factor of ten?

State briefly with reasons whether you expect standard electrode potentials for metal ion/metal electrodes to vary much with temperature.

$$\text{Cu}^{2+} (aq)|Cu, E^{\ominus} = 0 \cdot 337 \text{ V}.$$
$$\text{Ag}^{+} (aq)|Ag, E^{\ominus} = 0 \cdot 799 \text{ V}. \quad \text{Camb. Entrance}$$

7 What is meant in chemistry by the term *free energy*? Give two applications of free energy data.

Descirbe how you would determine experimentally the free energy value for the following reaction:

$$\text{Pb}^{2+}_{(aq)} + Zn_{(s)} \rightleftharpoons Zn^{2+}_{(aq)} + Pb_{(s)}.$$

For the above reaction, $\Delta H^{\ominus}_{298} = -154 \text{ kJ mol}^{-1}$ and $\Delta G^{\ominus}_{298} = -123 \text{ kJ mol}^{-1}$. Why do these values differ? Calculate $E^{\ominus}_{298}$ and a value for the equilibrium constant at 298 K for the reaction. $(R = 8 \cdot 314 \text{ J K}^{-1} \text{ mol}^{-1};$ $F = 9 \cdot 65 \times 10^4 \text{ C mol}^{-1}.)$ Oxford Schol. and Entrance (Physical Science)

8 How may a hydrogen electrode be constructed and set up to measure cell e.m.f.s? Under what conditions will the e.m.f. measured be a standard redox potential?

The standard redox potential of the half-cell:

$$[S_2O^{2-}_{6(aq)} + 4H^+_{(aq)}], 2H_2SO_{3(aq)} \mid Pt,$$

is $+0 \cdot 57$ V. How will this value be affected by a change in the hydrogen ion concentration?

The cell $Cu_{(s)} \mid Cu^{2+}_{(aq)}, Ag^+_{(aq)} \mid Ag_{(s)}$ was set up and the following e.m.f.s measured (at constant temperature), for varying silver ion concentrations:

Ag$^+$/M	E/V
0·20	+0·460
0·05	+0·425
0·007	+0·375
0·0002	+0·287

The concentration of the copper ions remained constant at 1 mol dm^{-3}. Deduce the equilibrium constant for the cell reaction.

Oxford Schol. and Entrance (Physical Science)

Equilibria in complex ion systems

19.1
The nature of complexes

It is not easy to give a short yet comprehensive definition of a complex but, for our purposes, it is sufficient to say that a complex (or co-ordination compound) is formed when one or more species with a lone-pair of electrons (either a negative ion or a neutral molecule) forms a co-ordinate bond with a positively charged metal ion (usually a transition metal ion).

For example, cyanide ions, CN^-, react with $Fe^{3+}(aq)$ ions to form a complex ion:

$$Fe^{3+}(aq) + 6CN^-(aq) \longrightarrow Fe(CN)_6{}^{3-}(aq)$$

The negative ion, or neutral molecule, which forms a co-ordinate bond with the metal ion is known as the **ligand**, and the number of lone-pairs on the ligands which become bonded to the metal ion is called the **co-ordination number** of the complex. The group comprising the metal ion and the ligands is called the **co-ordination sphere**.

A solution of aqueous ammonia reacts with aqueous copper(II) solutions to form a deep blue coloured solution:

$$Cu^{2+}(aq) + xNH_3(aq) \longrightarrow Cu(NH_3)_x{}^{2+}(aq)$$

In this example, the neutral ammonia molecule is acting as the ligand and complex ions of the general formula $Cu(NH_3)_x{}^{2+}(aq)$, with values of x ranging from 1 to 4, are formed. If copper(II) sulphate is used in the presence of an excess of aqueous ammonia, then the complex salt $Cu(NH_3)_4{}^{2+}SO_4{}^{2-}$ may be isolated and analysed. The co-ordination number of this complex salt is four since the sulphate ion lies outside the co-ordination sphere of the complex.

Complexes may be ions of either charge, e.g.

$$Fe^{3+}(aq) + 6F^-(aq) \longrightarrow FeF_6{}^{3-}(aq)$$
$$Ni^{2+}(aq) + 6NH_3(aq) \longrightarrow Ni(NH_3)_6{}^{2+}(aq)$$

or they may be neutral molecules which, being covalent compounds, are usually insoluble in aqueous solution:

$$Cu^{2+}(aq) + 2NH_2CH_2COO^-(aq) \longrightarrow Cu(NH_2CH_2COO)_2(s)$$

Complexes may also be formed between ligands and an uncharged metal atom. Carbon monoxide, for example, forms a series of carbonyls with neutral transition metal atoms,

$$Fe(s) + 5CO(g) \longrightarrow Fe(CO)_5(l)$$

but the bonding in these complexes is not of a simple co-ordinate type.

19.2
The formation of complexes

Complex formation can be treated as an extension of acid-base reactions. In an acid-base reaction a proton H^+ becomes detached from the acid and transferred to the base, with which it forms a co-ordinate bond. For example, in aqueous solution, a proton from hydrogen chloride is transferred to the water molecule:

$$HCl(aq) \longrightarrow H^+(aq) + Cl^-(aq)$$
$$H_2O(l) + H^+(aq) \longrightarrow H_3O^+(aq)$$

The structure of the resulting hydroxonium ion may be shown as:

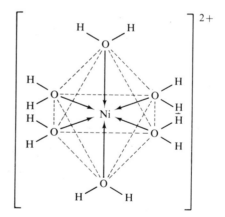

The lone-pair from the oxygen atom is accepted into the vacant $1s$ orbital of the proton.

In a similar way, other positive ions besides protons can bond with a lone-pair from a negative ion or a neutral molecule to form a complex. The nickel ion in aqueous solution is already complexed with water molecules which make efficient ligands:

$$Ni^{2+}(g) + 6H_2O(l) \longrightarrow Ni(H_2O)_6{}^{2+}(aq)$$

The structure of the hydrated nickel ion is shown below:

The lone-pairs on the oxygen atoms of the water ligands are accepted into the vacant $3d$, $4s$ and $4p$ orbitals of the nickel ion. It is the presence of this variety of vacant orbitals in transition metal ions which makes them so suitable for complex formation.

If the electron-pair donated by the ligand is equally shared between the metal ion and the ligand, then a truly covalent bond will result. In practice the bond is polarised and may have appreciable ionic character, on account of the different electronegativities of the metal and the donor atom. As each ligand bonds to the central metal ion, the latter will acquire a fraction of a negative charge which will effectively reduce its own positive charge.

Ligands may be treated as bases since they possess lone-pairs of electrons; but an important difference lies in the fact that whereas only one molecule of base will bond directly to a proton (whose vacant orbital can accommodate just two electrons), several ligands (usually a maximum of six, as in the hydrated nickel ion shown above) will form bonds with a metal ion. This is because a typical transition metal ion is much larger, in relative terms, than a proton and also because it has more vacant orbitals suitable for accepting the donated electrons from the ligands.

In aqueous solution, transition metal ions are generally complexed with six water molecules in the co-ordination sphere. The copper(II) ion is often shown co-ordinated with four water molecules, but it is thought that there are two more at a larger internuclear distance giving a distorted octahedral arrangement round the central ion.

$$Cu(H_2O)_4^{2+} \qquad Fe(H_2O)_6^{3+} \qquad Cr(H_2O)_6^{3+}$$

If a solution containing another ligand is added to an aqueous solution of a transition metal ion, the new ligands may replace water ligands in the co-ordination sphere. Thus a competition is set up between the two different ligands for the metal ion, in the same sort of way in which two different bases will compete for protons (p. 277).

The replacement of one ligand by another will occur in a stepwise manner:

$$Cu(H_2O)_4^{2+}(aq) \qquad + NH_3(aq) \rightleftharpoons Cu(H_2O)_3(NH_3)^{2+}(aq) + H_2O(l)$$
$$Cu(H_2O)_3(NH_3)^{2+}(aq) \quad + NH_3(aq) \rightleftharpoons Cu(H_2O)_2(NH_3)_2^{2+}(aq) + H_2O(l)$$
$$Cu(H_2O)_2(NH_3)_2^{2+}(aq) \quad + NH_3(aq) \rightleftharpoons Cu(H_2O)(NH_3)_3^{2+}(aq) \quad + H_2O(l)$$
$$Cu(H_2O)(NH_3)_3^{2+}(aq) \quad + NH_3(aq) \rightleftharpoons Cu(NH_3)_4^{2+}(aq) \qquad + H_2O(l)$$

The extent of replacement of one ligand by another depends upon the relative stabilities of the complexes formed but other factors are also involved. In the above example ammonia is a stronger ligand than water and on addition of ammonia solution the pale blue colour of the $Cu(H_2O)_4^{2+}(aq)$ complex ion gives way to the much deeper blue of the $Cu(NH_3)_4^{2+}(aq)$ complex ion. Transition metal complexes show a variety of colours and the colour change is generally taken as a first sign that ligand replacement has taken place. If concentrated hydrochloric acid is added to an aqueous solution of a cobalt(II) salt, the pink solution deepens in colour and eventually becomes deep blue. There are four stages involved in the stepwise replacement of the water molecules and consequently four equilibrium reactions may be written down. The overall reaction may be written:

$$\underset{\text{pink}}{Co(H_2O)_6^{2+}(aq)} + 4Cl^-(aq) \rightleftharpoons \underset{\text{blue}}{CoCl_4^{2-}(aq)} + 6H_2O(l)$$

Dilution of the blue solution, which contains the $CoCl_4^{2-}(aq)$ complex ion, results in the original pink colour being restored as the equilibrium position is driven over to the left.

Monodentate ligands

Ligands must contain an atom with a lone-pair of electrons which can form a co-ordinate bond with a metal ion; they may be negatively charged ions or neutral molecules. A ligand which can form only one co-ordinate bond is called a monodentate ligand and the most important

19.3
Ligands

of these are listed in Table 19.1.

Table 19.1 Some monodentate ligands

Ligand	Formula	Name
Water	H_2O	Aqua
Ammonia	NH_3	Ammine
Aliphatic amines	CH_3NH_2, $(CH_3)_2NH$ etc.	Methylamine etc.
Hydroxide ion	OH^-	Hydroxo
Oxide ion	O^{2-}	Oxo
Ethanoate ion	CH_3COO^-	Ethanato
Halide ion	F^-, Cl^-, Br^-, I^-	Fluoro, Chloro etc.
Cyanide ion	CN^-	Cyano

Polydentate ligands

A ligand which can simultaneously form more than one co-ordinate bond is called a polydentate ligand. A bidentate ligand can form two co-ordinate bonds, a tridentate ligand three co-ordinate bonds and so on. A widely used hexadentate ligand is the ion ethylenediamine tetra-acetate (EDTA) which has the formula:

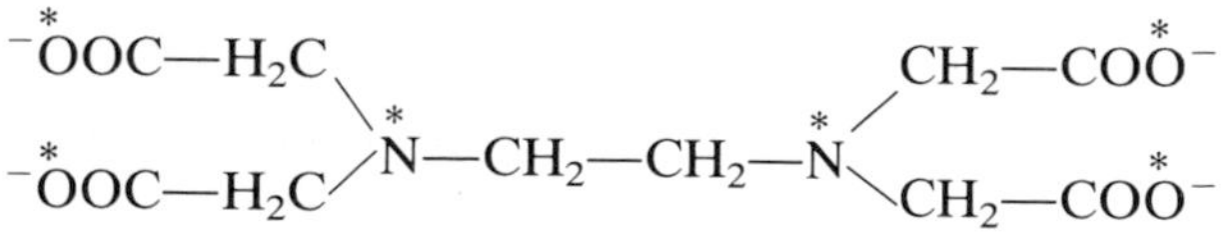

$$\begin{array}{c}
{}^*\!OOC{-}H_2C \\
{}^*\!OOC{-}H_2C
\end{array}\!\!>\!\overset{*}{N}{-}CH_2{-}CH_2{-}\overset{*}{N}\!<\!\!\begin{array}{c}
CH_2{-}\overset{*}{C}OO^- \\
CH_2{-}\overset{*}{C}OO^-
\end{array}$$

The bonding positions are indicated by an asterisk.

Chelating ligands

When a polydentate ligand simultaneously forms more than one bond with the same metal ion, a chelate* ring is formed. Nickel (II) ions, for example, react with 1,2-diaminoethane, $H_2N{-}CH_2{-}CH_2{-}NH_2$, which is abbreviated to en, to form a complex ion $Ni(en)_2^{2+}$ which has the following ring structure:

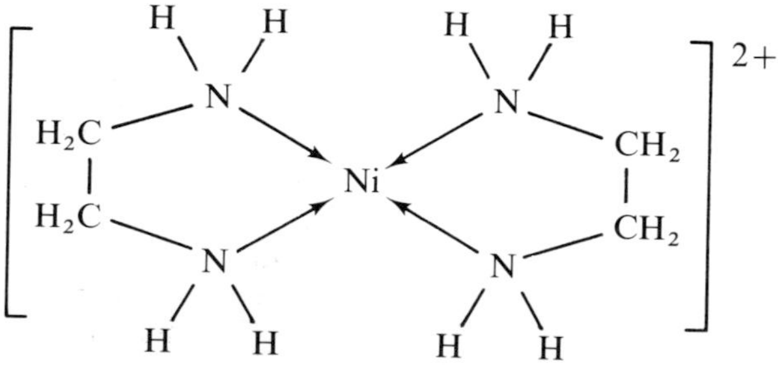

A polydentate ligand with many points of attachment can form several fused chelate rings. The $Ni(EDTA)^{2-}$ complex ion contains no less than five fused chelate rings:

* From the Greek word meaning a crab's claw

330

The formation of a chelate ring will often distort the bond angles in the ligand. The most stable rings are five-membered (including the metal ion) since this involves the minimum amount of distortion. Metal-EDTA complexes contain five-membered rings and are particularly stable.

19.4
The stability of complexes

A study of the energetic stability of complexes in solution is a study of dynamic equilibria.

Stability constants

As we have mentioned previously, the formation of a complex in aqueous solution does not occur in a single step but rather in a series of stepwise replacements. The initial step in the formation of the copper(II) tetrammine complex may be represented by the equation:

$$Cu(H_2O)_4^{2+}(aq) + NH_3(aq) \rightleftharpoons Cu(H_2O)_3NH_3^{2+}(aq) + H_2O(l)$$

The equilibrium constant, K_1, for this equilibrium is given by:

$$K_1 = \frac{[Cu(H_2O)_3NH_3^{2+}(aq)]_{eqm}}{[Cu(H_2O)_4^{2+}(aq)]_{eqm} [NH_3(aq)]_{eqm}}$$

Since the concentration of water is so high there is no significant change in this concentration during the setting up of the equilibrium and it is included in the constant K_1 (see p. 282).

The second step in the replacement of water ligands by ammonia can be represented as

$$Cu(H_2O)_3NH_3^{2+}(aq) + NH_3(aq) \rightleftharpoons Cu(H_2O)_2(NH_3)_2^{2+}(aq) + H_2O(l)$$

and

$$K_2 = \frac{[Cu(H_2O)_2(NH_3)_2^{2+}(aq)]_{eqm}}{[Cu(H_2O)_3NH_3^{2+}(aq)]_{eqm} [NH_3(aq)]_{eqm}}$$

The third and final steps follow in a similar manner with equilibrium constants of K_3 and K_4 respectively.

The overall process may be written as follows,

$$Cu(H_2O)_4{}^{2+}(aq) + 4NH_3(aq) \rightleftharpoons Cu(NH_3)_4{}^{2+}(aq) + 4H_2O(l)$$

for which the overall equilibrium constant is given by:

$$K = \frac{[Cu(NH_3)_4{}^{2+}(aq)]_{eqm}}{[Cu(H_2O)_4{}^{2+}(aq)]_{eqm} [NH_3(aq)]_{eqm}^4}$$

$K_1, K_2 \ldots$ etc. are called the **stepwise stability constants** and K is called the **overall stability constant** of the new complex. Clearly the larger the value of K the greater the stability of the complex. It is evident from this example that $K = K_1 K_2 K_3 K_4$, and in general:

$$K = K_1 K_2 K_3 \ldots K_n$$

The stepwise stability constants for the formation of the tetrammine copper(II) complex, at 298 K, are respectively:

$K_1 = 2 \cdot 0 \times 10^4$ mol^{-1} dm^3, $K_2 = 4 \cdot 2 \times 10^3$ mol^{-1} dm^3, $K_3 = 1 \cdot 0 \times 10^3$ mol^{-1} dm^3 and $K_4 = 1 \cdot 7 \times 10^2$ mol^{-1} dm^3.

This decrease which is commonly observed is partly due to a statistical factor; thus the chances of a water ligand being replaced are greater when there are four water molecules attached to the copper ion than when there are less than four. The overall stability constant, K, is $1 \cdot 4 \times 10^{13}$ mol^{-4} dm^{12} which means that the ammine complex is a very stable one compared with the aqua complex. It is usual to express stability constants in a logarithmic form, lg K. In this example lg $K = 13 \cdot 1$.

The stepwise and overall stability constants for some octahedral complexes containing the ligand 1,2-diaminoethane, en, are given in logarithmic form in Table 19.2. The table shows that complex stability varies in the manner:

$$Mn^{2+} < Fe^{2+} < Co^{2+} < Ni^{2+} < Cu^{2+} > Zn^{2+}$$

This same order is also observed for complexes with a wide range of ligands but there are some exceptions.

Table 19.2 Stability constants (in logarithmic form) for some M(en)$_3{}^{2+}$ complexes in aqueous solution at 298 K

Ligand		Mn^{2+}	Fe^{2+}	Co^{2+}	Ni^{2+}	Cu^{2+}	Zn^{2+}
en	lg K_1	2·7	4·3	5·9	7·5	10·6	5·7
	lg K_2	2·1	3·3	4·8	6·3	9·1	4·7
	lg K_3	0·9	2·0	3·1	4·3	−1·0	1·7
	lg K	5·7	9·6	13·8	18·1	18·7	12·1

The rather curious value of lg K_3 for the Cu^{2+} ion shows that there is some reluctance on the part of this ion to change from Cu(en)$_2{}^{2+}$(aq) to Cu(en)$_3{}^{2+}$(aq). The Cu^{2+} ion generally forms four strong square-planar bonds and two weaker ones, and this point was mentioned previously in connection with the aqua complex $Cu(H_2O)_4^{2+}$ (aq) (p. 329).

Some comments on stability constants

Consider the following two equilibria:

$$Ni(H_2O)_6^{2+}(aq) + 6NH_3(aq) \rightleftharpoons Ni(NH_3)_6^{2+}(aq) + 6H_2O(l)$$
$$Ni(H_2O)_6^{2+}(aq) + EDTA^{4-}(aq) \rightleftharpoons Ni(EDTA)^{2-}(aq) + 6H_2O(l)$$

The logarithms of the respective overall stability constants (at 298 K) for these two reactions are $\lg K' = 8\cdot9$ and $\lg K'' = 18\cdot6$. Clearly the EDTA complex of Ni^{2+} is far more stable in aqueous solution than is the hexammine complex (which is itself very stable). However, we must be very careful about comparing equilibrium constants whenever the stoichiometric coefficients of the replacing ligands are different as they are in these two examples, i.e. six and one respectively.

Suppose for the sake of argument that we assume that the concentrations of $Ni(H_2O)_6^{2+}(aq)$, $NH_3(aq)$ and $Ni(NH_3)_6^{2+}(aq)$ are respectively x, y and z mol dm^{-3}. Let us also assume that the concentrations of $Ni(H_2O)_6^{2+}(aq)$, $EDTA^{4-}(aq)$ and $Ni(EDTA)^{2-}(aq)$ are also respectively x, y and z mol dm^{-3}. If K' and K'' represent the overall stability constants for the formation of $Ni(NH_3)_6^{2+}(aq)$ and $Ni(EDTA)^{2-}(aq)$ respectively, then:

$$K' = \frac{z}{xy^6} \qquad K'' = \frac{z}{xy}$$

It follows that:

$$K'' = K'y^5 \text{ or } \lg K'' = \lg K' + 5 \lg y$$

If we now assume that y, the concentration of $NH_3(aq)$ and $EDTA^-(aq)$, is in the region $0\cdot1M$ at equilibrium (and this is a reasonable value) then $\lg K''$ would only need to be $3\cdot9$ compared with a value of $8\cdot9$ for $\lg K'$ to give a complex of equal stability. The fact that $\lg K''$ is actually $18\cdot6$ means that the EDTA complex is very stable indeed.

If the value of y is greater than $1M$ the effect would work in the opposite direction, but it would need to be $10M$ to swing it 5 units in the other direction. Solutions of concentration greater than $1M$ are seldom used, so this effect can be disregarded.

The effect of entropy changes on complex stability

Again let us use the two complex equilibria described in the above section, namely,

$$Ni(H_2O)_6^{2+}(aq) + 6NH_3(aq) \rightleftharpoons Ni(NH_3)_6^{2+}(aq) + 6H_2O(l)$$

and

$$Ni(H_2O)_6^{2+}(aq) + EDTA^{4-}(aq) \rightleftharpoons Ni(EDTA)^{2-}(aq) + 6H_2O(l)$$

The relevant thermodynamic data for these two reactions are given below for a temperature of 298 K:

Formation of	$\Delta H^\ominus$/kJ mol^{-1}	$\Delta S^\ominus$/kJ K^{-1} mol^{-1}	$T\Delta S^\ominus$/kJ mol^{-1}	$\Delta G^\ominus$/kJ mol^{-1}
$Ni(NH_3)_6^{2+}(aq)$	-79	$-0\cdot094$	-28	-51
$Ni(EDTA)^{2-}(aq)$	-35	$+0\cdot235$	$+70$	-105

The standard enthalpy change is more favourable for the formation of $Ni(NH_3)_6^{2+}(aq)$ than for the formation of $Ni(EDTA)^{2-}(aq)$, and it is the $T\Delta S^\ominus$ factor which is responsible for the latter complex being the more stable of the two in aqueous solution. The formation of the $Ni(NH_3)_6^{2+}(aq)$ complex involves no change in the number of particles, i.e. six ammonia molecules replace six water molecules in the aquo complex, hence the entropy change is small and negative. The formation of the $Ni(EDTA)^{2-}(aq)$ complex involves the replacement of six water molecules by one EDTA hexadentate ligand. There is a net increase in the number of particles as the second complex is formed, and the entropy change would be expected to be positive and reasonably large which it is.

It is generally true that polydentate ligands form more stable complexes with a given metal ion than unidentate ligands, and this entropy effect is often invoked as an explanation. Some EDTA complexes are so stable that further reaction is virtually impossible. For this reason EDTA is called a sequestering agent, meaning that it effectively removes metal ions from solution. It has been used successfully in the treatment of heavy metal poisoning, particularly lead poisoning ($\lg K$ for $Pb(EDTA)^{2-}(aq)$ is 18·0 at 298 K).

The stability of chelate complexes plays an important part in many biological processes. For example, the active parts of both the haemoglobin and chlorophyll molecules contain iron(II) and magnesium ions respectively, chelated to a complicated porphyrin ligand (fig. 19.1). An oxygen molecule can become attached to a vacant position on the iron(II) ion in the haemoglobin molecule and thus be transported round the body. However, strongly bonding ligands such as CN^- and CO will replace the oxygen molecule and so act as powerful poisons.

FIG. 19.1. *The 'active' part of haemoglobin*

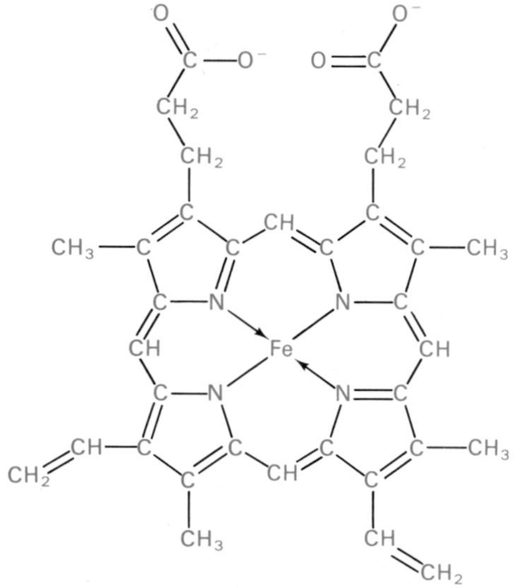

Complexometric titrations

A solution of metal ions can be titrated against a powerfully bonding ligand such as EDTA provided that an indicator can be found to detect the equivalence point. The indicator must form a complex with the metal ions that is both strongly coloured and stable enough to exist even when only a minute concentration of 'free' metal ions remain in solution. The metal-indicator complex must be less stable than the metal-EDTA complex so that, when the EDTA is in excess, it will replace the indicator ligands and the colour changes from that of the indicator-complex to that of the free indicator.

Common indicators in complexometric titrations are Eriochrome Black for Mg^{2+}, Ca^{2+} and Mn^{2+}; Pyrocatechol Violet for Bi^{3+} and Th^{4+}, and Murexide for Ca^{2+}, Ni^{2+} and Cu^{2+}.

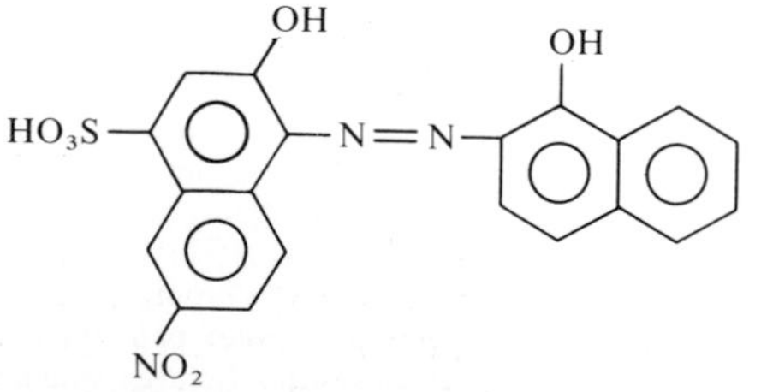

(Eriochrome Black)

Complexometric titrations are commonly used in industry as a means of estimating metal ion concentrations, for example, in determining the hardness of water samples.

19.5
The effect of
complexing on standard
electrode potentials

The equilibrium that is set up between a metal and its aqueous ions may be represented as:

$$M(s) \rightleftharpoons M^{n+}(aq) + ne^-$$

The effect of adding a ligand to the metal ion solution is to remove the aqueous ions, $M^{n+}(aq)$, from solution. The position of the equilibrium changes to produce more aqueous metal ions and, at the same time, more electrons. Therefore the metal electrode becomes more negative. This effect is illustrated in Table 19.3.

Table 19.3 The effect of complexing on standard electrode potentials at 298 K

Electrode Reaction	$E^{\ominus}/V$
$Cu^{2+}(aq) + 2e^- \rightleftharpoons Cu(s)$	+0·34
$Cu(NH_3)_4^{2+}(aq) + 2e^- \rightleftharpoons Cu(s) + 4NH_3(aq)$	−0·05
$Zn^{2+}(aq) + 2e^- \rightleftharpoons Zn(s)$	−0·76
$Zn(NH_3)_4^{2+}(aq) + 2e^- \rightleftharpoons Zn(s) + 4NH_3(aq)$	−1·03

This effect can be demonstrated practically. Thus if air is drawn through a mixture of copper turnings and a saturated aqueous solution of ammonia, the copper goes into solution as the $Cu(NH_3)_4^{2+}(aq)$ complex ion and a dark blue colour develops. This reaction occurs because the electrode potential of the complex system is lowered sufficiently for oxygen in the air to effect the oxidation.

The extent to which a given aqueous ion is removed from aqueous solution by different complexing agents will depend upon the stability constants of the complexes formed. It therefore follows that the higher the stability constant of the complex, the greater the reduction in the standard electrode potential of the system (but note what was said previously about comparing stability constants (p. 333)). This effect is illustrated in Table. 19.4.

Table 19.4 The effect of complexing on the standard electrode potential of silver at 298 K

Electrode Reaction	Lg K of complex	$E^{\ominus}/V$
$Ag^+(aq) + e^- \rightleftharpoons Ag(s)$	—	+0·80
$Ag(NH_3)_2^+(aq) + e^- \rightleftharpoons Ag(s) + 2NH_3(aq)$	7·2	+0·37
$Ag(CN)_2^-(aq) + e^- \rightleftharpoons Ag(s) + 2CN^-(aq)$	21·0	−0·38

In the case of an aqueous solution which contains ions of the same metal in different oxidation states, the effect of complexing depends upon the stabilities of the two complexes relative to their aquo complexes. The complex containing the metal in its higher oxidation state is generally the more stable and, if this is so, then complexing again makes the electrode potential more negative. The effect of different complexing agents on the Co^{3+}/Co^{2+} system is illustrated in Table 19.5.

Table 19.5 The effect of complexing on the Co^{3+}/Co^{2+} standard electrode potential at 298 K

Complexing Ligand	Lg K for Co^{3+}	Lg K for Co^{2+}	$E^{\ominus}/V$
H_2O	—	—	+1·84
$EDTA^{4-}$	36	16·3	+0·64
NH_3	33·7	4·4	−0·10
CN^-	—	—	−0·83

Co^{3+}(aq) is a sufficiently powerful oxidising agent to oxidise water to oxygen. On the other hand, the Co^{2+}-cyanide complex (which is actually $Co(CN)_5^{3-}$) reduces water to hydrogen.

The very stable complexes formed by the cyanide ion and many metal ions have important industrial applications. For example, gold may be oxidised by air in the presence of cyanide ions and thus removed from gold-bearing quartz by passing into solution as gold(I)

$$4Au(s) + 8CN^-(aq) + O_2(g) + 2H_2O(l) \rightarrow 4Au(CN)_2^-(aq) + 4OH^-(aq)$$

Whereas the standard electrode potential for the Au^+(aq)/Au(s) system is +1·70 V, the effect of complexing with cyanide ions reduces the electrode potential to +0·60 V. Addition of zinc to the solution containing the gold complex results in precipitation of the gold, since zinc forms a more stable complex with cyanide ions:

$$2Au(CN)_2^-(aq) + Zn(s) \longrightarrow 2Au(s) + Zn(CN)_4^{2-}(aq)$$

19.6
Questions on chapter 19

1 (a) What type of reaction is represented by the following equilibrium?
$$[M(H_2O)_6]^{n+} + mH_2O \rightleftharpoons [M(OH)_m(H_2O)_{6-m}]^{(n-m)+} + mH_3O^+$$
Discuss, with examples, how the position of the equilibrium will be influenced by the charge and size of the ion M^{n+} and show how you could promote the forward reaction.
(b) Discuss ligand substitution reactions of hydrated transition metal ions by showing, with one example in each case, how to bring about the following:
(i) ligand substitution with **no** change in coordination number;
(ii) ligand substitution with a change in coordination number;
(iii) ligand substitution which favours a change in oxidation state of the transition metal. JMB (Syllabus B)

2 Give an account of complex ion formation mentioning in your discussion the meaning of the following terms: (a) co-ordination number, (b) monodentate ligand, (c) chelating group, (d) stability constants.

3 An aqueous solution containing Fe^{3+}(aq) ions will oxidise an aqueous solution of an iodide to iodine. However, in the presence of fluoride ions, iron(II) solutions will reduce a solution of iodine to iodide ions. Explain these observations using the following redox data:
$$2Fe^{3+}(aq) + 2e^- \longrightarrow 2Fe^{2+}(aq) \qquad E^{\ominus} = +0·76 \text{ V}$$
$$2FeF_6^{3-}(aq) + 2e^- \longrightarrow 2FeF_6^{4-}(aq) \qquad E^{\ominus} = +0·4 \text{ V}$$
$$I_2(aq) + 2e^- \longrightarrow 2I^-(aq) \qquad E^{\ominus} = +0·54 \text{ V}$$

4 A solution containing Co^{3+}(aq) ions changes to a solution containing Co^{2+}(aq) ions and oxygen is liberated from the water at the same time. However, a solution containing Co^{2+}(aq) ions is readily oxidised to Co(III) if oxygen is bubbled through the solution in the presence of ammonia. Explain these observations using the following data:
$$2Co^{3+}(aq) + 2e^- \longrightarrow 2Co^{2+}(aq) \qquad E^{\ominus} = +1·82 \text{ V}$$
$$\tfrac{1}{2}O_2(g) + 2H^+(aq) + 2e^- \longrightarrow H_2O(l) \qquad E^{\ominus} = +1·23 \text{ V}$$
$$2Co(NH_3)_6^{3+}(aq) + 2e^- \longrightarrow 2Co(NH_3)_6^{2+}(aq) \qquad E^{\ominus} = +0·10 \text{ V}$$

Phase equilibria involving a single substance

20.1
Introduction

However strong the bonding that binds the individual particles together in a solid or liquid there are always some particles which possess sufficient energy to escape into the gaseous phase. If water is placed in an open vessel it evaporates and the loss of energetic molecules would normally result in a reduction of temperature. However, a constant temperature is maintained since the liquid absorbs heat from the surroundings and the average energy per mole of the sample remains unaltered.

Suppose now that the water vapour is not allowed to escape; this can be achieved by placing it in a closed vessel. If the vessel is previously evacuated and connected to a mercury manometer a pressure will develop. If only a tiny amount of water was originally used it may be completely converted into water vapour. However, if this is the case, more water can be added until some remains. At this stage the pressure reaches a maximum value and a state of **dynamic equilibrium** exists. At this point molecules of water are passing into the gaseous phase at the same rate at which molecules of water vapour are reforming liquid water. The maximum pressure developed by the water at a fixed temperature is called its **saturation vapour pressure**. All solids and liquids have their own characteristic saturation vapour pressures which increase as the temperature is raised. At room temperature the saturation vapour pressure may be too small to be detected—this is true of metals—or it may be quite appreciable—which is the case for ethoxyethane (ordinary ether) and is responsible for it being termed a volatile liquid.

The equilibrium between a liquid and its vapour

20.2
Equilibria between the various phases of a single substance

The equilibrium set up between a liquid and its saturated vapour can be expressed by the general equation:

$$\text{Liquid} \rightleftharpoons \text{Vapour}$$

This is an example of an heterogeneous equilibrium (p. 270) and the equilibrium constant, K_p, is simply the saturation vapour pressure of the liquid:

$$K_p = p_{\text{liquid}}$$

The conversion of a liquid into its vapour requires the absorption of heat, i.e. the above reaction is endothermic from left to right, thus the saturation vapour pressure of a liquid rises with an increase in temperature. A liquid boils when its **saturation vapour pressure** reaches the value exerted by the external pressure.

The equilibrium between a solid and its vapour

Most of what we have said about a liquid/vapour equilibrium applies in this case also. The saturation vapour pressure of a solid increases with an increase in temperature and, when it becomes equal to the saturation vapour pressure of the liquid form of the substance, solid, liquid and saturated vapour are in equilibrium. The saturation vapour pressure of a solid is generally called its **sublimation pressure**. In the case of a few solids, notably solid carbon dioxide, the sublimation pressure reaches atmospheric pressure on raising the temperature without the liquid state appearing. Such solids cannot be melted unless the pressure is greater than atmospheric pressure (p. 339).

The equilibrium between the solid, liquid and vapour states of a single substance

From what we have said in the previous paragraph, it is clear that the solid, liquid and vapour phases of a substance can only exist together at both a fixed temperature and a fixed pressure. For the system ice, water and water vapour these characteristic constants are 273·16 K and a pressure of $6·03 \times 10^{-3}$ atm. Any change of temperature or pressure, however slight, results in the disappearance of one of the phases.

The equilibrium between the solid and liquid states of a single substance

Application of pressure greater than the saturation vapour pressure of the liquid will prevent molecules leaving the liquid, i.e. there will be no vapour phase. If the temperature is now suitably lowered it is possible for the following equilibrium to be achieved:

$$\text{solid} \rightleftharpoons \text{liquid}$$

Since the solid state of a substance is more dense than its liquid form (a notable exception is the system ice/water) an increase in pressure will move the above equilibrium to the left. Since this is the direction in which heat is released, the increase in pressure will thus have the effect of raising the melting-point of the solid (for the ice/water system the reverse is true). Large pressures are needed to have a significant effect, since liquids are only slightly compressible.

The same basic kind of equilibrium may exist between solid allotropic forms of an element, e.g. rhombic and monoclinic sulphur (p. 340).

The phase diagram for the ice/water/water vapour system

This phase diagram is shown in fig. 20.1 (since the range of pressures covered is very large the diagram is not to scale). The line AB is the vapour pressure curve for ice (often called the sublimation curve) and line BC is the vapour pressure curve for water. At point B the vapour pressures of ice and water are the same (at a pressure of $6·03 \times 10^{-3}$ atm and a temperature of 273·16 K). This is called the **triple point** at which ice, water and water vapour are in equilibrium. Point C is the **critical point** (critical temperature 647 K and critical pressure 220 atm); above a temperature of 647 K water vapour cannot be converted into liquid water however great the applied pressure. Line *BE* represents the effect of pressure on the melting-point of ice. It slopes slightly from right to left in an upward direction, indicating that the melting-point of ice is lowered by an increase in pressure. Line *BD* represents the vapour pressure of water below its freezing point. It is a metastable condition, since water does not generally remain liquid below its freezing point. The phenomenon is referred to as supercooling.

FIG. 20.1. *The phase diagram for the ice/water/water vapour system (not to scale)*

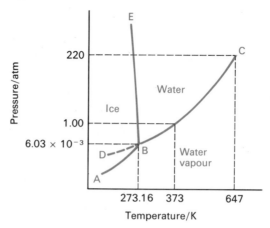

A phase diagram is useful in deciding the conditions under which different phases are in equilibrium. In an area only one phase exists, e.g. water vapour is the only stable phase below *ABC*. In this region, pressure and temperature can be varied independently over quite a range without altering the one-phase condition. Along a line two phases are in equilibrium, e.g. ice is in equilibrium with water vapour along the line *AB*. If the temperature is altered the pressure adjusts itself accordingly and vice-versa; in other words there is only one independent variable. At a point three phases are in equilibrium, e.g. point *B* in fig. 20.1. If either temperature or pressure is altered then one of the phases disappears—there are no independent variables.

The system is said to possess,

(a) two degrees of freedom in an area,
(b) one degree of freedom along a line,
(c) no degree of freedom at a point.

FIG. 20.2. *The phase diagram for carbon dioxide (not to scale)*

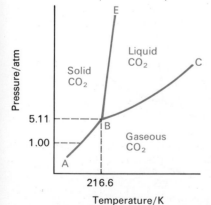

The phase diagram for carbon dioxide

This system is shown in fig. 20.2, and it is seen to be basically similar to the diagram for the ice/water/water vapour system. There are, however, two important differences: the triple point *B* is now above atmospheric pressure (it occurs at a pressure of 5·11 atm and a temperature of 216·6 K). This means that solid carbon dioxide will pass directly into carbon dioxide gas without ever becoming liquid, i.e. it sublimes, if it is allowed to warm up at pressures less than 5·11 atm. In other words, liquid carbon dioxide only exists at pressures greater than 5·11 atmospheres. The second difference is that the line *BE*, which represents the conditions under which solid and liquid carbon dioxide are in equilibrium, slopes from left to right in an upward direction. This is typical behaviour and means that an increase in pressure raises the melting-point of solid carbon dioxide.

339

The phase diagram for sulphur

Solid and liquid sulphur exhibit allotropy, i.e. the element exists in more than one form in the same physical state. In what follows, we shall discuss the allotropes of solid sulphur but not the allotropy of the liquid.

There are two crystalline forms of sulphur called rhombic (or α-) sulphur and monoclinic (or β-) sulphur. Each allotrope has a temperature range over which it is the more stable of the two, and a transition temperature exists (368·5 K) where the two allotropes are equally stable. Allotropy such as this is called enantiotropy:

$$\text{Rhombic sulphur} \underset{}{\overset{368·5\ K}{\rightleftharpoons}} \text{Monoclinic sulphur}$$
(stable < 368·5 K)　　　　　　　　　(stable > 368·5 K)

The phase diagram for sulphur is shown in fig. 20.3, and its interpretation follows logically from our previous discussion. The lines AB, BC and CD are respectively the vapour pressure curves for rhombic, monoclinic and liquid sulphur. Similarly the lines BE and CE represent the effect of pressure on the transition temperature of the two solid allotropes and the melting-point of monoclinic sulphur respectively. The slopes of these lines are much exaggerated but the direction of these slopes means that rhombic sulphur is denser than monoclinic sulphur, while the latter is denser than liquid sulphur.

Point B is the triple point for the three phases rhombic/monoclinic/vapour, while point C is the triple point for monoclinic/liquid/vapour. The dotted lines represent metastable equilibria. For example, the line BF represents the vapour pressure of rhombic sulphur above its transition temperature. Because the rate at which one solid allotrope changes into another at its transition temperature is very slow (a high activation energy barrier opposes solid transformations), fairly rapid heating of rhombic sulphur will allow the vapour pressure to rise as indicated by the line BF when it liquifies and the vapour pressure then follows the curve FCD. The dotted line EF shows the effect of pressure on the melting-point of metastable rhombic sulphur.

FIG. 20.3. *The phase diagram for sulphur (not to scale)*

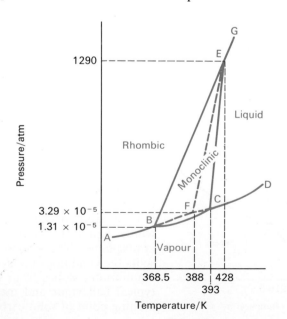

As fig. 20.3 shows, monoclinic sulphur is enclosed completely within the area *BCE* and this means that at pressures in excess of 1290 atm the cooling of liquid sulphur always results in the formation of rhombic sulphur.

20.3
Questions on chapter 20

1 (a) Sketch the pressure versus temperature phase diagram for the ice–water–water vapour system. Indicate the nature of the phases represented in the various parts of the diagram.

Use your diagram to explain the following:
  (i) The meaning of the term triple point.
  (ii) When ice is added to supercooled water, the water partially freezes and the vapour pressure and temperature of the system decrease.
(b) What is sublimation? What factors are necessary for a material to sublime at atmospheric pressure?   O Schol. and Entrance (Physical Science)

2 Sketch a simple phase diagram showing pressure as a function of temperature and indicate the regions occupied by the solid, liquid and vapour phases. Explain the meaning of the terms (a) the triple point, and (b) the critical point.

Why does the vapour pressure of a liquid increase with temperature?

The volume of a closed room is 60 m^3 and it is maintained at a constant temperature of 20°C. The partial pressure of water vapour in the room is initially 2.33×10^2 N m^{-2}. If a pan of water is then brought into the room calculate the mass of water that has evaporated when equilibrium is reached.

(Vapour pressure of water at 20°C = 2.33×10^3 N m^{-2}.)

O Schol. and Entrance (Physical Science)

Phase equilibria and solubility—part 1

21.1 Introduction

A solution is a homogeneous mixture of two or more substances, i.e. a solution is a one phase system. The extent to which one substance dissolves in another one to form a solution (its solubility) depends upon the nature of the substances involved. Solubility is influenced by temperature and pressure changes, although the latter is only of importance in the context of gases. In seeking an explanation of solubility trends we shall rely heavily on thermodynamic arguments, but before we embark upon these we must define some important terms.

There are a number of different types of solution but we shall restrict the discussion to the four most important ones which are set out below:

(a) gas in gas
(b) gas in liquid
(c) liquid in liquid
(d) solid in liquid

The component of a particular type of solution which is present in the greater quantity is referred to as the **solvent** while the other component(s) is called the **solute**. However, in the case of a solution formed from a solid and a liquid, the liquid is always referred to as the solvent even though there may be the odd occasion when the solid is present in the larger quantity.

A solution is unsaturated if it can still dissolve more solute at the same temperature; when it can dissolve no more of the solute at the same temperature it is said to be saturated. It is possible to obtain supersaturated solutions, i.e. solutions which contain more dissolved solute than they should at a particular temperature; such solutions are sometimes obtained when a hot concentrated solution of a solid in a liquid is cooled in the absence of traces of solids (even dust) so that nuclei on which crystallisation may commence are absent. In order to ensure that a particular solution is saturated but not supersaturated there must be a small amount of excess solute present. The distressing condition called the 'bends', from which divers may suffer if they surface from a great depth too rapidly, is caused by a supersaturated solution of nitrogen in the blood stream.

In order to give the term solubility a precise meaning we must specify the quantity of solute and solvent that are present in a saturated solution. There are several ways of expressing these concentrations, which are also used to specify the concentrations of unsaturated solutions, and the three most important ones are listed below:

(a) Concentration in terms of mol dm^{-3} (molarity)
This is the most common concentration unit and is employed in volumetric analysis. It refers to the number of moles of **solute** that dissolve in sufficient solvent to make 1 dm^3 (1000 cm^3) of **solution**.

(b) Concentration in terms of mol kg^{-1} (molality)
This concentration unit is defined to be the number of moles of **solute** that dissolve in 1 kg (1000 g) of **solvent**. Note that it refers to 1 kg of **solvent** and **not** 1 kg of the final **solution**.

(c) Concentration in terms of mole fractions
This concentration unit, unlike the previous two, is dimensionless. Suppose that n_A and n_B are the number of moles of solutes A and B respectively dissolved in n_C moles of solvent, then the mole fraction of each of the three components is as follows:

$$\text{Mole fraction of component A} = \frac{n_A}{n_A + n_B + n_C}$$

$$\text{Mole fraction of component B} = \frac{n_B}{n_A + n_B + n_C}$$

$$\text{Mole fraction of component C} = \frac{n_C}{n_A + n_B + n_C}$$

Note that the sum of the mole fractions of the components is unity.

Sometimes concentrations are expressed as mass of solute dissolved in a specified mass or a specified volume of solvent. However, for most purposes it is preferable to work in terms of numbers of molecules (moles) rather than in terms of mass of solute.

21.2 Gases in gases

Any number of gases have unlimited solubility in each other, provided they do not chemically combine, and they are said to be completely miscible in all proportions.

> We have already seen that the necessary criterion for change to occur is that there must be a decrease in free energy (p. 246). The change in free energy is related to changes in enthalpy and entropy by the equation,
>
> $$\Delta G = \Delta H - T\Delta S$$
>
> (we have omitted the superscripts indicating standard conditions, since we do not wish to be restricted to such conditions here).
>
> Consider the mixing of ideal gases where, by definition, there are no forces of attraction between the molecules. Then ΔH is zero and the entropy change ΔS must be positive in order that ΔG may be negative. There is obviously an increase in entropy when ideal gases mix—the 'mixed' state is clearly more disordered than the 'unmixed' state (p. 238). For real gases, ΔH is very close to zero, hence we can say that real gases mix because there is an increase in entropy when this happens:
>
> $$\Delta G = -T\Delta S \qquad (\Delta G \text{ is negative if } \Delta S \text{ is positive})$$

21.3 Gases in liquids

The solubility of a gas in a liquid is influenced by the nature of both gas and liquid and is also dependent upon the temperature and pressure. For example, methane is not very soluble in water but in a liquid hydrocarbon, with which it is chemically rather similar, the solubility is quite appreciable. Ammonia and hydrogen chloride are extremely soluble in water, the former principally because it can form hydrogen bonds and the latter because of chemical reaction to form hydrated ions. There is little likelihood of non-polar gases such as methane, oxygen and nitrogen interacting significantly with a highly polar solvent such as water and their solubilities are low.

343

Several ways of expressing the solubility of a gas in a liquid are current in chemical literature. One method is in terms of its **absorption coefficient** which is defined to be: the volume of gas, converted to s.t.p. (273 K and 1 atmosphere pressure), which will dissolve in unit volume of the solvent at a stated temperature and pressure. The absorption coefficients of a number of gases in water at three different temperatures are shown in Table 21.1.

Table 21.1 The absorption coefficients of some gases in water at three different temperatures and 1 atmosphere pressure

Gas	Temperature/K		
	273	293	323
Ammonia	1300	710	—
Hydrogen chloride	506	442	—
Oxygen	0·049	0·031	0·021
Nitrogen	0·024	0·016	0·011
Hydrogen	0·022	0·018	0·016

The effect of temperature on the solubility of a gas

All five gases in the above table show a decreasing solubility in water as the temperature is increased. This is true for most gases irrespective of the solvent they are dissolved in.

Suppose we consider a gas in contact with its saturated solution at a fixed temperature. The equilibrium can be represented as

$$\text{gas (in solvent)} \rightleftharpoons \text{gas (in gaseous phase)}$$
(a saturated solution)

The equilibrium will move to the right on increasing the temperature, i.e. the gas will become less soluble, provided that the change is endothermic in this direction. This is a direct consequence of Le Chatelier's Principle (p. 268). The change is thus exothermic in the opposite direction, consequently heat must be evolved when a gas dissolves in a solvent to produce a saturated solution, i.e. the gas molecules are solvated and energy of solvation is released in the process (p. 349).

A few examples are known where gas solubility increases with an increase in temperature, for example, the noble gases in hydrocarbon solvents, particularly in the temperature range 288 to 303 K. In such cases the process of forming a saturated solution must be endothermic.

The effect of pressure on the solubility of a gas

Consider again the equilibrium

$$\text{gas (in solvent)} \rightleftharpoons \text{gas (in gaseous phase)}$$
(a saturated solution)

Suppose the pressure is increased on the above system. Increase in pressure will have little effect on the volume of the saturated solution but it will have a considerable effect on the gas in contact with this liquid phase. Le Chatelier's Principle tells us that the effect of increasing the pressure will be to move the equilibrium in the direction which will reduce this pressure. The pressure can be reduced by more gas dissolving, i.e. a gas becomes more soluble in a particular solvent if the pressure is increased.

Let us pursue the problem a little further. The equilibrium between a gas and its saturated solution can obviously be expressed in terms of an equilibrium constant. Thus we may write:

$$K = \frac{[\text{Conc. of gas in gaseous phase}]_{eqm}}{[\text{Conc. of gas in saturated solution}]_{eqm}}$$

The concentration of gas in the gaseous phase may be represented by its pressure, while the concentration of gas in a saturated solution may be expressed as mass per unit volume or moles per unit volume etc. If m is the mass of gas dissolved in unit volume of solvent for a saturated solution and p is the gas pressure, we have

$$m = p/K$$

(clearly the value of K will depend upon which concentration units we choose). The above relationship was first stated by Henry and has subsequently become known as Henry's Law:

The mass of gas dissolved by a given volume of solvent, at a particular temperature, is proportional to the pressure of the gas with which it is in equilibrium.

Henry's Law is obeyed quite closely by gases of low solubility, provided that the pressure is not too high or the temperature too low, i.e. conditions which favour only small deviations from 'ideal' gas behaviour. Gases which interact strongly with the solvent to produce new species, e.g. ammonia and hydrogen chloride in water, do not obey Henry's Law at all well. This is not unexpected, since strong interaction with a solvent to form new species generally means that the gas has a high solubility in the particular solvent.

For gases which obey Henry's Law and, by implication, the gas laws it can easily be shown that the volume of gas dissolved by unit volume of solvent is independent of pressure. Thus suppose m g of gas dissolve in unit volume of solvent at a temperature of T K and p atmospheres pressure. If the pressure is doubled to $2p$ atmospheres, the temperature remaining unchanged, then by Henry's Law the mass of dissolved gas will increase to $2m$ g. Doubling the pressure halves the volume of a given mass of gas (Boyle's Law) but in this case the mass of gas increases from m g to $2m$ g. Hence the volume of gas dissolved by the solvent is independent of the pressure.

Gases which obey Henry's law quite closely when pure continue to do so when they are mixed, i.e. the presence of one gas in a particular solvent does not greatly influence the solubilities of others. We illustrate this point by means of the following example.

Example 1
A mixture of nitrogen and oxygen containing respectively 80% and 20% by volume of each gas is shaken with water at 273 K and 1 atmosphere pressure. Calculate the relative volumes of each gas that dissolve.

The absorption coefficients of nitrogen and oxygen under these conditions are respectively 0·024 and 0·049 (Table 21.1). The pressures of nitrogen and oxygen are respectively 0·8 atm and 0·2 atm (Dalton's Law of Partial Pressures). Thus the relative volumes of each gas that dissolve are:

$$\frac{\text{Volume of oxygen}}{\text{Volume of nitrogen}} = \frac{0.2 \times 0.049}{0.8 \times 0.024} = \frac{1}{2} \text{ (approx.)}$$

Approximately 33.3% of the dissolved gases is oxygen while the remaining 66.7% is nitrogen.

21.4 Liquids in liquids

Like gases, many pairs of liquids are completely miscible in all proportions (we restrict ourselves to pairs, since the discussion would otherwise become inordinately complex). However, in the case of liquid/liquid mixtures we cannot ignore the presence of intermolecular forces, as we did when we were dealing with gases, since it is the presence of these very forces which is responsible for the existence of the liquid state. Indeed, it is the difference in nature between the intermolecular forces in water (hydrogen-bonding) and in, say, tetrachloromethane (van der Waals' bonding) which is responsible for this pair of liquids being virtually immiscible.

Solutions of liquids in liquids will be discussed at some length in terms of vapour pressures in chapter 22, so we shall restrict ourselves to some general observations at this stage.

Pairs of liquids which are completely miscible in all proportions

Pairs of liquids such as benzene/methylbenzene ($C_6H_6/CH_3C_6H_5$), trichloromethane/propanone ($CHCl_3/CH_3COCH_3$) and carbon disulphide/propanone (CS_2/CH_3COCH_3) are completely miscible in all proportions. These three systems have been chosen since their enthalpies of mixing, ΔH, are respectively zero (or almost so), negative and positive. In discussing these three systems we shall let A and B represent molecules of the two constituents in the liquid mixture.

In the case of benzene and methylbenzene the intermolecular attractions A--A, B--B and A--B are similar in magnitude; consequently energy released in forming intermolecular bonds between A and B molecules just about balances the energy needed to break the intermolecular bonds between A and A molecules and B and B molecules. The enthalpy of mixing is therefore very close to zero. Pairs of liquids such as benzene and methylbenzene mix because there is an increase in disorder (entropy) when they do so.

Heat is evolved when trichloromethane and propanone are mixed, i.e. ΔH is negative. In this example a hydrogen bond is formed between the two components thus:

$$\begin{array}{c} H_3C \\ \diagdown \\ \diagup \\ H_3C \end{array} C = O \overset{\delta-}{\underset{}{}} \cdots \overset{\delta+}{\underset{}{}} H - C \begin{array}{c} \diagup Cl \\ - Cl \\ \diagdown Cl \end{array}$$

Hydrogen-bonding normally only exists when hydrogen is attached to the very electronegative fluorine, oxygen or nitrogen atoms (p. 90). However, the C—H bond in trichloromethane is extensively polarised by the presence of three chlorine atoms which exert a $-I$ inductive effect (p. 289), and the fractional positive charge residing on the hydrogen atom is large enough for hydrogen-bonding to take place with the oxygen atom of propanone.

There is always an increase of entropy when liquids mix, i.e. in the chemical system itself but not necessarily in the surroundings, hence ΔG which is given by the equation,

$$\Delta G = \Delta H - T\Delta S$$

will be negative (the signs of both ΔH and ΔS are favourable).

CAUTION: Explosions have been reported when trichloromethane and propanone are mixed. Experiments involving mixtures of these two compounds should not be attempted.

For pairs of liquids which show a negative enthalpy of mixing, A––B attractions are greater than the average of A––A and B––B attractions.

Heat is absorbed from the surroundings when a solution is formed by mixing carbon disulphide and propanone. This means that A––B attractions are smaller than the average of A––A and B––B attractions.

In effect, this unfavourable enthalpy of mixing means that this pair of liquids (and others which behave in a similar manner) are somewhat reluctant to form a solution. The fact that they do so means that the unfavourable enthalpy term is more than counterbalanced by the entropy of mixing.

Pairs of liquids which are immiscible and partially miscible

When the nature of the intermolecular bonding present in two liquids is sufficiently different in character, the two liquids are immiscible.[*] Such a pair of liquids is water and tetrachloromethane. Water is a highly polar hydrogen-bonded compound whereas tetrachloromethane is a non-polar liquid, with intermolecular bonding of the van der Waals' type (p. 93). The two liquids are completely different in character and thus virtually immiscible.

Certain pairs of liquids show a limited degree of solubility in each other, a typical example being ethoxyethane (ordinary ether) and water. If ether is gradually added to water it dissolves to a slight extent forming a saturated solution of ether in water. At this point, addition of more ether results in the formation of an upper layer which consists of a saturated solution of water in ether.

Other pairs of liquids behave in a similar manner to the ether/water system over a range of temperature, but show complete miscibility above or below a certain temperature. For example, hexane and phenylamine become completely miscible in all proportions above a temperature of 333 K, while triethylamine and water are only completely miscible below a temperature of 292 K.

21.5
Solids in liquids

The range of different types of solution formed from a solid solute and liquid solvent is wide. At the one extreme we have solutions of non-polar solutes in non-polar solvents, e.g. iodine in benzene, while at the other we are concerned with solutions of typically ionic solutes in a highly polar solvent, e.g. sodium chloride in water. We shall be concerned almost exclusively with the solubility of ionic compounds in water, but will briefly discuss the solubility of covalent solids in liquids.

* No two liquids are completely immiscible but many pairs approach this condition very closely.

Solubility of covalent solids in covalent liquids

In general, the type of interaction between a solid and a liquid which are both non-polar is similar to that between non-polar liquids. Typical examples include solutions of iodine in benzene, sulphur in carbon disulphide and paraffin wax in a liquid hydrocarbon. In each case solute-solute, solvent-solvent and solute-solvent interactions are similar in kind, i.e. if the enthalpy of solution is positive it is not prohibitively so.

> The entropy of solution for such systems would be expected to be very favourable, since the process of dissolving involves the breakdown of the 'order' present in a solid.

Iodine is not very soluble in water, since the solute-solute and solvent-solvent intermolecular forces are quite incompatible (van der Waals' and hydrogen-bonding respectively). Similarly, glucose is insoluble in a solvent such as benzene but soluble in water, with which it can engage in hydrogen-bonding.

Solubility of ionic compounds in water

FIG. 21.1. *Some solubility curves of ionic solids in water*

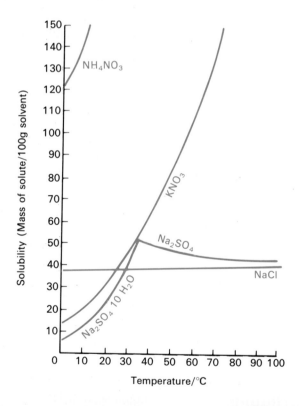

There is always a limit to the solubility of a solid in a liquid. Some ionic solids such as barium sulphate, $Ba^{2+}SO_4^{2-}$, are so slightly soluble in water that they may be regarded as insoluble, while others such as ammonium nitrate, $NH_4^+NO_3^-$, are exceedingly soluble. We shall be mainly concerned with trying to explain why it is that such a wide range of solubilities arises when typically ionic compounds are brought into contact with water. We shall restrict ourselves to the solvent water, since the rather unique properties of this common liquid make it one of the most effective solvents for producing solutions which contain ions.

(a) Variation of solubility with temperature

The variation of solubility of four ionic compounds with temperature is shown in fig. 21.1. In each case an increase in temperature leads to an increase in solubility, except for sodium sulphate above a temperature of 305 K (32°C). In this example the hydrate, $Na_2SO_4.10H_2O$, shows an increasing solubility until it changes into the anhydrous compound, Na_2SO_4, at 32°C; the solubility now decreases with an increase in temperature. By far the greatest proportion of ionic compounds become more soluble in water as the temperature is raised but there are a few, such as calcium ethanoate, which show the opposite kind of behaviour.

Consider the general equilibrium set up between a saturated solution and some undissolved solid solute at a fixed temperature:

$$\text{solid (in solvent)} \rightleftharpoons \text{solid solute}$$
(a saturated solution)

If the solid becomes more soluble as the temperature is raised, i.e. if the above equilibrium moves from right to left, then by Le Chatelier's Principle we can say that the dissolving of solid in its saturated solution is an endothermic process. Conversely the dissolving of a solute in its saturated solution is an exothermic process if the solubility decreases as the temperature is increased.

(b) A closer look at solubility

A completely quantitative discussion of solubility is very complicated and we shall do no more than point to a number of important factors that are involved in a comparative treatment.

Consider the process of a simple ionic compound, M^+X^-, dissolving in water which may be represented as

$$M^+X^-(s) + aq \longrightarrow M^+(aq) + X^-(aq) \tag{1}$$

This process may be broken down into two hypothetical stages as follows:

$$M^+X^-(s) \longrightarrow M^+(g) + X^-(g)$$
$$M^+(g) + X^-(g) + aq \longrightarrow M^+(aq) + X^-(aq)$$

In the first stage the lattice enthalpy is absorbed in separating the ions and effectively removing them to infinity, while in the second the hydration enthalpies of the two ions are released. The overall standard enthalpy change in equation (1)—the enthalpy of solution—is the algebraic sum of these two enthalpy changes.

Note that this enthalpy change refers to the formation of an aqueous solution containing each of the constituent ions at effective unit concentration (which we shall assume to be 1 mol dm^{-3}), i.e. it does not refer to the enthalpy change that takes place when a saturated solution is formed.

As an example, consider the following change at 298 K:

$$K^+Cl^-(s) + aq \longrightarrow K^+(aq) + Cl^-(aq)$$

The lattice enthalpy of potassium chloride is -711 kJ mol^{-1}, hence $+711$ kJ mol^{-1} is the energy needed to separate the ions completely. The standard enthalpies of hydration of the K^+ and Cl^- ions are respectively -305 kJ mol^{-1} and -384 kJ mol^{-1}, thus the total enthalpy of hydration is -689 kJ mol^{-1}. The standard enthalpy of solution of potassium chloride is thus $(+711 - 689)$ or $+22$ kJ mol^{-1}.

Potassium chloride is quite soluble with a solubility of 0·481 mol (100g H$_2$O)$^{-1}$ at 298 K; despite the fact that the process of solution is endothermic it is not prohibitively so.

Several important points emerge when other examples are analysed in a similar manner; these are listed below:

(a) The standard enthalpy of solution is the difference between two large energy terms, i.e. lattice enthalpies and hydration enthalpies. This means that errors in quoted values of enthalpies of solution may be quite large, particularly for enthalpies of solution which are numerically small (which they generally are).

(b) Although entropy must be considered (see later), it is likely that a strongly positive enthalpy of solution will be reflected in a low solubility. Silver chloride, for example, has a solubility of only $1·35 \times 10^{-6}$ mol (100g H$_2$O)$^{-1}$ and a standard enthalpy of solution of $+66$ kJ mol^{-1}.

(c) Standard enthalpies of hydration are large and negative, and this is a consequence of the highly polar nature of water molecules. They are able to form several strong ion-dipole bonds with every ion (negative and positive) which enable the high lattice enthalpy to be largely offset.

Less polar solvents than water cannot match this energy release when they solvate ions; consequently enthalpies of solution in such solvents are strongly positive and thus solution does not take place to any significant extent.

In order to achieve a favourable enthalpy of solution we require a low lattice enthalpy and a high enthalpy of hydration. Low lattice enthalpies are consistent with large singly-charged ions but these conditions are just those which give rise to lower hydration enthalpies. Similarly high lattice enthalpy and high hydration enthalpy are consistent with small highly-charged ions. There does not seem to be a way out of the impasse. However, lattice enthalpies are sensitive to relative ionic sizes and, other things being equal, tend to be largest when positive and negative ions are well-matched in size. On the other hand the sum of the hydration energies of anion and cation tends to be largest, other things being equal, when the ions are mismatched in size, the bulk of the hydration energy being provided by the smaller of the two ions. The steady decrease in solubility of the Group 1A chlorates(VII), and other salts containing large anions, as the size of the cation increases may be partially due to the operation of these two opposing factors. It is worth noting that the isolation of large complex ions from solution is facilitated by employing ions of an equal but opposite charge, and of approximately the same size as that of the complex ion. For example, the Ni(CN)$_5^{3-}$ complex ion can be precipitated from aqueous solution by employing the equally large Co(NH$_3$)$_6^{3+}$ complex ion.

In some cases the presence of covalent character is responsible for a solid having a higher lattice enthalpy than it would have if it were essentially ionic. In such cases the extra stability of the crystal lattice is often sufficient to cause insolubility. A typical example is silver chloride which we have previously noted (see p. 74 for the values of the calculated and experimental lattice energy of silver chloride).

We have previously mentioned that most ionic compounds become more soluble in water as the temperature is raised, yet an examination of data shows that many of these have a negative enthalpy of solution.

How can we explain these seemingly opposed facts in terms of Le Chatelier's Principle? The answer is that Le Chatelier's Principle can only be applied to an equilibrium situation, i.e. when solid is in contact with its saturated solution. Thus the dissolving of an ionic compound in its saturated solution is an endothermic process if solubility increases as the temperature is raised, irrespective of whether the enthalpy of solution is negative or positive. Can you explain why it is endothermic?

We conclude this section on solubility by noting a few points about entropies of solution.

The disruption of a crystal lattice to produce hypothetical gaseous ions will clearly lead to an iincrease in entropy (an increase in disorder). Hydration of the resulting gaseous ions results in a decrease of entropy, since order is imposed on the water molecules in the vicinity of the ions; in addition, the hydrated ions will suffer some restriction in their translational motion which makes its contribution to this decrease of entropy. Standard entropies of solution may thus be either positive or negative depending upon the relative magnitudes of these two opposing effects. In general, positive entropies of solution are favoured by large singly charged ions, since these ions will not impose as much order on the water molecules around them as small highly charged ones.

Whether an ionic solid dissolves in water to an appreciable extent, under standard conditions, is determined by the magnitude of $\Delta G^\ominus$, which is given by the usual expression:

$$\Delta G^\ominus = \Delta H^\ominus - T\Delta S^\ominus$$

Barium sulphate, for example, has a standard free energy of solution of $+49$ kJ mol^{-1} and is very insoluble. On the other hand, magnesium sulphate with a standard free energy of solution of -27 kJ mol^{-1} is very soluble.

Enough has been said to show that the subject of solubility is an extremely complex one. It is dangerous to disregard entropy effects and argue solely in terms of enthalpy changes. A difference of 76 kJ mol^{-1} in the standard free energies of solution of barium sulphate and magnesium sulphate is sufficient to account for the widely different solubilities of these two compounds; this is a small amount of energy by comparison with lattice enthalpies and hydration enthalpies. In other words, small differences in any one of the contributing factors is sufficient to determine the degree of solubility. Under the circumstances it is hardly surprising that it is not possible to lay down any hard and fast rules on this topic.

21.6
Solubility product

A solid in contact with its saturated solution is an example of dynamic equilibrium, and we should be able to discuss such a situation in terms of an equilibrium constant. In the case of a saturated solution containing an ionic compound, complications arise owing to ionic interference (p. 415). However, we can discuss equilibria of this type quantitatively, provided we restrict ourselves to sparingly soluble electrolytes such as barium sulphate and silver chloride. Solutions of sparingly soluble electrolytes are so dilute that the positive and negative ions in solution may be regarded as acting independently.

Consider a saturated solution of barium sulphate in equilibrium with some undissolved solid which may be represented by the equation below:

$$Ba^{2+}SO_4^{2-}(s) \rightleftharpoons Ba^{2+}(aq) + SO_4^{2-}(aq)$$

We may write down an equilibrium constant for this situation as follows:

$$K_c = \frac{[Ba^{2+}(aq)]_{eqm}[SO_4^{2-}(aq)]_{eqm}}{[Ba^{2+}SO_4^{2-}(s)]_{eqm}}$$

Since it does not matter how much solid barium sulphate is present (the concentration of a solid is simply its density), we can incorporate the denominator of the above equation into a new equilibrium constant,

$$K_s = [\text{Ba}^{2+}(\text{aq})]_{\text{eqm}}[\text{SO}_4{}^{2-}(\text{aq})]_{\text{eqm}}$$

where K_s is called the solubility product.

For a saturated solution of a sparingly soluble electrolyte, represented as C_cA_a, in contact with undissolved solid we may write:

$$C_cA_a(s) \rightleftharpoons cC^{a+}(\text{aq}) + aA^{c-}(\text{aq})$$

The solubility product, K_s, in this general case is given by the expression

$$K_s = [C^{a+}(\text{aq})]_{\text{eqm}}^{c}[A^{c-}(\text{aq})]_{\text{eqm}}^{a}$$

For example, the solubility product of bismuth sulphide, Bi_2S_3, is

$$K_s = [\text{Bi}^{3+}(\text{aq})]_{\text{eqm}}^{2}[\text{S}^{2-}(\text{aq})]_{\text{eqm}}^{3}$$

where the equilibrium concentration of each ion is expressed in terms of mol dm^{-3}.

The solubility product of a particular sparingly soluble electrolyte is constant provided the temperature remains unchanged. Values of solubility products are generally quoted for a temperature of 298 K and a selection of values is given in Table 21.2.

The connection between solubility and solubility product

There is clearly a connection between solubility and solubility product. The following two examples show how it is possible to calculate the solubility product for a sparingly soluble electrolyte from its solubility and vice-versa.

Table 21.2 The solubility products of some sparingly soluble electrolytes

Substance	Formula	K_s (298 K)*
Silver bromide	AgBr	$5 \cdot 0 \times 10^{-13}$ mol^2 dm^{-6}
Silver chloride	AgCl	$2 \cdot 0 \times 10^{-10}$ mol^2 dm^{-6}
Silver iodide	AgI	$8 \cdot 0 \times 10^{-17}$ mol^2 dm^{-6}
Barium sulphate	BaSO$_4$	$1 \cdot 0 \times 10^{-10}$ mol^2 dm^{-6}
Bismuth sulphide	Bi$_2$S$_3$	$1 \cdot 0 \times 10^{-97}$ mol^5 dm^{-15}
Copper(II) sulphide	CuS	$6 \cdot 3 \times 10^{-36}$ mol^2 dm^{-6}
Iron(II) hydroxide	Fe(OH)$_2$	$6 \cdot 0 \times 10^{-15}$ mol^3 dm^{-9}
Iron(III) hydroxide	Fe(OH)$_3$	$8 \cdot 0 \times 10^{-40}$ mol^4 dm^{-12}
Lead bromide	PbBr$_2$	$3 \cdot 9 \times 10^{-5}$ mol^3 dm^{-9}
Lead chloride	PbCl$_2$	$2 \cdot 0 \times 10^{-5}$ mol^3 dm^{-9}
Lead iodide	PbI$_2$	$7 \cdot 1 \times 10^{-9}$ mol^3 dm^{-9}
Lead sulphate	PbSO$_4$	$1 \cdot 6 \times 10^{-8}$ mol^2 dm^{-6}

* Although the literature values of a particular solubility product may show considerable variation, there is generally agreement about the order of magnitude.

Example 2
Calculate the solubility product of lead iodide, PbI_2, given that its solubility is 1.21×10^{-3} mol dm^{-3}.

Since x mole of PbI_2 will produce x mole of $Pb^{2+}(aq)$ and $2x$ mole of $I^-(aq)$ in solution, it follows that the equilibrium concentrations of $Pb^{2+}(aq)$ and $I^-(aq)$ are respectively 1.21×10^{-3} and 2.42×10^{-3} mol dm^{-3}.

$$K_s = [Pb^{2+}(aq)]_{eqm}[I^-(aq)]^2_{eqm}$$
$$= 1.21 \times 10^{-3} \times (2.42 \times 10^{-3})^2$$
$$= 7.1 \times 10^{-9} \text{ mol}^3 \text{ dm}^{-9}$$

Note that the concentration of $I^-(aq)$ is twice the concentration of $Pb^{2+}(aq)$, and that the former concentration must be squared as required by the equilibrium expression.

Example 3
Calculate the solubility of $PbCl_2$ in mol dm^{-3} given that the solubility product is 2.0×10^{-5} mol^3 dm^{-9}

Let x mol dm^{-3} be the equilibrium concentration of $Pb^{2+}(aq)$. Then $2x$ mol dm^{-3} will be the equilibrium concentration of $Cl^-(aq)$.

$$K_s = [Pb^{2+}(aq)]_{eqm}[Cl^-(aq)]^2_{eqm}$$
$$= x \cdot 4x^2 = 4x^3 = 2.0 \times 10^{-5}$$

thus

$$x = \sqrt[3]{\frac{2.0 \times 10^{-5}}{4}}$$
$$= 1.71 \times 10^{-2} \text{ mol dm}^{-3}$$

Since x or 1.71×10^{-2} mol dm^{-3} is the concentration of $Pb^{2+}(aq)$ it must also be the concentration of $PbCl_2$ that dissolves. Hence the solubility of $PbCl_2$ is

$$1.71 \times 10^{-2} \text{ mol dm}^{-3}$$

Note again that the concentrations of the two ions are not the same (one is twice the other), and that one of them is squared when substituted into the equilibrium expression.

Determination of solubility products

Since the concept of solubility product only applies to solutions of sparingly soluble electrolytes, sensitive methods will have to be employed to determine the concentrations of such solutions. There are two electrical methods that may be used to determine the solubilities of sparingly soluble electrolytes, from which solubility products can be calculated as shown in Example 2. The first method is based upon electrolytic conductivity and is discussed in chapter 23. The second method is based upon measurements of electrochemical cell e.m.f's and is described below.

As we have seen in chapter 18 the electrode potential of a metal depends upon the effective concentration of its metal ions with which it is in contact. Thus for the $Ag^+(aq)/Ag(s)$ system we may write

$$Ag^+(aq) + e^- \rightleftharpoons Ag(s) \qquad E^\ominus = +0.80 \text{ V}$$

Suppose we place a silver electrode into a saturated aqueous solution of silver chloride, then the electrode potential of silver will no longer be $+0.80$ V, since the concentration of aqueous silver ions is considerably lower than unit effective concentration. We can measure the e.m.f. of the cell comprising this system as one half-cell and a standard hydrogen electrode as the other, at 298 K. Suppose the e.m.f. of this cell is $+0.51$ V, then we can calculate the concentration of the silver chloride solution by applying the Nernst equation (p. 319), which in this case is

$$E = E^\ominus + 0.059 \lg[Ag^+(aq)]$$

Substituting the values of E and $E^\ominus$ into the above equation we obtain

$$+0.51 = +0.80 + 0.059 \lg[Ag^+(aq)]$$

hence

$$\lg[Ag^+(aq)] = -\frac{0.29}{0.059} = -4.915 = \bar{5}.085$$

$$[Ag^+(aq)] = \text{antilg } \bar{5}.085$$
$$= 1.22 \times 10^{-5} \text{ mol dm}^{-3}$$

The solubility of silver chloride is thus 1.22×10^{-5} mol dm^{-3} at 298 K. The solubility product for silver chloride is given by

$$K_s = [Ag^+(aq)]_{eqm}[Cl^-(aq)]_{eqm}$$

and since the $Ag^+(aq)$ and the $Cl^-(aq)$ concentrations are equal, we have

$$K_s = (1.22 \times 10^{-5})(1.22 \times 10^{-5})$$
$$= 1.49 \times 10^{-10} \text{ mol}^2 \text{ dm}^{-6}$$

The applications of solubility product

The conditions under which a sparingly soluble electrolyte will precipitate, or remain in solution, can be discussed in a quantitative manner by the application of the relevant solubility product. We shall employ both qualitative and quantitative ideas in the following discussions.

(a) The common ion effect
Consider the equilibrium set up when solid barium sulphate is in contact with its saturated solution:

$$BaSO_4(s) \rightleftharpoons Ba^{2+}(aq) + SO_4^{2-}(aq)$$

If a solution of either a soluble sulphate or a soluble barium salt is added to this equilibrium mixture, then the equilibrium position should move from right to left, i.e. the barium sulphate should become even less soluble in water. This phenomenon is known as the common ion effect, since both solid and added solution contain either identical anions or identical cations.

The common ion effect finds application in gravimetric analysis. For example, the concentration of an aqueous solution of a sulphate may be determined by precipitation as barium sulphate. After filtration, the precipitate is generally washed with dilute sulphuric acid rather than water prior to drying and weighing, in order to minimise loss of barium sulphate in these final stages.

Example 4
Calculate the solubility of barium sulphate in (a) water and (b) 0·1M H$_2$SO$_4$(aq), given that its solubility product is 1.0×10^{-10} mol^2 dm^{-6}.
We begin by writing down the equilibrium reaction:

$$BaSO_4(s) \rightleftharpoons Ba^{2-}(aq) + SO_4^{2-}(aq)$$

Since the solubility product is 1.0×10^{-10} mol^2 dm^{-6}, it follows that the concentrations of both $Ba^{2+}(aq)$ and $SO_4^{2-}(aq)$ are 10^{-5} mol dm^{-3}. The solubility of barium sulphate in water is thus 10^{-5} mol dm^{-3}.

In the presence of 0·1M H$_2$SO$_4$ the concentration of $Ba^{2+}(aq)$ will decrease from 10^{-5} to, say, x mol dm^{-3} (the common ion effect). The concentration of $SO_4^{2-}(aq)$ will be $(x + 10^{-1})$ mol dm^{-3}, but since we have seen that x will be less than 10^{-5} the concentration of $SO_4^{2-}(aq)$ may be taken to be 10^{-1} mol dm^{-3} (an error of about 0·01% which is clearly negligible). Since

$$K_s = [Ba^{2+}(aq)]_{eqm}[SO_4^{2-}(aq)]_{eqm}$$

we have

$$1 \cdot 0 \times 10^{-10} = 10^{-1}x$$

and

$$x = 10^{-9}$$

The solubility of barium sulphate in $0 \cdot 1M$ H_2SO_4 is thus 10^{-9} mol dm^{-3}, i.e. the solubility is reduced by a factor of $10^{-5}/10^{-9}$ or 10^4.

(b) Solubility product and complex ion formation

In some instances a sparingly soluble electrolyte can be taken into solution if a reagent is added with which it forms a complex ion. Sometimes the addition of a concentrated solution which contains a common ion will take the solid into solution when, as we have seen from the common ion effect (p. 354), it would be expected to reduce the solubility even more. There is nothing unusual in this behaviour once it is recognised that complex ion formation is taking place.

Consider, for example, the addition of concentrated hydrochloric acid (or a concentrated solution of any ionic chloride) to a saturated solution of lead chloride in contact with undissolved solid:

$$PbCl_2(s) \rightleftharpoons Pb^{2+}(aq) + 2Cl^-(aq)$$

The addition of $Cl^-(aq)$ would be expected to drive the equilibrium over to the left, i.e. make the lead chloride even less soluble. However, the large added concentration of $Cl^-(aq)$ combines with the $Pb^{2+}(aq)$ ions to form the $PbCl_4{}^{2-}$ (aq) complex ion:

$$Pb^{2+}(aq) + 4Cl^-(aq) \rightleftharpoons PbCl_4{}^{2-}(aq)$$

The concentration of $Pb^{2+}(aq)$ is reduced to such a low level that the solubility product of lead chloride is no longer exceeded and a solution results. The overall change can be represented by two competing equilibria:

$$PbCl_2(s) \rightleftharpoons Pb^{2+}(aq) + 2Cl^-(aq)$$
$$+$$
$$4Cl^-(aq) \text{ From conc. hydrochloric acid}$$
$$\updownarrow$$
$$PbCl_4{}^{2-}(aq)$$

The amphoteric behaviour of a number of metallic hydroxides is accounted for in a similar manner. For example, zinc hydroxide readily dissolves in the presence of an excess of sodium hydroxide solution:

$$Zn(OH)_2(s) \rightleftharpoons Zn^{2+}(aq) + 2OH^-(aq)$$
$$+$$
$$4OH^-(aq) \text{ From added } OH^-(aq)$$
$$\updownarrow$$
$$Zn(OH)_4{}^{2-}(aq)$$

(c) Solubility of sparingly soluble salts of weak acids in an aqueous solution of a strong acid

Consider the equilibrium set up between solid calcium phosphate (a salt of a strong base and a weak acid) and its saturated solution:

$$Ca_3(PO_4)_2(s) \rightleftharpoons 3Ca^{2+}(aq) + 2PO_4{}^{3-}(aq)$$

355

The equilibrium will move to the right, and the calcium phosphate dissolve if the phosphate ions, $PO_4^{3-}(aq)$, are removed. Since phosphoric acid, H_3PO_4, is a relatively weak acid, it follows that the phosphate ion, $PO_4^{3-}(aq)$, is a strong base, i.e. it will readily combine with $H^+(aq)$. The addition of a strong acid, e.g. dilute nitric acid, will provide $H^+(aq)$ which combine with $PO_4^{3-}(aq)$, and the removal of these latter ions will allow the equilibrium to move over to the right, i.e. the calcium phosphate dissolves. The changes involved can be represented by the following two equilibria:

$$Ca_3(PO_4)_2(s) \rightleftharpoons 3Ca^{2+}(aq) + 2PO_4^{3-}(aq)$$
$$+$$
$$2H^+(aq)$$
$$\updownarrow$$
$$2HPO_4^{2-}(aq)$$

The second of these two equilibria is followed by two more:

$$HPO_4^{2-}(aq) + H^+(aq) \rightleftharpoons H_2PO_4^-(aq)$$

and

$$H_2PO_4^-(aq) + H^+(aq) \rightleftharpoons H_3PO_4(aq)$$

A solution would not have formed if we had used dilute sulphuric acid, since calcium sulphate is only moderately soluble. All common metallic nitrates are soluble in water and this was the reason for stipulating dilute nitric acid.

It should be clear that an insoluble salt of a strong acid e.g. lead sulphate, $PbSO_4$, will not dissolve in dilute nitric acid since there can be no significant combination, in this example, between $SO_4^{2-}(aq)$ and $H^+(aq)$.

21.7 Partition of a solute between two immiscible solvents

When two immiscible liquids are in contact with each other and a solute is added, which is soluble in both liquids, then it is found that the solute partitions itself between the two liquids in a definite ratio. The numerical value of this ratio depends upon the particular immiscible liquids and solute which comprise the chemical system, and it varies with temperature. It is called the **partition or distribution coefficient**.

Suppose a solute A is dissolved in solvent 1 and a less dense immiscible solvent 2 is placed on top of this solution. Then there will be a movement of solute molecules across the interface between the two liquids. Because of the random motion of the dissolved solute molecules, some will pass back again into the lower solvent and, as time passes, the rate of this downward movement will increase. The result of this is that the net rate at which solute passes into the upper solvent will steadily decline, until a state of dynamic equilibrium is achieved. At this point the rates at which solute passes across the interface in an upward and downward direction are equal. If we represent the dissolved solute molecules as A(solv 1) and A(solv 2) in the two immiscible layers at equilibrium, we may write

$$A(\text{solv 1}) \rightleftharpoons A(\text{solv 2})$$

and

$$K_d = \frac{[A(\text{solv } 2)]_{\text{eqm}}}{[A(\text{solv } 1)]_{\text{eqm}}} \quad (2)$$

where K_d is the partition or distribution coefficient.

As we have stated above, the value of K_d depends upon the particular chemical system under investigation and it is temperature dependent. There are two important conditions that must be fulfilled if expression (2) above is to apply accurately to a particular system; these are given below:

(a) The two solutions which are in contact must be reasonably dilute. The reason for this restriction is that solvent-solute interaction will take place in each layer to different extents. This difference becomes less as the solutions become more dilute.

(b) The solute must be of the same molecular form in each solvent. The partition law does not apply, for example, to the system comprising hydrogen chloride, water and benzene; in water the dissolved species are $H^+(aq)$ and $Cl^-(aq)$ whereas in benzene the species are dissolved HCl molecules. Likewise a constant value of K_d is not obtained, for varying concentrations of dissolved solute, if it is essentially in the form of single molecules in one solvent but in an associated form in the other. We shall discuss such cases later (p. 358).

An example of a system which conforms closely to ideal conditions is the one comprising bromine (solute) partitioned between the two immiscible solvents tribromomethane, $CHBr_3$, and water. In both solvents the essential solute species is the Br_2 molecule. Table 21.3 shows that the partition coefficient for this system changes little over the range of concentrations listed.

Table 21.3 Partition coefficient of bromine between tribromomethane and water at 298 K

$[Br_2]_{\text{eqm}}$ (in $CHBr_3$)/mol dm^{-3}	0·500	1·000	1·500	2·000	2·500
$[Br_2]_{\text{eqm}}$ (in H_2O)/mol dm^{-3}	0·0075	0·015	0·022	0·029	0·036
$K_d = \dfrac{[Br_2(\text{in } CHBr_3)]_{\text{eqm}}}{[Br_2(\text{in } H_2O)]_{\text{eqm}}}$	66·7	66·7	68·1	68·9	69·4

It is worth noting that the expression given for the solubility of a gas in a solvent (p. 345) is, in effect, a partition coefficient, i.e. a solute (the gas) is partitioned between liquid and gaseous phases.

Solvent extraction

Partial removal of solute from one solvent into another immiscible one—or solvent extraction as it is called—is a useful technique that finds applications in organic chemistry.

For example, the problem might be to purify an organic acid which is present in aqueous solution together with ionic impurities. If the organic acid happens to be more soluble in ethoxyethane (ordinary ether) than in water, then it is shaken with ether in a separating funnel. After allowing the two layers to separate the ether layer is dried with a solid drying agent; the solution is then filtered to remove the drying agent and finally distilled to remove the ether. Ionic impurities remain in the aqueous solution. Ether is particularly useful since it not only dissolves a wide range of organic compounds but is easily removed at the end owing to its low boiling-point.

Solvent extraction is much more efficient if a given quantity of solvent is used in many small portions instead of one large amount. This is shown in the example below.

Example 5

At 291 K the partition coefficient of butanoic acid, $CH_3CH_2CH_2COOH$ between ethoxy-ethane (ether) and water is 3·5. Calculate the mass of butanoic acid extracted by shaking 100 cm^3 of water containing 10 g of butanoic acid with 100 cm^3 of ether. What would be the mass of butanoic acid removed in each of two further extractions using 100 cm^3 of ether each time? Compare the total mass of butanoic acid extracted in these three successive extractions with that in a single extraction using 300 cm^3 of ether. (O & C)

$$K_d = \frac{[CH_3CH_2CH_2COOH \text{ (in ether)}]_{eqm}}{[CH_3CH_2CH_2COOH \text{ (in water)}]_{eqm}} = 3·5$$

Let x g be the mass of acid extracted by 100 cm^3 of ether. The concentrations of acid in the ether and water are respectively $x/100$ g cm^{-3} and $(10 - x)/100$ g cm^{-3}. Thus

$$3·5 = \frac{x/100}{(10 - x)/100} = \frac{x}{(10 - x)}$$

On collecting terms we get

$$4·5x = 35 \qquad x = 7·78$$

The mass of acid extracted using one portion of 100 cm^3 of ether is 7·78 g.

The fraction of acid extracted is 7·78/10 or 0·778. Since $(10 - 7·78)$ or 2·22 g of acid is left in the water, a second extraction with 100 cm^3 portion of ether will extract

$$2·22 \times 0·778 \text{ or } 1·73 \text{ g of acid}$$

leaving $(2·22 - 1·73)$ or 0·49 g of acid in the water.

A third extraction will remove

$$0·49 \times 0·778 \text{ or } 0·38 \text{ g of acid}$$

The total mass of acid extracted using three separate portions of 100 cm^3 of ether is $(7·78 + 1·73 + 0·38)$ or 9·89 g, i.e. practically all the butanoic acid is extracted.

Let y g be the mass of acid extracted by one 300 cm^3 portion of ether. The concentrations of acid in the ether and water are now $y/300$ g cm^{-3} and $(10 - y)/100$ g cm^{-3}. Thus

$$3·5 = \frac{y/300}{(10 - y)/100} = \frac{y}{3(10 - y)}$$

On collecting terms we get

$$11·5y = 105 \qquad y = 9·13$$

The mass of acid extracted using one 300 cm^3 portion of ether is 9·13 g and this compares with 9·89 g when three separate 100 cm^3 portions were used.

It is always more efficient to use several small portions rather than one large portion of solvent for extraction.

Partition involving association or dissociation of the solute in one of the immiscible solvents

We have already mentioned that one of the conditions that must be met if the Partition Law is to be obeyed is that the solute must be of the same molecular form in both immiscible solvents.

358

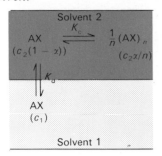

Suppose for example that a solute AX behaves normally in solvent 1 but is strongly associated into units having a relative molecular mass of n times its 'normal' value in solvent 2. The situation is depicted in fig. 21.2, where two equilibria are shown; one involves the simple partition of single molecules between the two solvents, and the other shows the association into larger molecules in the second solvent. It is important to realise that there cannot be any 'direct' equilibrium between single molecules in one solvent and associated molecules in the other, i.e. there must always be some single molecules in the second solvent.

If c_1 and c_2 are the concentrations of solute in solvent 1 and solvent 2 respectively, α is the degree of association of the solute in solvent 2, K_d is the partition coefficient, and K_c is the equilibrium constant for the association reaction in solvent 2, then we may write

$$K_d = \frac{c_2(1 - \alpha)}{c_1} \tag{3}$$

$$K_c = \frac{(c_2\alpha/n)^{1/n}}{c_2(1 - \alpha)} \tag{4}$$

From equation (4) we have

$$c_2(1 - \alpha) = \frac{(c_2\alpha/n)^{1/n}}{K_c}$$

Substituting for $c_2(1 - \alpha)$ in equation (3) we get

$$K_d = \frac{(c_2\alpha)^{1/n}}{K_c c_1 n^{1/n}}$$

Since K_d, K_c and $n^{1/n}$ are constants, we may write a new constant $K' = K_d K_c n^{1/n}$. The above expression now simplifies to

$$K' = \frac{(c_2\alpha)^{1/n}}{c_1} \tag{5}$$

If we may assume that the association of the solute in the second solvent is vitually complete, then α will be very close to unity. Under these circumstances it is clear that equation (5) can be written in the simpler form

$$K' = \frac{c_2^{1/n}}{c_1} \tag{6}$$

An alternative form of equation (6) is obtained by raising both sides to the power n:

$$K'' = K'^n = \frac{c_2}{c_1^n} \tag{7}$$

The reader may care to verify that equations (6) and (7) also apply if the solute exists as single molecules in the second solvent but dissociates into n particles in the first solvent. How then do we decide whether dissociation is taking place in solvent 1 or association is taking place in solvent 2? The answer is that chemical knowledge of the actual system under investigation generally allows the right choice to be made. For example, investigation of the partition of ethanoic acid (solute) between water (solvent 1) and benzene (solvent 2) gives results consistent with the expressions

$$K' = \frac{(\text{Conc. of ethanoic acid in benzene})^{1/2}}{\text{Conc. of ethanoic acid in water}}$$

or

$$K'' = \frac{\text{Conc. of ethanoic acid in benzene}}{(\text{Conc. of ethanoic acid in water})^2}$$

as we shall see in the example below. We have already seen in chapter 17 (p. 282) that an aqueous solution of ethanoic acid is only a very weak acid, i.e. it exists in water almost entirely in the form of single molecules. It therefore follows that ethanoic acid is associated into double molecules in the benzene (see p. 91) for the hydrogen-bonded structure of this dimer).

Example 6
Ethanoic acid associates into double molecules in benzene. Show that the following data support this view.

Conc. in water/mol dm^{-3} 0·04800   0·05500   0·07000
Conc. of benzene/mol dm^{-3} 0·00128   0·00168   0·00270

We can calculate the value of K' or K''. Let us choose the latter:

$$K'' = \frac{\text{Conc. of acid in benzene}}{(\text{Conc. of acid in water})^2}$$

Substituting the data into the above equation we get the following values for K'':

$$5\cdot56 \times 10^{-1}; \quad 5\cdot55 \times 10^{-1}, \quad 5\cdot51 \times 10^{-1}$$

The three values of K'' are reasonably constant and consistent with the association of ethanoic acid into double molecules in the benzene layer.

Suppose a system is being investigated and it is known that the solute exists as single molecules in solvent 1 and that it is associated into larger molecules in solvent 2. The value of n may be obtained by trial and error; there is a much less laborious way, however, of approaching the problem. If we take logarithms of equation (7) we get

$$\lg c_2 = n \lg c_1 + \lg K''$$

The above equation is of the form

$$y = mx + c \qquad \text{(equation of a straight line)}$$

Plotting the values of $\lg c_2$ along the y-axis and $\lg c_1$ along the x-axis will yield a straight line whose gradient is the value of n and whose intercept is $\lg K''$, from which K'' may be calculated.

In the above discussion we have assumed that the degree of association in one solvent (or dissociation in the other) is virtually complete, i.e. approaches the value of 1 quite closely. In some cases this approximation is not valid; however, in such cases it is still possible to calculate the value of n (and also α) but the details are beyond the scope of this book.

Investigation of complex ion formation using partition

A classic example of complex ion formation investigated by partition is that between iodine molecules and iodide ions in aqueous solution. The complex forming equilibrium is:

$$I_2(aq) + I^-(aq) \rightleftharpoons I_3^-(aq)$$

Assuming we know that I_3^- is the formula of the complex ion, we can determine the equilibrium constant for the above reaction (stability constant for the I_3^- complex) in the following manner.

Some iodine is partitioned, at a fixed temperature, between water and an immiscible organic solvent. Titration of known volumes of the water and organic layers with standard sodium thiosulphate solution will enable the partition coefficient of iodine between the two liquids to be determined. A second experiment is carried out in which iodine is now partitioned between an aqueous solution of potassium iodide of known concentration and the same organic solvent. Titration of known volumes of the aqueous potassium iodide solution and the organic solvent will enable the concentrations of iodine in both layers to be found. The 'total' concentration of iodine in the aqueous potassium iodide solution consists of the 'free' iodine, $I_2(aq)$, and the 'combined' iodine, $I_3^-(aq)$, and this total figure is what is obtained from the titration. The assumption is now made that the 'free' iodine partitions itself between the aqueous potassium iodide and the organic solvent in the same ratio as it does between the water and the organic solvent. It is thus possible to evaluate the concentrations of $I_2(aq)$, $I_3^-(aq)$ and $I^-(aq)$, and hence the stability constant of the $I_3^-(aq)$ complex. The example below shows how the calculation is carried out.

Example 7
The following results have been obtained for the equilibrium distribution of iodine between carbon disulphide and aqueous potassium iodide.

Initial concentration of KI = 0.200 mol dm^{-3}
Total concentration of I_2 in the aqueous layer (as I_2 and combined with I^-) = 0.020 mol dm^{-3}
Concentration of I_2 in CS_2 = 0.072 mol dm^{-3}

Given that the solubilities of I_2 in pure water and in CS_2 are 10^{-3} and 0.62 mol dm^{-3} respectively, determine the equilibrium constant for the reaction

$$I_2(aq) + I^-(aq) \rightleftharpoons I_3^-(aq) \qquad \text{(O \& C)}$$

The partition coefficient, K_d is given by

$$K_d = \frac{\text{Conc. of } I_2 \text{ in } H_2O}{\text{Conc. of } I_2 \text{ in } CS_2} = \frac{10^{-3}}{0.62} = 1.613 \times 10^{-3}$$

We can now use the partition coefficient to determine the concentration of the 'free' iodine in the aq KI as follows:

$$1.613 \times 10^{-3} = \frac{\text{Conc. of 'free' iodine in aq. KI}}{0.072}$$

hence

$$\text{Conc. of 'free' } I_2 \text{ in aq. KI} = 1.613 \times 10^{-3} \times 0.072$$
$$= 1.161 \times 10^{-4} \text{ mol } dm^{-3}$$

The concentration of 'combined' I_2 in the aq KI is equal to the 'total' I_2 minus the 'free' I_2, thus

$$\text{Conc. of 'combined' } I_2 \text{ in aq KI} = (0.020 - 1.161 \times 10^{-4})$$
$$= 1.988 \times 10^{-2} \text{ mol } dm^{-3}$$

The final concentration of $I^-(aq)$ is equal to the initial concentration of $I^-(aq)$ minus the concentration used in forming the I_3^-(complex), thus

$$\text{Final conc. of } I^-(aq) = (0.200 - 1.988 \times 10^{-2})$$
$$= 1.801 \times 10^{-1} \text{ mol } dm^{-3}$$

The equilibrium constant for the reaction is given by the expression

$$K_c = \frac{[I_3^-(aq)]_{eqm}}{[I_2(aq)]_{eqm}[I^-(aq)]_{eqm}}$$
$$= \frac{1.988 \times 10^{-2}}{(1.161 \times 10^{-4})(1.801 \times 10^{-1})}$$
$$= 9.51 \times 10^2 \text{ mol}^{-1} \text{ dm}^3$$

Another example of complex formation which has been investigated in a similar manner is that between aqueous ammonia and an aqueous solution of a copper(II) salt. Briefly, the method involves two separate experiments in which ammonia is partitioned between water and an organic solvent, and then between an aqueous solution of copper(II) sulphate and the same organic solvent. Having determined the partition coefficient of ammonia between water and the immiscible organic solvent, it is possible to determine the concentration of 'combined' ammonia in the copper(II) solution. Results consistent with the formula $Cu(NH_3)_4^{2+}(aq)$ are readily obtained.

21.8 Chromatographic methods

These methods, which are based on partition, are particularly valuable for separating complex mixtures. Four main methods are in general use:
(a) Column chromatography
(b) Paper chromatography
(c) Thin-layer chromatography (tlc)
(d) Gas-liquid chromatography (glc)
A fifth technique known as ion-exchange has much in common with chromatography and is also discussed.

Theoretical studies directed at understanding the mechanism of each type of chromatographic technique have been carried out. Details vary according to the type of stationary phase involved but, at a simple level, the underlying principle is the partition of the components in a mixture between the **stationary** and **mobile** phases.

The stationary phase may be an inert finely divided solid such as alumina (column chromatography), a layer of water adsorbed on the cellulose fibres of filter paper (paper chromatography), a thin layer of silica gel or alumina deposited on a glass plate (thin-layer chromatography), or a long-chain alkane coated onto an inert powered solid (gas-liquid chromatography). The mobile phase is a liquid (often a mixture of liquids), except in gas-liquid chromatography where a variety of gases have been employed, e.g. argon, hydrogen, nitrogen etc. The actual gas used is largely dictated by the type of instrumentation used to detect the separated components of a particular mixture.

Column chromatography

A typical piece of apparatus consists of a glass tube about 25 cm long and 2 cm in diameter, tapered at one end where it is joined to a glass tap and outlet tube (see fig. 21.3). A loose plug of glass wool is placed at the bottom of the tube, which is then partially filled with a liquid such as benzene. The stationary phase, e.g. alumina, is now carefully added until a column has been built up almost to the level of the liquid in the tube. A Buchner flask is fitted to the bottom of the column.

A solution of the mixture to be separated is carefully applied to the top of the alumina packing, the tap turned on, and a developing solvent added continuously. In time, the mixture will begin to separate and the process can be accelerated by connecting the Buchner flask to a water-pump.

If the components in the mixture are coloured there is no difficulty in keeping track of their movement down the column. In the case of colourless components which possess the property of fluorescence, their movement can be seen after illumination with ultra-violet light. Eventually one of the components will begin to emerge from the column and may be collected—a process known as **elution**. The process of elution is continued until all the components of the mixture have been collected in separate flasks. Since the rate at which a particular component travels down the column depends upon its solubility in the developing solvent and the degree to which it is adsorbed by the stationary phase, it follows that the first component to leave the column has the largest ratio

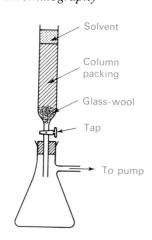

FIG. 21.3. *Column chromatography*

Solvent

Column packing

Glass-wool

Tap

To pump

$$\frac{\text{solubility in developing solvent}}{\text{affinity for the stationary phase}}$$

It may happen that some components stick tenaciously to the column and thus move very slowly, if at all. In such cases it is necessary to change to another solvent which allows a faster rate of travel.

A typical example of the use of column chromatography is the separation of the soluble pigments of spinach into the carotenes (orange) and the chlorophylls (green, and separable into their a and b components). This is achieved by placing a solution of the pigments in light petroleum onto the top of the column, and eluting the carotenes with benzene. The chlorophylls, which stick to the top of the column, can now be moved down the column and separated into their a and b components, by switching to a solvent consisting of butan-1-ol, ethanol and water.

Paper chromatography

This process is used for the identification of the components in a very small sample of a mixture, e.g. the amino acids produced as a result of hydrolysing protein material.

A thin pencil line is drawn near one edge of a rectangular piece of filter paper and parallel to it. The solid mixture, present in solution, is now applied as a small drop from a fine glass syringe along the pencil line, and near one vertical edge of the paper. If the problem is to identify the amino acids in a mixture, then small drops of separate solutions of pure amino acids, thought to be present in the mixture, are placed at intervals along the pencil line.

The filter paper is now hung vertically in a glass tank which contains the developing solvent, and the bottom edge of the filter paper is so positioned in the solvent that the pencil line is just clear of it. The tank is sealed with a lid to prevent the evaporation of solvent (see fig. 21.4).

FIG. 21.4. *Apparatus for paper chromatography*

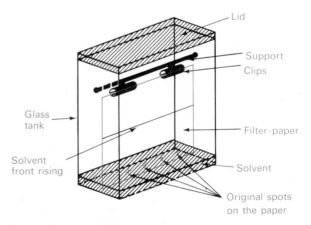

As the solvent travels up the paper it moves the components in the mixture, and the pure substances, at different rates. The paper is removed from the tank, and allowed to dry, once the solvent has almost reached the top of the paper. Since amino acids are colourless, their positions on the paper are located by spraying the whole paper with a solution of ninhydrin in propanone. After warming the paper in an oven set at a temperature of about 373 K the individual amino acids show up as blue-lilac spots.

363

Under carefully controlled conditions it is possible to characterise a particular compound separated from a mixture by its so-called R_F value. This is defined to be

$$\frac{\text{the distance moved by the pure substance}}{\text{the distance moved by the solvent front}}$$

Its value for a particular compound depends upon the solvent used and the temperature. From its definition above, it is clear that all R_F values are less than unity.

In some cases the use of one solvent may fail to separate two or more of the constituents present in the mixture, i.e. their R_F values under these conditions are almost identical. The answer to this problem is simple in theory. The paper is turned through 90° and the whole process repeated with a different solvent, the logic being that it is most unlikely that two or more components will have identical R_F values in a different solvent; this technique is called two-way chromatography. Figure 21.5 shows schematic representations of one- and two-way chromatography.

FIG. 21.5. *(a) Schematic representation of one-way chromatography (b) The R_F value of a pure compound (c) The principles of two-way chromatography*

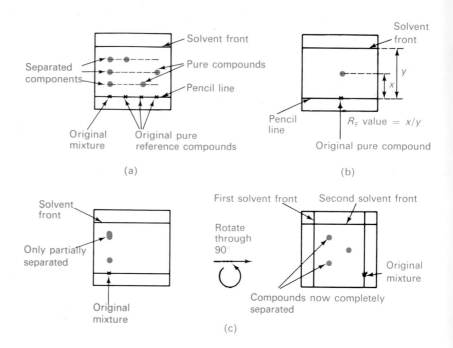

Thin-layer chromatography

Basically this process is similar to column chromatography but, like paper-chromatography, it is mainly used as a means of identifying the components in a very small sample of a solid mixture. Its advantage over paper chromatography stems from the fact that a variety of thin films, e.g. alumina, silica gel etc., can be used depending upon the nature of the substances to be separated. In addition, the thin-film is more compact than filter paper and this means that the sample spots are smaller and more concentrated. Useful separations can be achieved on thin-films deposited on microscope slides, so the process is a rapid one.

Brief experimental details for carrying out thin-layer chromatography are as follows. A slurry of silica gel, for example, is made in trichloromethane contained in an ordinary screw-cap bottle. A previously

364

cleaned microscope slide is immersed in the slurry and then carefully withdrawn using a pair of tweezers. The coated glass plate is now allowed to dry in an upright position and is then ready for use.

A typical experiment could be the separation of the nitration products of phenol (2-nitrophenol and 4-nitrophenol). Separate solutions of the mixed and the pure nitrophenols are made in ethanol. After marking the origin with two scratch marks, small spots of the three solutions (no wider than about 3–4 mm) are placed in a horizontal line by means of fine capillary tubes. The coated plate is now stood in a shallow depth of trichloromethane contained in a screw-cap bottle. When the solvent has risen about three-quarters of the way up the film, the plate is removed from the bottle and allowed to dry. The plate may now be developed by standing it in a beaker (covered with a watch-glass) containing a few crystals of iodine. In time, a dark spot appears where each compound is located on the plate (the position of 2-nitrophenol should be visible as a yellow spot before this final treatment).

Gas-liquid chromatography

This process is employed to separate mixtures of gases, liquids and volatile solids. A typical gas chromatograph is shown schematically in fig. 21.6.

FIG. 21.6. *A gas chromatograph*

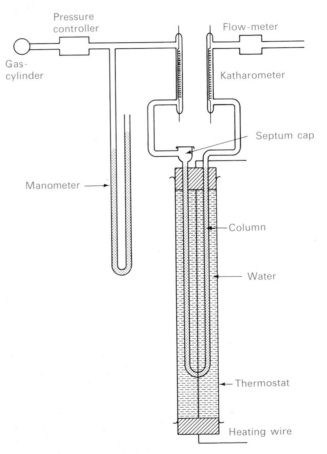

The stationary phase, which may be a non-volatile long-chain alkane deposited on an inert material, is packed into a long narrow column (typically 20 m long and 4 mm wide) bent into the shape of a U-tube. The mobile phase is a gas which will not interact with any of the components to be separated. It is referred to as the **carrier gas** and is frequently nitrogen. The progress of the carrier gas through the column is controlled by a valve (pressure controller) and monitored by a flow-meter.

The mixture to be separated is injected directly onto the top of the column through a self-sealing septum cap by means of a hypodermic syringe. As in the previously described chromatographic methods, partition of the components in the mixture occur to different extents between the stationary and mobile phases; they thus emerge from the column at different time intervals. In order to be able to cope with mixtures of differing degrees of volatility, provision is made for the column to be heated and its temperature thermostatically controlled.

One of the most common means of detecting the emerging components is by means of their thermal conductivities. The device used is called a **katharometer**. It consists of a metal block drilled to take two identical lengths of platinum resistance wire. The wires, which are heated electrically and eventually reach a steady temperature, are part of a Wheatstone bridge circuit which is balanced when carrier gas passes over both wires. Carrier gas alone passes over one wire whereas carrier gas together with emerging components pass over the other one. Since the thermal conductivities of the pure carrier gas and carrier gas/component mixture will be different, the temperature will change—and the bridge go out of balance—whenever a component passes out of the column. This out-of-balance signal is amplified and eventually recorded on a roll of paper.

FIG. 21.7. *A gas chromatogram of a mixture containing six components. The retention time for hex-1-ene is (B − A) min., and for heptane (C − A) min.*

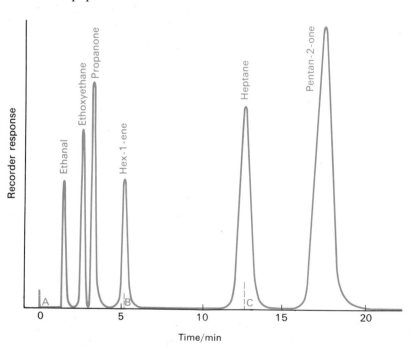

The time taken for a particular component to emerge from the column is called its **retention time**. Under carefully controlled conditions, i.e. for a particular carrier gas and flow rate, particular stationary phase and column temperature, the retention time may be used to identify a particular component. Care is needed, however, since two compounds may have the same retention time, and it is advisable to determine the retention times for two or more different types of stationary phase. A typical gas chromatogram is shown in fig. 21.7.

The area under a component peak is related to the quantity of that component. Thus if a known amount of the pure component is injected onto the column, in a separate experiment, it is possible to determine the quantity present in the mixture.

By channelling a portion of the components emerging from the column into a mass spectrometer it is possible to analyse very complex mixtures to a high degree of accuracy (p. 200).

Ion-exchange

The separation of ionic mixtures in solution can be achieved by using so-called ion-exchangers. These are complex polymeric materials which may be inorganic or organic in origin, although the latter are more common. The organic materials are manufactured in the form of small insoluble beads, and consist of a polymer chain composed of carbon atoms to which a number of acidic groups, e.g. $-SO_2OH$ and $-COOH$, or anionic groups, e.g. $-NH_3^+Cl^-$, are attached. The former types of material are called cationic-exchangers, e.g. $R-SO_2O^-H^+$ where R represents the polymer backbone, and the latter anionic exchangers, e.g. $R-NH_3^+Cl^-$.

Consider cationic-exchange: The cationic-exchanger is mixed with pure water and packed into a glass tube similar to the one used in column chromatography (fig. 21.3 p. 362). If a small amount of, say, copper(II) sulphate solution is placed on top of the column and water gradually added (as eluting agent), then a dilute solution of sulphuric acid will eventually emerge. The $Cu^{2+}(aq)$ ions replace an equivalent number of $H^+(aq)$ from the exchanger thus:

$$2R-SO_2O^-H^+ + Cu^{2+}(aq) \rightleftharpoons (R-SO_2O^-)_2Cu^{2+} + 2H^+(aq)$$

An immediate application which springs to mind is the estimation of the concentration of a solution of a salt, by titration of the eluted acid with standard alkali.

As the above equation shows, ion-exchange is a reversible process and this means that the original material can be recovered by washing it with dilute sulphuric acid (the $Cu^{2+}(aq)$ ions will now be eluted).

The versatility of the method is shown by the separation of the lanthanides in the +3 oxidation state. All the lanthanides form +3 ions, M^{3+}, whose ionic radii decrease progressively with increasing atomic number from lanthanum (La^{3+}, 0·125 nm) to lutetium (Lu^{3+}, 0·099 nm); this is the so-called lanthanide contraction. As the ionic radii decrease along this series of elements, the ability to form complex ions increases and this is the basis of their separation on an ion-exchange column.

A solution containing +3 lanthanide ions is placed at the top of a cation-exchange column and the lanthanide ions release an equivalent amount of hydrogen ions from the column. A solution of ammonium 1-hydroxy-1-methyl propanoate, $(CH_3)_2C(OH)COONH_4$, is slowly run down the column and the cations partition themselves between the ion-exchanger and the eluting solution (with which they form complexes). Since the smaller ions show a greater preference for the complexing solution, these ions are the first to emerge from the column. Fig. 21.8 shows some typical results, the order of elution being lutetium, ytterbium, thulium, erbium etc. A similar process has been used to separate and characterise the transuranium elements, their emergence from the column being monitored by a Geiger-Müller tube.

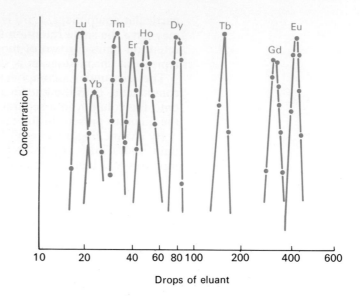

FIG. 21.8. *The separation of the lanthanides using a cation-exchanger*

A combination of anion- and cation-exchanger are used for the deionisation of water. An inorganic material (sodium aluminium silicate) manufactured under the trade name of Permutit is used in the softening of water ($Ca^{2+}(aq)$ and $Mg^{2+}(aq)$ ions are replaced by $Na^+(aq)$). For details of both these processes an inorganic textbook should be consulted.

21.9 Questions on chapter 21

1 Explain the term *solubility product* and state why the concept does not apply to aqueous solutions of sodium chloride and benzoic acid.

Derive an expression for the solubility product of iron(II) hydroxide, stating the units.

If the solubility product of iron(II) hydroxide has a numerical value of 10^{-15}, calculate the solubility in mol dm^{-3} of iron(II) hydroxide in
   (a) water,
   (b) 1 M sodium hydroxide,
   (c) 1 M ammonia.
(Dissociation constant K_b for ammonia = $1\cdot8 \times 10^{-5}$ mol dm^{-3}).   AEB

2 (a) (i) Write an equation for the equilibrium which may be presumed to exist between solid silver chloride and dissolved silver chloride in a saturated solution.
    (ii) Write an expression for the solubility product of silver chloride and state the conditions under which this applies.
    (iii) The solubility of silver chloride in water at 25°C is 1×10^{-5} mol dm^{-3}. Calculate the solubility product of silver chloride and state the units in which it is expressed.
  (b) The solubility product of lead(II) bromide at 25°C is $3\cdot9 \times 10^{-5}$ mol^3 dm^{-9}. Calculate the solubility in g dm^{-3} at 25°C of lead(II) bromide.
  (c) **Outline** a practicable method for determining the solubility of a named sparingly soluble ionic compound in water at 25°C.   AEB

3 Explain what is meant by *solubility product*.

The solubility of anhydrous calcium iodate(v) in water at 298 K is $3\cdot07$ g dm^{-3}. Calculate (a) its solubility product, (b) its solubility, (in g dm^{-3}), in an aqueous solution containing $0\cdot1$ mol dm^{-3} of sodium iodate(v).

Explain, with full experimental details, how you would determine the solubility of calcium iodate(v) in water at 298 K, emphasising any precautions you would have to take.   W

368

4 What do you understand by the term *solubility product*?
Write an expression for the solubility product of (a) lead(II) sulphide, (b) lead(II) chloride.
Explain **each** of the following observations as fully as you can.
  (i) When hydrogen sulphide is passed into an acidified aqueous solution of lead(II) nitrate and zinc nitrate, only lead(II) sulphide is precipitated.
  (ii) The solubility of lead(II) chloride in water decreases on addition of *dilute* hydrochloric acid but increases on addition of *concentrated* hydrochloric acid.
  (iii) When aqueous sodium hydroxide is gradually added to aqueous lead(II) nitrate, a white precipitate is formed initially but this dissolves in an excess of aqueous sodium hydroxide to give a colourless solution.
  (iv) Lead(IV) oxide dissolves in cold, concentrated hydrochloric acid to give a deep yellow solution from which a yellow solid separates on addition of a saturated aqueous solution of ammonium chloride.   C

5 Explain fully what is meant by the term *solubility product*, indicating the limitations on its use, and derive an expression for the solubility product of the salt A_3B_2 given that its solubility in water is s mol dm^{-3}.
  (a) Calculate the solubility of silver iodide (g dm^{-3}) in water given that the numerical value of its solubility product is 1.0×10^{-16}.
$$(AgI = 235)$$
  (b) Calculate the solubility product of lead iodide, stating the units in which it is expressed, given that the solubility in water is 6.04×10^{-1} g dm^{-3}.
$$(PbI_2 = 461)$$
  (c) If solid potassium iodide (which is freely soluble in water) is added in minute quantities at a time to a solution which is 0.02M in both lead(II) nitrate, $Pb(NO_3)_2$ and silver nitrate, $AgNO_3$,
    (i) which of the cations is precipitated first?
    (ii) what will be its concentration (mol dm^{-3}) when the second cation begins to be precipitated?   S

6 At a given temperature, the solubility of calcium phosphate, $Ca_3(PO_4)_2$, is x mol dm^{-3} and its solubility product has a value of y.
  (a) Derive a relationship between x and y and give the units of y.
  (b) A saturated solution of calcium phosphate was found to contain 10^{-5} mol dm^{-3} of PO_4^{3-}. Calculate the concentration of Ca^{2+} in the solution and hence determine the value of y.
  (c) Calculate the solubility in mol dm^{-3} of calcium phosphate in a 1.0 M solution of sodium phosphate, Na_3PO_4.
  (d) Explain why calcium phosphate is more soluble in 1.0 M hydrochloric acid than in water.
  (e) The solubility product of mercury(II) sulphide is 10^{-53} mol^2 dm^{-6}. Calculate the number of mercury(II) ions and of sulphide ions in one litre of a saturated solution of mercury(II) sulphide and comment on the values. Assume that the Avogadro constant is 10^{23}.
  (f) Comment on the problems involved in determining a value for the solubility product of a very insoluble salt such as mercury(II) sulphide.
                    L (S)

7 Describe, giving important experimental details, how you would determine the solubility in water of **one, named**, sparingly soluble compound.
The solubility product of silver chloride is 1.7×10^{-10} mol^2 dm^{-6}. Calculate its solubility, (in mol dm^{-3}), in ammonia solution of concentration 1 mol dm^{-3}, given that the equilibrium constant for the equilibrium
$$Ag^+ + 2NH_3 \rightleftharpoons Ag(NH_3)_2^+, \text{ is } 1.6 \times 10^7 \text{ mol}^{-2} \text{ dm}^6$$
State what approximations you make to facilitate your calculation.   W (S)

8 Define 'solubility product' and give examples of the applications of solubility products in analytical chemistry.
The solubility products of $BaSO_4$ and $SrSO_4$ are 1.1×10^{-10} and 2.8×10^{-7} mol^2 dm^{-6} respectively, at 298 K. A solution is 0.01 mol dm^{-3} in $BaCl_2$ and 0.01 mol dm^{-3} in $SrCl_2$. To this is gradually added concentrated sodium sulphate solution. Which cation precipitates first? Neglecting the effects

of dilution, calculate the concentration of this ion when the second cation starts to precipitate. How useful would the method be for separating Ba^{2+} and Sr^{2+}?

Oxford Schol. and Entrance

9 Discuss the factors which determine whether or not a solid will dissolve in a liquid.

Comment on the following: (a) Sucrose, $C_{12}H_{22}O_{11}$, will dissolve in water; the solution is not an electrolyte. (b) Barium sulphate, an ionic compound, is hardly soluble in water. (c) Hydrogen chloride is soluble in toluene, and also gives an aqueous solution which is a good electrolyte. (d) High polymers, though covalently bonded, are often not soluble in organic liquids. L (S)

10 (a) State the *partition law* for the distribution of a solute between two immiscible solvents. Under what conditions does the law apply?

(b) The following results were obtained in an experiment to determine the partition coefficient of ammonia between water and trichloromethane ($CHCl_3$):

10.0 cm^3 of the aqueous layer required 43.2 cm^3 of 0.250 mol dm^{-3} hydrochloric acid for neutralisation;

25.0 cm^3 of the trichloromethane layer required 21.6 cm^3 of 0.050 mol dm^{-3} hydrochloric acid for neutralisation.

Calculate the partition coefficient at the laboratory temperature.

(c) Why is ammonia much more soluble in water than in trichloromethane?

(d) When a 0.0250 mol dm^{-3} aqueous solution of copper(II) sulphate was allowed to reach equilibrium with an excess of ammonia and trichloromethane, the aqueous layer was found to contain 0.400 mol dm^{-3} of ammonia and the trichloromethane layer 0.0120 mol dm^{-3} of ammonia. Using your answer from (b), calculate

(i) the concentration of *free* ammonia in the aqueous layer,

(ii) the value of n in the formula of the complex ion $[Cu(NH_3)_n]^{2+}$, drawing attention to any assumptions which you make in your calculation. C

11 (a) State the law defining the equilibrium distribution of a solute between two immiscible solvents.

(b) (i) **Outline** the method you would use to determine the distribution coefficient of iodine between water and tetrachloromethane (carbon tetrachloride) at room temperature.

(ii) State how and why the distribution of iodine would be affected by the addition of potassium iodide to the aqueous layer in (i).

(c) Industrially, silver is extracted from molten lead using molten zinc which is insoluble in lead. The solubility of silver is 300 times greater in zinc than it is in an equal volume of lead. If 0.005 litre of molten zinc is added to a liquid solution of 2 g of silver in 0.1 litre of lead, calculate the percentage of silver extracted by the zinc when the system has attained equilibrium. AEB

12 (a) Define the *partition* (or *distribution*) *coefficient* of a solute between two immiscible solvents. Describe briefly how you would determine the partition coefficient for any system you care to select.

(b) The partition coefficient of a substance X between benzene and water is 8.0, X being more soluble in benzene than in water. Starting with 100 cm^3 of an aqueous solution containing 6.00 g of X, compare the amounts of X extracted (i) by shaking with one portion of 100 cm^3 of benzene, (ii) by shaking with two successive portions of 50 cm^3.

(c) Explain briefly the principles involved in any one chromatographic technique for the separation of the components of a mixture. O

13 When molecular iodine is dissolved in an aqueous solution of potassium iodide, the following equilibrium is established:

$$I_3^- \rightleftharpoons I_2 + I^-$$

[The equilibrium constant is 0.0015 mol dm^{-3} at 298 K.]

0.02 mole of iodine is dissolved in 500 cm^3 of a solution containing 0.2 mol dm^{-3} of potassium iodide at 298 K. Calculate the concentration of each ion present at equilibrium.

The above equilibrium may be investigated by partition between two solvents, a method which can sometimes be used to determine the molecularity of a compound in solution. Benzene carboxylic acid (benzoic acid) is more soluble in benzene than in water. In benzene it exists as double molecules. Explain how you could confirm this fact experimentally using a partition method, giving important experimental details and the relevant theoretical background. W (S)

14 What is the partition coefficient of a solute between two immiscible liquids? Give **two** examples (other than the one below) of cases where the partition law does not appear to be obeyed. Describe **one** practical use of partition other than chromatography.

Some iodine is dissolved in an aqueous solution of potassium iodide of concentration 0.102 mol dm^{-3}, and the solution is then shaken with tetrachloromethane (carbon tetrachloride) until equilibrium is attained (at 15°C). The amount of iodine at equilibrium, as determined by titration, is found to be 0.048 mol dm^{-3} in the aqueous layer and 0.089 mol dm^{-3} in the tetrachloromethane layer. The distribution coefficient of iodine between tetrachloromethane and water is 85. Calculate the equilibrium constant at 15°C for the reaction

$$I_3^-(aq) \rightleftharpoons I_2(aq) + I^-(aq). \qquad \text{O and C}$$

15 When a substance is added to a two phase liquid system it is generally distributed with different equilibrium concentrations in the two phases. Why is this so?

A certain amount of iodine was shaken with CS_2 and an aqueous solution containing 0.3 mol dm^{-3} of potassium iodide. By titrating with thiosulphate solution, it was found that the CS_2 phase contained 32.3 g dm^{-3} and the aqueous phase 1.14 g dm^{-3} of iodine. The distribution coefficient for iodine distributed between CS_2 and water is 585. Calculate the equilibrium constant for the reaction

$$I_2 + I^- \rightleftharpoons I_3^- \text{ at the prevailing temperature.} \quad \text{Camb. Entrance}$$

16 Discuss briefly the principles of chromatography.

Give brief details of the experimental methods used for column chromatography, thin layer chromatography and paper chromatography, giving a specific example of the use of each specialised technique.

Suggest analytical techniques (not necessarily chromatography) which would be most appropriate for quantitative analysis of the major components of **two** of the following samples:

(a) a mixture of nitrogen(II) oxide and ethane,
(b) a sample of North Sea oil,
(c) a mineral sample *brought back* to Earth from the Moon,
(d) a mineral sample *on* the planet Mars. O and C

Chapter 22

Phase equilibria and solutions—part 2

22.1
Introduction

In this chapter we shall be concerned mainly, although not exclusively, with the way in which the vapour pressure of a two-component liquid mixture is influenced by its composition and by the interaction between the two components present. The effect of vapour pressures on other relevant properties of such systems will be examined.

We begin by discussing two-component liquid mixtures in which both constituents are volatile. The problem will first of all be simplified by defining an 'ideal' solution as a model. The vapour pressure of a 'real' solution will be compared with that of the model system, and the type of behaviour exhibited will be seen to have a bearing on the technique·of fractional distillation—a process of immense industrial importance.

Two-component liquid mixtures, in which one of the components is non-volatile, will then be discussed, and the connection between the vapour pressure of this type of system and the so-called colligative properties* will be developed.

Finally some two-component phase equilibria will be described. These will comprise metal/metal systems, which are the province of the metallurgist, and salt/water systems. The latter are included to illustrate the extension of solubility curves (p. 348) at temperatures below the freezing-point of water.

22.2
Two-component liquid mixtures in which both components are volatile

The vapour pressure of an ideal solution

An ideal solution consisting of two components A and B is one in which the intermolecular attractions A--A, B--B and A--B are all equal. There is neither a volume change on mixing ($\Delta V_m = 0$), nor an enthalpy change on mixing ($\Delta H_m = 0$). We shall see later how real systems compare with such an ideal state.

From what we have said above, it follows that the tendency of molecules of A to escape from a solution of A and B and enter the vapour phase is the same as in the pure liquid A. The same is clearly true for the second component B. In a solution of A and B there are relatively fewer molecules of A than in the pure liquid A, consequently a simple relationship between the partial vapour pressure of A and its mole fraction in the mixture is expected. A similar relationship is clearly expected for the second component B. Such as relationship was first discovered by Raoult and is expressed in terms of a law (**Raoult's law**):

The partial vapour pressure of any volatile component of an ideal solution is equal to the vapour pressure of the pure component multiplied by the mole fraction of that component in the solution. This law can be expressed mathematically as follows:

Suppose p_A°, p_A and X_A are respectively the vapour pressure of pure

---

* Colligative properties are those which depend upon numbers of particles and not on their masses.

A, the partial vapour pressure of A in the solution, and the mole fraction of A in the solution, then

$$p_A = p_A^\circ X_A \tag{1}$$

Similarly for liquid B we have:

$$p_B = p_B^\circ X_B \tag{2}$$

The total vapour pressure of the solution (at a fixed temperature), p_T, will be the sum of the partial vapour pressures of A and B

$$p_T = p_A + p_B$$
$$p_T = p_A^\circ X_A + p_B^\circ X_B \tag{3}$$

(From the definition of mole fraction (p. 343) it follows that $X_A + X_B = 1$)

Fig. 22.1 shows the linear variation of total vapour pressure with mole fraction of the solution, the two dotted lines corresponding to the partial vapour pressures of the two components. We have assumed that component A is more volatile than component B, i.e. has the higher vapour pressure.

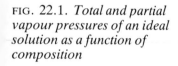

FIG. 22.1. *Total and partial vapour pressures of an ideal solution as a function of composition*

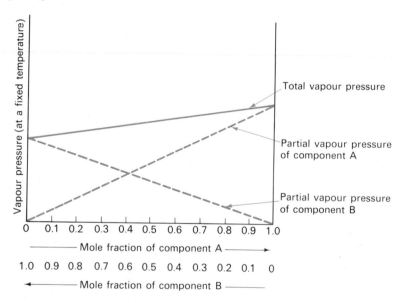

We must now enquire how the partial and total vapour pressures are related to the composition of the vapour. In fig. 22.1 the vapour pressure of pure A is greater than that of pure B. The vapour in contact with a solution of known composition would therefore be expected to be somewhat enriched in the more volatile component A. In other words, the mole fraction of A in the vapour would be expected to be higher than the mole fraction of A in the liquid in equilibrium with this vapour. We may examine the variation of vapour pressure with vapour composition in the following manner.

1 atm $\simeq 10^5$ Pa

Suppose that at a fixed temperature pure component A has a vapour pressure of 1×10^4 Pa while pure component B has a vapour pressure of 3×10^3 Pa. Consider a solution in which the mole fractions of A and

373

B are equal, i.e. $X_A = X_B = 0.5$. Then assuming that the vapour obeys Dalton's law of partial pressures (p. 143), we can calculate the mole fractions of the components in the vapour quite simply as follows:

Partial pressure of A is given by $p_A = p_A^o X_A$
$$= (1 \times 10^4)0.5$$
$$= 5 \times 10^3 \text{ Pa}$$

Partial pressure of B is given by $p_B = p_B^o X_B$
$$= (3 \times 10^3)0.5$$
$$= 1.5 \times 10^3 \text{ Pa}$$

Mole fraction of A in the vapour $= X'_A = \dfrac{p_A}{p_A + p_B}$

$$= \dfrac{5 \times 10^3}{6.5 \times 10^3} = 0.77$$

Mole fraction of B in the vapour $= X'_B = 1.00 - 0.77 = 0.23$

In the above example the enrichment of the vapour in component A is considerable ($X_A = 0.5$ while $X'_A = 0.77$). The mole fractions of the vapour, corresponding to other values of mole fractions of the liquid, may be calculated in a similar manner. The results are plotted in fig. 22.2, which also shows the vapour pressure/liquid composition curve. Inspection of fig. 22.2 shows that the composition of the vapour, in equilibrium with a liquid of known composition, can be readily obtained. Thus the horizontal line YZ shows that liquid (mole fraction $X_A = 0.5$) is in equilibrium with vapour (mole fraction $X'_A = 0.77$).

FIG. 22.2. *Vapour pressures of an ideal solution showing the compositions of (a) the liquid and (b) the vapour*

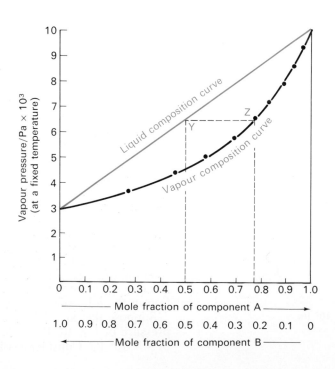

374

Boiling-point—Composition diagrams for ideal solutions

The fact that the vapour of an ideal solution is richer in the more volatile component, than the liquid with which it is in equilibrium, is the key to the process called fractional distillation. We now need to know how the boiling-points of a range of solutions, prepared from the same two volatile components, are related to the compositions of the liquid and vapour phases. Without going into details, this can readily be deduced if the variation of the vapour pressures of the two pure components with temperature is known.

The boiling-point—composition diagram for the system consisting of 2-methylpropan-1-ol, $CH_3CH(CH_3)CH_2OH$ (b.p. = 109°C) and propan-1-ol, $CH_3CH_2CH_2OH$ (b.p. = 82°C), which conforms very closely to ideal behaviour, is shown in fig. 22.3. The diagram is practically the mirror image of fig. 22.2 as expected, since propan-1-ol (component A) has the higher vapour pressure, at a fixed temperature, and thus the lower boiling-point at a fixed pressure (atmospheric).

FIG. 22.3. *Boiling-point-composition diagram for 2-methylpropan-1-ol and propan-1-ol at atmospheric pressure*

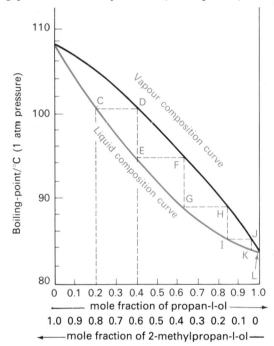

Consider a mixture of these two alcohols in which the mole fraction of propan-1-ol is 0·2 (see fig. 22.3). This mixture will boil at a temperature of about 100°C (point *C* on the diagram). This liquid will be in equilibrium with vapour which is richer in propan-1-ol (point *D*, which indicates a mole fraction of approximately 0·4). If this vapour were removed and the temperature reduced, then it would produce liquid of the same composition at point *E*. This liquid would now be in equilibrium with vapour still richer in propan-1-ol (point *F*). Removal of this vapour, followed by condensation would produce liquid (point *G*) of the same composition and, again, this liquid would be in equilibrium with vapour still richer in the more volatile of the two alcohols (point *H*). Repetition of this quite tedious process would eventually result in the separation of almost pure propan-1-ol from the liquid mixture.

Fortunately it is possible to carry out all these stages in a fractionating column as outlined in the next section.

Fractional distillation

The several stages described in the previous section for the separation of an almost ideal solution into its pure components can be carried out in one continuous operation. The process is called **fractional distillation** and it can be employed to separate liquid mixtures into their pure constituents, provided they have boiling-point—composition characteristics similar to the one shown in fig. 22.3. In practice, appreciable deviations from ideal behaviour can generally be tolerated (but see later p. 379).

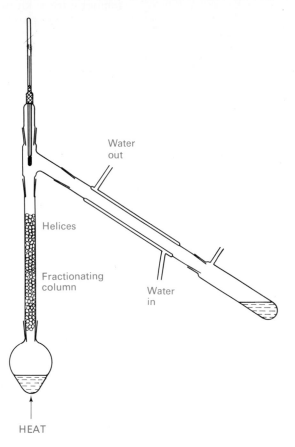

FIG. 22.4. *An apparatus for fractional distillation*

A typical laboratory apparatus for carrying out fractional distillation is shown in fig. 22.4. The key component of this apparatus is the fractionating column—a glass column packed with glass or stainless steel coils or beads. The column packing provides a large surface area so that good contact between descending liquid and ascending vapour is achieved. The mode of operation of a fractionating column can be understood by reference to fig. 22.5.

The boiling-point—composition diagram for the system 2-methylpropan-l-ol/propan-l-ol has been redrawn in fig. 22.5 with temperature decreasing along the y-axis. Note that this means that the upper curve now shows the liquid composition and the lower curve the vapour composition. If the liquid mixture is heated it will commence to boil at a temperature of 101°C (point C) producing vapour of composition represented by point D. This vapour condenses a little way up the column, producing liquid of the same composition (point E). This liquid

376

is in equilibrium with vapour of composition given by point *F*. If the sequence of upward steps is traced on the diagram, it will be seen that virtually pure propan-1-ol may be drawn off at the top of the column. Further consideration of fig. 22.5 shows that there are a number of liquid/vapour equilibria established at successively lower temperatures up the column, i.e. the so-called **tie-lines** *CD*, *EF* etc. In order to establish these equilibria, which are essential for a clean separation of the components in the mixture, it is necessary to have maximum contact between ascending vapour and descending liquid. This is the purpose of the packed column.

FIG. 22.5. *The mode of operation of a fractionating column*

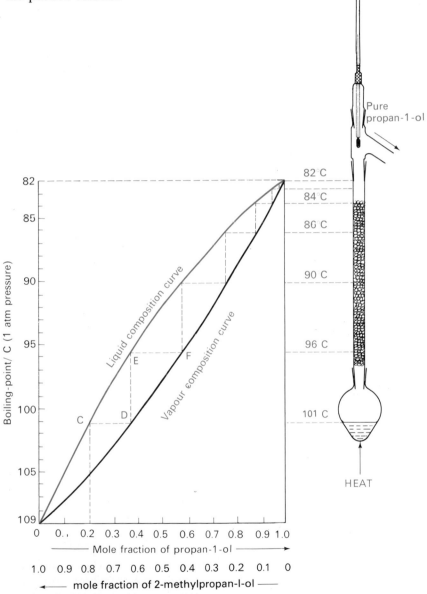

In practice the various equilibria established in the column are constantly changing during distillation. Removal of the more volatile constituent from the mixture means that the boiling-point of the mixture in the flask will steadily rise. Conditions in the column are therefore in a state of flux and slow distillation is essential if the system is to have time to continually re-adjust and so enable a good separation to be achieved.

One final point—the length of the fractionating column is largely dictated by the difference in boiling-points of the components to be separated, the smaller this difference the longer the column needed to achieve a good separation.

Fractionating columns used industrially may be anything up to 100 m in length. Essentially they are steel towers containing a series of trays, pipes and bubble caps. At each stage in the tower, ascending vapour is forced to pass through liquid contained in trays, provision being made for the overflow of liquid in one tray to another at a lower level. They are used, for example, in the distillation of liquid air and in the distillation of crude oil. In the latter process it would be an impossible task to separate the crude oil into the multitude of individual hydrocarbons it contains. In practice, a number of outlet pipes along the length of the fractionating column enable six principal fractions to be drawn off as the first stage of the refining process (consult an organic textbook for details). Fig. 22.6 illustrates the hardware used.

FIG. 22.6. *Distillation of crude oil*

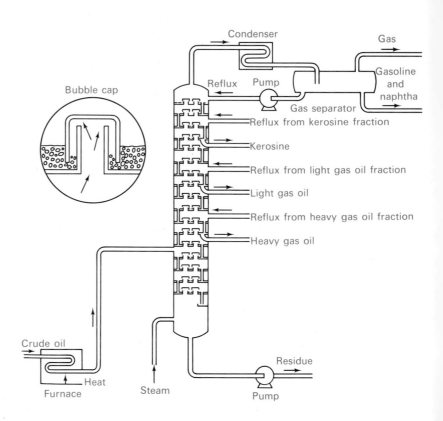

The vapour pressure of non-ideal solutions

Binary liquid mixtures such as 2-methylpropan-l-ol/propan-l-ol and benzene/methylbenzene obey Raoult's law quite closely over the complete concentration range. They are practically ideal solutions. Other pairs deviate to a greater or lesser extent and are said to be non-ideal solutions. These deviations may be of two distinct types, and are manifested in the total vapour pressure being either lower or higher than expected from Raoult's law. They are referred to as negative and positive deviations respectively.

378

FIG. 22.7. *(a) System showing slight negative deviations (b) System showing gross negative deviations*

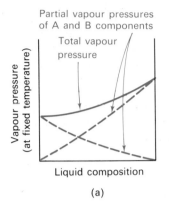

(a)

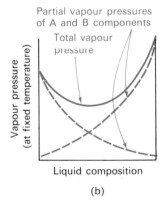

(b)

FIG. 22.8. *Boiling-point-composition diagram for trichloromethane and propanone at atmospheric pressure*

(a) Negative deviations from Raoult's law

Binary liquid mixtures which show this type of deviation have vapour pressure-composition curves similar to those shown in fig. 22.7(a) and (b). As in fig. 22.2 both liquid and vapour composition curves may be exhibited on the same diagram but only the former is shown.

Liquid mixtures which exhibit this sort of behaviour contain components A and B which show some degree of attraction for each other, i.e. A--B attractions are greater than the average of A--A and B--B attractions. There is both a decrease in volume and an evolution of heat when such solutions are formed. Methanol/water mixtures exhibit slight negative deviations as in fig. 22.7(a) and such mixtures can be separated by distillation. However, systems such as trichloromethane/propanone and nitric acid/water show gross negative deviations, a minimum occurring in their vapour pressure curves. Propanone hydrogen-bonds with trichloromethane (p. 346) while nitric acid ionises in water thus:

$$HNO_3(l) + H_2O(l) \longrightarrow H^+(aq) + NO_3^-(aq)$$

Liquid mixtures which have a minimum in their vapour pressure curves cannot be completely separated into their pure components by fractional distillation. This is explained below.

The boiling-point—composition curve for the binary liquid mixture trichloromethane/propanone is shown in fig. 22.8. A minimum in the vapour pressure-composition curve corresponds with a maximum in the boiling-point—composition diagram. This occurs at a composition corresponding to point Y on the diagram (approximately 0·65 mole fraction of trichloromethane). At this point both liquid and vapour have the same composition and such a mixture, known as an **azeotropic mixture** (or an azeotrope, or a maximum constant boiling-point mixture) will distil without any change in its composition. It behaves to all intents and purposes as a pure compound would, but it is clearly a mixture since its composition is dependent upon the pressure under which the distillation is carried out.

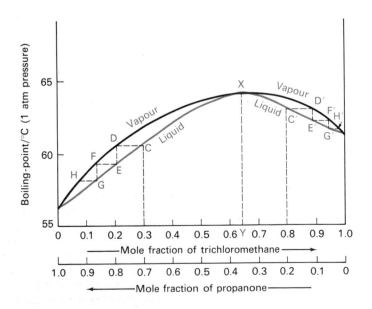

379

Consider a mixture of composition 0.3 mole fraction of trichloromethane (to the left of the vertical line XY in fig. 22.8). If this mixture is subjected to fractional distillation, virtually pure propanone may be obtained (equilibria represented by the tie-lines CD, EF, GH etc. will occur at increasing distances up the fractionating column). Removal of propanone will result in the mixture remaining in the flask becoming richer in trichloromethane; once the composition of the mixture in the flask corresponds to that of the azeotropic mixture it will distil unchanged, i.e. it will be impossible to distil over any more pure propanone. Similarly if a mixture of compositiion 0.8 mole fraction of trichloromethane (to the right of the vertical line XY in fig. 22.8) is fractionally distilled, it is possible to obtain virtually pure trichloromethane (equilibria represented by the tie-lines $C'D'$, $E'F'$, $G'H'$ etc. will be set up in the fractionating column). Removal of trichloromethane will gradually increase the mole fraction of propanone in the mixture in the flask. Once the composition of the azeotropic mixture is attained it will distil over unchanged.

It is therefore possible to obtain either pure propanone or pure trichloromethane from a mixture of these two liquids. In either case, however, a mixture is eventually left corresponding to the composition of the azeotrope. This then distils over unchanged. Probably the best way to visualise a system such as this is to regard the boiling-point composition diagram as two simple diagrams joined together along the line XY. It is not possible to move across this line since at this point both liquid and vapour have the same composition.

(b) Positive deviations from Raoult's law

Binary liquid mixtures which exhibit this type of behaviour have vapour pressure—composition curves similar to those shown in fig. 22.9(a) and (b). The fact that the total vapour pressure is greater than for an ideal solution, means that A‑‑B attractions are smaller than the average of A‑‑A and B‑‑B attractions. There is an increase in volume and an absorption of heat when such a solution is made. Tetrachloromethane/cyclohexane mixtures exhibit slight positive deviations as in fig. 22.9(a) and such mixtures can be separated by distillation. Gross positive deviations result in a maximum in the vapour pressure—composition diagram (fig. 22.9(b)). Pairs of liquids which exhibit such behaviour include carbon disulphide/propanone, benzene/ethanol and water/ethanol. Such mixtures cannot be separated completely by fractional distillation as is explained below.

Fig. 22.10 shows the boiling-point—composition diagram for the system benzene/ethanol. This binary liquid mixture forms an azeotrope—this time a minimum boiling-point mixture—containing approximately 0.55 mole fraction of ethanol at atmospheric pressure.

Fractional distillation of a binary liquid mixture which exhibits a maximum vapour pressure, and hence a minimum boiling-point, gives a distillate which has the composition of the azeotrope. The liquid remaining in the flask, at the end of the distillation, will be one of the pure components. For example, in fig. 22.10, any mixture of composition to the left of the vertical line XY will result in pure benzene remaining in the distillation flask, once the azeotrope has been distilled. Similarly pure ethanol accumulates in the distillation flask if any mixture of composition to the right of the vertical line XY is distilled. Once again, the behaviour of a liquid mixture of any composition can readily be

FIG. 22.9. (a) System showing slight positive deviations
(b) System showing gross positive deviations

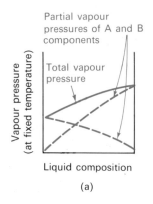

Partial vapour pressures of A and B components

Total vapour pressure

Vapour pressure (at fixed temperature)

Liquid composition

(a)

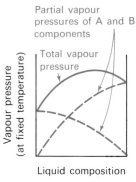

Partial vapour pressures of A and B components

Total vapour pressure

Vapour pressure (at fixed temperature)

Liquid composition

(b)

understood, if the diagram is visualised in terms of two simple diagrams joined together along the line XY.

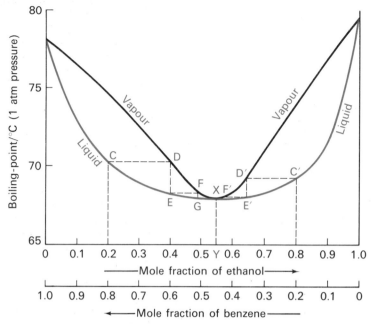

FIG. 22.10. *Boiling-point-composition diagram for benzene and ethanol at atmospheric pressure*

Ethanol and water form a minimum boiling point mixture containing 96% by mass of ethanol. Fractional distillation of aqueous solutions of ethanol thus produce a distillate of this composition, known as rectified spirit. Removal of the remaining 4% of water can be achieved by standing over calcium oxide for several hours followed by distillation. The product is now known as absolute ethanol.

Immiscible liquid mixtures

When positive deviations become sufficiently large, partial miscibility and, in the extreme, almost complete immiscibility can result (p. 347). We shall be concerned here with binary liquid mixtures which to all intents and purposes may be assumed to be immiscible, e.g. nitrobenzene and water.

Consider a mixture of nitrobenzene and water placed in a container, from which all the air has been removed, and connected to a manometer. At a fixed temperature, this system will produce a vapour pressure which is independent of the relative amounts of the two components present (agitation of the mixture is necessary to enable the denser of the two liquids—in this case nitrobenzene—to establish equilibrium with its own vapour). In general, the total vapour pressure exhibited by a mixture of immiscible liquids is equal to the sum of their separate saturation vapour pressures.

A liquid boils when its vapour pressure reaches the value exerted by the external pressure. Thus the total vapour pressure exerted by an immiscible mixture of liquids will reach atmospheric pressure at a temperature below the boiling-point of the most volatile constituent. For example, a mixture of water (b.p. = 100°C) and nitrobenzene (b.p. = 210°C) will boil at a temperature below 100°C. This is illustrated in fig. 22.11.

The vapour pressure curves for nitrobenzene and water are shown in fig. 22.11. It will be seen that the total vapour pressure reaches atmospheric pressure at a temperature of about 98°C, and this is the temperature at which the mixture boils.

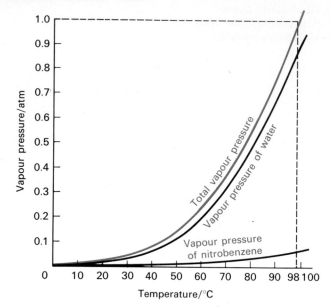

It is possible to calculate the ratio of the masses of the two liquids in the distillate in the following manner. Suppose p_w and p_n are the vapour pressures of water and nitrobenzene respectively at the boiling-point of this mixture, and n_w and n_n are respectively the number of moles of water and nitrobenzene in the vapour, we have

$$\frac{p_n}{p_w} = \frac{n_n}{n_w}$$

Since the number of moles of each constituent is simply the mass (in grams) divided by the relative molecular mass (expressed in grams), we may write

$$\frac{p_n}{p_w} = \frac{w_n/M_n}{w_w/M_w}$$

or

$$\frac{w_n}{w_w} = \frac{p_n M_n}{p_w M_w}$$

Distillation of immiscible liquids in which one of the constituents is water is called steam distillation. The technique has been used to purify liquids with high boiling-points, which would decompose if distilled in the ordinary manner at atmospheric pressure. However, vacuum distillation is a better and cleaner method of achieving the same result. Steam distillation is a useful method of extracting oils from material of vegetable origin, e.g. in perfumery.

22.3
Two-component mixtures comprising a volatile solvent and a non-volatile solute

The addition of a non-volatile solute (one that exerts a negligible vapour pressure) to a volatile solvent results in a lowering of the vapour pressure. In a simple kind of way this may be thought of as arising because the presence of the non-volatile solute will inevitably result in some decrease in concentration of the solvent. Raoult's law (p. 372) is again applicable to this type of situation provided that the solution approximates closely to an ideal one.

Suppose that p° is the vapour pressure of a pure solvent at some constant temperature, and that this reduces to p when a certain quantity of non-volatile solute is added to it. Then by Raoult's law we have:

$$p = p^\circ X_{\text{solvent}}$$

Although the above equation applies only to ideal solutions, non-ideal solutions obey the relationship quite well provided they are dilute.

If X_{solute} is the mole fraction of the non-volatile solute in the solution we may write:

$$p = p^\circ(1 - X_{\text{solute}})$$

Rearranging the above equation we get,

$$p^\circ - p = p^\circ X_{\text{solute}}$$

and dividing both sides by p° gives:

$$\frac{p^\circ - p}{p^\circ} = X_{\text{solute}} \tag{4}$$

If n_1 and n_2 are the number of moles of non-volatile solute and solvent respectively in a solution, then equation (4) may be written as:

$$\frac{p^\circ - p}{p^\circ} = \frac{n_1}{n_1 + n_2} \tag{5}$$

Now provided that we restrict ourselves to dilute solutions, it follows that $(n_1 + n_2)$ will approximate to n_2 and no significant error will be introduced by rewriting equation (5) in the approximate form:

$$\frac{p^\circ - p}{p^\circ} \simeq \frac{n_1}{n_2} \tag{6}$$

Equation (5)—or in the approximate form equation (6)—provides the key to the problem of determining the relative molecular mass of a non-volatile solute in solution. Continuing with equation (5), suppose w_1 and w_2 are the respective masses of non-volatile solute and solvent in a solution. Similarly M_1 and M_2 are their respective relative molecular masses. Then we may write:

$$\frac{p^\circ - p}{p^\circ} = \frac{w_1/M_1}{(w_1/M_1 + w_2/M_2)} \tag{7}$$

The relative lowering of the vapour pressure, i.e. $(p^\circ - p)/p^\circ$, may be determined by measuring the vapour pressures of pure solvent and solution. The masses w_1 and w_2 are known; thus if the relative molecular mass of the solvent, M_2 is known, the relative molecular mass of the solute, M_1, may be calculated. Using the approximate expression (equation 6), it may readily be shown that the relative molecular mass of the solute, M_1, is given by the expression:

383

$$M_1 \simeq \frac{w_1 M_2 p^{\circ}}{w_2(p^{\circ} - p)} \qquad (8)$$

The relative lowering of the vapour pressure of a solvent by the addition of a non-volatile solute is directly proportional to the number of moles of solute present in a fixed amount of solvent (provided that the solution may at all times be taken to be a dilute one), i.e. the relative lowering of the vapour pressure depends upon the number of solute particles present in a fixed quantity of solvent and not on their mass. Properties which depend upon numbers of particles and not on their actual masses are referred to as **colligative properties**. We shall now see that the elevation of the boiling-point and the depression of the freezing-point of a volatile solvent on addition of a non-volatile solute are also colligative properties; so too is osmotic pressure. All three properties are directly related to relative lowering of vapour pressure, and provide convenient means of determining relative molecular masses of non-volatile solutes in solution.

Boiling-points of solutions comprising a volatile solvent and a non-volatile solute

The vapour pressure of a pure solvent rises as the temperature is increased, and the liquid boils when its vapour pressure reaches the value exerted by the external pressure. The vapour pressure of a dilute solution of a non-volatile solute dissolved in a volatile solvent behaves in a similar manner; but at each temperature the vapour pressure of the dilute solution is always lower than that of the pure solvent. In fact, the vapour pressure curves for both pure solvent and dilute solution are practically parallel. Fig. 22.12 shows the type of vapour pressure—temperature curves characteristic of a pure solvent and two dilute solutions of different concentration.

FIG. 22.12. *Vapour pressure curves for a volatile solvent and for dilute solutions containing a non-volatile solute*

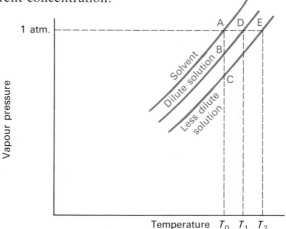

Over the small temperature range of interest the lines BD and CE, shown in fig. 22.12, may be taken to be almost straight. Since we have previously mentioned that they are practically parallel to each other, it follows that ABD and ACE approximate very closely to similar triangles. Thus

k = a constant

$$\frac{AD}{AB} = \frac{AE}{AC} = k$$

384

Considering the triangle ABD, the side AD represents the elevation of boiling-point, ΔT, and the side AB the lowering of the vapour pressure $(p^\circ - p)$. We may therefore write

$$\frac{\Delta T}{(p^\circ - p)} = k$$

i.e. the elevation of boiling-point is directly proportional to the lowering of the vapour pressure.

If we substitute $\Delta T/k$ for $(p^\circ - p)$ in equation (8), the following expression is obtained:

$$M_1 = \frac{kw_1 M_2 p^\circ}{w_2 \Delta T} \quad \text{and} \quad \Delta T = \frac{kw_1 M_2 p^\circ}{M_1 w_2} \tag{9}$$

For a given solvent the relative molecular mass, M_2, is a constant and the vapour pressure p° is one atmosphere. We may therefore incorporate both of these values in a new constant k', and simplify equation (9) to:

$k' = kM_2 p^\circ$

$$\Delta T = \frac{k' w_1}{M_1 w_2} \tag{10}$$

It is customary to choose 1 kg (1000 g) of solvent for w_2. Since w_1/M_1 is the number of moles of solute, then $w_1/(M_1 1000)$ is the number of moles of solute in 1000 g of solvent, i.e. the molality, m, of the solution (see p. 343). The final equation is therefore

$$\Delta T = k_b m \tag{11}$$

The constant k_b in equation (11) is known as the molal boiling-point elevation constant (or the ebullioscopic constant). It is the elevation in boiling-point that occurs when 1 mole of non-volatile solute is dissolved in 1000 g* of solvent. Its value is independent of the non-volatile solute but varies from solvent to solvent. Table 22.1 lists the values of the ebullioscopic constants for a number of solvents.

Table 22.1 The ebullioscopic constants for some solvents

Substance	Formula	Ebullioscopic constant k_b/K mol^{-1} kg
Benzene	C_6H_6	2·53
Cyclohexane	C_6H_{12}	2·79
Ethanoic acid	CH_3COOH	3·07
Ethoxyethane	$C_2H_5OC_2H_5$	2·11
Propanone	CH_3COCH_3	1·71
Tetrachloromethane	CCl_4	5·02
Trichloromethane	$CHCl_3$	3·63
Water	H_2O	0·52

* Sometimes the ebullioscopic constant is quoted for 100 g of solvent. This constant is ten times the value of the molal boiling-point elevation constant.

(a) *Determination of the relative molecular mass of a non-volatile solute in solution by the elevation of boiling-point method*

The principle of the method involves determining the boiling-points of the pure solvent and a dilute solution. Knowing the concentration of the solution and the value of the ebullioscopic constant for the solvent, it is a simple matter to calculate the relative molecular mass of the solute.

The determination of the relative molecular mass may be carried out by the Landsberger method illustrated in fig. 22.13. A small amount of the pure solvent is placed in the graduated vessel and a supply of vapour from the boiling solvent is passed through it. After some time the solvent in the vessel boils and the steady maximum temperature is recorded, using a thermometer graduated in tenths of a degree. Excess solvent vapour passes through a small hole in the bulbous part of the apparatus and is condensed and collected.

FIG. 22.13. *Landsberger's method of determining the elevation of boiling-point*

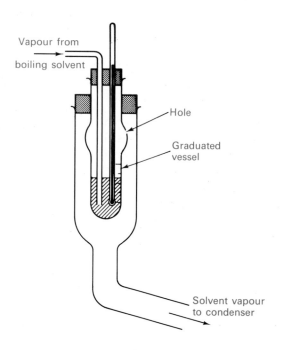

Vapour from boiling solvent

Hole

Graduated vessel

Solvent vapour to condenser

Some of the solvent is removed from the vessel and a small weighed amount of non-volatile solute is dissolved in that which remains. Vapour from the boiling solvent is now passed through the dilute solution. Eventually the maximum temperature is recorded at the same time as the supply of vapour from the boiling solvent is interrupted (it is necessary to stop the flow of solvent vapour when a reading is taken since it will continue to condense, giving up its latent heat in order to boil the solution).

The thermometer and vapour inlet tube are removed, taking care to shake off any liquid adhering, and the volume of solution recorded. It is generally accurate enough to assume that the volume of solution is the same as the volume of solvent, hence the mass of solvent can readily be calculated if its density is known.

Further results may be obtained by reconnecting the supply of boiling solvent vapour and recording the elevation of boiling-points for progressively more dilute solutions.

386

Example 1

The boiling-point of pure propanone is 56·2°C. A solution of 0·81 g of a non-volatile solute in 10 g of propanone boiled at 58·5°C. What is the relative molecular mass of the solute if the ebullioscopic constant for propanone is 1·71 K mol⁻¹ kg?

0·81 g of solute in 10 g of propanone gives an elevation of (58·5 − 56·2) or 2·3°C
R.M.M. (in g) of solute in 1000 g of propanone gives an elevation of 1·71° C
We can deduce the relative molecular mass by simple proportion thus:

$$R.M.M. = 0{\cdot}81 \times \frac{1000}{10} \times \frac{1{\cdot}71}{2{\cdot}3}$$

$$= 60.(2) = 60$$

Notice that the elevation in boiling-point is quite small (2·3°C) and the accuracy is likely to be no greater than ±0·1°C. An overall accuracy no greater than about 5–10% is expected in the value of the relative molecular mass. This is no real disadvantage, for the method is only used to determine the molecular formula of the solute in solution once its empirical formula (determined with much more precision) is known.

Freezing-points of solutions comprising a volatile solvent and a non-volatile solute.

The vapour pressure curve of a solvent shows an abrupt change in direction once it solidifies (p. 339). This is shown in fig. 22.14 which also includes the vapour pressure curves for two dilute solutions. Provided pure solid solvent separates from a dilute solution when it freezes, the presence of a non-volatile solute results in a lowering of the freezing-point of the solvent. It is for this reason that common salt is applied to icy roads and ethane-1,2,diol (ethylene glycol) is added to car radiators in the winter months.

FIG. 22.14. *The relationship between lowering of vapour pressure and the depression of freezing-point*

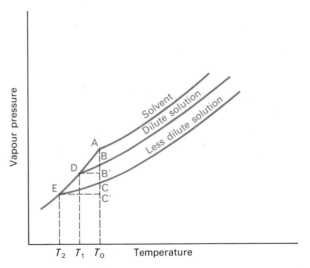

Since *ABD* and *ACE* may be taken to approximate very closely to similar triangles we may write:

$$\frac{AB}{BD} = \frac{AC}{CE} \tag{12}$$

Also *BB'D* and *CC'E* are similar triangles, hence

$$\frac{BD}{B'D} = \frac{CE}{C'E} \quad \text{and} \quad BD = \frac{B'D.CE}{C'E}$$

387

Substituting for BD in equation (12) and simplifying we obtain:

$$\frac{B'D}{AB} = \frac{C'E}{AC} = k$$

Considering the triangles ABD and $AB'D$, the side $B'D$ represents the depression of freezing-point, ΔT, and the side AB the lowering of vapour pressure ($p^\circ - p$). We can therefore write

$$\frac{\Delta T}{(p^\circ - p)} = k \tag{13}$$

i.e. the depression of freezing point is directly proportional to the lowering of the vapour pressure.

Since equation (13) is similar to the one derived for the elevation of boiling-point (p. 385), except that the values of the constant k are clearly different in both cases, it follows that the rest of the argument is the same. The final equation is therefore

$$\Delta T = k_f m \tag{14}$$

The constant k_f in equation (14) is known as the molal freezing-point depression constant. It is the depression in freezing-point that takes place when 1 mole of non-volatile solute is dissolved in 1000 g of solvent, and is sometimes referred to as the cryoscopic constant*. Like the ebullioscopic constant its value is independent of the non-volatile solute but varies from solvent to solvent. Table 22.2 lists the values of the cryoscopic constants for some solvents.

Table 22.2 The cryoscopic constants for some solvents

Substance	Formula	Cryoscopic constant k_f/K mol^{-1} kg
Benzene	C_6H_6	5·12
Cyclohexane	C_6H_{12}	20·1
Ethanoic acid	CH_3COOH	3·90
Tetrachloromethane	CCl_4	30
Water	H_2O	1·86

(b) *Determination of the relative molecular mass of a non-volatile solute in solution by the depression of freezing-point method*
The method involves the determination of the freezing-point of the pure solvent, and the depression of freezing-point that takes place when a known amount of non-volatile solute is dissolved in a fixed quantity of the solvent. Knowing the value of the appropriate cryoscopic constant, the relative molecular mass of the solute can readily be calculated. The experimental details are as follows.

A known quantity of solvent is placed in the inner glass vessel (see fig. 22.15) which is then cooled by immersing it in an appropriate

* Sometimes the cryoscopic constant is quoted for 100 g of solvent. This constant is ten times the value of the molal freezing-point depression constant.

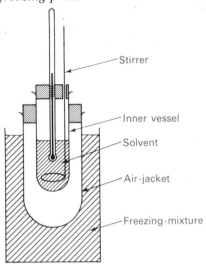

FIG. 22.15. *Apparatus for determining the depression of freezing-point*

Stirrer

Inner vessel

Solvent

Air-jacket

Freezing-mixture

freezing-mixture. Once the solvent has solidified the vessel is removed and the solvent just melted by warming in the hand. It is then placed in the air-jacket and re-immersed in the freezing-mixture. The solvent is stirred and the presence of the air-jacket ensures that the temperature drops slowly. The temperature may drop below the freezing-point of the solvent without any solid appearing—a phenomenon known as super-cooling. However, steady stirring will ensure that the temperature eventually rises and remains constant for some time at the freezing-point of the solvent. This temperature is recorded using a thermometer graduated in tenths of a degree.

The apparatus is now removed from the freezing-mixture and the solvent just melted. A known mass of solute is added, the inner vessel replaced in the air-jacket and the whole assembly placed in the freezing-mixture once again. The solution is stirred and the temperature recorded when solid first appears. Several results may be obtained with the same solution by melting followed by refreezing.

Example 2
The freezing-point of cyclohexane is 6·5°C. A solution of 0·65 g of naphthalene in 19·2 g of cyclohexane begins to freeze at a temperature of 1·2°C. What is the relative molecular mass of naphthalene? The cryoscopic constant for cyclohexane is 20·1 K mol^{-1} kg.
0·65 g of naphthalene in 19·2 g of cyclohexane gives a depression of 5·3°C
R.M.M (in g) of naphthalene in 1000 g of cyclohexane gives a depression of 20·1°C
We can deduce the relative molecular mass of naphthalene by simple proportion thus:

$$\text{R.M.M} = 0.65 \times \frac{1000}{19.2} \times \frac{20.1}{5.3}$$
$$= 128.(4) = 128$$

Relative molecular masses determined by depression of freezing-point are more accurate than those determined by the elevation of boiling-point, since the cryoscopic constant for a particular solvent is generally larger than its ebullioscopic constant. For example, k_b and k_f for cyclohexane are respectively 2·79 and 20·1 K mol^{-1} kg.

Camphor, a solid of m.p. = 179·5°C, has a cryoscopic constant of 40·0 K mol^{-1} kg, thus quite large depressions may be obtained by using molten camphor as the solvent.

The melting-point of camphor is first determined. Then a known mass of solute is melted with a larger known amount of camphor and the mixture allowed to solidify. It is then ground to a powder and the melting-point of the mixture determined (the temperature is recorded once the last trace of solid has melted). Relative molecular masses of sufficient accuracy can be achieved using a thermometer graduated in degrees only. The process is known as Rast's method.

Osmosis and osmotic pressure

When a solution is separated from its pure solvent by a **semi-permeable membrane**, i.e. a membrane which allows solvent to pass through it but not solute, there is a flow of solvent into the solution. This process is known as **osmosis**. The phenomenon is also observed whenever two solutions, made from the same solvent but of different concentrations, are separated from each other by a semi-permeable membrane. In such cases solvent flows from the less concentrated solution into the more concentrated one. The first recorded observations of osmosis were made by the Abbé Nollet in 1748, who employed a pig's bladder as a semi-permeable membrane.

Despite the wide range of materials that have been used as membranes none has been found to be completely non-permeable to solute particles. In addition, a membrane which approaches the ideal of semi-permeability for solutions which contain large solute particles may be quite useless for solutions which contain smaller solute particles. The choice of membrane is largely governed by the size and nature of the solute particles in solution.

The passage of solvent from one solution (or pure solvent) into a more concentrated one through a semi-permeable membrane gives rise to an osmotic pressure, which can be demonstrated quite easily.

A small porous pot is immersed in a warm aqueous solution of gelatine and propane-1,2,3-triol (glycerine) for several hours. It is then removed and, on standing, a gel is formed in the pores which acts as a semi-permeable membrane. The porous pot is then filled with a concentrated solution of sucrose and fitted with a rubber bung carrying a long piece of glass tubing (see fig. 22.16(a)). The pot is then placed in water. After a little while liquid begins to move up the glass tubing and, provided the membrane does not rupture, eventually reaches a stationary level (see fig. 22.16(b)). The column of liquid of height h exerts just sufficient pressure to prevent further osmosis, and this is equal to the osmotic pressure of the final diluted solution (not the initial more concentrated one).

FIG. 22.16. *Apparatus to demonstrate osmotic pressure, (a) at the start of the experiment, (b) at the end of the experiment*

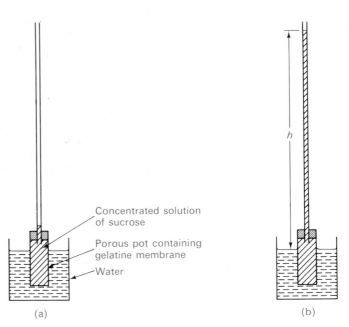

Concentrated solution of sucrose

Porous pot containing gelatine membrane

Water

(a)

(b)

As we shall see later, the best definition of **osmotic pressure** is '**that pressure which must be applied to a solution to prevent osmosis, when the solution is separated from its pure solvent by a semi-permeable membrane**'

(a) The Laws of Osmotic Pressure

Much of the early pioneering work was done by Pfeffer from about 1880 onwards. He used membranes of copper(II) hexacyanoferrate(II), $Cu_2Fe(CN)_6$, deposited in the sides of fine-grained porous pots. The pots were filled with a solution of copper(II) sulphate and immersed in a solution of potassium hexacyanoferrate(II). Diffusion of the $Cu^{2+}(aq)$ and $Fe(CN)_6^{4-}(aq)$ ions from opposite sides of the walls of the pots led to the formation of membranes where the two oppositely charged ions met.

Pfeffer experimented with dilute sucrose solutions of varying concentration and was able to observe osmotic pressures up to about 25 atmospheres with his apparatus. Some typical results are shown in Table 22.3.

Table 22.3 Some of Pfeffer's results on the osmotic pressures of sucrose solutions in water at 293 K

Concentration of sucrose solution/mol dm^{-3}	Observed osmotic pressure/atm
0·1	2·59
0·2	5·06
0·5	12·75
1·0	26·64

From results such as those given in Table 22.3 and from others obtained by carrying out experiments at different temperatures, Pfeffer was able to show that dilute solutions of sucrose obeyed the following rules quite closely:
(a) The osmotic pressure of a solution is directly proportional to its concentration if the temperature remains constant.
(b) The osmotic pressure of a solution of fixed concentration is directly proportional to the absolute temperature.

In the period from about 1900–1923 improvements in the technique of measuring osmotic pressures were made. Berkeley and Hartley, for example, designed a piece of apparatus which was capable of measuring osmotic pressures of several hundred atmospheres. Equally important was the fact that their apparatus, which is illustrated in fig. 22.17, allowed measurements to be obtained much more quickly than previous methods allowed.

FIG. 22.17. *Berkeley and Hartley's method of measuring osmotic pressure*

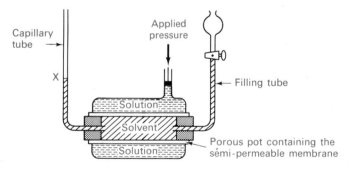

The apparatus consisted of a porous pot whose pores were sealed with a copper(II) hexacyanoferrate(II) semi-permeable membrane. The pot was enclosed in a metal jacket which was provided with a side-arm through which pressure could be applied.

The porous pot was filled with the solvent until it reached a known level X in the capillary tube. The dilute solution under investigation was placed in the outer jacket and the whole apparatus put in a thermostat at a fixed temperature. The movement of solvent through the semi-permeable membrane into the outer jacket was prevented by applying pressure to the solution. Just sufficient pressure was applied to keep the level of solvent in the capillary tube at the fixed mark X; this pressure is the osmotic pressure of the solution.

Using Pfeffer's data on dilute solutions, van't Hoff, in 1886, was able to show that there was a close parallel between the pressure exerted by a gas and the osmotic pressure developed by a dilute solution. Thus the ideal gas equation

$$pV = nRT$$

may be replaced by a similar equation:

II = osmotic pressure

$$\Pi V = nRT \tag{15}$$

The value of R in equation (15) is very close to the accepted value of R for an ideal gas, i.e. $8 \cdot 314$ J K^{-1} mol^{-1}.

Provided we are dealing with dilute solutions, osmotic pressure measurements may be used to determine the relative molecular masses of non-volatile solutes in solution as the following example shows.

Example 3
A solution of a sugar containing 10 g dm^{-3} was found to give an osmotic pressure of 0·703 atm at 20°C. Calculate the relative molecular mass of the sugar given that $R = 8\cdot314$ J K^{-1} mol^{-1}.
We must convert the pressure into N m^{-2}, the volume into m^3 and the temperature into the absolute scale. 1 atmosphere is 101 325 N m^{-2}, 1 dm^3 is 10^{-3} m^3 and 20°C is 293 K.
Substituting into equation (15) we have,

$$0\cdot703 \times 101\ 325 \times 10^{-3} = n \times 8\cdot314 \times 293$$

hence

$$n = \frac{0\cdot703 \times 101\ 325 \times 10^{-3}}{8\cdot314 \times 293}$$
$$= 0\cdot0292$$

10 g of the sugar is 0·0292 mole, thus the relative molecular mass is given by,

$$\text{R.M.M} = \frac{10}{0\cdot0292} = 342\cdot(5) = 343$$

(b) Theories of semi-permeability
One of the earliest theories put forward to explain the action of a semi-permeable membrane supposed that the membrane acted as a sieve, allowing the smaller solvent molecules to pass through it while, at the same time, acting as a barrier to the larger solute molecules. This theory may well explain the high permeability of some biological membranes which are composed of lipid (fat) molecules, punctuated with water-filled pores. However, it is clearly impossible to explain the semi-permeability of membranes which act as a barrier to small solute molecules in terms of a simple sieve action.

Another theory was based on selective solubility. For example, a membrane which dissolved water but not solute would be able to transport water from a dilute solution into a more concentrated one, while remaining impermeable to the solute particles.

Yet a third theory discusses semi-permeability in terms of differences in vapour pressure at both sides of the membrane. If the pores of the

392

membrane can be considered to be free from liquid, then vapour would be expected to diffuse from solvent side (where the vapour pressure is higher) to solution side (where the vapour pressure is lower).

While the above three simple theories go some way towards explaining the semi-permeability of specific membranes there is still a great deal to be learnt about the way in which they function.

(c) Osmosis in biological systems

Animal cells are surrounded by a plasma membrane which is semi-permeable. If such a cell is immersed in water or a dilute salt solution, whose concentration is lower than the cell fluid, then water will pass into the cell which consequently swells. Conversely cells immersed in a more concentrated salt solution shrink as water passes out of the cell into the concentrated salt solution.

This is easily demonstrated using an egg which has had the outer shell removed by treatment with dilute hydrochloric acid. When placed in water it will float and expand as water passes through the external membrane. In a concentrated solution of salt, however, it sinks and shrinks as water passes out of the egg into the surrounding solution.

If animal cells are to preserve their shape they must either exist in an isotonic environment, i.e. an environment in which the concentrations of the fluids both inside and outside the cells are such that no osmosis occurs, or there must be some mechanism which allows the cells to adapt to the surrounding fluid. For example, there are species of the Amoeba which live in both salt and fresh water environments. The species which lives in salt water has cell fluid which is isotonic with its surrounding medium and thus no osmosis occurs. On the other hand, the fresh water variety has to contend with the continual passage of water into the cell itself. This species has what amounts to a pump (the contractile vacuole) which periodically discharges this water and so prevents the animal bursting.

(d) Reverse osmosis

If just sufficient pressure is applied to a solution which is separated from its pure solvent by a semi-permeable membrane no osmosis occurs. If the applied pressure is now increased, solvent from the solution is forced to pass through the semi-permeable membrane into the pure solvent, i.e. pure solvent is recovered from the solution. This process, called **reverse osmosis**, has been considered as a means of obtaining potable water from sea water. Unfortunately the process is as yet uneconomic in terms of energy expenditure.

(e) Determination of relative molecular masses of polymers by osmosis

In general, membranes which approach the ideal state of complete semi-permeability can be prepared provided that the solute particles are not too small. Ideally the process of osmosis is best suited to the determination of the relative molecular masses of polymers in solution. Osmometry—as the technique is called—has been employed successfully in the determination of relative molecular masses in the range from 10^4 to 5×10^5. It is worth noting that neither the elevation of boiling-point nor the depression of freezing-point methods are satisfactory, since the temperature changes are much too small to be measured accurately.

393

Example 4

A solution of a polymer containing 5 g dm^{-3} was found to give an osmotic pressure of 600 N m^{-2} (4·5 mm Hg) at 15°C. Calculate the relative molecular mass of the polymer.

The osmotic pressure is small but using a manometer fluid much less dense than mercury it can be measured with some accuracy.

We use the equation,

$$\Pi V = nRT$$

and proceed as in Example 3.

$$600 \times 10^{-3} = n \times 8\cdot314 \times 288$$

hence

$$n = \frac{600 \times 10^{-3}}{8\cdot314 \times 288} = 2\cdot5 \times 10^{-4} \text{ mol}$$

5 g of polymer is $2\cdot5 \times 10^{-4}$ mole, thus the relative molecular mass is given by,

$$\text{R.M.M.} = \frac{5}{2\cdot5 \times 10^{-4}} = 2\cdot0 \times 10^4$$

Polymers are composed of molecules of different chain lengths, hence the relative molecular mass is an 'average' value.

Association and dissociation of the solute in solution

The three methods based on colligative properties, namely elevation of boiling-point, depression of freezing-point and osmotic pressure, may be used to investigate association and dissociation in solution. For example, the relative molecular mass of ethanoic acid in benzene, determined by the depression of freezing-point method, is approximately twice that expected for the formula CH_3COOH. It thus exists as a dimer in this solvent (see p. 91 for the hydrogen-bonded structure of this compound).

Electrolytes such as barium chloride, $Ba^{2+}(Cl^-)_2$, which are essentially ionic in the solid state, would be expected to exhibit colligative properties consistent with the total number of ions in solution. For example, barium chloride might be expected to give three times the elevation of boiling-point expected for the 'single' substance $BaCl_2$ in aqueous solution. In other words, barium chloride would be expected to have a relative molecular mass in solution equal to one third of that corresponding to the formula $BaCl_2$.

Observations carried out on solutions of electrolytes confirm that these ideas are essentially correct, although the number of 'effective' ions always tends to be lower than the number of 'actual' ions present. This discrepancy, which is due to ionic interferences (see chapter 23, p. 415), becomes less and less as the solution is progressively diluted.

22.4
Some two component phase equilibria

We shall be concerned in the final section of this chapter with the behaviour exhibited by two component liquid mixtures when they are cooled and solid allowed to separate. Detailed studies of such systems, when the components are metallic, are carried out by metallurgists. Such work is important in research devoted to the development of alloys with certain specific properties. We shall restrict ourselves to two different types of behaviour (although it should be appreciated that there are more types). The examination will be a brief one.

The types of system we shall look at can be summarised as below:
(a) Type I. Only pure components separate out as solid on cooling.
(b) Type II. A solid compound separates out on cooling.

Type I. Pure components separate out as solid on cooling

A typical phase diagram for such a system is shown in fig. 22.18. It refers to bismuth—cadmium mixtures of varying compositions.

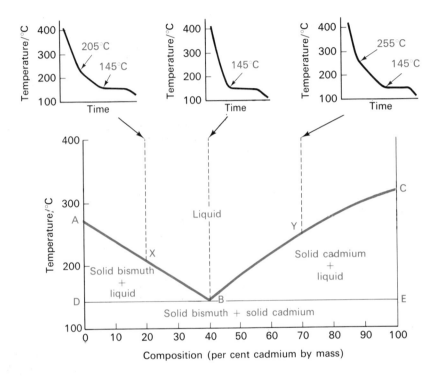

FIG. 22.18. *The phase diagram for a bismuth-cadmium mixture. The three cooling curves above the diagram are for mixtures containing respectively 20, 40 and 70% by mass of cadmium*

Point *A* is the melting-point of pure bismuth and the line *AB* represents the lowering of the melting-point of bismuth in the presence of successively larger amounts of cadmium. Point *C* is the melting-point of cadmium and line *CB* shows the effect of successively larger amounts of bismuth on the melting-point of cadmium.

The complete phase diagram may be obtained by taking a number of different mixtures which are then melted and allowed to cool. A cooling curve is plotted for each mixture and the temperature at which a solid phase begins to separate is indicated by a change of slope. Three cooling curves are shown at the top of fig. 22.18 and refer to mixtures containing respectively 20, 40 and 70% by mass of cadmium. The cooling curve on the extreme left shows a change of slope at 205°C, when solid bismuth begins to separate (point *X* on the phase diagram), and an even larger one at a temperature corresponding to point *B* when the whole mixture solidifies. The other two cooling curves are interpreted in a similar manner. The separation of a solid from a liquid results in the liberation of latent heat and this is the reason why the rate of cooling is partially or wholly arrested for some time.

Consideration of the phase diagram shows that above ABC the system is liquid. In the area ABD solid bismuth is present in the liquid mixture, while in the area BCE solid cadmium exists in the liquid mixture. Below the line DBE the entire system is a solid mixture of bismuth and cadmium. The temperature corresponding to point B is called the **eutectic point** and is the lowest temperature that can be obtained before complete solidification takes place. The mixture of composition represented by point B (in this example 60% bismuth and 40% cadmium by mass) is called the **eutectic mixture.**

Microscopic examination of the solid obtained by cooling the liquid of eutectic composition reveals the presence of two types of crystals, i.e. the eutectic is definitely a mixture and not a compound. If liquid of composition represented by point X is gradually cooled, microscopic examination of the resulting solid reveals the presence of crystals of bismuth embedded in a finely divided eutectic mixture. Similar examination of the solid resulting from the slow solidification of liquid of composition represented by point Y shows the presence of crystals of cadmium interspersed in a fine eutectic mixture.

Salt-water systems exhibit the same sort of behaviour and this is illustrated in fig. 22.19 for a solution of potassium iodide in water. It is clearly impossible to extend the diagram to 100% potassium iodide at normal atmospheric pressure, but this is only a detail.

FIG. 22.19. *The phase diagram for potassium iodide-water mixtures*

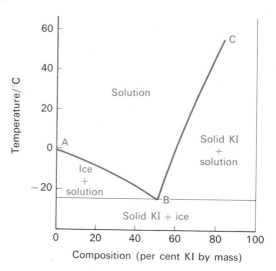

Point A is the melting-point of ice (or the freezing-point of water) and the line AB represents the lowering of the melting-point of ice (or the freezing-point of water) by the addition of potassium iodide. The line BC represents the solubility curve of potassium iodide in water. The interpretation of the diagram is exactly similar to the previous case. Point B is the eutectic point ($-23°C$) and is the lowest temperature at which a solution of potassium iodide can exist.

Other two component systems which exhibit the same kind of behaviour include lead/tin, gold/thallium, benzene/chloromethane.

Type II. A solid compound separates on cooling

An example of a system which shows this type of behaviour is calcium—magnesium and the phase diagram for this mixture is shown in fig. 22.20.

FIG. 22.20. *The phase diagram for a calcium-magnesium mixture*

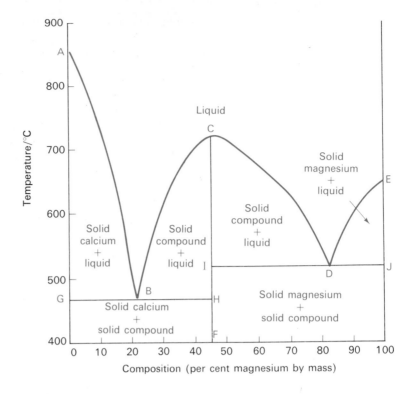

FIG. 22.21. *The phase diagram for the system Fe₂Cl₆—H₂O*

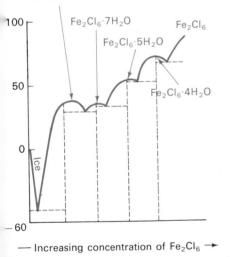

Point *A* is the melting-point of pure calcium and the line *AB* represents the lowering of the melting-point of calcium by the addition of increasing amounts of magnesium. Similarly point *E* is the melting-point of pure magnesium and the line *ED* shows the effect of increasing amounts of calcium on the melting-point of magnesium.

The maximum at point *C* corresponds to the formation of a compound of composition 55·5% calcium and 44·5% magnesium (a mole ratio Ca/Mg of 3/4). The formula of this compound is therefore Ca_3Mg_4—a curious stoichiometry. The lines *CB* and *CD* represent the effect of increasing amounts of calcium and magnesium respectively on the melting-point of this compound.

The whole phase diagram is best visualised in terms of two separate systems. To the left of the vertical line *CF* the diagram refers to the system Ca/Ca_3Mg_4, while to the right the corresponding system is Mg/Ca_3Mg_4. Below the line *GH* the system is completely solid and consists of a mixture of calcium and the compound. Similarly solid magnesium and solid compound exist below the line *IJ*.

Salts that form a number of hydrates exhibit this sort of behaviour, each maximum in the diagram corresponding to the formation of a definite hydrate. As in the case of the previous example they are best considered in terms of a number of composite phase diagrams. Fig. 22.21 shows the Fe_2Cl_6/water system, the maxima in the diagram corresponding to four definite hydrates.

Other two component systems which exhibit compound formation include magnesium/zinc, gold/tellurium and phenol/urea (carbamide).

22.5
Questions on chapter 22

1 Explain, with the aid of suitable boiling point/composition curves, what is meant by (a) *fractional distillation*, (b) *an azeotropic mixture*.

Predict, giving reasons, whether a benzene/methylbenzene mixture shows ideal or non-ideal behaviour.

Use the data in the table below to plot accurate temperature/composition curves to illustrate the equilibria between the liquid and vapour phases for mixtures of benzene and methylbenzene.

Mol fraction of benzene in liquid	0	0·05	0·20	0·40	0·60	0·80	1·0
Boiling point of liquid/°C	111	107·5	99·5	93·0	88·0	83·5	80·1
Mol fraction of benzene in vapour in equilibrium with boiling liquid	0	0·18	0·49	0·70	0·83	0·93	1·0

   (i) What is the mol fraction of benzene in the vapour above a boiling liquid mixture in which the mol fraction of benzene is 0·50?
   (ii) How many separate simple distillations would be needed to obtain a distillate containing a mol fraction of benzene of over 0·95 from a liquid mixture in which the mol fraction of benzene is initially 0·50?

C (Overseas)

2 (a) At atmospheric pressure tetrachloromethane boils at 77°C, and tin(IV) chloride at 114°C. A liquid mixture of these two compounds can be separated by fractional distillation.
    (i) Draw rough graphs to show (1) how you would expect the total vapour pressure of such a mixture to depend on its composition at a given temperature, (2) the relation between the boiling-point at a fixed pressure and the composition of the liquid and vapour phases.
    (ii) On the basis of the diagram in your answer to (i) (2), explain the process of fractional distillation.
   (b) Sulphur can exist as rhombic sulphur, monoclinic sulphur, liquid sulphur and sulphur vapour, according to the temperature and pressure. Sketch a pressure—temperature diagram to illustrate this statement, and state clearly the significance of the lines on your diagram.   O

3 Discuss the origin of the various possible types of curve arising when the vapour pressure of a mixture of two miscible volatile solvents is plotted against the composition expressed in mole fraction. You should refer to the ideal case and also to two non-ideal cases.

   For each case mentioned above, draw also the boiling point/composition curve. In the ideal case, show how the curve may be used to explain the process of fractional distillation; and for one non-ideal case, show how the curve may be used to explain the existence of an azeotropic mixture.   JMB (Syllabus B)

4 Explain, using a vapour pressure—composition diagram and giving important practical details, how two ideally miscible liquids may be separated by fractional distillation.

   Show, concisely, using another diagram, how the fractional distillation of an ethanol-water mixture differs from the one you have described.

   Methanol-ethanol solutions are almost ideal. If the vapour pressures of methanol and ethanol at 335 K are $8·1 \times 10^4$ and $4·5 \times 10^4$ N m^{-2} respectively, calculate the volume composition of the vapour over a mixture of 64 g of methanol and 46 g of ethanol at 335 K, assuming ideality.   W

5 At standard atmospheric pressure nitric acid (boiling point 87°C) and water form a constant boiling mixture, having a boiling point of 122°C and composition 65% by mass of nitric acid.
   (a) Define the term constant boiling mixture.
   (b) Sketch and label fully the boiling point/composition diagram for nitric acid and water.
   (c) State Raoult's law. Explain what is meant by the statement that a nitric acid/water mixture shows negative deviation from the law.
   (d) What changes take place when nitric acid is added to water?
   (e) State qualitatively what happens to the temperature and composition of the residual mixture during the distillation of a nitric acid/water mixture containing initially 20% by mass of nitric acid.
   (f) Name a pair of liquids which give a positive deviation from Raoult's law

and another pair which obey Raoult's law very closely.　AEB

6　State and explain Raoult's Law of Vapour Pressure Lowering.

By sketching vapour pressure/composition and boiling-point/composition curves, describe and explain what happens during the course of the distillation of an ideal mixture of two perfectly miscible liquids. What happens when deviations from Raoult's Law occur?

Methanoic acid (formic acid), b.p. 101°C, and water are miscible in all proportions and the boiling-point/composition curve shows a maximum for a mixture containing 73% by weight of methanoic acid. Propan-1-ol, b.p. 97°C, and water are also completely miscible yet the corresponding curve shows a minimum for a mixture containing 80% by weight of the alcohol. What is the result of the fractional distillation of separate aqueous solutions containing 50% by weight of these compounds?　S

7　Two liquids, A and B, are miscible over the whole range of composition and may be treated as ideal (obeying Raoult's Law). At 350 K the vapour pressure of pure A is 24·0 kPa (180 mmHg) and of pure B is 12·0 kPa (90 mmHg). A mixture of 60% A and 40% B is distilled at this temperature; what is the pressure in a closed distillation apparatus from which air is excluded? A small amount of the distillate is collected and redistilled at 350 K; what is the composition of the second distillate?

Describe and explain the features which may appear when two liquids which are not ideal and which do not react chemically are mixed and distilled.

Camb. Entrance

8　(a) What is meant by (i) *saturated vapour pressure*, (ii) *partial pressure*?

(b) Explain the process of *steam distillation*, and state why it is used in the preparation of phenylamine (aniline) from nitrobenzene.

(c) The vapour pressures of pure water and pure phenylamine at the stated temperatures are:

Temperature in °C	85	90	95	100	105
Vapour Pressure of water in kPa	57·9	70·1	84·5	101·3	120·5
in mmHg	434	526	634	760	906
Vapour Pressure of phenylamine in kPa	3·0	3·9	4·9	6·1	7·3
in mmHg	22·9	29·2	36·5	45·7	55·0

With the aid of suitable graphs

(i) find to the nearest °C the temperature at which a mixture of phenylamine and water will steam distil under standard atmospheric pressure.

(ii) Calculate the percentage by mass of phenylamine which would be expected to be present in the distillate.　AEB

9　Bromobenzene and water are immiscible liquids.

(a) Sketch curves to show how the vapour pressure of the system varies with (i) composition, (ii) temperature, showing in each case the contributions made by separate components, and explain briefly what is illustrated. Derive an expression relating the ratio of the masses of the components obtained on distillation to their molar masses (relative molecular masses) and vapour pressures.

Explain why steam distillation is important in the laboratory.

(b) Describe briefly how the behaviour of the system would differ if the two liquids were miscible.

(c) An aromatic compound distils in steam at 98·2°C under 731·9 mmHg pressure. The ratio of the mass of compound to mass of water in the distillate is 0·188. Calculate its molar mass (relative molecular mass) given that the saturated vapour pressure of water at 98·2°C is 712·4 mmHg, and state the units in which it is expressed.

($H_2O = 18$)

Note: the non-SI unit of pressure, mmHg, is used in this calculation.　S

10　(i) Draw a diagram of an apparatus suitable for the purification of a liquid by steam distillation and indicate what further operations are necessary to obtain a pure sample of the liquid. What advantages may this process have over simple distillation of the liquid as a method of purification?

(ii) Chlorobenzene distils in steam, when the atmospheric pressure 101·0 kPa, at 364 K, at which temperature the vapour pressure o chlorobenzene is 29·0 kPa. What mass of distillate would contain 10·0 g o chlorobenzene?

(iii) Give a critical summary of other methods which can be used for th purification of compounds having relative molecular masses of <300.

11 (a) The elevation in boiling point of a solvent by a solute is known as *colligative property*. What is meant by this term. Give **two** other examples o colligative properties.

(b) How would you determine the molar mass of a compound in a name solvent by an experiment involving one of these properties? Give concis experimental details.

(c) State with reasons, why measurement of the elevation of boiling point coul be used with an aqueous solution, to obtain the molar mass of urea (ca bamide), but not ethanoic (acetic) acid.

(d) A solution of 2·8 g of cadmium iodide (CdI_2) in 20 g of water boiled at temperature which was 0·20 K higher than the boiling point of pure wate under the same conditions of pressure. Calculate the molar mass of th solute, and comment on the result. (The boiling point elevation constant fo water is 0·52 K kg/mol). AE

12 (a) (i) Describe with the aid of a sketch an apparatus for the determination the elevation of the boiling-point of a solvent by a solute at a know concentration.

(ii) Explain how such measurements can be used to estimate the relativ molecular mass of the solute.

(iii) State briefly what factors or circumstances could give rise to a mislea ing answer in such a determination of a relative molecular mass.

(b) 0·600 g of iron(III) chloride are introduced into a vessel of internal volum 200 cm^3. The vessel is evacuated and closed, and then heated to 600 K. Th iron(III) chloride completely vaporizes, and the pressure in the vessel found to be 4·60 × 10^4 Pa. Calculate the relative molecular mass of iron(II chloride in the gaseous state at this temperature, and comment briefly o your result.

13 State carefully Raoult's Law of Vapour Pressure Lowering.

(a) *Vapour pressure lowering, boiling-point elevation, freezing-point depres sion and osmotic pressure are all closely related properties of solution* Comment and indicate why different methods for determining mola masses (relative molecular masses), which are based on them, are pre ferred according to the circumstances.

(b) Describe how you would determine the molar mass of a compound in th laboratory by measuring
  either the elevation of the boiling-point
    or the depression of the freezing-point
of a suitable solvent on forming a solution of the compound. Include th relevant vapour pressure/temperature graph in your account and link briefly to Raoult's Law.

(c) A 4·0% solution of ribitol in water (i.e. 4 g in 100 g of solution) has th same boiling-point as a 4·5% solution of glucose ($C_6H_{12}O_6$ = 180) i water. Calculate the molar mass of ribitol.

(d) If the apparent degree of ionization of potassium chloride (KCl = 74·5 in water at 290 K is 0·86, calculate the mass of potassium chloride whic must be made up to 1 dm^3 (1 l) of aqueous solution to have the sam osmotic pressure as the solution of glucose at that temperature.

14 (a) Explain what is meant by *the freezing point depression constant* for solvent.

(b) Explain **each** of the following observations:
  (i) 0·1 mole of sodium chloride depresses the freezing point of a give mass of water, twice as much as does 0·1 mole of glucose.
  (ii) 0·1 mole of ethanoic acid depresses the freezing point of a given mas

of benzene, half as much as does 0·1 mole of naphthalene, (a solid hydrocarbon).

    (iii) 0·1 mole of aluminium chloride, $(AlCl_3)$, depresses the freezing point of a given mass of benzene, half as much as does 0·1 mole of naphthalene.

(c) If a mixture of two immiscible liquids is boiled, the boiling point is below that of each of the liquids.

    (i) Explain why this is so.

    (ii) When the two liquids are water and nitrobenzene, the mixture boils at 372 K and the vapour pressures at this temperature are 97·7 kN m^{-2} (water) and 3·6 kN m^{-2} (nitrobenzene). Calculate the percentage of nitrobenzene in the vapour. (The relative molecular masses of water and nitrobenzene are 18 and 123 respectively.)

    (iii) What separation process used in organic chemistry utilises this principle?   W

15 (a) Assuming that the vapour pressure of water at any constant temperature is reduced on adding an involatile solute such as sugar by an amount proportional to the concentration of the solute, show with the aid of a diagram that the elevation of the boiling-point of such a solution is proportional to the concentration of the solute.

(b) Give a labelled sketch of the apparatus you would use to determine the depression of the freezing-point of a solution of sugar.

(c) The freezing-point of pure benzene is 5·533°C. The freezing-point of a solution of 6·40 g of naphthalene $(C_{10}H_8)$ in 1 000 g of benzene is 5·277°C, while that of a solution of 15·25 g of benzoic acid, $C_7H_6O_2$, in 1 000 g of benzene is 5·175°C. What conclusions can you draw from this information?

                                                       O

16 Describe how you would determine the relative molecular mass of a solute by measuring the depression in freezing-point of water. State clearly the precautions you would take to ensure accuracy.

When 0·775 g of white phosphorus was dissolved in 50·0 g of benzene, the solution froze at 4·89°C. What value do these data give for the relative molecular mass of white phosphorus? What explanation can you offer for this value?

[Freezing point of pure benzene = 5·53°C. Cryoscopic (freezing point) constant for benzene = 5·12°C mol^{-1} kg, i.e. 5·12°C mol^{-1} for 1000 g of solvent.]

                                                       C (Overseas)

17 Explain what is understood by the term "colligative property". Name three such properties and describe the practical determination of one such property. Explain how such a measurement can be used to determine relative molecular mass. Give two instances when the method selected would give unexpected results.

A 0·10 M solution of $HgCl_2$ in water freezes at $-0·186$°C whereas a 0·10 M solution of $Hg(NO_3)_2$ freezes at $-0·558$°C. Deduce the structural states of these two compounds. Given that the melting point of mercury(II) chloride is 280°C what is the predominant type of bonding in this compound? (The freezing point depression constant for water is 1·86 K for 1 mole of solute in 1 litre of solution.)   JMB (Syllabus B)

18 *Relative molecular mass* may be calculated from measurements of *colligative properties*

    (a) Explain the terms in italics.

    (b) What conditions must be met for the determination of relative molecular mass by these methods.

    (c) Show graphically the relationships between lowering of vapour pressure, elevation of boiling point and depression of freezing point.

    (d) What is meant by the term *molar depression of freezing point*?

An aqueous solution of a solute **X** boils at 0·13°C higher than the boiling point of water. Calculate (i) the freezing-point, (ii) the osmotic pressure at 13°C, of the solution. (Molar depression of freezing-point = 1·85°C/1000 g; molar elevation of boiling-point = 0·52°C/1000 g).   AEB

19 Discuss the principles underlying methods for determining relative molecular masses (molecular weights) in solution. What methods would you consider most suitable for measuring the relative molecular masses of sucrose, polystyrene and benzoic acid in solution? Give reasons for your choice and state what factors influence your choice of solvent in each case.　　Oxford Schol. and Entrance

20 (a) What is meant by the terms *colligative property* and *osmosis*? State how the osmotic pressure of a system is related to (i) the concentration of the solute (ii) temperature and (iii) the relative molecular mass of the solute.

(b) Describe a method used to give a fairly accurate determination of the relative molecular mass of a solute using osmosis. What kind of solutes are best suited for such a determination?

(c) An aqueous solution of a substance **X** at 27°C exhibited an osmotic pressure of 779 kPa (7·69 atmospheres). What would be the freezing point of the same solution at standard atmospheric pressure? The molecular depression constant for water is 1·86 K kg/mol. (**X** is known not to associate or dissociate in aqueous solution).　　　　　　　　　　　　　AEB

21 (a) Explain what is meant by a *colligative property*. Name **three** properties **apart** from osmosis which can be classified as colligative. For each of these properties give the relationship that exists between the property and the concentration of the solution.

(b) The following table gives values of osmotic pressure at 25°C for various concentrations of naphthalene in methylbenzene (toluene).

Plot the osmotic pressure of these solutions against concentration, and use the graph to calculate the relative molar mass of naphthalene.

Concentration/g dm^{-3}	Osmotic Pressure/kPa
0·5	10·0
1·0	20·0
1·5	28·0
2·0	37·0
2·5	46·0
3·0	56·0
4·0	74·0
5·0	92·0
6·0	110·0

(c)

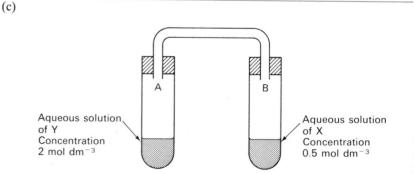

Aqueous solution
of Y
Concentration
2 mol dm^{-3}

Aqueous solution
of X
Concentration
0.5 mol dm^{-3}

The above apparatus was allowed to stand for some time at a constant temperature. Assuming the solutes **X** and **Y** to be in the same monomolecular condition in the water.

(i) what would be observed in both tubes during this time? Explain why,

(ii) what difference if any would be observed if the experiment was repeated at a higher temperature?　　　　　　　　　　　　　AEB

22 What do you understand by the terms *allotropy*, *alloy* and *eutectic mixture*? How could you distinguish between a eutectic mixture and a pure substance?

A mixture of 27% gold and 73% thallium melts and freezes at 131°C as if it were a pure substance. Draw on graph paper the phase diagram for the gold-thallium system, labelling each of the main areas.

Sketch the temperature–time curves that would be obtained on allowing the following mixtures to cool to room temperature from 1000°C:

	(a)	(b)
gold	60%	27%
thallium	40%	73%

Describe what happens in the system during the cooling of mixture (a).

O and C

Chapter 23

Ions in solution

23.1
Metallic conductors and electrolysis

The science of electrochemistry dates from the discovery by Volta in 1800 of a method of producing a direct current of electricity. The so-called Voltaic cell consisted of alternate discs of copper and zinc separated from each other by pieces of card saturated with an aqueous solution of sodium chloride. With this new invention a means was at hand to study the effect of electricity on matter.

Metals and a few other substances like graphite, conduct electricity quite well but no chemical change appears to take place and they are accordingly classed as **metallic conductors**.

However, the majority of solids allow no electricity to pass whatsoever, but in some instances conduction of electricity takes place if the solid is allowed to get damp. This seems strange, since water can be shown to be a poor conductor of electricity. Further investigation shows that many solutions of inorganic solids conduct, although in all cases less efficiently than metals. A typical piece of apparatus which is useful for investigating the conduction of electricity in aqueous solution is shown in fig. 23.1.

FIG. 23.1. *Conduction of electricity through aqueous solutions*

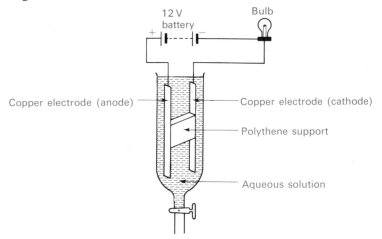

As a result of applying a potential difference across two metallic electrodes placed in a variety of aqueous solutions, the following rules can be drawn up.

(a) Aqueous solutions of mineral acids, alkalis and salts readily conduct electricity. They are called **strong electrolytes**.

(b) Insoluble salts and most organic substances, whether pure or in aqueous solution, do not allow a current to pass and are called **non-electrolytes**.

With a more sensitive piece of apparatus than that shown in fig. 23.1 it is possible to differentiate between weak and strong electrolytes. Typical **weak electrolytes** are aqueous solutions of ammonia and ethanoic acid.

Whenever a direct current passes through an aqueous solution, a chemical change at the electrodes is invariably found to take place as well. This truth was first established by Nicholson and Carlisle (1800) who were able to show that impure water was decomposed into hydrogen and oxygen. However, the products of electrolysis are often unexpected. The results of some typical experiments are shown in Table 23.1.

Table 23.1 Electrolysis of some aqueous solutions of salts

Electrolyte	Anode	Cathode	At the anode	At the cathode	Any other changes
$CuSO_4(aq)$	Pt	Pt	O_2 liberated	Cu deposited	$H_2SO_4(aq)$ formed at the anode
$CuSO_4(aq)$	Cu	Cu	Cu removed	Cu deposited	No other change
$KI(aq)$	Pt	Pt	I_2 liberated	H_2 liberated	$KOH(aq)$ formed at the cathode
$NaCl(aq)$	Pt	Pt	O_2 and Cl_2 liberated	H_2 liberated	More Cl_2 relative to O_2 as the conc. increases
$ZnI_2(aq)$	Pt	Pt	I_2 liberated	Zn deposited Some H_2 liberated	Conc. of solution decreases

Summarising the results of these and many other experiments, it is clear that the solvent water has an important part to play, not only in making the solid salts conducting, but also in the actual chemical changes taking place.
(a) Salts like ZnI_2, whose aqueous solutions decompose into the elements of which the salt is formed, are comparatively rare.
(b) Soluble salts of metals below zinc in the electrode potential series (p. 313) tend to liberate the metal at the cathode.
(c) Soluble iodides always liberate iodine at the anode, while most other soluble salts give rise to oxygen at the anode. Chlorides tend to give oxygen at the anode if they are dilute but increasing amounts of chlorine (relative to oxygen) are produced as the concentration of the solution increases.
(d) Salts of metals above zinc in the electrode potential series almost always liberate hydrogen and oxygen.
(e) Metallic anodes more reactive than platinum tend to pass into solution instead of oxygen being produced.

Faraday's work on electrolysis

As a result of some careful measurements of the quantities of substances liberated by electrolysis under different conditions, Faraday in 1830 was able to summarise his findings in terms of two laws:
(a) The First Law. The mass of a substance liberated at an electrode during electrolysis is always proportional to the quantity of electricity passed.

The quantity of electricity is measured in coulombs and is equal to the product of current (in amperes) and time (in seconds).
(b) The Second Law. In modern form this law can be stated as follows: The same quantity of electicity is required to liberate 1 mole of atoms of any univalent element during electrolysis.

405

During the electrolysis of aqueous silver nitrate solution, for example, 107·87 g of silver (1 mole) are deposited on the cathode by the passage of 96 500 coulombs of electricity. The Faraday constant, F, may be taken as:

$$F = 96\ 500 \text{ C mol}^{-1}$$

It requires 2 faradays of electricity to deposit one mole of a divalent metal, such as copper, during electrolysis. Similarly the quantity of electricity needed to deposit one mole of a trivalent metal, such as aluminium, is equal to 3 faradays.

The Ionic Theory

The ultimate particles concerned in electrochemical decompositions were first called 'ions' by Faraday although he had no idea of their nature. It was not until the year 1887 when the Swedish chemist Arrhenius proposed his ionic theory that an acceptable explanation of electrolysis was given. Previously Clausius (1857) had postulated the existence of a small concentration of ions in an electrolyte; Arrhenius took the bold step and suggested that strong electrolytes existed entirely as ions. There is a good deal of experimental evidence to support this view, some of which is listed below:

(a) Colligative properties of dilute solutions of electrolytes are easily explained on the basis of complete ionisation (p. 394).
(b) The almost constant values for the enthalpies of neutralisation of an aqueous solution of a strong acid by a strong base find a ready explanation in terms of the fundamental reaction:

$$H^+(aq) + OH^-(aq) \longrightarrow H_2O(l) \qquad \text{(p. 221)}$$

(c) X-ray analysis has shown that many salts such as sodium chloride are completely ionised in the solid state (p. 177).

In terms of the ionic theory, the deposition of 1 mole of silver during electrolysis of an aqeuous solution of silver nitrate is simply represented as:

$$\underset{\text{1 mole}}{Ag^+(aq)} + \underset{\text{1 mole}}{e^-} \longrightarrow \underset{\text{1 mole}}{Ag(s)}$$

It does not matter whether the substance undergoing electrolysis is in aqueous solution or simply in the fused (molten) state. For example, the formation of 1 mole of sodium during the electrolysis of fused sodium chloride may be represented:

$$\underset{\text{1 mole}}{Na^+} + \underset{\text{1 mole}}{e^-} \longrightarrow \underset{\text{1 mole}}{Na(s)}$$

In general, the discharge of 1 mole of ions, M^{n+}, may be written:

$$\underset{\text{1 mole}}{M^{n+}} + \underset{n \text{ moles}}{ne^-} \longrightarrow \underset{\text{1 mole}}{M(s)}$$

Thus n faradays of electricity are required to deposit 1 mole of an n-valent element.

The situation is similar if we are considering the discharge of anions at the anode. For example,

$$2I^-(aq) \longrightarrow I_2(aq) + 2e^-$$

$$\underset{\text{2 moles}}{\hphantom{2I^-(aq)}} \quad \underset{\text{1 mole}}{\hphantom{I_2(aq)}} \quad \underset{\text{2 moles}}{\hphantom{2e^-}}$$

One mole of univalent ions, electrons and atoms contain the Avogadro constant of 'particles' and the charge on a univalent ion is the same in magnitude as the charge on the electron; it therefore follows that the relationship between the Faraday constant, F, the Avogadro constant, L, and the electronic charge, e, is

$$F = Le$$

Mechanism of electrolysis in terms of the Ionic Theory

Ions which are free to move, and which therefore transfer electric charge, can be produced in three ways.
(a) Breaking up an ionic crystal by melting it.
(b) Breaking up an ionic crystal by dissolving it in water.
(c) Allowing non-ionic substances to react with water.

If the energy released by the reaction of a non-ionic substance with water is greater than the energy required to break the bonds and separate the ions formed, the substance may ionise on contact with water. Gaseous hydrogen chloride is an example:

$$HCl(g) + H_2O(l) \longrightarrow H^+(aq) + Cl^-(aq)$$

An aqueous solution of ammonia is a weak electrolyte (p. 286) and may be represented by the equilibrium:

$$NH_3(g) + H_2O(l) \rightleftharpoons NH_4^+(aq) + OH^-(aq)$$

Although pure water is itself a very weak electrolyte, the following equilibrium is nevertheless of importance because electrolysis is generally carried out in aqueous solution:

$$H_2O(l) \rightleftharpoons H^+(aq) + OH^-(aq)$$

When a liquid containing positive and negative ions is electrolysed with direct current, the flow of electrons from the battery—which can be thought of as an 'electron pump'—is available at the cathode to convert cations to the free metal (or hydrogen). At the anode, negatively charged anions give up their electrons to the positive electrode. Thus an electric current flows, but it is important to realise that no one electron ever passes right round the circuit from one electrode to the other. The situation is depicted in fig. 23.2. In the electrolysis of fused sodium chloride, for example, electrons supplied to the cathode are used to convert sodium ions, Na^+, to sodium, Na, while chloride ions, Cl^-, supply electrons to the anode and produce chlorine molecules, Cl_2:

$$2Na^+ + 2e^- \longrightarrow 2Na(s) \quad \text{(at the cathode)}$$
$$2Cl^- \longrightarrow Cl_2(g) + 2e^- \quad \text{(at the anode)}$$

FIG. 23.2. *The mechanism of electrolysis*

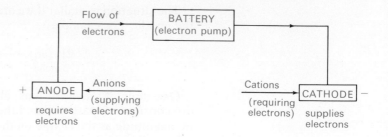

Discharge of ions at the cathode

If the electrolyte is a fused salt there is only one type of cation present and this is discharged at the cathode (see the example above).

Complications arise if there are two or more different types of cation present, but in general the cation which is discharged is the one that can gain electrons most readily. For aqueous solutions which contain ions of comparable concentrations the ease of discharge of a particular cation is determined by its position in the electrode potential series (p. 313).

$$
\begin{array}{l}
K^+(aq) \\
Ca^{2+}(aq) \\
Na^+(aq) \\
Mg^{2+}(aq) \\
Al^{3+}(aq) \\
Zn^{2+}(aq) \\
Fe^{2+}(aq) \\
Sn^{2+}(aq) \\
Pb^{2+}(aq) \\
H^+(aq) \\
Cu^{2+}(aq) \\
Ag^+(aq)
\end{array}
\qquad
\begin{array}{c}
\text{Increasing ease of discharge} \\
\text{at the cathode}
\end{array}
$$

The production of hydrogen at the cathode merits further consideration. Electrolysis of aqueous solutions of salts containing metals above zinc in the electrode potential series almost always give rise to hydrogen at the cathode, despite the fact that only about 1 in every 5×10^8 molecules of water is split up into ions. The mechanism of this process is still a subject of controversy, although the common explanation is as follows:

$$2H^+(aq) + 2e^- \longrightarrow 2H(g)$$
$$2H(g) \longrightarrow H_2(g)$$

As the discharge of hydrogen ions occurs, more water molecules continue to ionise to maintain the constant but minute concentration of $H^+(aq)$. However, it has been suggested that direct interaction between water molecules and electrons may occur at the cathode, particularly with alkaline solutions in which the concentration of $H^+(aq)$ ions is even lower:

$$2H_2O(l) + 2e^- \longrightarrow H_2(g) + 2OH^-(aq)$$

Two factors tend in specific cases to interfere with the simple order of discharge of aqueous cations given by the electrode potential series.

(a) The concentration effect

When one metallic ion $X^{m+}(aq)$ is present in a much higher concentration than another metallic ion $Y^{n+}(aq)$, then $X^{m+}(aq)$ may be discharged at the same time as $Y^{n+}(aq)$, even though under normal conditions only $Y^{n+}(aq)$ would be discharged. For example, brass (an alloy of copper and zinc) may be deposited from aqueous solution by using an electrolyte which contains a much higher concentration of $Zn^{2+}(aq)$ ions than $Cu^{2+}(aq)$ ions.

(b) The overvoltage effect

Hydrogen ions are discharged much as one would expect in terms of the electrode potential series at a bright platinum surface; but at other electrodes—notably lead, zinc, tin, aluminium and especially mercury—they are not discharged so readily as the position of $H^+(aq)$ in this series indicates. The phenomenon is called the overvoltage effect since a higher potential difference is needed to discharge hydrogen ions at these metallic cathodes than is the case when a bright platinum electrode is used. It has been suggested that the effect is due to a high energy of activation (p. 445) for one or both of the reactions:

$$2H^+(aq) + 2e^- \longrightarrow 2H(g)$$
$$2H(g) \longrightarrow H_2(g)$$

Platinum is known to be a very effective catalyst for the recombination of hydrogen atoms.

Electrode processes at the anode

If two or more anions are present in aqueous solution they are discharged selectively in the order shown:

$I^-(aq)$
$OH^-(aq)$
$Cl^-(aq)$ Increasing ease of discharge
$NO_3{}^-(aq)$ at the anode
$SO_4{}^{2-}(aq)$
other large anions

For example, in the electrolysis of an aqueous solution of potassium iodide (Table 23.1, p. 405), iodide ions are discharged in preference to hydroxyl ions.

    As in the case of hydrogen evolution at the cathode, (p. 408) there is still much uncertainty regarding the mechanism of oxygen production at the anode, although the favourite explanation is as follows:

$$4OH^-(aq) \longrightarrow 4OH + 4e^-$$
$$4OH \longrightarrow 2H_2O(l) + O_2(g)$$

The discharge of hydroxyl ions will result in the ionisation of more molecules of water to maintain the minute concentration (p. 279). Once again an alternative mechanism has been proposed for oxygen evolution, especially from acidic solutions where the $OH^-(aq)$ concentration is considerably lower than the minute value in neutral solutions. Direct involvement of water molecules at the cathode are envisaged and the overall process may be represented as:

$$2H_2O(l) \longrightarrow 4H^+(aq) + O_2(g) + 4e^-$$

(a) The concentration effect

This is important for anions, especially in the electrolysis of aqueous solutions of chlorides. For example, electrolysis of a very dilute solution of sodium chloride results in the liberation of oxygen at the anode. Increasing the concentration of the solution results in the evolution of chlorine at the expense of oxygen.

(b) The overvoltage effect

The discharge of oxygen in the electrolysis of alkaline solutions takes place less readily than expected, even at bright platinum electrodes, but otherwise the overvoltage effect at the anode is not important.

(c) Alternative electrode process at the anode

If the atoms of the anode can lose electrons and pass into solution as hydrated ions more easily than the hydrated anions can be discharged they will do so. Thus in the electrolysis of a dilute solution of copper(II) sulphate the reaction at the copper anode is

$$2Cu(s) \longrightarrow 2Cu^{2+}(aq) + 4e^-$$

rather than

$$4OH^-(aq) \longrightarrow 2H_2O(l) + O_2(g) + 4e^-$$

The industrial importance of electrolytic methods

Only a brief summary of some of the many industrial processes which involve electrolysis can be given in this book. More details are available in textbooks of inorganic chemistry.

(a) Electrolysis of fused halides is the method used for extracting the metals in Groups 1A and 2A of the Periodic Table.

(b) Aluminium is manufactured by the electrolysis of purified aluminium oxide, Al_2O_3, dissolved in fused cryolite, Na_3AlF_6.

(c) Chlorine and sodium hydroxide solution are obtained by the electrolysis of a concentrated solution of sodium chloride. A flowing mercury cathode is employed into which sodium ions discharge (hydrogen ions have a high overvoltage at a mercury surface (p. 409)). The sodium amalgam is passed into water when sodium hydroxide solution and hydrogen are produced.

(d) Copper is purified by making it the anode of an electrolytic cell in which a pure strip of copper is used as the cathode and the electrolyte is an aqueous solution of copper(II) sulphate. Copper is effectively transferred from the anode to the cathode. An 'anode sludge' containing silver and gold accumulates and helps make the process economically viable.

(e) A variety of metals can be used for electroplating, e.g. chromium, silver and nickel. The article to be plated is the cathode, the anode is the pure plating metal, and the electrolyte is a dilute aqueous solution containing the plating metal ions. The plating metal goes into solution as hydrated ions as electrolysis proceeds and these ions are discharged onto the cathode.

23.2
The conductivity of solutions of electrolytes

I = current (in amperes)
V = potential difference (in volts)
R = resistance (in ohms)

Solutions of electrolytes, like metallic conductors, obey Ohm's law, i.e. the current (in amperes) passing through a solution of a particular electrolyte is given by the equation:

$$I = \frac{V}{R}$$

The resistance of an electrolyte is directly proportional to the length, l, of liquid through which the current passes, and inversely proportional to its area of cross section, A. It therefore follows that

$$R = \frac{\rho l}{A} \qquad (1)$$

where ρ is the constant of proportionality and is called the **resistivity**. It is a constant for an aqueous solution of a given electrolyte of fixed concentration at a particular temperature. In the basic SI units, resistivity has the units of ohm metre (Ω m).

In practice it is found to be more convenient to use the reciprocal of the resistivity to characterise the electrical behaviour of a particular electrolyte. This is called the **electrolytic conductivity** (or simply conductivity) and is denoted by the symbol κ.

$$\kappa = 1/\rho \qquad (2)$$

The basic SI units of electrolytic conductivity are thus ohm^{-1} metre^{-1} (Ω^{-1} m^{-1}), although it is still common to express this quantity in terms of the centimetre as a unit of length. The connection between the two units is:

$$100 \ \Omega^{-1} \ m^{-1} = 1 \ \Omega^{-1} \ cm^{-1}$$

Measurement of electrolytic conductivity

The first studies of this type employed direct current and it was soon noticed that complications arose. For example, electrolysis could lead to the removal of ions from solution and therefore to an increase in resistance. In addition, the release of gases at the electrodes (notably hydrogen and oxygen) often gave rise to an increase in resistance in the vicinity of the electrodes. These complications disappeared when an alternating-current was used, and nowadays it is customary to use a-c frequencies within the range 1000-4000 hertz. The rapid rate at which the electrodes change polarity ensures that equilibrium conditions obtain with no electrolysis taking place.

A typical bridge circuit used for determining the conductivity of an electrolyte solution is shown in fig. 23.3. A known volume of the liquid under test is placed in a conductivity cell (described later) which is then thermostated (generally at 298 K). The sliding contact is then moved along a metre wire (AB) until minimum noise is detected in the headphones. If R_{cell} is the resistance of the solution, R_{box} is the selected resistance in the resistance box, and X is the balance point, then

$$\frac{R_{cell}}{R_{box}} = \frac{BX}{AX}$$

FIG. 23.3. *A bridge circuit for measuring conductivity*

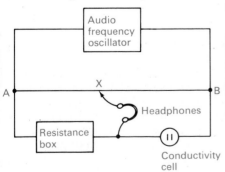

411

$$R_{cell} = \frac{R_{box} \cdot BX}{AX}$$

In practice the value of R_{box} selected is such as to allow a balance point to be obtained near the mid-point of the metre wire, since errors in the ratio BX/AX are thereby minimised. A refinement which is needed in accurate work is the inclusion of a variable condenser in parallel with the resistance box. This is adjusted to minimise the capacitance effect of the conductivity cell and results in a sharper balance point being achieved.

The conductivity cell itself consists of a glass vessel fitted with platinum disc electrodes which are a fixed distance apart. Before use, the electrodes are plated with a coating of platinum black and this helps to achieve the essential reversible conditions at the electrodes. Fig. 23.4(a) and (b) illustrates two types of cell which may be used depending upon whether the solution in question has a low or high conductivity.

FIG. 23.4. *Types of conductivity cell (a) for a low conductivity solution, (b) for a high conductivity solution*

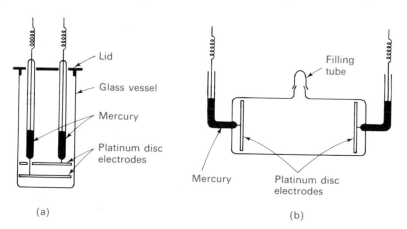

(a)

(b)

Accurate conductivity measurements require the use of very pure water for making up the electrolyte solutions. Ordinary distilled or deionised water absorbs carbon dioxide from the atmosphere which reacts with the water producing a few ions. These must be removed by passing the water through an ion exchange resin (p. 367) before it is used. The glass conductivity cell should be 'steamed out' before use and should be constructed from so-called 'insoluble' glass or better, from fused silica.

The resistance of a solution measured as described above can be converted into conductivity by the application of equations (1) and (2), (p. 411):

$$\kappa = \frac{1}{R_{cell}} \times \frac{l}{A} \qquad (3)$$

However, we need to know the dimensions of the conductivity cell in order to calculate the value of l/A (the cell constant). These measurements are difficult to obtain with a high degree of accuracy, so the cell constant is determined by an indirect method. The resistance of a solution of known conductivity is measured at 298 K, and one of the standard solutions used for this purpose is a 0·1M solution of potassium chloride.

Example 1

The resistance of a 0·1M solution of potassium chloride is found to be 29·0 Ω. If the conductivity of this solution of potassium chloride is 1·29 Ω^{-1} m^{-1} (1·29 × 10^{-2} Ω^{-1} cm^{-1}) determine the cell constant. Using the same cell, the resistance of a 0·1M solution of silver nitrate was 34·9 Ω. Calculate the conductivity of this solution.

From equation (3) the cell constant, (l/A) is given by:

$$l/A = R_{cell} \cdot \kappa$$

Substituting into the above equation we have

$$l/A = 29·0 \times 1·29 = 37·4$$

The cell constant is 37·4 m^{-1} (or 3·74 × 10^{-1} cm^{-1})
For the 0·1M solution of silver nitrate we have

$$\kappa = \frac{37·4}{34·9} = 1·07$$

The conductivity of the silver nitrate solution is 1·07 Ω^{-1} m^{-1} (or 1·07 × 10^{-2} Ω^{-1} cm^{-1})

Variation of conductivity with temperature and concentration

The conductivity of a solution of an electrolyte increases as the temperature is raised. For example, the conductivities of a 0·1M solution of potassium chloride are 1·12 × 10^{-2} Ω^{-1} cm^{-1} and 1·29 × 10^{-2} Ω^{-1} cm^{-1} respectively at 291 and 298 K. We shall not be concerned with temperature effects, except to stress that measurements must be taken at a fixed temperature which, as we have mentioned previously, is generally 298 K.

The conductivity of a particular electrolyte at a fixed temperature depends upon
(a) the number of ions present in unit volume of solution,
(b) the speed at which the ions oscillate in step with the applied alternating voltage.

Some typical conductivity values, at a temperature of 298 K, are given in Table 23.2 for aqueous solutions of potassium chloride (a strong electrolyte) and ethanoic acid (a weak electrolyte). It will be seen

Table 23.2 The conductivities of aqueous solutions of potassium chloride and ethanoic acid at 298 K

Concentration c/mol dm^{-3}	Conductivity, $\kappa \times 10^3/\Omega^{-1}$ cm^{-1}	
	KCl	CH$_3$COOH
0·0005	0·0739	0·0339
0·001	0·147	0·0492
0·005	0·718	0·115
0·01	1·41	0·163
0·02	2·77	0·232
0·05	6·67	0·370
0·10	12·9	—
0·20	24·8	—
0·50	58·6	—
1·00	112	—

that in both examples the conductivity increases with an increase in concentration. However, the two electrolytes behave differently in that their conductivities which are widely different at a concentration of, say, 0·05 mol dm^{-3} are roughly comparable when diluted one hundred-fold to give solutions of concentration 0·0005 mol dm^{-3}.

(a) Variation of conductivity with concentration for a strong electrolyte

It is impossible to plot all the conductivity values, given in Table 23.2, for solutions of potassium chloride on one graph since the values are wide-ranging. However,, fig. 23.5 shows the values plotted for solutions of concentration within the range 0·01 to 1·00 mol dm^{-3}.

Potassium chloride exists in the form of ions as a solid and thus there are one hundred times as many ions in unit volume of solution at a concentration of 1·0 mol dm^{-3} as there are at a concentration of 0·01 mol dm^{-3}. However, the conductivity of M potassium chloride solution is not one hundred times the conductivity of a 0·01M solution

FIG. 23.5. *Variation of conductivity for solutions of potassium chloride at 298 K*

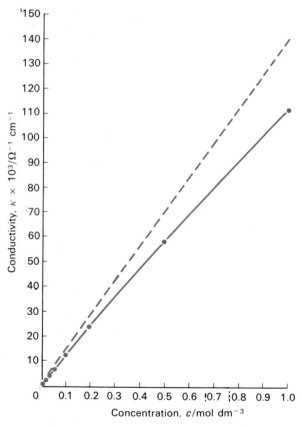

i.e. it is not 1·41 × 10^{-3} × 10^2 or 1·41 × 10^{-1} Ω^{-1} cm^{-1} but 1·12 × 10^{-1} Ω^{-1} cm^{-1}. Thus the conductivity of the more concentrated solution is less than expected simply on the basis of the number of ions in unit volume of solution.

The solid line in fig. 23.5 shows the actual variation of conductivity with concentration, while the dotted line shows what would happen if conductivity depended solely on numbers of ions present in unit volume of solution. The divergence between the two lines gets larger as the concentration of the solution increases. This general behaviour is typical of solutions of all strong electrolytes; indeed, concentrated solutions of very soluble electrolytes often show a maximum conductivity value which then decreases on further concentration of the solution, e.g. calcium chloride solutions.

What then is the reason for solutions of strong electrolytes, particularly more concentrated ones, having lower conductivities than might reasonably be expected simply on the basis of ionic concentration? The answer lies in the fact that ionic distances become less as concentration increases. This results in greater and greater attraction between positive and negative ions with a consequent reduction in ionic speeds. The phenomenon is called **ionic interference**.

It is reasonable to assume that the distribution of ions in a solution of a strong electrolyte is not a purely random one. A given ion would be expected to be surrounded at close quarters by rather more ions of opposite than of the same charge (in the extreme case this is exactly what is observed in the symmetrical structure of an ionic solid). This is depicted in fig. 23.6(a) which shows a symmetrical ionic atmosphere surrounding a single positive ion. When an electric field is applied ions of opposite charge migrate in opposite directions and the ionic atmosphere becomes asymmetric. The reason for this asymmetry is due to the fact that it takes a finite time to establish an ionic atmosphere around a moving ion, which continually builds up in front and decays behind it. A direct consequence of this asymmetry, which is depicted in fig. 23.6(b), is that there is a net attraction on the ion in question, tending to shift it back again to its central position and hence its motion is retarded.

FIG. 23.6. *(a) A symmetrical ionic atmosphere surrounding a stationary positive ion (b) An asymmetrical ionic atmosphere induced by the application of an electric field*

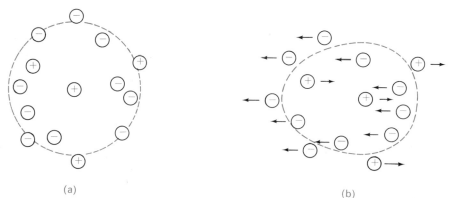

(a)   (b)

(b) Variation of conductivity with concentration for a weak electrolyte
The conductivity data shown in Table 23.2 for solutions of ethanoic acid are plotted graphically in fig. 23.7. How can we explain the shape of this curve? We certainly cannot invoke ionic interference as we did in the previous example, since a weak electrolyte produces very few ions in solution. In fact application of equation (7) (p. 285) shows that the degree of ionisation of 0·01M ethanoic acid solution is 0·042 or 4·2%, while the application of the more accurate equation (6)* (p. 284) gives a value of 0·125 or 12·5% for the degree of ionisation of 0·001M ethanoic acid solution. As the concentration of ethanoic acid increases there is an increase in the number of ethanoic acid molecules but a smaller proportion of them are ionised as we have just seen. The variation in conductivity of a weak electrolyte with concentration is due to an interplay of these two opposing effects. Pure ethanoic acid is an exceedingly poor conductor of electricity so we may anticipate that the extension of fig. 23.7 would show the conductivity reaching a maximum value followed by a steady decrease.

* The accurate equation must be used here since $(1 - \alpha)$ no longer approximates to the value 1.

415

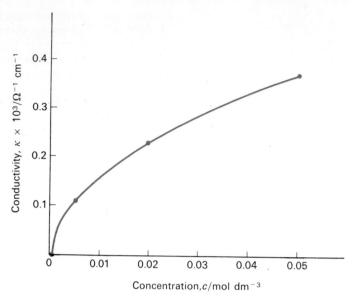

Molar conductivity

When comparing the conductivities of solutions of electrolytes, it is clearly advantageous to have a measure which refers to equal amounts of the solute. Such a measure is obtained by dividing the electrolytic conductivity (which up until now we have called simply the conductivity) by the concentration of the solution. It is called the **molar conductivity** and is denoted by the symbol Λ. Thus

$$\text{Molar conductivity } (\Lambda) = \frac{\text{Electrolytic conductivity } (\kappa)}{\text{Concentration } (c)}$$

In the basic SI units the concentration must be expressed in units of mol m^{-3} and since the corresponding units of electrolytic conductivity are Ω^{-1} m^{-1} it follows that molar conductivity has the units Ω^{-1} m^2 mol^{-1}. If electrolytic conductivity is expressed in terms of Ω^{-1} cm^{-1} then the concentration must be given in units of mol cm^{-3}. In this case the units of molar conductivity are Ω^{-1} cm^2 mol^{-1}. The connection between the two sets of units is as follows:

$$\Omega^{-1} \text{ m}^2 \text{ mol}^{-1} = 10^4 \ \Omega^{-1} \text{ cm}^2 \text{ mol}^{-1}$$

The molar conductivities of solutions of potassium chloride and ethanoic acid are given in Table 23.3 and are plotted graphically in fig. 23.8. These values have been obtained from the electrolytic conductivity and concentration data listed in Table 23.2. A little care is needed when converting electrolytic conductivity into molar conductivity and we illustrate this with an example.

A solution of potassium chloride of concentration 0·0005 mol dm^{-3} has an electrolytic conductivity of 0·0739 $\times$ 10^{-3} Ω^{-1} cm^{-1}. The concentration is 0·0005 mole in 1000 cm^3 or 5 $\times$ 10^{-7} mol cm^{-3}. If we divide the electrolytic conductivity by this concentration we obtain:

$$\Lambda = \frac{0 \cdot 0739 \times 10^{-3}}{5 \times 10^{-7}} = 147 \cdot 8$$

The molar conductivity of this solution is 147·8 Ω^{-1} cm^2 mol^{-1} which in SI units becomes 147·8 × 10^{-4} or 1·478 × 10^{-2} Ω^{-1} m^2 mol^{-1}. The other values are obtained in a similar manner.

Table 23.3 The molar conductivities of aqueous solutions of potassium chloride and ethanoic acid at 298 K

Concentration $c \times 10^3$/mol cm^{-3}	Molar conductivity, Λ/Ω^{-1} cm^2 mol^{-1}	
	KCl	CH$_3$COOH
0·0005	147·8	67·8
0·001	147·0	49·2
0·005	143·6	23·0
0·01	141·0	16·3
0·02	138·5	11·6
0·05	133·4	7·4
0·10	129·0	—
0·20	124·0	—
0·50	117·2	—
1·00	112·0	—

FIG. 23.8. *Variation of molar conductivity for solutions of potassium chloride and ethanoic acid at 298 K*

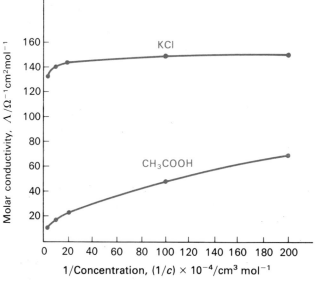

Note that the molar conductivity values in fig. 23.8 are plotted against the reciprocal of the concentration. The graph for potassium chloride shows that a limiting value of molar conductivity is being approached as the concentration diminishes. However, it is difficult to determine the value from a graph such as this. Ethanoic acid, on the other hand, behaves completely differently with molar conductivity increasing steadily as the solution gets more and more dilute. All strong electrolytes exhibit the same kind of behaviour as potassium chloride; similarly, weak electrolytes behave very much like ethanoic acid.

In order to determine the limiting value for the molar conductivity of a strong electrolyte, we make use of an empirical equation first discovered by Kohlrausch, namely

417

Λ = molar conductivity
Λ° = molar conductivity at zero concentration
b = constant
c = concentration

$$\Lambda = \Lambda^\circ - b\sqrt{c}$$

As can be seen from fig. 23.9 this equation holds quite well for potassium chloride solutions of low concentration (between A and B). This linear part of the graph can be extrapolated to cut the y-axis at a value of $150.0 \ \Omega^{-1} \ cm^2 \ mol^{-1}$, which is the molar conductivity at zero concentration, Λ°, for potassium chloride. Between the points B and C the graph begins to level off, and the divergence from linearity over this region can be judged by extrapolating the line AB backwards to D (shown as the dotted line). Solutions of other strong electrolytes exhibit the same kind of behaviour.

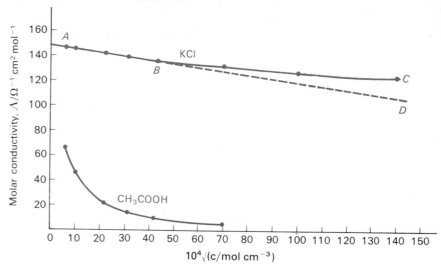

FIG. 23.9. *Graphs of molar conductivity against $\sqrt{concentration}$ for solutions of potassium chloride and ethanoic acid at 298 K*

Solutions of weak electrolytes, such as ethanoic acid, do not obey the Kohlrausch equation because even in dilute solution there is a significant increase in degree of ionisation with further dilution.

Kohlrausch's Law of Independent Migration of Ions

By 1875, Kohlrausch had obtained the molar conductivities at zero concentration for a number of strong electrolytes. He observed that the difference between these values for pairs of salts having a common ion was close to a constant value. This is illustrated in Table. 23.4.

Table 23.4 Some molar conductivities at zero concentration, $\Lambda^\circ/\Omega^{-1} \ cm^2 \ mol^{-1}$ (298 K)

	KI (150·4)	KCl (149·9)	Difference 0·5
	NaI (126·9)	NaCl (126·5)	Difference 0·4
Difference	23·5	23·4	

He concluded that the molar conductivity at zero concentration for a strong electrolyte could be expressed as the sum of separate contributions from the constituent ions:

$$\Lambda^\circ = \Lambda^\circ_+ + \Lambda^{\circ*}_- \tag{4}$$

* The above equation applies to a 1:1 electrolyte. For an electrolyte such as $BaCl_2$ the equation becomes $\Lambda^\circ = \Lambda^\circ_+ + 2\Lambda^\circ_-$ but we shall not be concerned with such cases in this book.

We have seen that it is not possible to determine the molar conductivity at zero concentration for a weak electrolyte directly. However, this value can be obtained by applying Kohlrausch's Law. For example, the molar conductivity at zero concentration for ethanoic acid may be obtained by determining Λ° values for sodium ethanoate, sodium chloride, and hydrochloric acid (all strong electrolytes). The simple calculation is shown below, the numerical values of Λ° being those at a temperature of 298 K.

$$\Lambda^\circ\ (CH_3COOH) = \Lambda^\circ\ (CH_3COONa) + \Lambda^\circ\ (HCl) - \Lambda^\circ\ (NaCl)$$
$$\qquad\qquad\quad (91\cdot0) \qquad\qquad + (426\cdot2) \quad - (126\cdot5)$$
$$\Lambda^\circ\ (CH_3COOH) = 390\cdot7\ \Omega^{-1}\ cm^2\ mol^{-1}$$

The application of molar conductivity measurements

We conclude this section by describing a number of applications of conductivity measurements.

(a) *Determination of the ionisation constant of a weak electrolyte*
The ratio Λ/Λ°, where Λ is the molar conductivity of a weak electrolyte at a particular concentration, may be identified with the degree of ionisation, α, of the weak electrolyte:

$$\alpha = \Lambda/\Lambda^\circ$$

We have seen in chapter 17 (p. 284) that the acid dissociation (ionisation) constant of a weak acid is given by the equation:

$$K_a = \frac{\alpha^2}{(1-\alpha)V}$$
$$= \frac{\alpha^2 c}{(1-\alpha)} \qquad (5)$$

$c = 1/V$,
i.e., conc. in mol dm^{-3}

Substituting for α in equation (5) we have:

$$K_a = \frac{(\Lambda/\Lambda^\circ)^2 c}{(1 - \Lambda/\Lambda^\circ)}$$

(b) *Ionic interference in solutions of strong electrolytes*
The ratio Λ/Λ° for solutions of strong electrolytes cannot be identified with degree of ionisation, since many strong electrolytes already exist in the form of ions in the solid state (p. 65). This ratio, which is also represented by the symbol α, is a measure of the degree of ionic interference at a particular concentration. Values obtained from conductivity measurements agree with those based on colligative properties (p. 394).

(c) *The ionic product for water*
The ionic product for water, K_w, was used in chapter 17 as a means of defining a pH scale. The value of this constant may be determined in the following manner.

Using a sensitive bridge circuit (p. 411), the electrolytic conductivity of extremely pure water is determined at a temperature of 298 K. From this value, the molar conductivity of water is now calculated. By the application of Kohlrausch's Law, the molar conductivity at zero concentration is calculated and the degree of ionisation of water obtained. It is now a simple matter to evaluate $[H^+(aq)]_{eqm} = [OH^-(aq)]_{eqm}$ and hence the value of K_w.

The calculation is shown below. Electrolytic conductivity, $\kappa = 5\cdot5 \times 10^{-8}\ \Omega^{-1}\ cm^{-1}$ (298 K). Since 1 mole of water has a volume of 18 cm^3, its concentration is 1/18 mol cm^{-3}. The molar conductivity of water is given by the equation:

$$\Lambda = \frac{\kappa}{c} = \frac{5\cdot5 \times 10^{-8}}{1/18}$$
$$= 9\cdot9 \times 10^{-7}\ \Omega^{-1}\ cm^2\ mol^{-1}$$

$$\Lambda^\circ\ (H_2O) = \Lambda^\circ\ (NaOH) + \Lambda^\circ\ (HCl)\ -\ \Lambda^\circ\ (NaCl)$$
$$(248\cdot4) + (426\cdot2) - (126\cdot5)$$
$$\Lambda^\circ\ (H_2O) = 548\cdot1\ \Omega^{-1}\ cm^2\ mol^{-1}$$

The degree of ionisation of water, α, is given by the equation:

$$\alpha = \frac{\Lambda}{\Lambda^\circ} = \frac{9\cdot9 \times 10^{-7}}{548\cdot1} = 1\cdot806 \times 10^{-9}$$

We have now to be rather careful. The concentration term, c, in an equilibrium expression is given in units of mol dm^{-3}. Since there is 1/18 mole of water in 1 cm^3, there will be 1000/18 moles in 1 dm^3.

$$[H^+(aq)]_{eqm} = [OH^-(aq)]_{eqm} = \alpha c$$
$$= \frac{1\cdot806 \times 10^{-9} \times 1000}{18}$$
$$= 1\cdot003 \times 10^{-7}\ mol\ dm^{-3}$$

The ionic product for water, K_w, is given by the equation:

$$K_w = [H^+(aq)]_{eqm}[OH^-(aq)]_{eqm}$$
$$= (1\cdot003 \times 10^{-7})^2$$
$$= 10^{-14}\ mol^2\ dm^{-6}\ (at\ 298\ K)$$

(*d*) *The determination of solubility products*
We illustrate the procedure by using silver chloride as an example.

The electroytic conductivity of a saturated solution of silver chloride is $1\cdot887 \times 10^{-6}\ \Omega^{-1}\ cm^{-1}$ at 298 K. In view of this small value we must subtract the electrolytic conductivity of pure water ($5\cdot5 \times 10^{-8}\ \Omega^{-1}\ cm^{-1}$). The corrected conductivity is thus $1\cdot832 \times 10^{-6}\ \Omega^{-1}\ cm^{-1}$.

If the solubility of silver chloride is x mol dm^{-3} at 298 K, then it will be $10^{-3}x$ mol cm^{-3}. Since the substance is so very slightly soluble we may assume that the calculated value of Λ approximates very closely to Λ°, i.e. ionic interference in the saturated solution is negligible.

$$\Lambda^\circ = \frac{1\cdot832 \times 10^{-6}}{10^{-3}x}$$

But

$$\Lambda^\circ\ (AgCl) = \Lambda^\circ\ (AgNO_3) + \Lambda^\circ\ (KCl) - \Lambda^\circ\ (KNO_3)$$
$$(133\cdot4) + (149\cdot9) - (145\cdot0)$$
$$\Lambda^\circ\ (AgCl) = 138\cdot3\ \Omega^{-1}\ cm^2\ mol^{-1}$$

We therefore have

$$\frac{1\cdot832 \times 10^{-6}}{10^{-3}x} = 138\cdot3$$

On rearranging the above expression we get

$$x = \frac{1\cdot832 \times 10^{-3}}{138\cdot3}$$
$$= 1\cdot325 \times 10^{-5}$$

From the definition of solubility product (p. 352) we may write:

$$K_s = [Ag^+(aq)]_{eqm}[Cl^-(aq)]_{eqm}$$
$$= (1\cdot325 \times 10^{-5})^2$$
$$= 1\cdot76 \times 10^{-10} \text{ mol}^2 \text{ dm}^{-6}$$

23.3
Conductimetric titrations

FIG. 23.10. *Conductivity cell with dipping electrodes*

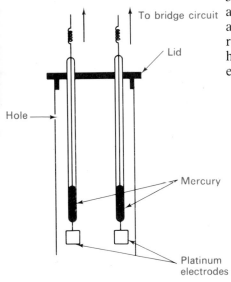

FIG. 23.11. *Apparatus used for studying changes of conductivity*

Reactions which show a significant change in electrolytic conductivity as they proceed can conveniently be studied by monitoring these changes, using a conductivity cell with dipping electrodes, previously coated with platinum black, (see fig. 23.10). We illustrate the method by considering the conductimetric titration of both a strong and a weak acid with a strong base.

A convenient volume of an aqueous solution of a strong acid (say, 50 cm³ 0.1M HCl) is placed in a beaker containing the conductivity cell, and the solution is stirred magnetically (see fig. 23.11). Portions of an aqueous solution of a strong base (say, M NaOH) are added and the resistance of the solution measured after each addition. The reason for having the acid much more dilute than the alkali is to minimise dilution effects, which would tend to mask any conductivity changes. The

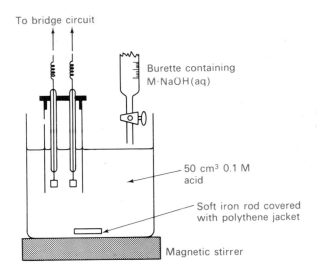

reciprocal of the measured resistance is proportional to the electrolytic conductivity (see equation (3), p. 412) so it is not essential to calculate the conductivity values at each stage of the titration.

The changes in conductivity (1/Resistance) for this titration are shown in the upper graph of fig. 23.12. As alkali is added the highly mobile $H^+(aq)$ ions of the acid are replaced by the slower $Na^+(aq)$ ions of the alkali, since virtually unionised water is produced. The conductivity falls in a linear manner along the line AB. After the end-point, the conductivity rises sharply along the line BC as additional $Na^+(aq)$ and $OH^-(aq)$ ions are added. The sharp rise in conductivity in this region is primarily due to the presence of the highly mobile $OH^-(aq)$ ions. The slope of the line BC is not as great as that of the line AB, since $OH^-(aq)$ ions are not as mobile as $H^+(aq)$ ions.

The lower graph of fig. 23.12 refers to the titration of a 0.1M solution of a weak acid (say, ethanoic acid). The conductivity rises along the line

$A'B'$ as weakly ionised ethanoic acid is replaced by increasing amounts of sodium ethanoate, i.e. $Na^+(aq)$ and $CH_3COO^-(aq)$ ions. After the end-point the conductivity rises sharply along the line $B'C'$ and this is primarily due to the presence of the highly mobile $OH^-(aq)$ ions that are added.

The interested reader might like to predict the shapes of the titration curves for the combinations strong acid/weak alkali and weak acid/weak alkali.

FIG. 23.12. *Changes in conductivity when (a) 0·1 M HCl is titrated with M NaOH (b) 0·1 M CH₃COOH is titrated with M NaOH*

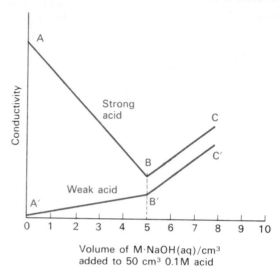

Volume of M·NaOH(aq)/cm³
added to 50 cm³ 0.1M acid

23.4
Questions on chapter 23

1 (a) State Faraday's laws of electrolysis. Describe an experiment to illustrate the second law.
   (b) Calculate the time in minutes necessary for a current of 10 A to deposit 1 g of copper from an aqueous solution of copper(II) sulphate.
   (c) Show why the ratio of the masses of copper and sodium deposited, under the appropriate conditions, by the same quantity of electricity is 1·38.
   (d) Calculate the charge on an electron given that the Avogadro constant is $6·02 \times 10^{23}$/mol. AEB

2 Describe the lattice structure of crystalline sodium chloride and show how it accounts for the characteristic physical properties of the substance.
   What products can be obtained from the electrolysis of aqueous sodium chloride under different conditions? Account for any differences as far as you can, writing equations for the reactions that occur.
   When a current of 0·25 A was passed through dilute aqueous sodium chloride for 32 min 10 s, 56 cm³ (corrected to s.t.p.) of gas was evolved at the cathode. Calculate the Faraday constant from these data and hence derive a value for the Avogadro constant, obtaining any other data from a Data Book. C (Overseas)

3 (a) Define the terms (i) *electrolytic conductivity (specific conductance)*, (ii) *molar conductivity* of an aqueous solution of an electrolyte.
   (b) The electrolytic conductivity of a saturated aqueous solution of thallium(ı) chloride, TlCl, at 25°C is $2·40 \times 10^{-3}$ ohm^{-1} cm^{-1}. The molar conductivities at infinite dilution (zero concentration) of thallium(ı) hydroxide, sodium hydroxide and sodium chloride are 273, 248 and 126 ohm^{-1} cm^2 mol^{-1} respectively. Estimate (i) the molar conductivity of thallium(ı) chloride, (ii) the solubility of TlCl in water at 25°C in mol dm^{-3}.
   (c) Explain concisely the *principles* involved in the purification of water by ion exchange for use in laboratory experiments. O

4 (a) What is meant by the terms *electrolytic conductivity* and *molar conductivity*? Describe how you would determine the conductivity of a strong electrolyte of known concentration.

Show graphically how the molar conductivity will vary with dilution for
  (i) strong electrolytes,
  (ii) weak electrolytes.
(b) The molar conductivity of a solution of sodium chloride containing 12·5 g
    solid in 1 kg of water was found to be $96·2 \times 10^4$ S m^2 mol^{-1}
    ($96·2$ Ω^{-1} cm^2 mol^{-1}) at 298 K, whereas the corresponding value for in-
    finite dilution was $12·6 \times 10^5$ S m^2 mol^{-1} (126 Ω^{-1} cm^2 mol^{-1}).
      Calculate the freezing point of this solution. (The molar mass of sodium
    chloride is 58·5). The molar freezing point depression constant for water is
    1·86 K kg mol^{-1}). AEB

5 (a) Describe concisely, but including essential experimental detail, how the
       electrolytic conductivity of a solution of an electrolyte can be determined.
   (b) What is a buffer solution? If you were asked to prepare a set of buffer
       solutions with pH values between 4·5 and 5·5, what kind of substances or
       solutions would you require? Indicate the steps you would take to prepare
       such buffer solutions. (A qualitative account only is required.)
   (c) State one use which is made of ion exchange (either in the laboratory or in
       everyday life), and explain the principles underlying this use. O

6 Explain the following terms:
   (a) *electrolytic (specific) conductivity*,
   (b) Λ, *molar conductivity* (**or** *equivalent conductivity*),
   (c) Λ_∞, *molar conductivity* (**or** *equivalent conductivity*) *at infinite dilution*
       (zero concentration).
   The Λ_∞ values at 25°C for some electrolytes are as follows:

	Λ_∞/S cm^2 mol^{-1}
CH$_3$COONa	91
HNO$_3$	421
NaNO$_3$	122
KNO$_3$	145

   [Ω denotes ohm; S = Ω^{-1}, where S denotes the siemen.]
   Calculate Λ_∞ at 25°C for
   (i) CH$_3$COOK, and
   (ii) CH$_3$COOH.
   What is *Ostwald's dilution law*? Illustrate your answer by reference to ethanoic acid.
     If the value of Λ for 0·016 M CH$_3$COOH (aq) is 13·0 S cm^2 mol^{-1} at 25°C,
   what is the dissociation constant of the acid at this temperature? C (Overseas)

7 Sketch graphs to show how the molar electrical conductances of weak and strong
   electrolytes depend on the square root of the concentration. Briefly account for
   what is observed, explain the significance of using *molar* conductance and state
   the units in which concentration is measured here.
     Derive an expression relating the degree of ionization (dissociation) of a *very
   weak* acid to its acidity constant (acid ionization or dissociation constant) and
   molar concentration.
     Without giving experimental details, show briefly how the validity of the
   expression might be tested experimentally and the results displayed graphically.
     Evaluate the molar conductance at zero concentration for potassium chloride,
   potassium ethanoate, hydrochloric acid and *ethanoic (acetic) acid*, using the
   following data collected at 18°C.

Concentration/mol m^{-3}	Molar conductance $\Lambda c/\Omega^{-1}$ cm^2 mol^{-1}		
	KCl	CH$_3$COOK	HCl
1·00	130	99·0	374
9·00	127	95·0	368
25·00	123	91·0	363
81·00	117	85·0	352
100	115	83·0	350

S

8 (a) Explain what you understand by *solubility product* and indicate the conditions under which the principle can be applied. Discuss its application to one of the group tests in the systematic analysis of cations.

(b) Calculate the solubility product of silver chloride, AgCl, from the following data.

The electrolytic (specific) conductivity of a saturated solution of silver chloride in pure water at 25°C is $3 \cdot 41 \times 10^{-6}$ ohm^{-1} cm^{-1}, and the electrolytic conductivity of the water alone is $1 \cdot 60 \times 10^{-6}$ ohm^{-1}cm^{-1}. The salts listed below have the following molar conductivities at infinite dilution (zero concentration) at 25°C:

AgNO$_3$ 133·4 ohm^{-1} cm^2 mol^{-1};
KNO$_3$ 145·0 ohm^{-1} cm^2 mol^{-1};
KCl   149·9 ohm^{-1} cm^2 mol^{-1}.

Mention briefly any assumptions you have made in calculating your answer.

O (S)

Kinetics

Rates of chemical reactions

24.1 Introduction

We have been careful to point out in previous chapters that reactions which are favoured in the thermodynamic sense may not take place at a sufficient rate to be detectable. For example, carbon and oxygen are thermodynamically less stable than carbon dioxide at 298 K, yet coke does not spontaneously catch fire in air. Similarly, diamond is thermodynamically less stable than graphite, but at 298 K the rate of conversion of diamond to graphite is, happily, infinitesimally small. Thermodynamics has nothing to say about the **rate** at which reactants can be transformed into products; this is the province of chemical kinetics and is the subject matter of the following two chapters.

The study of chemical kinetics (reaction rates) is clearly of importance to the industrial chemist who is concerned with producing a particular product as economically as possible. Consider for example the synthesis of methanol from carbon monoxide and hydrogen:

$$CO(g) + 2H_2(g) \longrightarrow CH_3OH(l)$$

This reaction is thermodynamically feasible at a temperature of 298 K but in practice the reaction proceeds infinitesimally slowly at this temperature. In order to achieve a satisfactory rate for this reaction the chemist would consider the possibilities of increasing the concentration of the reactants and increasing the temperature, for these are variables that increase the number of encounters between reactant molecules. As a last resort he might search for a suitable catalyst to enable an easier reaction path to be followed and thus allow a lower temperature to be employed. By carrying out experiments along these lines the optimum conditions for the synthesis of methanol have been discovered; they are a temperature of about 400°C (673 K), a pressure in the region of 300 atmospheres and a catalyst containing zinc oxide and chromium(III) oxide.

24.2 Increasing the rate of chemical reactions

Temperature, concentration of reacting substances, radiation (e.g. the effect of light on silver salts, which is the basis of photography), surface area of reacting solids and the presence of a catalyst are all found by experiment to influence the rate at which reactions take place. For example, the rate at which magnesium reacts which dilute hydrochloric acid can be investigated using the apparatus shown in fig. 24.1. The most convenient way to follow the rate of this reaction is by measuring the volume of hydrogen produced at given intervals of time.

Fig. 24.2 indicates the type of curves obtained when volume of hydrogen evolved is plotted against the time, the inset table attached showing the experimental conditions. It is clear from the results that the rate of this reaction is increased by increasing the surface area of the magnesium, increasing the concentration of the acid, and by raising the

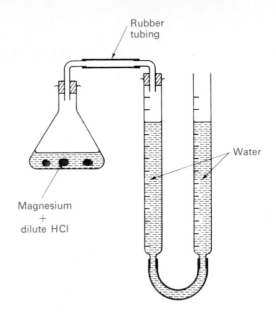

FIG. 24.1. *Apparatus used to study the rate of reaction between magnesium and dilute HCl*

Rubber tubing

Water

Magnesium
+
dilute HCl

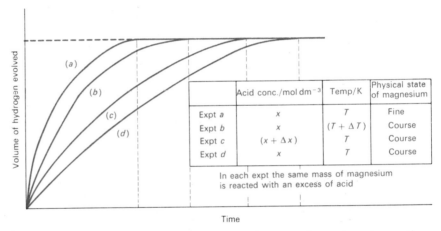

FIG. 24.2. *Typical results for the reaction between magnesium and dilute HCl*

Volume of hydrogen evolved

(a)

(b)

(c)

(d)

	Acid conc./mol dm^{-3}	Temp/K	Physical state of magnesium
Expt a	x	T	Fine
Expt b	x	$(T + \Delta T)$	Course
Expt c	$(x + \Delta x)$	T	Course
Expt d	x	T	Course

In each expt the same mass of magnesium is reacted with an excess of acid

Time

temperature. For any given experiment, the rate of reaction at a particular time can be obtained by drawing a tangent to the curve at this point; the gradient of this tangent will be the rate of reaction at this time (units volume time^{-1}) and is given by the usual calculus notation $d(\text{volume})/dt$.

Drawing a tangent to a curve is an operation which is subject to appreciable error, and we shall see that in some instances it is possible to derive mathematical equations from which reaction rates can be calculated more accurately.

24.3
Orders of reaction

Consider the alkaline hydrolysis of ethyl ethanoate, the stoichiometric equation for this reaction being:

$$CH_3COOC_2H_5 + OH^-(aq) \longrightarrow CH_3COO^-(aq) + C_2H_5OH$$

The rate of this reaction, at a fixed temperature, can be expressed in several ways; either as a decrease in the concentration of ethyl ethanoate or hydroxyl ions, i.e.

$$\frac{-d[CH_3COOC_2H_5]}{dt} = \frac{-d[OH^-(aq)]}{dt}$$

or as an increase in the concentration of ethanoate ions or ethanol, i.e.

$$\frac{d[CH_3COO^-(aq)]}{dt} = \frac{d[C_2H_5OH]}{dt}$$

Note that the increase in concentration of both ethanoate ions and ethanol are equal to the decrease in concentration of both ethyl ethanoate and hydroxyl ions.

Careful experiments carried out on this chemical reaction have established that its rate is proportional to the concentrations of both ethyl ethanoate and hydroxyl ions, i.e.

$$\frac{-d[CH_3COOC_2H_5]}{dt} \propto [CH_3COOC_2H_5][OH^-(aq)]$$

or

$$\frac{-d[CH_3COOC_2H_5]}{dt} = k[CH_3COOC_2H_5][OH^-(aq)] \qquad (1)$$

Since the experimentally determined rate equation is found to depend on the concentrations of ethyl ethanoate and hydroxyl ions (both raised to the power one), it is said to be a second order reaction (first order with respect to each of the reactants). The constant of proportionality, k, in equation (1) is called the rate coefficient or **rate constant**. It is a constant for a particular reaction at a fixed temperature. Knowing the rate constant and order of reaction, we can calculate the rate of reaction at any concentration at that temperature.

It might appear from the above discussion that the form of the rate equation could be deduced from the stoichiometric equation for the reaction, but this is seldom true. For example, the hydrolysis of 2-chloro-2-methylpropane, $(CH_3)_3CCl$, by aqueous alkali may be represented by the stoichiometric equation:

$$CH_3-\underset{\underset{CH_3}{|}}{\overset{\overset{CH_3}{|}}{C}}-Cl + OH^-(aq) \longrightarrow CH_3-\underset{\underset{CH_3}{|}}{\overset{\overset{CH_3}{|}}{C}}-OH + Cl^-(aq) \qquad (2)$$

This reaction is not a second order reaction as might be supposed in view of the form of the above stoichiometric equation. It is found to be a first order reaction (first order with respect to the concentration of 2-chloro-2-methylpropane and zero order with respect to the concentration of aqueous alkali, i.e. the concentration of the alkali does not appear in the rate equation):

$$\frac{-d[(CH_3)_3CCl]}{dt} = k[(CH_3)_3CCl]$$

Similarly the oxidation of aqueous hydriodic acid by hydrogen peroxide is not a fifth order reaction as the unwary might suppose in view of the stoichiometric equation:

$$H_2O_2 + 2H^+(aq) + 2I^-(aq) \longrightarrow 2H_2O + I_2 \qquad (3)$$

This reaction is second order (first order with respect to the concentrations of both hydrogen peroxide and aqueous iodide ions) as the experimentally determined rate equation shows:

$$\frac{-d[H_2O_2]}{dt} = \frac{d[I_2]}{dt} = k[H_2O_2][I^-(aq)]$$

Fractional orders are known; thus the decomposition of ethanal vapour at a high temperature is represented by a deceptively simple stoichiometric equation:

$$CH_3CHO(g) \longrightarrow CH_4(g) + CO(g) \qquad (4)$$

However, the rate equation is,

$$\frac{-d[CH_3CHO]}{dt} = \frac{d[CH_4]}{dt} = k[CH_3CHO]^{3/2}$$

which shows that the order is 1·5.

As a final example consider the synthesis of hydrogen bromide from its elements:

$$H_2(g) + Br_2(g) \longrightarrow 2HBr(g) \qquad (5)$$

This seemingly simple reaction has a complicated rate equation:

$$\frac{d[HBr]}{dt} = \frac{k'[H_2][Br_2]^{1/2}}{1 + k''([HBr]/[Br_2])}$$

It is not easy to visualise what is meant by the order of a reaction with a rate equation of such complexity. However, at the start of the reaction the denominator of the rate equation will be close to unity and, under these conditions, the order will approximate to 1·5 (first order with respect to hydrogen and 0·5 order with respect to bromine).

We shall examine some of the above reactions in detail later, to see what can be deduced from their orders. Suffice it to say at this stage that **the order of a reaction is an experimentally determined quantity, which is seldom in any way connected with the stoichiometry of the particular reaction.** In the few instances where the rate equation is what one would predict on the basis of the stoichiometric equation, then these reactions proceed in a simple step, i.e. they are elementary reactions.

Experimental methods of determining orders of reactions

The first stage in determining the order of a particular reaction is to set up the reaction in a thermostat at a selected temperature and then monitor the changes of concentration as the reaction proceeds. Since the stoichiometric equation for the reaction will be known, it is only necessary to follow the changes of concentration of one substance, either one of the reactants or one of the products. Several methods for monitoring the course of a chemical reaction which is essentially complete over a period of minutes or hours are available as outlined below.

429

The techniques for following very fast reactions, i.e. those which are completed within seconds or indeed milliseconds, are outside the scope of this book.

(a) Volumetric analysis

In this method a known volume of the reaction mixture is removed at various times and rapidly added to a larger volume of ice-cold water which hopefully effectively stops the reaction or at least slows it down considerably. The concentration of one of the substances present in the reaction mixture is now determined by a suitable titration.

This method could be used to follow the rate of hydrolysis of 2-chloro-2-methylpropane (equation (2), p. 428), the alkali being titrated with standard acid.

(b) Pressure changes

If a reaction carried out in the gaseous phase results in a change of pressure, it can be followed by recording the pressure at various intervals of time. This method could be used to study the thermal decomposition of ethanal vapour at high temperature (equation (4), p. 429), since the reaction proceeds with an increase in pressure.

(c) Conductivity

If the electrolytic conductivity changes sufficiently during the course of a reaction then this is often a convenient method of analysis. In the hydrolysis of 2-chloro-2-methylpropane by aqueous alkali (equation (2), p. 428) there will be a decrease in conductivity as the reaction proceeds since the faster moving $OH^-(aq)$ ions are replaced by the slower $Cl^-(aq)$ ions ($\Lambda^o_- = 198\cdot6\ \Omega^{-1}\ cm^2\ mol^{-1}$ for $OH^-(aq)$ and $\Lambda^o_- = 76\cdot4\ \Omega^{-1}\ cm^2\ mol^{-1}$ for $Cl^-(aq)$, both at 298 K).

(d) Spectroscopy

If one of the species in a reaction absorbs radiation in a convenient part of the electromagnetic spectrum, then the reaction may be followed by recording the changes in the intensity of absorption. In the oxidation of aqueous hydriodic acid by hydrogen peroxide (equation (3), p. 429) only iodine absorbs in the visible region and the intensity of the absorption will increase as reaction proceeds.

Similarly, in the direct synthesis of hydrogen bromide (equation (5), p. 429) bromine absorbs visible light and the intensity of this absorption will fall as reaction proceeds. Note that this gas phase reaction cannot be studied by method (b) above since no pressure change occurs in the reaction.

(e) Mass spectrometry and chromatography

In both these methods the basic idea is to remove portions of the reaction mixture at varying time intervals, and then to use the techniques either singly or in concert to analyse the samples (p. 196, and p. 362).

First order reactions

Let us suppose that a first order reaction may be represented by the general equation:

$$A \longrightarrow Products$$

Since the reaction is first order we may write:

$$\frac{-d[A]}{dt} = k[A]$$

If the initial concentration of A is represented as a (in appropriate concentration units) and $(a - x)$ when the reaction has been in progress for a time t, then we may write

$$\frac{-d(a - x)}{dt} = k(a - x)$$

or

$$\frac{dx}{dt} = k(a - x)$$

On rearranging the above equation and integrating we have,

$$\int \frac{dx}{(a - x)} = \int k \, dt$$

or

$$-\ln (a - x) = kt + \text{const.}$$

The integration constant can be evaluated by noting that $x = 0$ when $t = 0$, thus:

$$-\ln a = \text{const.}$$

The complete equation thus becomes:

$$\ln a - \ln (a - x) = kt \tag{6}$$

Thus using calculus we have:

$$\ln \frac{a}{(a - x)} = kt \tag{7}$$

If we rearrange equation (7) it becomes:

$$t = \frac{1}{k} \ln \frac{a}{(a - x)} \tag{8}$$

A particular reaction is first order if a plot of $\ln a/(a - x)$ against t gives a straight line. Alternatively it is often less time consuming to plot $\ln (a - x)$ against t (see equation (6) above). In this latter case the straight line has a negative gradient and the intercept is $\ln a$; both methods of procedure are shown in fig. 24.3. Note that in each case the gradient is the value of the rate constant k which, for a first order reaction, has units of time^{-1} (this is clear from the form of equation (8)).

Let $t_{1/2}$ be the time needed for a first order reaction to be half completed, then the value of x will be $a/2$. Substituting into equation (8) we get,

431

$$t_{1/2} = \frac{1}{k} \ln \left(a/\tfrac{1}{2}a\right) = \frac{1}{k} \ln 2$$

or

$$t_{1/2} = \frac{0 \cdot 6932}{k} \tag{9}$$

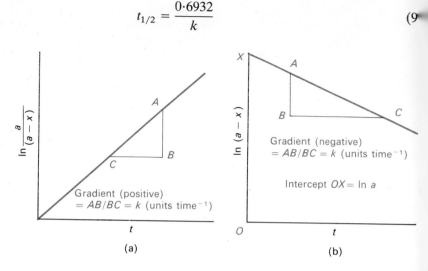

FIG. 24.3. *Two ways of plotting the data for a first order reaction*

(a) *ln* $\dfrac{a}{(a-x)}$ *plotted against* t

(b) *ln* (a − x) *plotted against* t

Note that the scales used along the y-axis are different in each case

As equation (9) shows, the time taken for the reaction to be half completed $t_{1/2}$, or the half-life as it is often called, is independent of the initial concentration a. Showing that the half-life of a particular reaction is independent of the initial concentration amounts to proving that the reaction is first order. This means that the time taken for a first order reaction to proceed from 0 to 50% conversion is exactly the same as the time needed to proceed from 50 to 75% conversion and so on. Simple inspection of experimental data will often allow a quick decision to be made as to whether a reaction is first order or not.

Typical examples of first order reactions include radioactive decay (p. 53) and the reactions represented by the stoichiometric equations below:

$$\begin{array}{c} CH_2 \\ \triangle \\ H_2C \quad CH_2 \end{array} \longrightarrow CH_3-CH=CH_2$$

$$2N_2O_5(g) \longrightarrow 4NO_2(g) + O_2(g)$$

$$(CH_3)_3C-Cl + OH^-(aq) \longrightarrow (CH_3)_3C-OH + Cl^-(aq)$$

Example 1
A gas phase decomposition was monitored by measuring the total pressure p in the reaction vessel:

time/s	0	300	600	900	1200
p/atm	0·367	0·526	0·604	0·662	0·691

The pressure eventually reached an asymptotic value of 0·736 atm. Show that the reaction is first order and estimate the times at which the reaction could be said to be 50% complete and 75% complete. What is the rate constant for this reaction?

Since the final pressure is double the initial pressure, within experimental error, it follows that the reactant gas decomposes into two other gases. The reaction may be represented as below, where the initial pressure of the reactant and the pressures of each constituent in the gaseous mixture are given at time t.

432

	A	→	B	+ C
Initial pressure/atm	0·367		0	0
Pressure/atm of each constituent at time t	$(0·367 - x)$		x	x

The pressure of the gas mixture at time t is given by the expression $(0·367 - x) + x + x$ or $(0·367 + x)$ atm.

We can calculate the values of x at varying times by equating this total pressure to the numerical values given in the original data; hence we can calculate the pressure of the reactant gas at varying times. For example, at time $t = 300$ we have $(0·367 + x) = 0·526$, which gives a value of $x = 0·159$. The corresponding pressure of gas A is thus $(0·367 - x) = (0·367 - 0·159)$ or $0·208$ atm. The remaining three values can be obtained in a similar manner and are as set out below.

time/s	0	300	600	900	1200
p/atm of reactant gas	0·367	0·208	0·130	0·072	0·043

Fig. 24.4 shows the graph obtained by plotting the pressure of the reactant gas A against time. As can be seen from the graph the time taken for the reaction to be 50% complete is about 375 s and the time taken for the reaction to be 75% complete is about 750 s. Since the time needed to go from 50 to 75% completion is 375 s, i.e. the same as the time for the reaction to go from 0 to 50% completion, the reaction is first order.

FIG. 24.4. *Graph of pressure of reactant gas A against time*

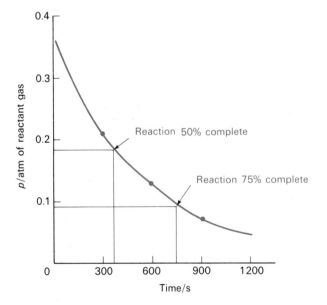

We may calculate the rate constant, k, for the reaction by employing the equation:

$$t_{1/2} = \frac{0·6932}{k}$$

Substituting $t_{1/2}$ into this equation gives a value of k equal to $1·848 \times 10^{-3}$ s^{-1}.

The problem could have been approached slightly differently. Thus the initial pressure of the reactant gas A, p_o, is the concentration term a in equation (7), p. 431. Similarly the pressure of the reactant gas A at time t, which may be represented as p_t is the concentration term $(a - x)$. Rewriting equation (7) in the form applicable to this example we have:

$$\ln (p_o/p_t) = kt \tag{10}$$

The values of p_o/p_t and $\ln (p_o/p_t)$ at varying times are given below.

time/s	0	300	600	900	1200
p_o/p_t	1	1·764	2·823	5·097	8·535
$\ln (p_o/p_t)$	0	0·568	1·038	1·629	2·144

Fig. 24.5 shows the graph obtained by plotting $\ln(p_o/p_t)$ against t. The gradient of the straight line graph, AB/BC, has a value of $0.8/450$ or 1.778×10^{-3}. The value of the rate constant, k, is thus $1.778 \times 10^{-3}\,s^{-1}$, which is likely to be a more accurate value than the previously estimated value of $1.848 \times 10^{-3}\,s^{-1}$ in view of the errors associated with drawing a curve. The value of $t_{1/2}$ is obtained from the equation:

$$t_{1/2} = \frac{0.6932}{k} = \frac{0.6932}{1.778 \times 10^{-3}} = 390\,s$$

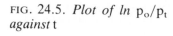

FIG. 24.5. *Plot of ln* p_o/p_t *against* t

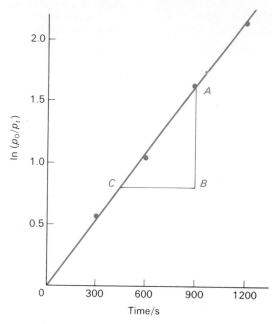

Second order reactions

Let us suppose that a second order reaction may be represented by the general equation:

$$A + B \longrightarrow Products$$

Since the reaction is second order we may write:

$$\frac{-d[A]}{dt} = k[A][B]$$

The initial concentrations of A and B may be the same or different. Let us assume that they are the same, since the integration of the rate equation is then simpler to handle, and that they may be represented as a (in appropriate concentration units). After the reaction has been in progress for a time t both concentrations will be $(a - x)$. We may therefore write,

$$\frac{-d(a - x)}{dt} = k(a - x)^2$$

or

$$\frac{dx}{dt} = k(a - x)^2$$

434

On rearranging the above equation and integrating we have,

$$\int \frac{dx}{(a-x)^2} = \int k\,dt$$

$$\text{or} \quad \frac{1}{(a-x)} = kt + \text{const.}$$

The integration constant is found by noting that $x = 0$ when $t = 0$, thus:

$$\frac{1}{a} = \text{const.}$$

The complete equation therefore becomes,

$$\frac{1}{(a-x)} - \frac{1}{a} = kt$$

Thus,

$$t = \frac{1}{k}\frac{x}{a(a-x)} \tag{11}$$

It is necessary to be rather careful in defining the conditions under which equation (11) applies to second order reactions. The reaction must involve the consumption of equal concentrations of both reactants A and B as it proceeds, unless there is only one reactant (see later), and the initial concentrations of A and B must be the same. With these restrictions in mind, a particular reaction is second order if a plot of $x/a(a-x)$ against t gives a straight line. The gradient of the graph is the value of the rate constant k which, in this case, has units of $conc^{-1}\ time^{-1}$.

The half-life of a second order reaction to which equation (11) applies can easily be found by substituting $x = a/2$. The final expression is,

$$t_{1/2} = \frac{1}{ka} \tag{12}$$

i.e. the half-life is inversely proportional to the initial concentration.

Typical examples of second order reactions which conform to equations (11) and (12) are the following.

$$CH_3COOC_2H_5 + OH^-(aq) \longrightarrow CH_3COO^-(aq) + C_2H_5OH$$
$$CH_3I + OH^-(aq) \longrightarrow CH_3OH + I^-(aq)$$
$$2NOCl(g) \longrightarrow 2NO(g) + Cl_2(g)$$

Note that in the last example there is only one reactant and the rate equation is accordingly written:

$$\frac{-d[NOCl]}{dt} = k[NOCl]^2$$

At the risk of repetition, equations (11) and (12) apply to the first two examples, only if equal concentrations of each reactant are employed.

Example 2
A solution of methyl ethanoate and sodium hydroxide, in a thermostat at 298 K, is initially 0·01 M with respect to both compounds. Titration of samples with standard acid at various times gives the following results:

435

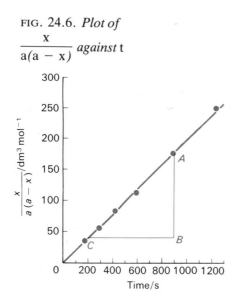

FIG. 24.6. *Plot of*

$$\frac{x}{a(a-x)} \text{ against } t$$

y-axis: $\dfrac{x}{a\,(a-x)}$/dm^3 mol^{-1}

x-axis: Time/s

time/s	0	180	300	420	600	900	1260
[OH$^-$(aq)]/mmol dm^{-3}	10·0	7·40	6·34	5·50	4·64	3·63	2·88

Show that the reaction is second order and calculate the rate constant.
The stoichiometric equation for the reaction is:

$$CH_3COOCH_3 + OH^-(aq) \longrightarrow CH_3COO^-(aq) + CH_3OH$$

Since equal concentrations of ester and alkali, initially of the same concentration, are consumed during the reaction we can apply equation (11) (p. 435). We need to calculate x, the amount of alkali consumed, and the expression $x/a(a-x)$ at the various times. These values are set out below.

time/s	0	180	300	420	600	900	1260
$(a-x)$/mmol dm^{-3}	10.0	7.40	6.34	5.50	4.64	3.63	2.88
x/mmol dm^{-3}	0	2.60	3.66	4.50	5.36	6.37	7.12
$\dfrac{x}{a(a-x)}$ / dm^3 mol^{-1}	0	35.1	57.7	81.8	115.1	175.5	247.2

The values of $x/a(a-x)$ plotted against t are shown in fig. 24.6. Since the graph is a straight line, the reaction must be second order. The gradient of the graph, AB/BC, has a value of $136/700$ or $1·94 \times 10^{-1}$. The value of the rate constant is thus $1·94 \times 10^{-1}$ dm^3 mol^{-1} s^{-1}.

Determination of orders of reaction by the initial rate method

As we have mentioned previously (p. 429) the oxidation of aqueous hydriodic acid (or an acidified solution of an iodide) by hydrogen peroxide is a second order reaction. This reaction has the stoichiometric equation as given below:

$$H_2O_2 + 2H^+(aq) + 2I^-(aq) \longrightarrow 2H_2O + I_2$$

This is an example of a second order reaction in which the amount of iodide ions consumed is twice as great as that of the hydrogen peroxide. Here we have an example where equation (11) (p. 435) is not applicable. How then do we determine the order of this reaction?

One convenient method is to allow the reaction to take place in the presence of a small amount of sodium thiosulphate solution and starch. By comparison with the reaction being studied, the rate of reaction between iodine and sodium thiosulphate is very fast, so that the characteristic blue colour that iodine produces with starch does not appear until the small concentration of sodium thiosulphate has been used up. The reciprocal of the time taken for a small but definite fraction of iodine to be produced will be directly proportional to the average rate of reaction over this period. This, in turn, will approximate quite closely to the initial rate of the reaction.

A series of experiments is carried out in which the initial acid and iodide concentrations are kept constant but the hydrogen peroxide concentration is varied. The times taken for a definite but small amount of iodine to form in each case is recorded. If the reciprocals of these times are denoted by $1/t$ then we may write:

m, n and *p* are the orders with respect to each reactant

$$\text{Initial rate } (\propto 1/t) = k[H_2O_2]_{\text{init}}^m[I^-(aq)]_{\text{init}}^n[H^+(aq)]_{\text{init}}^p$$

Since the initial concentrations of iodide ions and acid do not change in the experiments, the above equation simplifies to:

436

$$\text{Initial rate } (\propto 1/t) = k'[H_2O_2]_{\text{init}}^m$$

Experimental results show that the initial rate is directly proportional to the initial concentration of hydrogen peroxide, i.e. $m = 1$ and the reaction is first order with respect to hydrogen peroxide.

Further series of experiments in which the initial concentrations of hydrogen peroxide and acid are kept constant but the iodide ion concentration is varied show that $n = 1$, i.e. the reaction is also first order with respect to iodide ions. Finally by varying the initial acid concentration but keeping the concentrations of hydrogen peroxide and iodide ions constant it can be established that $p = 0$ within quite wide limits, i.e. the rate of reaction is zero order with respect to H^+(aq) ions.

Since the overall order is the sum of the separate orders the reaction is second order, i.e. $m + n + p = 2$.

Before leaving this initial rate method of determining orders of reaction, it is worth noting that the experimental results given in Example 2 (p. 435) only allow the overall order of the alkaline hydrolysis of methyl ethanoate to be determined. In order to establish that the reaction is first order with respect to each reactant, the initial rate method could be employed.

24.4
Reaction mechanisms

One of the main reasons for determining the order of a reaction is to see whether the experimentally determined order sheds any light on the detailed mechanism by which the reaction occurs. We consider four examples below.

The alkaline hydrolysis of 2-chloro-2-methylpropane

This reaction may be represented by the stoichiometric equation:

$$\underset{\overset{|}{CH_3}}{\overset{\overset{CH_3}{|}}{CH_3-C-Cl}} + OH^-(aq) \longrightarrow \underset{\overset{|}{CH_3}}{\overset{\overset{CH_3}{|}}{CH_3-C-OH}} + Cl^-(aq)$$

Experiment has established that the rate of this reaction is first order with respect to the organic chloride but zero order with respect to the OH^-(aq) ions, i.e. the concentration of OH^-(aq) ions does not influence the speed of the reaction. There must of course be a minimum concentration of OH^-(aq) ions present in order to obtain the product. Can we suggest a reaction mechanism which is consistent with these facts?

Well, we can at once rule out a one-stage reaction involving both reactants, since this would imply a reaction whose rate depended upon the concentrations of both reactants, i.e. a second order reaction. The only logical conclusion seems to be that the reaction must take place in stages. If the first stage only involves the organic chloride, and possibly the solvent, and is slow by comparison with the other stage(s), which involves the organic intermediate(s) and the hydroxide ions, then this slow stage would control the rate of reaction and first order kinetics would be expected. The suggested mechanism for this reaction is:

$$CH_3-\overset{\overset{\displaystyle CH_3}{|}}{\underset{\underset{\displaystyle CH_3}{|}}{C}}-Cl \xrightarrow{\text{slow}} CH_3-\overset{\overset{\displaystyle CH_3}{|}}{\underset{\underset{\displaystyle CH_3}{|}}{C^+}} + Cl^-$$

$$CH_3-\overset{\overset{\displaystyle CH_3}{|}}{\underset{\underset{\displaystyle CH_3}{|}}{C^+}} + OH^- \xrightarrow{\text{fast}} CH_3-\overset{\overset{\displaystyle CH_3}{|}}{\underset{\underset{\displaystyle CH_3}{|}}{C}}-OH$$

Although the kinetic results are consistent with the above mechanism they do not prove it. Indeed, reaction mechanisms can never be proved, since they are the result of reasoned theorising and are not facts. However, we can find another piece of evidence which lends support to this mechanism.

The reaction is normally carried out in aqueous ethanol as solvent, the ethanol being necessary in order to dissolve the organic chloride. However, if separate experiments are performed in a solvent mixture which is progressively richer in water, then the reaction rate is observed to increase. This is consistent with the fact that water is a more polar solvent than ethanol, i.e. the greater the polar nature of a solvent, the less difficult the process of ionisation becomes. In fact the ions in the above reaction scheme should really be shown solvated by solvent molecules.

The oxidation of an acidified solution of an iodide with hydrogen peroxide

The stoichiometric equation for this reaction is:

$$H_2O_2 + 2H^+(aq) + 2I^-(aq) \longrightarrow 2H_2O + I_2$$

As we have stated previously, this reaction is found to be first order with respect to both hydrogen peroxide and iodide ions but zero order with respect to hydrogen ions (within quite wide limits). The rate controlling step in this reaction must be the first step and involve one H_2O_2 molecule and one $I^-(aq)$ ion. We therefore write down a plausible elementary reaction consistent with these facts:

$$H_2O_2 + I^-(aq) \xrightarrow{\text{slow}} H_2O + IO^-(aq)$$

Since HIO is a weak acid a reasonable suggestion for the first fast step would be:

$$H^+(aq) + IO^-(aq) \xrightarrow{\text{fast}} HIO$$

The final step or steps must be consistent with the overall stoichiometry of the reaction and can only be guessed. A possible one-step process might be:

$$HIO + H^+(aq) + I^-(aq) \xrightarrow{\text{fast}} H_2O + I_2$$

The above mechanism may or may not be the correct one; all we can say with certainty is that the above reaction sequence is consistent with the experimental kinetic data.

The iodination of propanone in aqueous acid solution

Propanone reacts with iodine in aqueous acid solution according to the stoichiometric equation:

$$CH_3COCH_3 + I_2 \longrightarrow CH_2ICOCH_3 + H^+(aq) + I^-(aq)$$

The reaction is catalysed by $H^+(aq)$ ions and it will be noted that these are produced in the reaction, i.e. the reaction is autocatalytic. In order that the build up of concentration of $H^+(aq)$ produced during the reaction does not interfere with the kinetics of the reaction to any significant extent, the initial concentration of $H^+(aq)$ ions used is deliberately made much larger, in the relative sense, than the maximum concentration of $H^+(aq)$ ions produced in the reaction. Under these conditions the acid concentration remains effectively constant.

A convenient way of following the kinetics of this reaction is to use the initial rate method (p. 436), noting the times for a small but definite amount of iodine to disappear. A series of experiments is carried out in which the concentration of propanone, iodine and acid are varied in turn. The results of such experiments show that the kinetics of this reaction are in accordance with the rate equation:

$$\frac{-d[I_2]}{dt} = k[CH_3COCH_3][I_2]^0[H^+(aq)]$$

The reaction therefore has an overall order of 2, being first order with respect to both propanone and $H^+(aq)$ ions and zero order with respect to iodine.

A possible reaction mechanism involves the slow formation of the 'enol' form of propanone in which one molecule of propanone and one $H^+(aq)$ ion participate, and which results in the regeneration of the catalyst as shown below:

$$CH_3-\underset{\underset{O}{\parallel}}{C}-CH_3 + H^+ \rightarrow CH_3-\underset{\underset{^+OH}{\parallel}}{C}-CH_3 \rightarrow$$

'keto' form

$$CH_2{=}\underset{\underset{OH}{|}}{C}-CH_3 + H^+ \quad \text{(slow step)}$$

'enol' form

By comparison, the subsequent step(s) is/are fast and is written below as a single process:

$$CH_2{=}\underset{\underset{OH}{|}}{C}-CH_3 + I_2 \rightarrow CH_2I-\underset{\underset{O}{\parallel}}{C}-CH_3 + H^+ + I^- \quad \text{(fast step)}$$

The decomposition of ethanal vapour at high temperature
This reaction has the deceptively simple stoichiometric equation:

439

$$CH_3CHO(g) \longrightarrow CH_4(g) + CO(g)$$

Kinetic studies of the reaction have shown that its rate equation is,

$$\frac{-d[CH_3CHO]}{dt} = \frac{d[CH_4]}{dt} = k[CH_3CHO]^{3/2}$$

which shows that the order is 1·5.

This reaction is an example of a chain reaction (see chapter 25) in which the idea of a slow step in a reaction sequence controlling the overall rate, and which we have employed in the previous three examples in this section, is no longer tenable. The mechanism of the above reaction is thought to be:

$$CH_3CHO \xrightarrow{k_1} CH_3\cdot + \cdot CHO \qquad \text{Initiation}$$

$$CH_3CHO + CH_3\cdot \xrightarrow{k_2} CH_4 + CH_3\dot{C}O \qquad \text{Propagation}$$

$$CH_3\dot{C}O \xrightarrow{k_3} CH_3\cdot + CO \qquad \text{Propagation}$$

$$2CH_3\cdot \xrightarrow{k_4} C_2H_6 \qquad \text{Termination}$$

The first step (initiation) involves the splitting of a carbon-carbon bond in such a way that both fragments produced have an unpaired electron (denoted by a dot in the above equations). These fragments are called free radicals.

The next step in the reaction sequence involves the reaction between an ethanal molecule and a methyl radical to produce methane and a $CH_3\dot{C}O$ radical, which then decomposes in the next propagation step to regenerate a methyl radical and produce carbon monoxide. The products of the overall reaction, i.e. methane and carbon monoxide are formed in the two propagation steps. In fact these two propagation reactions would proceed indefinitely with no overall consumption of the small initial concentration of methyl radicals produced in the initiation step, were it not for the fact that occasionally two methyl radicals collide and produce ethane. The removal of methyl radicals in this manner is called termination. Chain termination is much less frequent than chain propagation, since ethanal vapour is present in much larger concentration than the methyl radicals; collision between an ethanal molecule and a methyl radical is therefore a more probable event than collision between two methyl radicals.

If we cannot explain a chain reaction in terms of a slow step controlling the overall reaction rate, then we must make some other simplifying assumption. The assumption that is generally made is that the rate of formation of a particular free radical is equal to its rate of removal. Methyl radicals are produced in the initiation step and in the second propagation step; they are removed in the first propagation step and in the termination process. Hence we may write:

Rate of formation of $CH_3\cdot = k_1[CH_3CHO] + k_3[CH_3\dot{C}O]$
Rate of removal of $CH_3\cdot \quad = k_2[CH_3CHO][CH_3\cdot] + 2k_4[CH_3\cdot]^2$

Assuming that both these rates are equal we have,

$$k_1[CH_3CHO] + k_3[CH_3\dot{C}O] = k_2[CH_3CHO][CH_3\cdot] + 2k_4[CH_3\cdot]^2 \qquad (13)$$

The radical CH_3CO is produced in the first propagation step and removed in the second. Hence in a similar manner we may write:

$$k_2[CH_3CHO][CH_3\cdot] = k_3[CH_3\dot{C}O] \qquad (14)$$

From equations (13) and (14) it can easily be shown that:

$$[CH_3\cdot] = \left(\frac{k_1}{2k_4}\right)^{1/2}[CH_3CHO]^{1/2}$$

Now the overall rate of formation of methane is given by the rate equation:

$$\frac{d[CH_4]}{dt} = k_2[CH_3CHO][CH_3\cdot]$$

440

Substituting into this equation the concentration of methyl radicals found in the previous expression we get,

$$\frac{d[CH_4]}{dt} = k_2 \left(\frac{k_1}{2k_4}\right)^{1/2} [CH_3CHO]^{3/2}$$

or

$$\frac{d[CH_4]}{dt} = k[CH_3CHO]^{3/2} \text{ where } k = k_2 \left(\frac{k_1}{2k_4}\right)^{1/2}$$

The proposed mechanism has been able to account for an order of 1·5 for this reaction.

It is possible to propose a satisfactory mechanism for the synthesis of hydrogen bromide from its elements, which is represented by a complex rate equation (p. 429), in terms of a chain reaction. The details of this are, however, outside the scope of this book.

24.5
A practical application of chemical kinetics

During the past few years a growing concern has been expressed about air pollution, and one of the most publicised examples is the occurrence of so-called 'photochemical smog', which can afflict cities with high traffic densities on warm sunny days in the summer.

The main source of this type of pollution is the exhaust emission from cars and lorries, which contains oxides of nitrogen (mainly NO and NO_2) together with traces of hydrocarbons from uncombusted petrol. The chain of events starts with the decomposition of nitrogen dioxide induced by radiation of wavelength less than about 430 nm which reaches the earth from the sun:

v = frequency of the radiation and $h\nu$ = the energy of one quantum of this radiation

$$NO_2 \xrightarrow{h\nu} NO + O\cdot$$

The oxygen atoms produced in the above reaction then initiate furthur reactions which result in the formation of ozone and the regeneration of nitrogen dioxide thus:

$$O\cdot + O_2 \longrightarrow O_3$$
$$O_3 + NO \longrightarrow O_2 + NO_2$$

Once oxygen atoms have been formed they enter into a complex reaction sequence in which hydrocarbons (from the uncombusted petrol) are oxidised to a variety of products, including aldehydes and ketones, some of which are known to be powerful irritants.

This type of pollution is particularly severe in Los Angeles where special geographical conditions combine to produce temperature inversion layers, i.e. cooler air trapped under warm air, which prevent vertical circulation of the air and hence a build up of pollutants at ground level. On a typical summer sunny day in this city there is a steady reduction in visibility as a white and sometimes brown haze appears gradually throughout the day. Towards evening as the intensity of the sunlight diminishes the smog disperses.

The type of pollution described above is localised and much more concern is now being expressed about possible contamination of the stratosphere, a layer in the atmosphere extending from a vertical height of about 16 km to about 45 km. It is in this region where a belt of ozone is situated which acts as a very efficient filter, preventing dangerous short wavelength ultra-violet radiation reaching the earth.

Laboratory experiments have shown that nitrogen oxide combines very rapidly with ozone and there is thus the possibility that nitrogen oxide emission from the exhaust systems of supersonic aircraft, which normally operate in the stratosphere, might be slowly depleting the concentration of this ozone layer.

Another threat to this ozone layer is probably posed by the use of two fluorochloromethanes, i.e. $CFCl_3$ and CF_2Cl_2, which are used in aerosol sprays and as refrigerants. It is ironic that the non-toxic properties of these two compounds, which make them so useful, mean that they are not destroyed in the lower atmosphere and they tend to collect in the stratosphere. Laboratory experiments have shown that short wavelength ultra-violet radiation of the type present in the stratosphere, and which the ozone layer effectively filters out, decompose these compounds forming chlorine atoms, which then rapidly react with ozone.

Some American sources suggest that the ozone layer could be depleted by about 7% by continual release of fluorocarbons at the rates of the early 1970s. While there is not universal acceptance of this figure, the general consensus of opinion would seem to be that the use of the compounds $CFCl_3$ and CF_2Cl_2 should be severely restricted and indeed there is now EEC legislation to restrict their use. Some would argue that their use should be prohibited altogether.

Questions on this chapter are at the end of chapter 25.

Activation energy, catalysis and chain reactions

25.1
Collision theory of gas phase reactions

Z = collision frequency
d = collision diameter
n = number of molecules per unit volume
$\bar{c}$ = average speed of molecules

In the previous chapter we saw how the rates of a number of different reactions were related to the concentrations of the reactants. It is now time to develop some theoretical ideas in order to see if we can account for the rates of chemical reactions. Gas phase reactions are the simplest to handle and we shall restrict the initial discussion to such systems although, as we shall see, the general ideas which emerge can be extended to other systems.

The simplest theory—called the collision theory—is based on the reasonable assumption that chemical reactions take place as a result of molecular collisions. As we saw in chapter 9 (p. 123), a formula can be derived which allows the collision frequency to be calculated for the molecules of a single gas, namely:

$$Z = \frac{\sqrt{2}\pi d^2 n^2 \bar{c}}{2} \tag{1}$$

Substituting values of d and $\bar{c}$ for oxygen into the above formula gives a collision frequency of approximately 10^{35} collisions m^{-3} s^{-1} at s.t.p.

Calculations show that collision frequencies seldom differ by a factor of more than about 10^2 for most gases at the same temperature and pressure, and that collision frequencies are not very sensitive to temperature changes. Indeed this same order of magnitude for collision frequency is also observed for collisions between molecules of two different gases, although needless to say the relevant equation needed now is a modified version of equation (1).

Collision frequencies in the region of 10^{33}–10^{35} collisions m^{-3} s^{-1} are so enormous that every gas phase reaction would be essentially complete in a very small fraction of a second if every collision leads to reaction. The inference therefore is that only molecules which collide with an energy greater than a certain critical energy, called the **activation energy**, are able to react.

The kinetic theory of gases (p. 119) shows that the speeds and hence the kinetic energies of gas molecules cover a wide range. Some have very small, others have intermediate, while a very few have very high energies. It is not difficult to see in a qualitative manner why this should be so for, suppose that at a particular instant all the molecules in a gas had the same kinetic energy, random collisions would tend to speed up some molecules and slow others down and in no time at all a complete range of speeds, and hence kinetic energies, would result. The spread of energies follows the Maxwell distribution shown for two different temperatures in fig. 25.1 (note that these curves are similar to the distribution of molecular speed curves (see fig. 9.3 (p. 120)) except that the initial rise is convex instead of concave, the reason being that kinetic energy is proportional to the square of the speed).

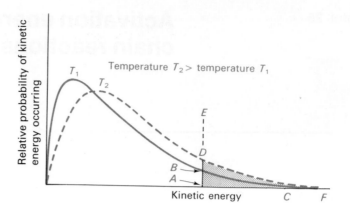

FIG. 25.1. *The Maxwell distribution of kinetic energies at two different temperatures*

The number of molecules having energies greater than E will be represented by the area (ABC) at temperature T_1 and by the area (ADF) at temperature T_2, i.e. the number of molecules having high energies increases markedly as the temperature rises. The result can be expressed mathematically by the equation:

$$n = n_o e^{-E/RT} \qquad (2)$$

where n_o is the total number of molecules and n is the number of molecules with energies greater than E.

Since n/n_o is the fraction of molecules with energies equal to or greater than some critical activation energy E_a, then this fraction is also equal to the fraction of collisions that lead to reaction, i.e.

$$\text{Fraction of effective collisions} = e^{-E_a/RT}$$

The collision theory thus predicts that the rate of a gas phase reaction is given by the expression:

$$\text{Rate of reaction} = Ze^{-E_a/RT} \qquad (3)$$

For a second order gaseous reaction the rate of reaction is given by the expression:

k = rate constant
n_A = number of molecules of A in unit volume
n_B = number of molecules of B in unit volume

$$\text{Rate of reaction} = kn_A n_B \qquad (4)$$

Combining both equations we have,

$$k = \frac{Z}{n_A n_B} e^{-E_a/RT}$$

or

$$k = Z^{\circ} e^{-E_a/RT} \qquad (5)$$

where $Z^{\circ} = Z/n_A n_B$ and is called the **collision number**.

In order to test the validity of equation (5) we need to compare the experimentally determined value of the rate constant, k, of a reaction at a given temperature with the calculated value. The calculated value is obtained from the calculated value of Z° and the experimentally determined activation energy, E_a, (p. 445).

For a few reactions, including some that take place in solution, the agreement is good; however there are many reactions where rates differ by many orders of magnitude from those predicted by theory. In these cases an arbitrary constant P is incorporated into equation (5) and we write:

$$k = PZ^{\circ} e^{-E_a/RT} \tag{6}$$

It is customary to refer to P as the steric factor thus implying that the orientation of collision between two molecules is of importance in some reactions.

25.2 Determination of activation energy

Before the collision theory had been developed Arrhenius discovered that the variation in rate constant, k, with temperature, for a given reaction, could be expressed by the equation,

$$k = Ae^{-E_a/RT} \tag{7}$$

where the constant A is equal to PZ° in the collision theory. The constant A is called the frequency factor or the pre-exponential factor.

When expressed logarithmically the Arrhenius equation is:

$$\ln k = \text{const.} - E_a/RT$$

A plot of $\ln k$ against the reciprocal of the temperature K $(1/T)$ should be a straight line, whose slope is $-E_a/R$ from which the energy of activation, E_a, can be evaluated. Thus in order to determine the energy of activation for a reaction it is necessary to determine the values of the rate constant, k, at a number of different temperatures; the example below shows the procedure for calculating E_a.

Example 1
The rate constant for the dissociation of hydrogen iodide varies with temperature as follows:

T/K	$10^5 k/\text{dm}^3 \, \text{mol}^{-1} \, \text{s}^{-1}$
678	4·233
727	43·07
761	177·8
816	1432
855	5012

Determine the activation energy for this reaction

The values of $\ln k$ plotted against the reciprocal of the temperature K $(1/T)$ are shown in fig. 25.2. The gradient of the graph is AB/BC and has a value of $7\cdot4 \times 10^3/0\cdot33$ which is the value of E_a/R. Since R has a value of $8\cdot314$ J mol^{-1} K^{-1} the activation energy for the reaction is given by:

$$E_a = \frac{7\cdot4 \times 10^3 \times 8\cdot314}{0\cdot33}$$
$$= 186\,400 \text{ J mol}^{-1}$$
$$= 186\cdot4 \text{ kJ mol}^{-1}$$

The Arrhenius plots for the dissociation of hydrogen iodide give a very good straight-line fit, the activation energy for the reaction $(186\cdot4$ kJ mol$^{-1})$ being the energy barrier which has to be surmounted before reaction can proceed (fig. 25.3(a)).

445

FIG. 25.2. *The Arrhenius plot for the dissociation of hydrogen iodide*

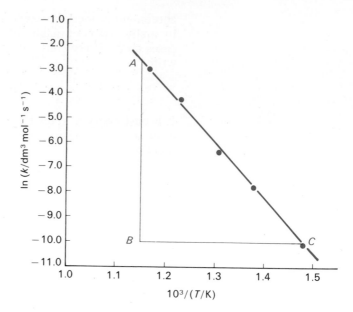

FIG. 25.3. *(a) Energy of activation for the reaction:*
$2HI(g) \rightarrow H_2(g) + I_2(g)$
(b) Energy of activation for the reaction:
$H_2(g) + I_2(g) \rightarrow 2HI(g)$

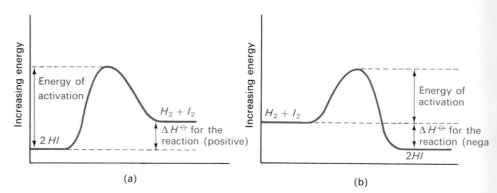

Note that the reverse reaction which is exothermic (fig. 25.3(b)) has an activation energy lower than the above value by an amount $\Delta H^{\ominus}$ (25·9 kJ mol^{-1}), i.e. it is about $186·4 - 25·9$ or 160·5 kJ mol^{-1}.

A crude analogy that might help in appreciating the concept of activation energy is the problem of how to transfer water over an intervening barrier from a higher to a lower level (analogous to an exothermic reaction). The obvious solution is to use a siphoning tube; once some of the water has been raised over the barrier and thus energy expended, the rest flows spontaneously from the higher to the lower potential energy level.

For reactions that take place in stages, e.g. the oxidation of an acidified solution of an iodide with hydrogen peroxide, the experimentally determined activation energy is that for the rate-determining step, i.e. the slowest step.

25.3
The effect of temperature on reaction rates

The Arrhenius equation (equation (7), p. 445) is obeyed by many chemical reactions, i.e. a plot of ln k against the reciprocal of the temperature K $(1/T)$ gives a straight line from which the energy of activation may be calculated. This implies that the pre-exponential factor A is not sensitive to temperature changes as we previously noted

(p. 443). It therefore follows that the variation in rate of a chemical reaction is largely controlled by the factor $e^{-E_a/RT}$.

Consider a reaction which has an activation energy of about 52 kJ mol^{-1}, and suppose we wish to compare the rates of this reaction at room temperature (about 298 K) and at a temperature about 10° higher (say 308 K). Then we substitute the appropriate quantities into the exponential factor $e^{-E_a/RT}$. At 298 K we have

$$e^{-E_a/RT} = e^{-(52\times10^3)/(8\cdot314\times298)}$$
$$= 7\cdot67 \times 10^{-10}$$

Similarly at 308 K we get

$$e^{-E_a/RT} = e^{-(52\times10^3)/(8\cdot314\times308)}$$
$$= 15\cdot18 \times 10^{-10}$$

The reaction speeds up by a factor very close to 2 for a 10° rise in temperature from 298 to 308 K. For the same temperature range, a reaction with an activation energy of 150 kJ mol^{-1} speeds up by a factor of about 7.

Small temperature changes are seen to have a pronounced influence on the rates of chemical reactions. Reactions with large activation energies are slower than those with smaller activation energies at the same temperature, but a rise in temperature has a more significant effect on reactions with high activation energies as we have seen above.

25.4 Catalysis

There is another possible way of increasing the exponential factor $e^{-E_a/RT}$ apart from increasing the temperature; a reduction of E_a would have the same effect as an increase in temperature. Catalysts enable E_a to be reduced.

The essential features of catalysts are as follows:
(a) They become temporarily involved in a reaction, providing an alternative reaction path of lower activation energy than that for the uncatalysed reaction. They are reformed at the end of the reaction.
(b) They catalyse both forward and backward reactions to the same extent in a reversible reaction and thus have no effect on the equilibrium constant.
(c) They are generally highly specific, i.e. a particular catalyst for one reaction is not necessarily a catalyst for another reaction.

It is convenient to discuss catalysis under two distinct headings, namely homogeneous catalysis (the catalyst and the reacting system constituting one phase only) and heterogeneous catalysis (the catalyst and the reacting system constituting two distinct phases).

Homogeneous catalysis

The essential features of this type of catalysis can be explained in terms of the reaction between aqueous solutions of iodide and persulphate ions:

$$2I^-(aq) + S_2O_8^{2-}(aq) \longrightarrow I_2 + 2SO_4^{2-}(aq)$$

The above reaction can be studied by the initial rate method, similar to that described for the reaction between an acidified solution of an iodide and hydrogen peroxide (p. 436). The addition of a small amount of an aqueous solution containing either Fe^{3+}(aq) or Fe^{2+}(aq) considerably quickens the rate of reaction.

A plausible though possibly oversimplified explanation of this catalysis might be:

$$2Fe^{3+}(aq) + 2I^-(aq) \longrightarrow 2Fe^{2+}(aq) + I_2$$
$$2Fe^{2+}(aq) + S_2O_8^{2-}(aq) \longrightarrow 2Fe^{3+}(aq) + 2SO_4^{2-}(aq)$$

It is known that both the above reactions take place rapidly, and it would be expected that two reactions between ions of opposite charge would be faster than one reaction between ions of the same type of charge. Transition metal ions are particularly effective catalysts for such types of reaction, since they can easily change their oxidation states.

It is often possible to depict catalysis in terms of the formation of an intermediate compound in the following manner:

$$A + Catalyst \longrightarrow A\text{—}Catalyst$$
$$A\text{—}Catalyst + B \longrightarrow A\text{—}B + Catalyst$$

The larger activation energy for the uncatalysed reaction,

$$A + B \longrightarrow A\text{—}B$$

is, in this case, replaced by two smaller activation energies for the catalysed reaction. This is illustrated in fig. 25.4.

FIG. *25.4. A catalyst lowers the energy of activation*

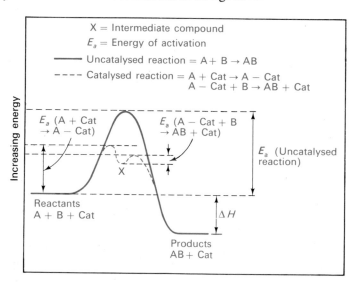

Heterogeneous catalysis

Catalysis at a solid surface is of enormous industrial importance. Examples of such reactions include the Haber process (p. 265) and the Contact process (p. 267). The wide range of products obtained from crude oil are, at some stage in their manufacture, dependent upon reactions involving solid catalysts.

Catalysis at a solid metal surface involves the formation of bonds between reactant molecules and the surface catalyst atoms (the transition metals are particularly effective as catalysts since they have d electrons in addition to s electrons which can be utilised in bonding); this has the effect of increasing the concentration of the reactants at the catalyst surface and also of weakening the bonds in the reactant molecules, i.e. the activation energy is lowered.

As an example, consider the hydrogenation of ethene at a nickel surface which takes place readily at a temperature of about 400 K, the overall reaction being:

$$H_2C{=}CH_2(g) + H_2(g) \longrightarrow CH_3{-}CH_3(g)$$

Both ethene and hydrogen can bond to surface nickel atoms by a process called chemisorption. By this means, bonds approaching the strength of a conventional covalent bond are formed; indeed, it is thought that the majority of the hydrogen is bonded to the catalyst as single atoms.

The mechanism is visualised in terms of a collision between a hydrogen atom on the catalyst surface and one end of a bonded ethene molecule. If the collision is favourable the fragment $CH_3{-}CH_2{-}$ is formed which remains bonded to the nickel surface at one point. Collision with another favourably placed hydrogen atom then forms $CH_3{-}CH_3$ which breaks free from the metal surface. The sequence of events is shown in fig. 25.5.

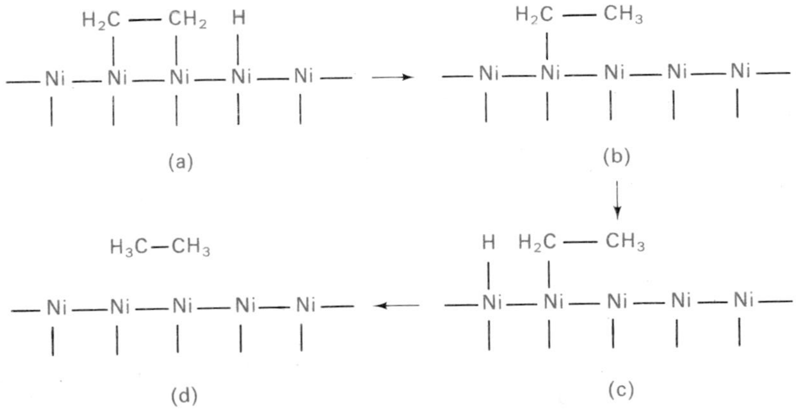

FIG. 25.5. *Catalytic hydrogenation of ethene*

Catalysis by enzymes

Enzymes are essentially proteins which are responsible for catalysing reactions which occur in living matter. They function best at about body temperature, i.e. at about 37°C, rapidly becoming inactive if the temperature rises above 50–60°C owing to partial breakdown of their structures. Reactions catalysed by enzymes do not, therefore, conform to the Arrhenius equation (p. 445). Since these substances are comprised of very large molecules which form colloidal solutions with water, they fall somewhere between the truly homogeneous systems on the one hand and the completely heterogeneous ones on the other.

Enzymes vary widely in their specificity. For example, the enzyme pepsin will only hydrolyse a peptide link ($-NH-CO-$) if an aromatic ring is present in a given position adjacent to this linkage. On the other hand the lipase group of enzymes will catalyse the hydrolysis of almost any ester.

449

Enzymes are far more efficient than any catalyst system devised by man, since they generally lower the activation energy of a reaction in which they participate quite dramatically. For example, the decomposition of hydrogen peroxide into water and oxygen which may be represented thus,

$$2H_2O_2(aq) \longrightarrow 2H_2O(l) + O_2(g)$$

has an activation energy of about 75 kJ mol^{-1}. The reaction may be catalysed by colloidal platinum which lowers the activation energy to about 50 kJ mol^{-1}; on the other hand the energy of activation is lowered to a value of about 21 kJ mol^{-1} when the enzyme catalase is used.

The overall shape of the enzyme molecule is particularly important; it seems as though the molecule in which reaction is to occur (the substrate) must fit exactly into a cavity in the enzyme structure (analogous to the fitting of a key into a lock). Reaction occurs within this complex of enzyme and substrate resulting in the release of a new molecule.

25.5
Chain reactions

In the previous chapter we noted that the thermal decomposition of ethanal, which has an order of 1·5, could be explained in terms of a chain reaction (p. 440). We now examine in a little more detail some of the essential features of such reactions. We shall concentrate on those initiated by radiation in the visible and near ultra-violet part of the electromagnetic spectrum—so-called photochemical chain reactions.

If a mixture of hydrogen and chlorine in equimolar concentrations is placed in a stoppered polythene bottle and then irradiated with light from a photofloodlight, combination of the two gases takes place with explosive violence to produce hydrogen chloride:

$$H_2(g) + Cl_2(g) \longrightarrow 2HCl(g)$$

Experimentally it is found that light of wavelength 400 nm initiates the reaction which is equivalent to an energy of 299 kJ mol^{-1}. Since the bond dissociation energy of the Cl_2 molecule is 242 kJ mol^{-1} while that of H_2 is 436 kJ mol^{-1} it is the Cl—Cl bond which is broken in this process thus:

$$Cl—Cl \longrightarrow Cl\cdot + Cl\cdot$$

The breaking of a bond in such a manner that the two fragments produced have an unpaired electron is called **homolysis**. It should be noted that alternative bond fission to produce, in this case Cl^+ and Cl^-, is ruled out since it requires the input of very much more energy than the former process.

The next step in the sequence is reaction between a chlorine atom and a hydrogen molecule:

$$Cl\cdot + H—H \longrightarrow H—Cl + H\cdot$$

This step produces a hydrogen atom which can continue the chain thus:

$$H\cdot + Cl—Cl \longrightarrow H—Cl + Cl\cdot$$

The process can then be repeated over and over again until either the chlorine atoms or the hydrogen atoms are removed by recombination. It is known that chain termination takes place more frequently by combination of chlorine atoms than by removal of hydrogen atoms. The presence of a third body (represented by M) is necessary to carry away the excess energy when recombination occurs. It can be a molecule in the gas phase but, more often, it is the wall of the vessel:

$$Cl\cdot + Cl\cdot + M \longrightarrow Cl—Cl + M$$

The complete reaction is thus represented by the following steps:

$$Cl—Cl \longrightarrow Cl\cdot + Cl\cdot \qquad \text{Initiation}$$
$$\left.\begin{array}{l} Cl\cdot + H—H \longrightarrow H—Cl + H\cdot \\ H\cdot + Cl—Cl \longrightarrow H—Cl + Cl\cdot \end{array}\right\} \text{Propagation}$$
$$Cl\cdot + Cl\cdot + M \longrightarrow Cl—Cl + M \qquad \text{Termination}$$

The essential requirements for a fast chain reaction are rapid propagation steps. The first propagation step in the hydrogen/chlorine reaction is slightly endothermic and the second very exothermic, both having relatively low activation energies. The extent of the propagation steps, compared with the final termination stage, can be gauged from the fact that under the most favourable conditions 10^6 molecules of product can be obtained by the initial absorption of one photon of radiation by the chlorine molecule. This is referred to as the **quantum yield**. A high quantum yield in a photochemical reaction is indicative of a chain mechanism.

Similar to the hydrogen/chlorine reaction in general features is the reaction between excess methane and chlorine, which may be initiated photochemically, or by heating to a temperature of about 600 K; the overall reaction is:

$$CH_4(g) + Cl_2(g) \longrightarrow CH_3Cl(g) + HCl(g)$$

The basic steps in this reaction are as follows:

$$Cl—Cl \longrightarrow Cl\cdot + Cl\cdot \qquad \text{Initiation}$$
$$\left.\begin{array}{l} Cl\cdot + CH_4 \longrightarrow H—Cl + CH_3\cdot \\ CH_3\cdot + Cl—Cl \longrightarrow CH_3Cl + Cl\cdot \end{array}\right\} \text{Propagation}$$
$$\left.\begin{array}{l} Cl\cdot + Cl\cdot \longrightarrow Cl—Cl \\ Cl\cdot + CH_3\cdot \longrightarrow CH_3Cl \\ CH_3\cdot + CH_3\cdot \longrightarrow C_2H_6 \end{array}\right\} \text{Termination}$$

In the presence of excess chlorine further chain reactions are possible to give successively CH_2Cl_2, $CHCl_3$ and CCl_4 by similar types of mechanism.

Chain reactions involving free radicals, e.g. $CH_3\cdot$, or atoms, e.g. $Cl\cdot$, are speeded up by adding another substance that yields free radicals readily. For example, tetramethyl-lead vapour, $Pb(CH_3)_4$, when added to the system decomposes thermally to give methyl radicals:

$$Pb(CH_3)_4 \longrightarrow Pb + 4CH_3\cdot$$

These methyl radicals can enter into a chain propagation reaction and so speed it up, e.g. the second of the two propagation reactions above.

451

Other substances, particularly nitrogen oxide (which contains an unpaired electron), may slow some chain reactions down and thus act as inhibitors. These substances lower the concentration of free radicals by combining with them, thereby restricting the chain propagation reactions, e.g.

$$CH_3 \cdot + \cdot NO \longrightarrow CH_3NO$$

25.6
The transition state theory of reaction rates

FIG. 25.6. *The energy profile for the reaction:*

$$H—H + Cl \cdot \rightarrow H \cdot + H—Cl$$

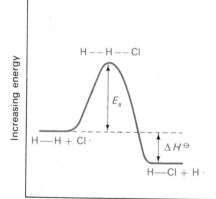

We began this chapter by discussing the collision theory of gas phase reactions; we now end it by having a brief look at the transition state theory of reaction rates.

This theory is more sophisticated than collision theory and in particular directs attention on the processes that are likely to occur just prior to reaction. Chemical reactions are assumed to take place via a transition state in which reactants come together as an activated complex in a particular orientation. The essential idea in this theory is that bond-breaking and bond-making are not instantaneous processes but occur continuously and simultaneously. For instance, the reaction between a hydrogen molecule and a chlorine atom is considered to occur by a gradual stretching of the H—H bond when two favourably orientated species come together to form the activated complex, which occurs at the maximum of an energy profile (see fig. 25.6). The H--Cl bond in the complex gradually shortens and the complex decomposes to give an H—Cl molecule:

$$H—H + CI \cdot \rightleftharpoons H--H--CI \longrightarrow H \cdot + H—Cl$$

activated complex
in equilibrium with
the reactants

This theory, when fully developed, leads to an equation of the form:

$$k = \text{const. } e^{-\Delta G_c/RT}$$

The term ΔG_c is the free energy change in forming the activated complex from the reactants; but, since $\Delta G_c = \Delta H_c - T\Delta S_c$ (see chapter 15) where ΔH_c and ΔS_c are respectively the enthalpy and entropy changes in forming the activated complex from the reactants, the equation may be written:

$$k = \text{const. } e^{\Delta S_c/R} \cdot e^{-\Delta H_c/RT}$$
$$\simeq \text{const. } e^{\Delta S_c/R} \cdot e^{-E_a/RT}$$

Thus the constant A in the Arrhenius equation is replaced by the term:

$$\text{const. } e^{\Delta S_c/R}$$

For gas reactions involving two reactants the change in entropy ΔS_c in forming the activated complex is usually negative and often large, since the formation of the complex involves the association of molecules and hence an increase in order.

Its advantage over the collision theory lies in the fact that transition state theory gives a reasonable quantitative agreement with experiment for simple reacting systems, and even for more complicated systems the exponential entropy factor is able to account for variations in rate—even if only in a qualitative manner.

25.7
Questions on chapters 24 and 25

1 (a) Define the term *order of a reaction*.
   (b) In an experimental study of a reaction in solution between two compounds A and B, the following information was obtained for the *initial* rate of the reaction:

Initial rate in mol dm^{-3} s^{-1}	Initial concentrations of A and B in mol dm^{-3}	
	A	*B*
$1 \cdot 0 \times 10^{-4}$	$1 \cdot 0 \times 10^{-1}$	$1 \cdot 2 \times 10^{-1}$
$4 \cdot 0 \times 10^{-4}$	$2 \cdot 0 \times 10^{-1}$	$1 \cdot 2 \times 10^{-1}$
$8 \cdot 0 \times 10^{-4}$	$2 \cdot 0 \times 10^{-1}$	$2 \cdot 4 \times 10^{-1}$

What is the order of the reaction with respect to (i) the reactant A, (ii) the reactant B?

(c) The decomposition of benzene diazonium chloride in aqueous solution is a reaction of the first order which proceeds according to the equation
$$C_6H_5N_2Cl = C_6H_5Cl + N_2(g).$$
A certain solution of benzene diazonium chloride contains initially an amount of this compound which gives 80 cm^3 of nitrogen on complete decomposition. It is found that, at 30°C, 40 cm^3 of nitrogen are evolved in 40 minutes. How long after the start of the decomposition will 70 cm^3 of nitrogen have been evolved? [All volumes of nitrogen refer to the same temperature and pressure.]

O

2   Explain what you understand by the terms *rate equation, rate constant* and *order of reaction*.

Give one possible reason why the rate equation cannot be deduced from the over-all stoichiometric equation.

For a reaction of your own choice studied during the course, describe how you determined the rate of a reaction in the laboratory and explain how the rate constant was obtained from your observations.

Using two reaction vessels of differing surface material, the reaction between A and B in the gaseous phase was investigated. The initial rate refers to the rate of removal of A (mol m^{-3} s^{-1}) in the experimental data set out below. Deduce the rate equation in each case and assuming that no errors have occured, offer an explanation for what is found.

	$[A]$/mol m^{-3}	$[B]$/mol m^{-3}	Relative rate initially
Experiment 1	0·15	0·15	1
	0·30	0·15	4
	0·15	0·30	2
Experiment 2	0·15	0·15	1
	0·30	0·30	4
	0·60	0·30	16

S

3   (a) Describe what is meant by the *order of a chemical reaction*.

(b) The following data represent the activity of the radioactive isotope $^{223}_{87}Fr$, the longest lived isotope of francium.

Time in min	0	5	10	15	20	25	30	35	40
Activity in counts/min	680	575	495	425	355	300	265	230	210

   (i) Use these data to plot a graph.
   (ii) From the graph determine the time for the count to drop
      (1) to half the first value;
      (2) from three-quarters to three-eighths of the first value;
      (3) from two-thirds to one-third of the first value.
   (iii) Determine the half-life of the radioactive isotope $^{223}_{87}Fr$.
   (iv) Of what order is this decay process? Give your reasoning.
   (v) Calculate the rate constant for the radioactive decay.

(c) The reaction between an aqueous solution of hydrogen peroxide and an aqueous solution of an iodide, in the presence of an acid, can be represented by the equation
$$H_2O_2(aq) + 2I^-(aq) + 2H^+(aq) \rightleftharpoons 2H_2O(l) + I_2(aq)$$
Explain why, using the above information, it is wrong to state that the order of the forward reaction is 5.   AEB (Syllabus II)

4   Explain the meaning of the terms (a) *rate (velocity) constant*, (b) *order of reaction*.

How is a knowledge of the order of a reaction useful to the chemist?

The reaction between propanone and iodine in acidic solution is first order with respect to both propanone and hydrogen ion and zero order with respect\ ˺ iodine. Write a rate equation for the reaction and describe in outline how you would attempt to verify this statement.

Explain the following observations as fully as you can.

(i) When a piece of silver wire is added to 2 mol dm^{-3} aqueous hydrogen peroxide, oxygen is slowly evolved but, when colloidal silver is added to 2 mol dm^{-3} aqueous hydrogen peroxide, oxygen is rapidly evolved.

(ii) The rate of the reaction between potassium manganate(VII) and ethanedioic acid in sulphuric acid solution gradually increases at first and then decreases.

C (Overseas)

5 Define, or explain, the terms *rate* and *order of a reaction*.

(a) Describe and explain carefully in terms of molecular collision theory the effect of temperature change on the rate of a homogeneous gaseous reaction.

(b) For the exothermic gaseous reaction

$$A_2(g) \longrightarrow 2A(g)$$

sketch graphs to show the course of the decomposition (dissociation) by plotting the concentrations of A_2 and A against time.

Using the same scale on another graph, plot the concentration of A_2 against time to show what happens in similar experiments in which

(i) the temperature is raised,

(ii) the pressure is lowered,

(iii) a suitable catalyst is introduced.

Add a brief statement in each case to clarify your answer.

(c) The decomposition of nitrogen pentoxide (N_2O_5) dissolved in tetra-chloromethane (carbon tetrachloride) at 45°C is *first* order. Using the concentrations of nitrogen pentoxide and the times given, estimate by a graphical method the rate constant for the decomposition, stating the units in which it is expressed.

time/s	0	250	500	750	1000	1500	2000	2500
concentration/ mol dm^{-3}	2·33	1·95	1·68	1·42	1·25	0·95	0·70	0·50

S

6 Phenylethane can be chlorinated in two different positions in the alkane side chain and the alkaline hydrolysis of the products has been studied in aqueous ethanol as solvent.

A solution 0·1 M with respect to both 1-chloro-2-phenylethane and sodium hydroxide is allowed to react at 30°C. Portions of the mixture are withdrawn at definite times and titrated with a standard acid solution.

Time/days	Titre/cm^3
0·01	23·2
2·0	18·3
4·0	15·1
6·0	12·7
8·0	11·0
10·0	9·7

The experiment was repeated using 1-chloro-1-phenylethane with the following results:

Time/hours	Titre/cm^3
0·1	21·2
2·0	18·7
4·0	16·3
6·0	14·2
8·0	12·6
10·0	11·0

By plotting appropriate graphs determine whether the reactions are first order or second order. On the basis of your conclusions, propose tentative mechanisms for both the reactions.

N (S)

7 Trichloromethane (chloroform) is hydrolysed by sodium hydroxide solution according to the equation:

$$2CHCl_3 + 7OH^- \longrightarrow CO + HCO_2^- + 6Cl^- + 4H_2O$$

The reaction is first order with respect to both trichloromethane and hydroxide ion.

(a) Outline a procedure by which the reaction could be 'followed' with a view to determining the rate constant.

(b) What is the ratio between the initial rate of the reaction and the rate when half of the hydroxide ions have reacted if the initial concentrations are both 0.1 M?

(c) What is the rate of chloride ion formation when the trichloromethane is reacting at a rate of 2×10^{-4} mol s^{-1}?

N (S)

8 The rates of chemical reactions may be affected by such factors as (a) temperature, (b) pressure or concentrations of reactants, (c) physical state of the reactants, (d) presence of catalysts.

Discuss the physico-chemical principles involved in THREE of these factors and illustrate your answer by reference to TWO examples in each case. L

9 When strips of cobalt foil are rotated rapidly at a constant rate in sodium peroxodisulphate(VI) solution, there is a slow reaction and the cobalt dissolves. The reaction can be followed by removing and weighing the foil at intervals.

Experiment 1 at 0·5°C		Experiment 2 at 13·5°C		Experiment 3 at 25°C	
Time/min	Mass/mg	Time/min	Mass/mg	Time/min	Mass/mg
0	130	0	130	0	130
20	120	10	115	4	116
60	98	15	106	8	103
80	86	30	86	12	87

Determine a rate constant k at each temperature as a loss in mass per minute (mg min^{-1}), and then determine the activation energy E of the reaction using the relationship

$$k \propto 10^{-E/2 \cdot 3RT}$$

What differences in the results would you predict if the cobalt foil were NOT rotated? N

10 Write an account of the importance of activation energy in the study of chemical reactions.

You may refer to one or more of the following aspects of the subject:
(i) its experimental determination,
(ii) the magnitude of its values,
(iii) its connection with catalysis. N

11 (a) Explain what is meant by the terms (i) *velocity constant* (or *rate constant*), (ii) *activation energy*.

(b) Nitrogen dioxide decomposes in the gaseous phase into nitrogen oxide and oxygen. Suggest an experimental method for determining the rate of this reaction.

(c) The following results were obtained for the velocity constant k of this reaction:

$k/dm^3 \, mol^{-1} \, s^{-1}$	Temperature/K
3·16	650
28·2	730
158	800
1120	900
5010	1000

Use this information to find the activation energy for this reaction. O (S)

12 State **three** characteristics of catalysis.

By means of an appropriate example in each case, show what you understand by
(a) heterogeneous catalysis,
(b) a negative catalyst (an inhibitor),
(c) enzyme catalysis.

How is enzyme catalysis affected by temperature changes?

The first few drops of potassium permanganate solution titrated into an acidified oxalic acid solution above 60°C are not immediately decolourised although subsequent drops are immediately decolourised. Explain why this occurs and state how you would use manganese(II) sulphate (manganous sulphate) to confirm your explanation.

C

13 Peroxodisulphate ions, $S_2O_8^{2-}$, oxidize iodide ions to iodine, as follows:
$$S_2O_8^{2-} + 2I^- \longrightarrow I_2 + 2SO_4^{2-}$$
The reaction, with reagents of suitable concentrations, can proceed at a reasonable rate at room temperature.

Describe

(a) how you would find the order of this reaction with respect to each reactant, and

(b) how you would show that Fe^{3+} ions catalyze the reaction.

Some catalysts are said to act by lowering the activation energy of a reaction. How would you investigate whether this is so in the case of this catalyzed reaction?

L

14 Write an essay on the catalysis of reactions. Your answer should include an explanation of how catalysts work.

N

15 Define clearly what is meant by catalysis and illustrate the importance of catalysts in industrial and biochemical processes. How would you investigate the effectiveness of metal ions in catalysing the following reaction:
$$OCl^-(aq.) \longrightarrow Cl^-(aq.) + \tfrac{1}{2}O_2(g)? \quad \text{Oxford Schol. and Entrance}$$

16 Suggest how the rate equation for the reaction:
$$BrO_{3(aq)}^- + 5Br_{(aq)}^- + 6H_{(aq)}^+ \longrightarrow 3Br_{2(aq)} + 3H_2O_{(l)}$$
may be determined experimentally.

The rate constant for the dissociation of hydrogen iodide varies with temperature as follows:

T/K	$k/10^{-5}$ dm^3 mol^{-1} s^{-1}
678	4·233
727	43·07
761	177·8
816	1432
855	5012

Determine the activation energy for this reaction. ($R = 8\cdot3$ J mol^{-1} K^{-1}.)

What will be the effect on the activation energy of (i) continuously removing the iodine vapour formed, and (ii) adding a homogenous catalyst?

Oxford Schol. and Entrance (Physical Science)

17 Describe, with experimental details, how you would determine the rate of the acid-catalysed hydrolysis of methyl ethanoate (acetate) at 40°C.

Account for each of the following observations.

(a) The reaction $S_2O_8^{2-}(aq) + 2I^-(aq) = 2SO_4^{2-}(aq) + I_2(aq)$ is first order with respect to $S_2O_8^{2-}$ and also with respect to I^-.

(b) At room temperature, a mixture of hydrogen and chlorine will not react until irradiated with ultraviolet light.

(c) The calculated collision frequency between gas molecules is usually some 10^{10} times greater than the measured rate of reaction between the same molecules.

(d) The collision frequency between gas molecules is proportional to the square root of the thermodynamic temperature but a rise in temperature of 10 K often increases the rate of a reaction between gases by a factor of about two.

O and C

18 Ammonia decomposes on a hot tungsten surface to dinitrogen (nitrogen) and dihydrogen (hydrogen). The following data were found using a tungsten filament at 1400 K:

Initial pressure of ammonia, kPa	35·3	17·3	7·73
Half-life of reaction, s	456	222	100

(The half-life is the time taken for the pressure of ammonia to drop to half its initial value.)

Find the order of the reaction and the rate constant and then answer the following questions:

(a) Is the decomposition reaction occurring at a significant rate in the gas phase?

(b) Would you expect the reaction order to remain the same at the very end of an experiment, when very little ammonia remains?

(c) How strongly are ammonia, dinitrogen (nitrogen) and dihydrogen (hydrogen) adsorbed on tungsten?

Camb. Entrance

19 What do you understand by the terms *rate expression* and *reaction order*? Suggest one method each for measuring the rates of the following reactions:

(a) The hydrolysis of ethyl ethanoate (ethyl acetate).

(b) The oxidation of hydriodic acid by hydrogen peroxide.

The equilibrium between two isomers, A and B can be represented:

$$A \underset{k_2}{\overset{k_1}{\rightleftharpoons}} B$$

where k_1 and k_2 are rate constants for the forward and reverse reactions, respectively. Starting with a non-equilibrium mixture of concentrations $[A]_0 = a$ and $[B]_0 = b$ it was found that x moles of A had reacted after time t. Give an expression for the rate, dx/dt, and hence show that the integrated rate expression is

$$\ln\left(\frac{p}{p - x}\right) = (k_1 + k_2)t$$

where

$$p = \frac{k_1 a - k_2 b}{k_1 + k_2}.$$

After 69·3 minutes $x = p/2$; calculate k_1 and k_2 if the equilibrium constant $K = 4$.

Camb. Entrance

20 Explain what is meant by the terms *activation energy* and *rate constant* of a chemical reaction. What would be the appropriate units to express a rate constant of (a) a first-order reaction and (b) a second-order reaction? Indicate briefly how you would determine the order of a reaction? Is it possible to have (a) a zero order reaction, (b) a sixth order reaction?

A rate constant determined initially at 303 K was found to have increased by a factor of 5 at 314·4 K and by a further factor of 5 at 327·0 K. What is the activation energy for this reaction?

Oxford Schol. and Entrance

Appendix I

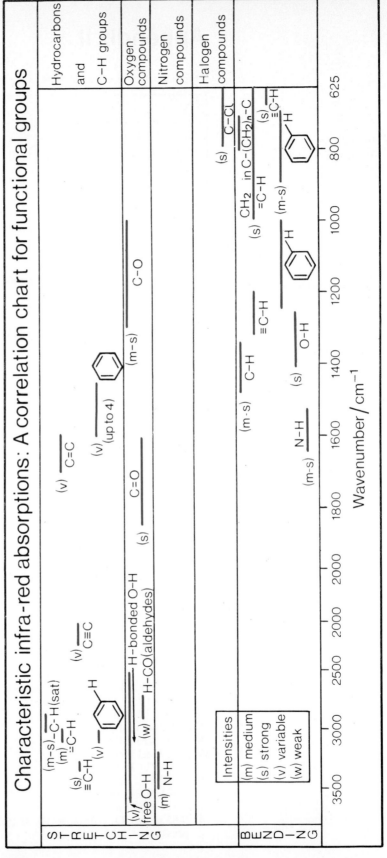

Characteristic infra-red absorptions: A correlation chart for functional groups

STRETCHING

Hydrocarbons and C–H groups
- (m-s) –C–H(sat)
- (m) =C–H
- (s) ≡C–H
- (v) =C–H
- (v) C≡C
- (v) C=C

Oxygen compounds
- (v) free O–H
- (v) H-bonded O–H
- (w) H–CO(aldehydes)
- (s) C=O
- (v) (up to 4)
- (m-s) C–O

Nitrogen compounds
- (m) N–H

BENDING

Intensities
- (m) medium
- (s) strong
- (v) variable
- (w) weak

- (m-s) C–H
- (s) N–H
- (s) ≡C–H
- (m-s) O–H
- (s) ≡C–H
- (s) CH₂ in C–(CH₂)ₙ–C
- (s) =C–H
- (m-s) =C–H

Halogen compounds
- (s) C–Cl

Wavenumber / cm⁻¹

3500 3000 2500 2000 2000 1800 1600 1400 1200 1000 800 625

459

Appendix II

Typical Values of Chemical Shifts/τ for Hydrogen Atoms in Organic Compounds

Type of Proton	Chemical Shift Range/τ
$\underline{CH_3}$—CH_2—	8.9–9.2
—CH_2—$\underline{CH_2}$—CH_2—	8.3–8.8
$\underline{CH_3}$—CH_2—X (X=Cl, Br, OH)	8.1–8.9
$\underline{CH_3}$—C— $\parallel$ O	7.4–8.0
CH_3—$\underline{CH_2}$—X (X=Cl, Br, OH)	6.5–6.7

Answers to questions

CHAPTER 3
7. (a) 13·5 eV (b) 122·2 nm (c) 0·053 nm

CHAPTER 4
2. 0·0245 nm

CHAPTER 5
7. $4·27 \times 10^3$ disintegrations s^{-1}
9. 66 years

CHAPTERS 9 & 10
1. 482 m s^{-1}
4. 4·00
5. (a) 88·9 (b) 758 K
6. (b) (iii) 1·254
7. (i) 3991 J (ii) 250 m s^{-1} (iii) 2:1 (iv) 120 s
8. $1·93 \times 10^3$ m s^{-1}

CHAPTER 11
7. (b) $6·0213 \times 10^{23}$ mol^{-1}
8. RMM = 58·7, Radius of metal atom = 0·1245 nm

CHAPTER 13
2. C_8H_8
11. (i) $2·477 \times 10^{-28}$ m^3 (ii) $6·023 \times 10^{23}$ mol^{-1}
12. 0·267 nm, 0·191 nm, 0·310 nm

CHAPTER 14
1. $\Delta H^{\ominus} = -1368·1$ kJ mol^{-1}
2. (i) −120 kJ mol^{-1} (ii) −206 kJ mol^{-1}
3. 74·6% CH_4, 25·4% C_2H_6
4. −1430 kJ mol^{-1}
5. −312 kJ mol^{-1}
8. −359 kJ mol^{-1}
10. −1890 kJ mol^{-1}

CHAPTER 15
5. −190·6 kJ mol^{-1}

CHAPTER 16
1. (b) 0·577 mole
4. 0·85 mole CO, 1·70 moles H_2, 0·15 mole CH_3OH, pressure = $3·33 \times 10^4$ kPa

5. (c) (i) 2 moles   (ii) 1/3 and 2/3
6. (c) (i) 0·25   (ii) 40%   (iii) $2·69 \times 10^4$ N m^{-2}
7. $2·43 \times 10^5$ N m^{-2}
9. (i) 0·1528   (ii) $9·55 \times 10^3$ Pa
11. 0·236 atm

CHAPTER 17

4. (b) 8·18 cm^3
5. 2·38
8. (i) 3   (ii) 5
10. (a) 13   (b) (i) $1·148 \times 10^{-2}$ (ii) $1·33 \times 10^{-5}$ mol dm^{-3}
11. (b) 2·62
13. 7·5
15. (i) 2·73   (ii) 3·87   (iii) 3·00   (a) 0·64   (b) 13·36

CHAPTER 18

7. $E^{\ominus} = 0·637$ V, $K_c = 3·64 \times 10^{21}$

CHAPTER 20

2. 930 g

CHAPTER 21

1. (a) $6·30 \times 10^{-6}$ mol dm^{-3} (b) 10^{-15} mol dm^{-3}
   (c) $5·55 \times 10^{-11}$ mol dm^{-3}
2. (a) (iii) 1×10^{-10} mol^2 dm^{-6} (b) 7·84 g dm^{-3}
3. (a) $1·95 \times 10^{-6}$ mol^3 dm^{-9} (b) 0·0761 g dm^{-3}
5. (a) $2·35 \times 10^{-6}$ g dm^{-3} (b) $8·99 \times 10^{-9}$ mol^3 dm^{-9}
   (c) $[Ag^+] = 1·49 \times 10^{-13}$ mol dm^{-3}
6. (a) $y = 108x^5$ (b) $1·5 \times 10^{-5}$ mol dm^{-3}, $3·38 \times 10^{-25}$ mol^5 dm^{-15}
   (c) $2·32 \times 10^{-9}$ mol dm^{-3} (e) $3·16 \times 10^{-4}$
7. 0·50 mol dm^{-3}
8. $3·93 \times 10^{-6}$ mol dm^{-3}
10. (b) 25   (d) (i) 0·3 mol dm^{-3} (ii) 4
11. (c) 93·75%
12. (b) (i) 5·33 g   (ii) 5·76 g
13. $[I_3^-] = 0·0397$ mol dm^{-3}, $[I_2] = 0·0003$ mole dm^{-3}
   $[I^-] = 0·1603$ mol dm^{-3}
14. $1·23 \times 10^{-3}$ mol dm^{-3}
15. 66·4

CHAPTER 22

4. 78·3% CH_3OH, 21·7% C_2H_5OH
7. 85·7% A, 14·3% B
9. (c) 123·6
10. (ii) 13·97 g
11. (d) 364 g
12. (b) 325
13. (c) 160 g   (d) 10·01 g
14. (c) (ii) 20·11%
16. 124
18. (d) (i) −0·4625°C   (ii) 5·87 atm
20. (c) −0·58°C

CHAPTER 23

1. (b) 5·07 min   (d) $1·603 \times 10^{-19}$ C
2. 96 500 C mol^{-1}, $6·02 \times 10^{23}$ mol^{-1}
3. (b) (i) 151 Ω^{-1} cm^2 mol^{-1} (ii) $1·59 \times 10^{-2}$ mol dm^{-3}
4. −0·70°C
6. (c) (i) 114 S cm^2 mol^{-1} (ii) 390 S cm^2 mol^{-1}
8. $1·71 \times 10^{-10}$ mol^2 dm^{-6}

CHAPTERS 24 & 25

1. (c) 120 min
18. zero order, $k = 3·87 \times 10^{-2}$ kPa s^{-1}
19. $k_1 = 8 \times 10^{-3}$ min^{-1}, $k_2 = 2 \times 10^{-3}$ min^{-1}
20. 112 kJ mol^{-1}

Index

Table of Relative Atomic Masses to Four Significant Figures

(Scaled to the relative atomic mass $^{12}C = 12$ exactly)

Values quoted in the table, unless marked * or †, are reliable to at least ±1 in the fourth significant figure. A number in parentheses denotes the atomic mass number of the isotope of longest known half-life.

ATOMIC NUMBER	NAME	SYMBOL	RELATIVE ATOMIC MASS
1	Hydrogen	H	1·008
2	Helium	He	4·003
3	Lithium	Li	6·941*†
4	Beryllium	Be	9·012
5	Boron	B	10·81†
6	Carbon	C	12·01
7	Nitrogen	N	14·01
8	Oxygen	O	16·00
9	Fluorine	F	19·00
10	Neon	Ne	20·18
11	Sodium	Na	22·99
12	Magnesium	Mg	24·31
13	Aluminium	Al	26·98
14	Silicon	Si	28·09
15	Phosphorus	P	30·97
16	Sulphur	S	32·06†
17	Chlorine	Cl	35·45
18	Argon	Ar	39·95
19	Potassium	K	39·10
20	Calcium	Ca	40·08†
21	Scandium	Sc	44·96
22	Titanium	Ti	47·90*
23	Vanadium	V	50·94
24	Chromium	Cr	52·00
25	Manganese	Mn	54·94
26	Iron	Fe	55·85
27	Cobalt	Co	58·93
28	Nickel	Ni	58·70
29	Copper	Cu	63·55
30	Zinc	Zn	65·38
31	Gallium	Ga	69·72
32	Germanium	Ge	72·59*
33	Arsenic	As	74·92
34	Selenium	Se	78·96*
35	Bromine	Br	79·90
36	Krypton	Kr	83·80
37	Rubidium	Rb	85·47
38	Strontium	Sr	87·62†
39	Yttrium	Y	88·91
40	Zirconium	Zr	91·22
41	Niobium	Nb	92·91
42	Molybdenum	Mo	95·94*
43	Technetium	Tc	(98)
44	Ruthenium	Ru	101·1
45	Rhodium	Rh	102·9
46	Palladium	Pd	106·4
47	Silver	Ag	107·9
48	Cadmium	Cd	112·4
49	Indium	In	114·8
50	Tin	Sn	118·7
51	Antimony	Sb	121·8
52	Tellurium	Te	127·6

* Values so marked are reliable to ±3 in the fourth significant figure.
† Values so marked may differ from the atomic weights of the relevant elements in some naturally occurring samples because of a variation in the relative abundance of the isotopes.

This is a 'Table of Atomic Weights to Four Significant Figures,' published by the International Union of Pure and Applied Chemistry and reproduced with their permission.

Atomic Number	Name	Symbol	Relative Atomic Mass
53	Iodine	I	126·9
54	Xenon	Xe	131·3
55	Caesium	Cs	132·9
56	Barium	Ba	137·3
57	Lanthanum	La	138·9
58	Cerium	Ce	140·1
59	Praseodymium	Pr	140·9
60	Neodymium	Nd	144·2
61	Promethium	Pm	(145)
62	Samarium	Sm	150·4
63	Europium	Eu	152·0
64	Gadolinium	Gd	157·3
65	Terbium	Tb	158·9
66	Dysprosium	Dy	162·5
67	Holmium	Ho	164·9
68	Erbium	Er	167·3
69	Thulium	Tm	168·9
70	Ytterbium	Yb	173·0
71	Lutetium	Lu	175·0
72	Hafnium	Hf	178·5
73	Tantalum	Ta	180·9
74	Wolfram (Tungsten)	W	183·9
75	Rhenium	Re	186·2
76	Osmium	Os	190·2
77	Iridium	Ir	192·2
78	Platinum	Pt	195·1
79	Gold	Au	197·0
80	Mercury	Hg	200·6
81	Thallium	Tl	204·4
82	Lead	Pb	207·2†
83	Bismuth	Bi	209·0
84	Polonium	Po	(209)
85	Astatine	At	(210)
86	Radon	Rn	(222)
87	Francium	Fr	(223)
88	Radium	Ra	(226)
89	Actinium	Ac	(227)
90	Thorium	Th	232·0
91	Protactinium	Pa	(231)
92	Uranium	U	238†
93	Neptunium	Np	(237)
94	Plutonium	Pu	(244)
95	Americium	Am	(243)
96	Curium	Cm	(247)
97	Berkelium	Bk	(247)
98	Californium	Cf	(251)
99	Einsteinium	Es	(252)
100	Fermium	Fm	(257)
101	Mendelevium	Md	(258)
102	Nobelium	No	(259)
103	Lawrencium	Lr	(260)

Logarithms

	0	1	2	3	4	5	6	7	8	9	1	2	3	4	5	6	7	8	9
10	·0000	0043	0086	0128	0170	0212	0253	0294	0334	0374	4	8	12	17	21	25	29	33	37
11	·0414	0453	0492	0531	0569	0607	0645	0682	0719	0755	4	8	11	15	19	23	26	30	34
12	·0792	0828	0864	0899	0934	0969	1004	1038	1072	1106	3	7	10	14	17	21	24	28	31
13	·1139	1173	1206	1239	1271	1303	1335	1367	1399	1430	3	6	10	13	16	19	23	26	29
14	·1461	1492	1523	1553	1584	1614	1644	1673	1703	1732	3	6	9	12	15	18	21	24	27
15	·1761	1790	1818	1847	1875	1903	1931	1959	1987	2014	3	6	8	11	14	17	20	22	25
16	·2041	2068	2095	2122	2148	2175	2201	2227	2253	2279	3	5	8	11	13	16	18	21	24
17	·2304	2330	2355	2380	2405	2430	2455	2480	2504	2529	2	5	7	10	12	15	17	20	22
18	·2553	2577	2601	2625	2648	2672	2695	2618	2742	2765	2	5	7	9	12	14	16	19	21
19	·2788	2810	2833	2856	2878	2900	2923	2945	2967	2989	2	4	7	9	11	13	16	18	20
20	·3010	3032	3054	3075	3096	3118	3139	3160	3181	3201	2	4	6	8	11	13	15	17	19
21	·3222	3243	3263	3284	3304	3324	3345	3365	3385	3404	2	4	6	8	10	12	14	16	18
22	·3424	3444	3464	3483	3502	3522	3541	3560	3579	3598	2	4	6	8	10	12	14	15	17
23	·3617	3636	3655	3674	3692	3711	3729	3747	3766	3784	2	4	6	7	9	11	13	15	17
24	·3802	3820	3838	3856	3874	3892	3909	3927	3945	3962	2	4	5	7	9	11	12	14	16
25	·3979	3997	4014	4031	4048	4065	4082	4099	4116	4133	2	3	5	7	9	10	12	14	15
26	·4150	4166	4183	4200	4216	4232	4249	4265	4281	4298	2	3	5	7	8	10	11	13	15
27	·4314	4330	4346	4362	4378	4393	4409	4425	4440	4456	2	3	5	6	8	9	11	13	14
28	·4472	4487	4502	4518	4533	4548	4564	4579	4594	4609	2	3	5	6	8	9	11	12	14
29	·4624	4639	4654	4669	4683	4698	4713	4728	4742	4757	1	3	4	6	7	9	10	12	13
30	·4771	4786	4800	4814	4829	4843	4857	4871	4886	4900	1	3	4	6	7	9	10	11	13
31	·4914	4928	4942	4955	4969	4983	4997	5011	5024	5038	1	3	4	6	7	8	10	11	12
32	·5051	5065	5079	5092	5105	5119	5132	5145	5159	5172	1	3	4	5	7	8	9	11	12
33	·5185	5198	5211	5224	5237	5250	5263	5276	5289	5302	1	3	4	5	6	8	9	10	12
34	·5315	5328	5340	5353	5366	5378	5391	5403	5416	5428	1	3	4	5	6	8	9	10	11
35	·5441	5453	5465	5478	5490	5502	5514	5527	5539	5551	1	2	4	5	6	7	9	10	11
36	·5563	5575	5587	5599	5611	5623	5635	5647	5658	5670	1	2	4	5	6	7	8	10	11
37	·5682	5694	5705	5717	5729	5740	5752	5763	5775	5786	1	2	3	5	6	7	8	9	10
38	·5798	5809	5821	5832	5843	5855	5866	5877	5888	5899	1	2	3	5	6	7	8	9	10
39	·5911	5922	5933	5944	5955	5966	5977	5988	5999	6010	1	2	3	4	5	7	8	9	10
40	·6021	6031	6042	6053	6064	6075	6085	6096	6107	6117	1	2	3	4	5	6	8	9	10
41	·6128	6138	6149	6160	6170	6180	6191	6201	6212	6222	1	2	3	4	5	6	7	8	9
42	·6232	6243	6253	6263	6274	6284	6294	6304	6314	6325	1	2	3	4	5	6	7	9	9
43	·6335	6345	6355	6365	6375	6385	6395	6405	6415	6425	1	2	3	4	5	6	7	8	9
44	·6435	6444	6454	6464	6474	6484	6493	6503	6513	6522	1	2	3	4	5	6	7	8	9
45	·6532	6542	6551	6561	6571	6580	6590	6599	6609	6618	1	2	3	4	5	6	7	8	9
46	·6628	6637	6646	6656	6665	6675	6684	6693	6702	6712	1	2	3	4	5	6	7	7	8
47	·6721	6730	6739	6749	6758	6767	6776	6785	6794	6803	1	2	3	4	5	5	6	7	8
48	·6812	6821	6830	6839	6848	6857	6866	6875	6884	6893	1	2	3	4	4	5	6	7	8
49	·6902	6911	6920	6928	6937	6946	6955	6964	6972	6981	1	2	3	4	4	5	6	7	8
50	·6990	6998	7007	7016	7024	7033	7042	7050	7059	7067	1	2	3	3	4	5	6	7	8
51	·7076	7084	7093	7101	7110	7118	7126	7135	7143	7152	1	2	3	3	4	5	6	7	8
52	·7160	7168	7177	7185	7193	7202	7210	7218	7226	7235	1	2	2	3	4	5	6	7	7
53	·7243	7251	7259	7267	7275	7284	7292	7300	7308	7316	1	2	2	3	4	5	6	6	7
54	·7324	7332	7340	7348	7356	7364	7372	7380	7388	7396	1	2	2	3	4	5	6	6	7

Logarithms

	0	1	2	3	4	5	6	7	8	9	1	2	3	4	5	6	7	8	9
55	·7404	7412	7419	7427	7435	7443	7451	7459	7466	7474	1	2	2	3	4	5	5	6	7
56	·7482	7490	7497	7505	7513	7520	7528	7536	7543	7551	1	2	2	3	4	5	5	6	7
57	·7559	7566	7574	7582	7589	7597	7604	7612	7619	7627	1	2	2	3	4	5	5	6	7
58	·7634	7642	7649	7657	7664	7672	7679	7686	7694	7701	1	1	2	3	4	4	5	6	7
59	·7709	7716	7723	7731	7738	7745	7752	7760	7767	7774	1	1	2	3	4	4	5	6	7
60	·7782	7789	7796	7803	7810	7818	7825	7832	7839	7846	1	1	2	3	4	4	5	6	6
61	·7853	7860	7868	7875	7882	7889	7896	7903	7910	7917	1	1	2	3	4	4	5	6	6
62	·7924	7931	7938	7945	7952	7959	7966	7973	7980	7987	1	1	2	3	3	4	5	6	6
63	·7993	8000	8007	8014	8021	8028	8035	8041	8048	8055	1	1	2	3	3	4	5	5	6
64	·8062	8069	8075	8082	8089	8096	8102	8109	8116	8122	1	1	2	3	3	4	5	5	6
65	·8129	8136	8142	8149	8156	8162	8169	8176	8182	8189	1	1	2	3	3	4	5	5	6
66	·8195	8202	8209	8215	8222	8228	8235	8241	8248	8254	1	1	2	3	3	4	5	5	6
67	·8261	8267	8274	8280	8287	8293	8299	8306	8312	8319	1	1	2	3	3	4	5	5	6
68	·8325	8331	8338	8344	8351	8357	8363	8370	8376	8382	1	1	2	3	3	4	4	5	6
69	·8388	8395	8401	8407	8414	8420	8426	8432	8439	8445	1	1	2	2	3	4	4	5	6
70	·8451	8457	8463	8470	8476	8482	8488	8494	8500	8506	1	1	2	2	3	4	4	5	6
71	·8513	8519	8525	8531	8537	8543	8549	8555	8561	8567	1	1	2	2	3	4	4	5	5
72	·8573	8579	8585	8591	8597	8603	8609	8615	8621	8627	1	1	2	2	3	4	4	5	5
73	·8633	8639	8645	8651	8657	8663	8669	8675	8681	8686	1	1	2	2	3	4	4	5	5
74	·8692	8698	8704	8710	8716	8722	8727	8733	8739	8745	1	1	2	2	3	4	4	5	5
75	·8751	8756	8762	8768	8774	8779	8785	8791	8797	8802	1	1	2	2	3	3	4	5	5
76	·8808	8814	8820	8825	8831	8837	8842	8848	8854	8859	1	1	2	2	3	3	4	5	5
77	·8865	8871	8876	8882	8887	8893	8899	8904	8910	8915	1	1	2	2	3	3	4	4	5
78	·8921	8927	8932	8938	8943	8949	8954	8960	8965	8971	1	1	2	2	3	3	4	4	5
79	·8976	8982	8987	8993	8998	9004	9009	9015	9020	9025	1	1	2	2	3	3	4	4	5
80	·9031	9036	9042	9047	9053	9058	9063	9069	9074	9079	1	1	2	2	3	3	4	4	5
81	·9085	9090	9096	9101	9106	9112	9117	9122	9128	9133	1	1	2	2	3	3	4	4	5
82	·9138	9143	9149	9154	9159	9165	9170	9175	9180	9186	1	1	2	2	3	3	4	4	5
83	·9191	9196	9201	9206	9212	9217	9222	9227	9232	9238	1	1	2	2	3	3	4	4	5
84	·9243	9248	9253	9258	9263	9269	9274	9279	9284	9289	1	1	2	2	3	3	4	4	5
85	·9294	9299	9304	9309	9315	9320	9325	9330	9335	9340	1	1	2	2	3	3	4	4	5
86	·9345	9350	9355	9360	9365	9370	9375	9380	9385	9390	1	1	2	2	3	3	4	4	5
87	·9395	9400	9405	9410	9415	9420	9425	9430	9435	9440	0	1	1	2	2	3	3	4	4
88	·9445	9450	9455	9460	9465	9469	9474	9479	9484	9489	0	1	1	2	2	3	3	4	4
89	·9494	9499	9504	9509	9513	9518	9523	9528	9533	9538	0	1	1	2	2	3	3	4	4
90	·9542	9547	9552	9557	9562	9566	9571	9576	9581	9586	0	1	1	2	2	3	3	4	4
91	·9590	9595	9600	9605	9609	9614	9619	9624	9628	9633	0	1	1	2	2	3	3	4	4
92	·9638	9643	9647	9652	9657	9661	9666	9671	9675	9680	0	1	1	2	2	3	3	4	4
93	·9685	9689	9694	9699	9703	9708	9713	9717	9722	9727	0	1	1	2	2	3	3	4	4
94	·9731	9736	9741	9745	9750	9754	9759	9763	9768	9773	0	1	1	2	2	3	3	4	4
95	·9777	9782	9786	9791	9795	9800	9805	9809	9814	9818	0	1	1	2	2	3	3	4	4
96	·9823	9827	9832	9836	9841	9845	9850	9854	9859	9863	0	1	1	2	2	3	3	4	4
97	·9868	9872	9877	9881	9886	9890	9894	9899	9903	9908	0	1	1	2	2	3	3	4	4
98	·9912	9917	9921	9926	9930	9934	9939	9943	9948	9952	0	1	1	2	2	3	3	4	4
99	·9956	9961	9965	9969	9974	9978	9983	9987	9991	9996	0	1	1	2	2	3	3	3	4

Antilogarithms

	0	1	2	3	4	5	6	7	8	9	1	2	3	4	5	6	7	8	9
·00	1000	1002	1005	1007	1009	1012	1014	1016	1019	1021	0	0	1	1	1	1	2	2	2
·01	1023	1026	1028	1030	1033	1035	1038	1040	1042	1045	0	0	1	1	1	1	2	2	2
·02	1047	1050	1052	1054	1057	1059	1062	1064	1067	1069	0	0	1	1	1	1	2	2	2
·03	1072	1074	1076	1079	1081	1084	1086	1089	1091	1094	0	0	1	1	1	1	2	2	2
·04	1096	1099	1102	1104	1107	1109	1112	1114	1117	1119	0	1	1	1	1	2	2	2	2
·05	1122	1125	1127	1130	1132	1135	1138	1140	1143	1146	0	1	1	1	1	2	2	2	2
·06	1148	1151	1153	1156	1159	1161	1164	1167	1169	1172	0	1	1	1	1	2	2	2	2
·07	1175	1178	1180	1183	1186	1189	1191	1194	1197	1199	0	1	1	1	1	2	2	2	2
·08	1202	1205	1208	1211	1213	1216	1219	1222	1225	1227	0	1	1	1	1	2	2	2	3
·09	1230	1233	1236	1239	1242	1245	1247	1250	1253	1256	0	1	1	1	1	2	2	2	3
·10	1259	1262	1265	1268	1271	1274	1276	1279	1282	1285	0	1	1	1	1	2	2	2	3
·11	1288	1291	1294	1297	1300	1303	1306	1309	1312	1315	0	1	1	1	2	2	2	2	3
·12	1318	1321	1324	1327	1330	1334	1337	1340	1343	1346	0	1	1	1	2	2	2	3	3
·13	1349	1352	1355	1358	1361	1365	1368	1371	1374	1377	0	1	1	1	2	2	2	3	3
·14	1380	1384	1387	1390	1393	1396	1400	1403	1406	1409	0	1	1	1	2	2	2	3	3
·15	1413	1416	1419	1422	1426	1429	1432	1435	1439	1442	0	1	1	1	2	2	2	3	3
·16	1445	1449	1452	1455	1459	1462	1466	1469	1472	1476	0	1	1	1	2	2	2	3	3
·17	1479	1483	1486	1489	1493	1496	1500	1503	1507	1510	0	1	1	1	2	2	2	3	3
·18	1514	1517	1521	1524	1528	1531	1535	1538	1542	1545	0	1	1	1	2	2	2	3	3
·19	1549	1552	1556	1560	1563	1567	1570	1574	1578	1581	0	1	1	1	2	2	3	3	3
·20	1585	1589	1592	1596	1600	1603	1607	1611	1614	1618	0	1	1	1	2	2	3	3	3
·21	1622	1626	1629	1633	1637	1641	1644	1648	1652	1656	0	1	1	2	2	2	3	3	3
·22	1660	1663	1667	1671	1675	1679	1683	1687	1690	1694	0	1	1	2	2	2	3	3	3
·23	1698	1702	1706	1710	1714	1718	1722	1726	1730	1734	0	1	1	2	2	2	3	3	4
·24	1738	1742	1746	1750	1754	1758	1762	1766	1770	1774	0	1	1	2	2	2	3	3	4
·25	1778	1782	1786	1791	1795	1799	1803	1807	1811	1816	0	1	1	2	2	2	3	3	4
·26	1820	1824	1828	1832	1837	1841	1845	1849	1854	1858	0	1	1	2	2	3	3	3	4
·27	1862	1866	1871	1875	1879	1884	1888	1892	1897	1901	0	1	1	2	2	3	3	3	4
·28	1905	1910	1914	1919	1923	1928	1932	1936	1941	1945	0	1	1	2	2	3	3	4	4
·29	1950	1954	1959	1963	1968	1972	1977	1982	1986	1991	0	1	1	2	2	3	3	4	4
·30	1995	2000	2004	2009	2014	2018	2023	2028	2032	2037	0	1	1	2	2	3	3	4	4
·31	2042	2046	2051	2056	2061	2065	2070	2075	2080	2084	0	1	1	2	2	3	3	4	4
·32	2089	2094	2099	2104	2109	2113	2113	2123	2128	2133	0	1	1	2	2	3	3	4	4
·33	2138	2143	2148	2153	2158	2163	2168	2173	2178	2183	0	1	1	2	2	3	3	4	4
·34	2188	2193	2198	2203	2208	2213	2218	2223	2228	2234	1	1	2	2	3	3	4	4	5
·35	2239	2244	2249	2254	2259	2265	2270	2275	2280	2286	1	1	2	2	3	3	4	4	5
·36	2291	2296	2301	2307	2312	2317	2323	2328	2333	2339	1	1	2	2	3	3	4	4	5
·37	2344	2350	2355	2360	2366	2371	2377	2382	2388	2393	1	1	2	2	3	3	4	4	5
·38	2399	2404	2410	2415	2421	2427	2432	2438	2443	2449	1	1	2	2	3	3	4	4	5
·39	2455	2460	2466	2472	2477	2483	2489	2495	2500	2506	1	1	2	2	3	3	4	5	5
·40	2512	2518	2523	2529	2535	2541	2547	2553	2559	2564	1	1	2	2	3	4	4	5	5
·41	2570	2576	2582	2588	2594	2600	2606	2612	2618	2524	1	1	2	2	3	4	4	5	5
·42	2630	2636	2642	2649	2655	2661	2667	2673	2679	2685	1	1	2	2	3	4	4	5	6
·43	2692	2698	2704	2710	2716	2723	2729	2735	2742	2748	1	1	2	3	3	4	4	5	6
·44	2754	2761	2767	2773	2780	2786	2793	2799	2805	2812	1	1	2	3	3	4	4	5	6
·45	2818	2825	2831	2838	2844	2851	2858	2864	2871	2877	1	1	2	3	3	4	5	5	6
·46	2884	2891	2897	2904	2911	2917	2924	2931	2938	2944	1	1	2	3	3	4	5	5	6
·47	2951	2958	2965	2972	2979	2985	2992	2999	3006	3013	1	1	2	3	3	4	5	5	6
·48	3020	3027	3034	3041	3048	3055	3062	3069	3076	3083	1	1	2	3	4	4	5	6	6
·49	3090	3097	3105	3112	3119	3126	3133	3141	3148	3155	1	1	2	3	4	4	5	6	6

Antilogarithms

	0	1	2	3	4	5	6	7	8	9	1	2	3	4	5	6	7	8	9
·50	3162	3170	3177	3184	3192	3199	3206	3214	3221	3228	1	1	2	3	4	4	5	6	7
·51	3236	3243	3251	3258	3266	3273	3281	3289	3296	3304	1	2	2	3	4	5	5	6	7
·52	3311	3319	3327	3334	3342	3350	3357	3365	3373	3381	1	2	2	3	4	5	5	6	7
·53	3388	3396	3404	3412	3420	3428	3436	3443	3451	3459	1	2	2	3	4	5	6	6	7
·54	3467	3475	3483	3491	3499	3508	3516	3524	3532	3540	1	2	2	3	4	5	6	6	7
·55	3548	3556	3565	3573	3581	3589	3597	3606	3614	3622	1	2	2	3	4	5	6	7	7
·56	3631	3639	3648	3656	3664	3673	3681	3690	3698	3707	1	2	3	3	4	5	6	7	8
·57	3715	3724	3733	3741	3750	3758	3767	3776	3784	3793	1	2	3	3	4	5	6	7	8
·58	3802	3811	3819	3828	3837	3846	3855	3864	3873	3882	1	2	3	4	4	5	6	7	8
·59	3890	3899	3908	3917	3926	3936	3945	3954	3963	3972	1	2	3	4	5	5	6	7	8
·60	3981	3990	3999	4009	4018	4027	4036	4046	4055	4064	1	2	3	4	5	6	6	7	8
·61	4074	4083	4093	4102	4111	4121	4130	4140	4150	4159	1	2	3	4	5	6	7	8	9
·62	4169	4178	4188	4198	4207	4217	4227	4236	4246	4256	1	2	3	4	5	6	7	8	9
·63	4266	4276	4285	4295	4305	4315	4325	4335	4345	4355	1	2	3	4	5	6	7	8	9
·64	4365	4375	4385	4395	4406	4416	4426	4436	4446	4457	1	2	3	4	5	6	7	8	9
·65	4467	4477	4487	4498	4508	4519	4529	4539	4550	4560	1	2	3	4	5	6	7	8	9
·66	4571	4581	4592	4603	4613	4624	4634	4645	4656	4667	1	2	3	4	5	6	7	9	10
·67	4677	4688	4699	4710	4721	4732	4742	4753	4764	4775	1	2	3	4	5	7	8	9	10
·68	4786	4797	4808	4819	4831	4842	4853	4864	4875	4887	1	2	3	4	6	7	8	9	10
·69	4898	4909	4920	4932	4943	4955	4966	4977	4989	5000	1	2	3	5	6	7	8	9	10
·70	5012	5023	5035	5047	5058	5070	5082	5093	5105	5117	1	2	4	5	6	7	8	9	11
·71	5129	5140	5152	5164	5176	5188	5200	5212	5224	5236	1	2	4	5	6	7	8	10	11
·72	5248	5260	5272	5284	5297	5309	5321	5333	5346	5358	1	2	4	5	6	7	9	10	11
·73	5370	5383	5395	5408	5420	5433	5445	5458	5470	5483	1	3	4	5	6	8	9	10	11
·74	5495	5508	5521	5534	5546	5559	5572	5585	5598	5610	1	3	4	5	6	8	9	10	12
·75	5623	5636	5649	5662	5675	5689	5702	5715	5728	5741	1	3	4	5	7	8	9	10	12
·76	5754	5768	5781	5794	5808	5821	5834	5848	5861	5875	1	3	4	5	7	8	9	11	12
·77	5888	5902	5916	5929	5943	5957	5970	5984	5998	6012	1	3	4	5	7	8	10	11	12
·78	6026	6039	6053	6067	6081	6095	6109	6124	6138	6152	1	3	4	6	7	8	10	11	13
·79	6166	6180	6194	6209	6223	6237	6252	6266	6281	6295	1	3	4	6	7	9	10	11	13
·80	6310	6324	6339	6353	6368	6383	6397	6412	6427	6442	1	3	4	6	7	9	10	12	13
·81	6457	6471	6486	6501	6516	6531	6546	6561	6577	6592	2	3	5	6	8	9	11	12	14
·82	6607	6622	6637	6653	6668	6683	6699	6714	6730	6745	2	3	5	6	8	9	11	12	14
·83	6761	6776	6792	6808	6823	6839	6855	6871	6887	6902	2	3	5	6	8	9	11	13	14
·84	6918	6934	6950	6966	6982	6998	7015	7031	7047	7063	2	3	5	6	8	10	11	13	15
·85	7079	7096	7112	7129	7145	7161	7178	7194	7211	7228	2	3	5	7	8	10	12	13	15
·86	7344	7261	7278	7295	7311	7328	7345	7362	7379	7396	2	3	5	7	8	10	12	13	15
·87	7413	7430	7447	7464	7482	7499	7516	7534	7551	7568	2	3	5	7	9	10	12	14	16
·88	7586	7603	7621	7638	7656	7674	7691	7709	7727	7745	2	4	5	7	9	11	12	14	16
·89	7762	7780	7798	7816	7834	7852	7870	7889	7907	7925	2	4	5	7	9	11	13	14	16
·90	7943	7962	7980	7998	8017	8035	8054	8072	8091	8110	2	4	6	7	9	11	13	15	17
·91	8128	8147	8166	8185	8204	8222	8241	8260	8279	8299	2	4	6	8	9	11	13	15	17
·92	8318	8337	8356	8375	8395	8414	8433	8453	8472	8492	2	4	6	8	10	12	14	15	17
·93	8511	8531	8551	8570	8590	8610	8630	8650	8670	8690	2	4	6	8	10	12	14	16	18
·94	8710	8730	8750	8770	8790	8810	8831	8851	8872	8892	2	4	6	8	10	12	14	16	18
·95	8913	8933	8954	8974	8995	9016	9036	9057	9078	9099	2	4	6	8	10	12	15	17	19
·96	9120	9141	9162	9183	9204	9226	9247	9268	9290	9311	2	4	6	8	11	13	15	17	19
·97	9333	9354	9376	9397	9419	9441	9462	9484	9506	9528	2	4	7	9	11	13	15	17	20
·98	9550	9572	9594	9616	9638	9661	9683	9705	9727	9750	2	4	7	9	11	13	16	18	20
·99	9772	9795	9817	9840	9863	9886	9908	9931	9954	9977	2	5	7	9	11	14	16	18	20